Fundamentals of Mechanisms and Machines

Dhanesh N. Manik

Fundamentals of Mechanisms and Machines

Springer

Dhanesh N. Manik
Department of Mechanical Engineering
Indian Institute of Technology Bombay
Mumbai, Maharashtra, India

ISBN 978-981-97-1812-2 ISBN 978-981-97-1810-8 (eBook)
https://doi.org/10.1007/978-981-97-1810-8

This Springer imprint is published by the registered company Springer Nature Singapore Pte Ltd.
The registered company address is: 152 Beach Road, #21-01/04 Gateway East, Singapore 189721, Singapore

Paper in this product is recyclable.

Dedicated to my children: Chandani and Sandesh

Preface

This book is mainly focused on providing the fundamentals of machines and mechanisms, which basically deals with the kinematics and dynamics of machines; many times, the contents of this book are also studied as theory of machines. This is one of the most important core subjects of mechanical engineering that has a greater futuristic role in the design of machines, especially electric and hybrid vehicles and robots. When machines are expected to be designed from the viewpoints of energy efficiency, quietness, sustainability, global warming, etc., there will be a paradigm shift in the design of machines that has to be supported. Since kinematics and dynamics of machinery is a prerequisite to a course on machine design, most of the scientific aspects behind machine design are studied in this course. Therefore, it is very important to provide strong fundamentals to mechanical engineering students in this area. All these factors motivated me to take up writing this book, to present the material with proper flow and in a systematic manner, keeping in view the demands expected of it.

As the title of the book suggests, this book is all about mechanisms and machines. A mechanism represents the basic structure of a component of the machine that can do a specific job, whereas the machine contains several mechanisms that are configured to make a machine do its intended job. Since there are many different ways a mechanism can be configured and there are many mechanisms in a machine, there is a tremendous opportunity to build a better machine of our choice. The mechanisms also have many common features that can be easily recognized so we are using the most relevant mechanism for a particular application.

Although mechanisms and machines have thousands of configurations, there is a common thread that needs to be recognized so it becomes easier to design any of them with the same set of rules. This is the special feature of this book in which the same set of rules are consistently applied throughout for designing any mechanism or machine. Therefore, the same mechanism or similar mechanisms are used in many parts of the book, both in the chapters as well as in the exercises, for presenting different points of view required for their study.

Another important feature of this book is the use of block diagrams for representing the kinematic pairs of mechanisms. Although neat sketches are provided while discussing the mechanisms, there are limitations to be able to clearly identify the kinematic pairs of the mechanism, which is very important for further analysis. This aspect of the block diagram representation of the mechanisms will also be useful in obtaining inversions of the kinematic chain from which they evolved. Block diagrams of mechanisms are provided for the exercise problems also.

Most parts of this book have evolved based on the undergraduate and postgraduate courses in this area that I have taught at IIT Bombay for the past 30 years. I have also independently taught a basic dynamic course at Auburn University, USA, as a graduate student, using vector-based methods. Therefore, it has undergone many useful changes over a period of time, based on the feedback from students. The above course, taught over a period of 42 weeks with tutorials, quizzes exams, their solutions, etc., did generate inputs for this book. Hence, sufficient care has been taken to write this book so the student can easily comprehend the basics of machines and mechanisms through independent reading, supplemented by classroom teaching. Practicing engineers can also benefit using this as a reference book as there are many practical applications illustrated throughout this book. In short, providing an easy learning curve is an important feature of this book.

Another important feature of the book is the use of vector approach throughout. Some students are little hesitant with vector-based methods, but a systematic approach of using vectors will definitely make it easier for students. In addition, it will help them to move to the next higher level of extending the knowledge acquired in these books to applications in robotics. Initially, graphical solutions are provided along with vector-based solutions, so there is a smooth transition in learning from graphical methods to vector-based methods.

Chapter 1 discusses the basics of mechanisms. Kinematics pairs, kinematic chains, and their inversions are discussed with reference to some of the known mechanisms. In addition, Gruebler's criterion for determining the degrees of freedom is discussed accounting for various types of mechanisms used in practice. The position analysis of mechanisms is introduced that will be helpful in the kinematic and kinetic analysis of mechanisms in later chapters. Block diagram representation is used for defining the kinematic pairs in mechanisms that will help in having physical insight into the construction of any mechanism. Some of the mechanisms discussed in this chapter will be discussed in later chapters for analyzing them from different points of view. The exercise problems of this chapter will also be carried forward into later chapters.

Chapter 2 covers the fundamentals of kinematics. It begins with the discussion of various coordinate systems: Cartesian, spherical, and cylindrical. The basics of obtaining the derivative of a vector in moving coordinates are explained in a general context that will be applied to solve a variety of problems throughout the text, including this chapter. The above discussion is extended to relating the velocity between two points of a rigid body in plane motion of known angular velocity. Similarly, acceleration between two points of a rigid body in plane motion with known angular velocity and acceleration is discussed that can be used to obtain the kinematic analysis of a vast number of mechanisms. However, there are specific mechanisms in which a linear motion is occurring along a link that is oscillating. In such cases, a combination of fixed and moving coordinate systems will have to be used for obtaining the kinematic analysis. An alternate analysis using graphical analysis is also presented in many places so vector analysis techniques can be clearly understood and appreciated. The advantage of vector analysis is that they are amenable to using computers and are also much more accurate than graphical solutions. In addition, the vector approach is useful for synthesizing new mechanisms.

Chapter 3 is concerned with spur gears and Chap. 4 with helical, bevel, and worm gears. Since design and manufacture of gears is an exhaustive topic, only relevant portions related to these gears are discussed that will help integrate them into practical mechanisms. Chapter 5 deals with gear trains that have useful application in many mechanisms.

Chapter 6 discusses cam-follower mechanisms that are very useful in controlling specific time-based movements in a mechanism. The challenge lies in obtaining accurate timing in a smooth manner. Hence, determining the profile of a cam for a specific application is the most important aspect of cam design. Various aspects of obtaining the cam profile are discussed that will result in a smoothly operating cam.

Chapter 7 is concerned with the synthesis of mechanisms. Although the earlier chapters are concerned with mechanisms that are already designed, it is possible to obtain new mechanism configurations that may overcome the existing limitations of some mechanisms; they can also be designed subject to specific kinematics expected of them. Generally, position synthesis is focused upon wherein the mechanism is synthesized based on some possible positions taken by some of the constituent members. Some of the known mechanisms that were kinematically analyzed in Chap. 2 are synthesized in the exercise problems that would guarantee some performance criteria based on expected kinematics from some of the links.

Chapter 8 discusses the relationship between forces and moments to accelerations (linear and angular) of a rigid body in plane motion. The equations so derived are used for obtaining the torque output from slider crank mechanisms used in IC engines, as they have been traditionally used for force analysis. However, the vector analysis is general enough to be applied to any mechanism in plane motion. With some modification, the same analysis can be extended to a general three-dimensional motion of rigid bodies, which is demonstrated in Chap. 12.

Chapter 9 deals with balancing of rotary and reciprocating mechanisms. Since unbalance in machines invariably results in excessive vibration, it is important to either balance machines with existing unbalance or minimize the unbalance during design. Rotary balancing on single and multiple planes is perhaps one of the easily achievable tasks. However, reciprocating unbalance is not very easy to solve and hence the concept of using multi-cylinders is discussed that will not only reduce the unbalance but also result in uniform torque output from such machines.

Chapter 10 is concerned about speed control of machines using governors. Although governors were initially used with steam engines that have now become extinct, the basic concept of speed control is still relevant, and various modifications of the earlier design of governors are discussed. The following governors are discussed: Portal, Proell, and Hartnell. Speed control by using some of the concepts presented in this chapter can be useful in designing hybrid and electric vehicles.

Chapter 11 is concerned about vibration from machines. Using the basic concepts of single-degree of freedom systems, multi-degree of freedom systems, and continuous systems like shafts, a wide number of practical applications are discussed; both free and forced vibration are discussed from a practical viewpoint. The kinematic and kinetic analysis presented in the earlier chapters plays a major role in analyzing vibration in machines.

Chapter 12, although named as gyroscopes, deals with general three-dimensional motion. Gyroscopes are mainly used for navigation, but the gyroscopic forces, in general, become important in many practical applications. The analysis presented is general enough to account for all aspects of gyroscopic effects for any application. In addition, the analysis presented in this chapter can also be used for force analysis of three-dimensional mechanisms.

Regarding the illustrations used in this book, I have personally created all the illustrations using vector graphics, in addition to typing the entire manuscript. So, I could go back and forth between the illustrations and the corresponding text till they matched perfectly; I would either modify the text or the illustration. Hence, although it has taken more time than usual to prepare the manuscript, it has helped in seamless integration of the text and illustrations.

The book is organized in such a way that information related to machines and mechanisms is logically presented that can be applied to solve many practical problems. The sequence is such that all the chapters can be sequentially studied without having to go back and forth. The physics of dynamic systems is emphasized throughout so that the reader is not just lost in mathematics. Therefore, the information presented here will be very useful for designing many practical systems. After going through this book, many other reference books and journal papers in this area can be easily studied.

Mumbai, India
December 2023

Dhanesh N. Manik

Acknowledgements I thank the Director, IIT Bombay, for granting sabbatical leave during the academic year 2018–2019 and for the valuable support to undertake the preparation of the manuscript. Most of the groundwork for preparing the manuscript was laid during the above sabbatical leave.

I thank my undergraduate and postgraduate students at IIT Bombay who have taken courses related to this subject. The classroom discussions and the feedback received by these students have been invaluable in preparing the manuscript. The reader is encouraged to contact me at dnmanik@gmail.com. Answers to the exercise problems, computer programs and errata are available at https://sites.google.com/view/dnmanik/home

Contents

About the Author

Dhanesh N. Manik is Professor in the Department of Mechanical Engineering, Indian Institute of Technology Bombay. He has done his bachelors in Mechanical Engineering from Mysore University (SJ College of Engineering) in 1982 and has done his M.E. and Ph.D. in Mechanical Engineering from IISc Bangalore, India and Auburn University, USA in 1985 and 1991, respectively. His areas of expertise are vibration, acoustics, control systems, and dynamics. He has more than 30 years teaching experience, covering a diverse set of courses in the fields of design engineering, mechanisms, and machines. He has authored/coauthored three textbooks and numerous scientific articles.

Chapter 1
Basics of Mechanisms

In this book, we are embarking on a journey to study the details of most machines that are used in practice around us. The title of this book is Fundamentals of Mechanisms and Machines, which would basically consist of understanding how machines are built to do various tasks. This would require analyzing the kinematics and dynamics of machines, which is basically computing accelerations (linear and angular) and then relating them to forces and moments. But we are going to study many details about machines that would go beyond the above computations. The first step in this journey is to learn how the basic configuration of a machine can be arrived at. Although there are various types of machines, it is possible to come up with a scientific analysis to find a common ground in all these machines. This would not only help in understanding existing machines but also help design new machines that would have never been built thus far.

In this chapter, we shall discuss the basics of mechanisms. Before we build machines, we must be sure of how the relative motion of various components of the machine is to be configured, without going into the details of forces and moments that would act on them during their operation. This basic configuration through which relative motion between various elements can be ascertained is called a mechanism. First, we try understanding the basic building blocks of a mechanism and the manner in which they are connected through various types of connections known as kinematic pairs. We then build kinematic chains and figure out various mechanisms that can be obtained through inversion by changing the fixed frame of reference assigned to one of the links. Although we shall deal with three-dimensional motions while discussing gyroscopes in Chap. 12, in this book, we are generally interested in planar mechanisms, and hence in this chapter, we shall discuss various planar mechanisms that are widely used. Then we try to compute the degrees of freedom of these mechanisms. We shall discuss metrics that decide the nature of a four-bar mechanism by deciding the fixed link based on its geometric dimensions. Finally, we carry out the position analysis of some of the mechanisms that will help us in the later chapters to conduct kinematic and kinetic analysis of mechanisms.

D. N. Manik, *Fundamentals of Mechanisms and Machines*,
https://doi.org/10.1007/978-981-97-1810-8_1

1.1 Basic Definitions

The following definitions are to be understood first:

Mechanism

A mechanism is a combination of rigid bodies capable of relative motion or it could even be a combination of several mechanisms.

Machine

A machine is a device used for transferring and transforming motion and force (power, torque) from the source to the load. It can be defined as a device that receives energy and uses it to perform some desired work. In this process, the motion has to be transformed between the source and load. The nature of such transformations in a machine depends on the type of available input and the desired output required for an operation. As shown in Table 1.1, there are various possibilities of input and output in a machine. Hence, several mechanisms may be involved in the operating of a machine. So, we need different types of mechanisms within a machine to achieve such transformations.

Table 1.1 can be discussed as follows:

(a) The input motion is in the form of continuous rotation, but the desired output motion is a rectilinear motion.
(b) The input motion is a rectilinear motion, and the desired output motion is a continuous rotary motion.
(c) The input motion is a continuous rotary motion, and the desired output motion is oscillatory along a straight line.
(d) The input motion is oscillating along a straight line and the output motion is a continuous rotary motion.
(e) The input motion is continuous rotary and the output motion is oscillatory rotary.
(f) The input motion is continuous rotary and the output motion is intermittent.

Let us first define the following terms:

Kinematics

The subject that deals with the geometric aspects of relative motion, without any reference to the force that is causing the motion is called kinematics. Hence, kinematics is the subject which deals with only geometric aspects of motion without any consideration of force. In many applications, especially in sophisticated instruments, obtaining the correct motion is very important and the forces involved in operating them can be negligible. Hence, strength is of secondary importance in such applications where kinematics plays a major role. For the study of kinematics, a machine may be referred to as a mechanism, which is a combination of interconnected rigid bodies capable of relative motion. This aspect is mainly discussed in Chap. 2.

Table 1.1 Various types of input and output

	Input	**Output**
(a)	Rotary	Linear
(b)	Linear	Rotary
(c)	Rotary	Oscillating linear
(d)	Oscillating linear	Rotary
(e)	Rotary	Angular oscillation
(f)	Rotary	Intermittent

Dynamics

When forces acting on various elements of a machine are accounted, for then the study is called dynamics. Dynamics is further divided into *statics* when the masses of the elements are neglected and *kinetics* when the masses are accounted for. Whether one has to use statics or kinetics or a combination of both will depend on the nature of the problem as we shall see in later chapters.

Machine Versus Mechanism

For example, consider the slider-crank mechanism of an IC engine. For studying the kinematics of this mechanism, we are only interested in the dimensions of the crank and connecting rod that will give the total distance traveled by the piston. Therefore, we ignore the gas forces, friction forces, unbalanced forces, etc. Hence, we call it a slider-crank mechanism and not a slider-crank machine for obvious reasons. Therefore, mechanisms or machines must be contextually addressed.

Link

Each part or a component of a mechanism that has relative motion with another is known as a link or element. The links could be rigid or flexible, but we shall be generally concerned about rigid links in this book. Each link can be constructed from various parts but as long as it has no relative motion with respect to its constituent parts, it can be viewed as a single entity. For example, a piston is constructed from several parts but it is denoted as a single link.

1.2 Examples of Mechanisms

If you look around, you can easily recognize many mechanisms. Even an automobile which we all take for granted has many mechanisms as shown in Table 1.2.

Table 1.2 lists 33 mechanisms that are present in a modern automobile, carrying out various functions that are necessary for operating the automobile; some of them are interfaced with many electrical and electronic systems. Each of them is distinctly different and some of them are too simple and some very complex; some of them may be rarely used and some of them always used, and there could be some more mechanisms missing from the list. But we all drive these automobiles without giving much thought to the various mechanisms that support many operations. The main objective of Table 1.1 is to present various mechanisms in an automobile that are also representative of many types of mechanisms. Although each of them might appear to be slightly different, it is possible to recognize some common features of all these mechanisms in many machines. This is necessary because we should not only be aware of the existing mechanisms but also be able to design new mechanisms for a specific application. Therefore, the main objective of this chapter is to recognize the characteristic features of these mechanisms so they can be subjected to a systematic study.

Table 1.2 Mechanisms in an automobile

Slider crank (engine)	Airbag	Screw jack
Steering	Air inlet control	Bonnet cover opening
Clutch	Temperature control	Bonnet cover restraining
Brake	Seat adjustment (horizontal)	Trunk cover restraining
Accelerator	Seat adjustment (inclination)	Handbrake
Gearbox	Seat adjustment (vertical)	Side rearview mirrors
Differential	Seat belt	Door-activated cabin lights
Wiper	Sun visor	Gear shifting mechanism
Door	Cabin mirror	Mobile stand
Door lock	Petrol tank cover	High beam flasher
Window glass	Trunk cover opening	Windshield cleaner
Choke	Turn indicator	Belt drive

1.3 Kinematic Pairs

The behavior of a mechanism mainly depends on the interconnections between their links and the interconnection between two links, when their relative motion is completely (consistently) constrained, they form a kinematic pair; the two members are known as pairing elements. The nature of kinematic pairs is very important for understanding kinematics, and hence a thorough understanding of them is important. Hence, we shall first discuss various kinematic pairs and the nature of relative motion permitted in them. Although we are going to discuss the degree of freedom of a mechanism later, it is important to note the degree of freedom in the context of kinematic pairs is the number of coordinates needed to define relative movement between them.

In every kinematic pair, there will be relative movement and hence some coordinates are required to describe the relative movement. These coordinates are referred to as the degrees of freedom and are also known as pair variables. A free unconnected rigid body has six degrees of freedom in a three-dimensional space; three translational and three rotational. Hence, when a free rigid body is connected to another free rigid body through a kinematic pair, the original degrees of freedom are naturally curtailed. Similar curtailment of degrees of freedom results while forming kinematic pairs.

Kinematic pairs can be broadly classified as (1) lower pairs, (2) higher pairs, and (3) wrapping pair. In a lower kinematic pair, the two connected bodies have area contact. On the other hand, in the case of a higher pair, the contact between the two bodies has either a line or a point contact. And, in the case of a wrapping pair, as in the case of a belt or chain drive, one body completely wraps over the other. However, lower and higher pairs are most common.

1.3.1 Lower Pairs

Revolute pair

Figure 1.1 shows a revolute pair in which a shaft and bearing are connected. This basically is the representation of a simple hinge joint that is commonly used in connecting links. The shaft is cylindrical that goes into the hole of the bearing. For the sake of illustration, we can assume that the bearing is fixed. The shaft cannot translate in any of the three directions that are restrained by the revolute pair. The only possible relative motion is for the shaft to rotate about the vertical axis. Hence the pair variable θ represents the relative motion between the shaft and bearing. It should be noted that both the shaft and bearing have a common axis that is fixed in space. A revolute pair is designated by “R”.

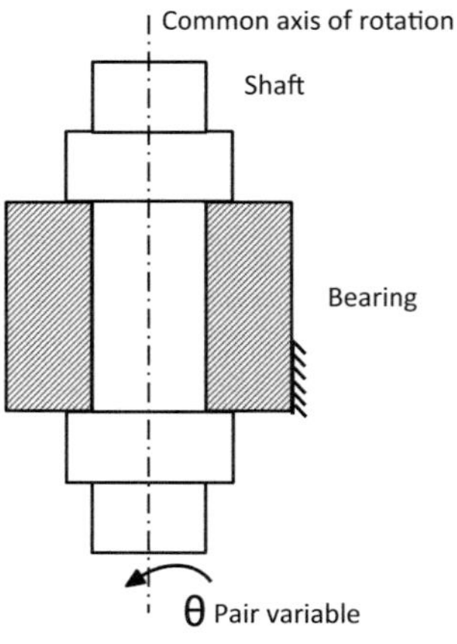

Fig. 1.1 Revolute pair

Prismatic pair

Figure 1.2 shows a rectangular prism sliding inside another rectangular prism, which is called a prismatic pair. In this prismatic pair, if body 2 is fixed, then body 1 is restrained to move only along the vertical direction. Hence, no other translation or rotational motion of body 1 is possible. Therefore, y representing the relative sliding motion between bodies 1 and 2, is the pair variable and hence a prismatic pair also has a single degree of freedom similar to a revolute pair. However, the fundamental difference between a prismatic and revolute pair is that a prismatic pair doesn't have a definite axis for motion, but only a direction that is vertical. Similarly, a piston moving inside the cylinder of an IC engine forms a prismatic pair as it is constrained to move along the cylinder. A prismatic pair is designed by "P".

Screw Pair

In the screw pair shown in Fig. 1.3, the screw moves inside a nut. Hence, if the nut is fixed, the only relative motion of the screw possible is its rotation θ about the axis of the nut. But as the screw rotates, it also moves along the axis of the nut. Although

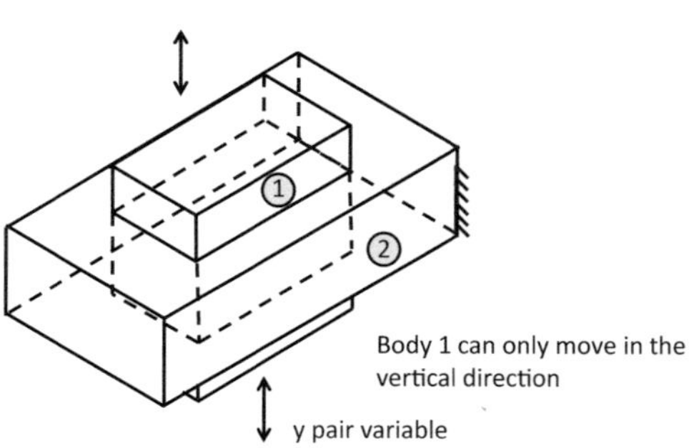

Fig. 1.2 Prismatic pair

Piston moving inside a cylinder forms a prismatic pair

the screw can rotate and translate, the kinematic pair has a single degree of freedom since the rotational and translation degree of freedom are related by $\frac{\Delta\theta}{2\pi} = \frac{\Delta s}{L}$. (L, lead of the screw, is the distance moved for one rotation.)

Spherical Pair

A spherical pair, also known as ball and socket joint, is shown in Fig. 1.4. There is a spherical ball in body 2 and a spherical cavity in socket 1. This type of kinematic pair is known as a spherical pair. If body 1 is fixed, body 2 will lose all its translational degrees of freedom. However, it will have all the three rotational degrees of freedom, represented by three independent angles: α, β, and γ.

Cylindrical Pair

A cylindrical pair is shown in Fig. 1.5. In this kinematic pair, a cylindrical body passes through a cylindrical hole of a plate. If the plate is held fixed, then the cylindrical body can rotate about the vertical axis and also translate in the vertical direction. Hence, a cylindrical pair has two independent degrees of freedom with pair variables y and θ.

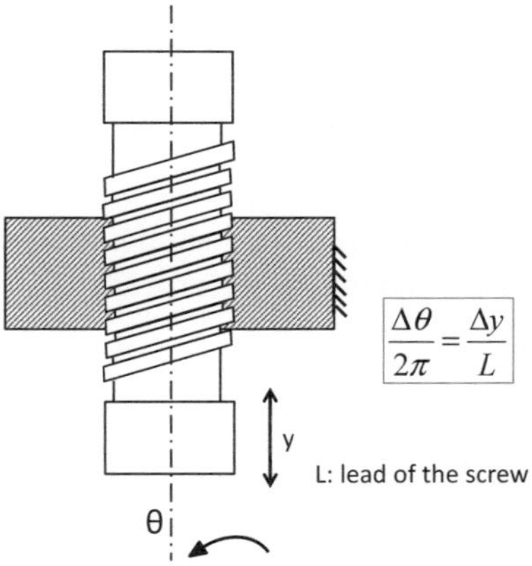

Fig. 1.3 Screw pair

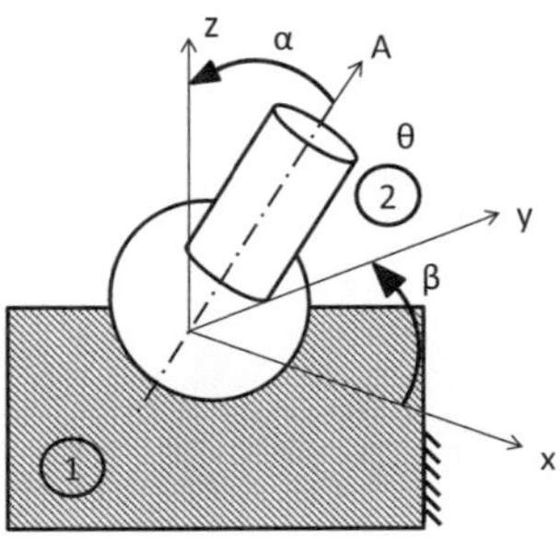

Fig. 1.4 Spherical pair

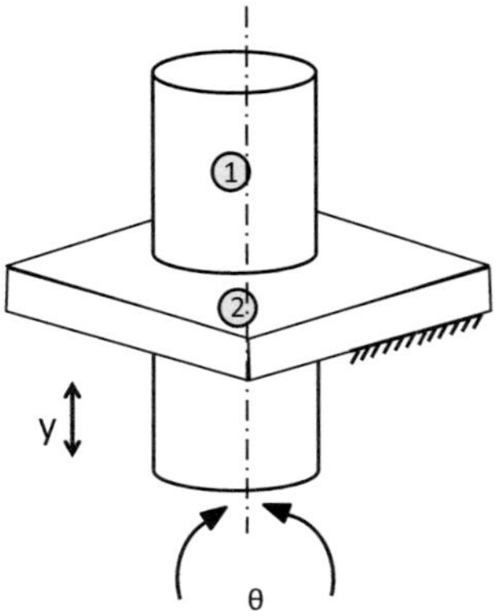

Fig. 1.5 Cylindrical pair: two degrees of freedom, s and θ (that are independent)

Planar Pair

A planar pair is shown in Fig. 1.6. Here, the body can have translation along the X- and Y-axes and can also rotate about the z-axis. x, y, and θ are the independent pair variables.

Pin-in-Slot Pair

In many mechanisms, a very common connection is a pin-in-slot pair as shown in Fig. 1.7. As shown, it allows two links to rotate and slide with each other. Link 2 is having a slot and link 1 is connected to link 2 by means of a pin that is not counted here. We can have a similar arrangement with a prismatic pair between a sliding piston and link 2 and a revolute pair between link 1 and the piston.

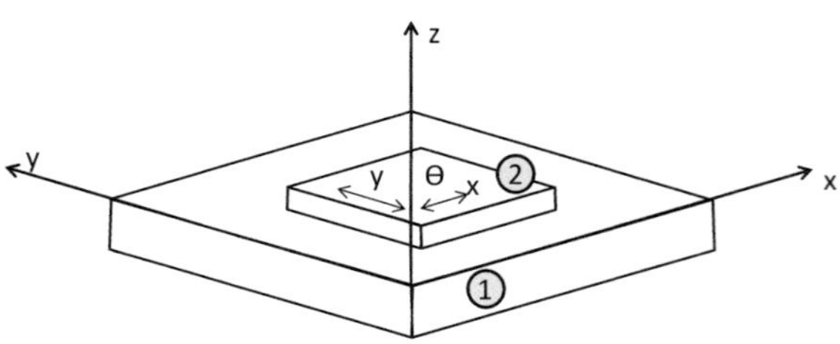

Fig. 1.6 Planar pair: three degrees of freedom, x and y, and θ

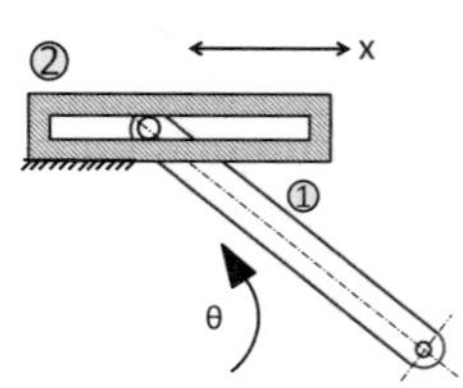

Fig. 1.7 Pin-in-slot pair

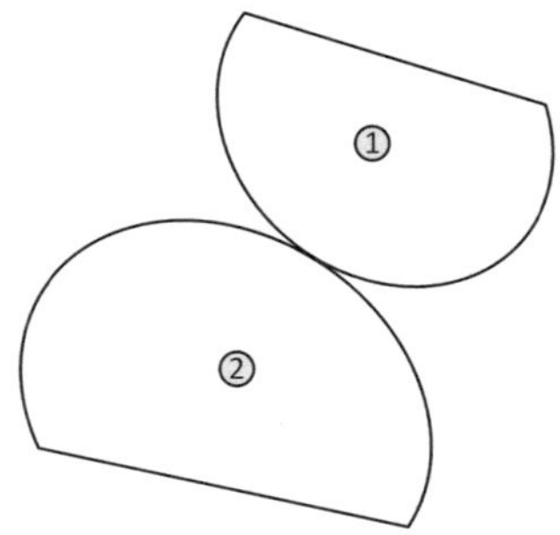

Fig. 1.8 Higher pairs

1.3.2 Higher Pairs

A set of rigid bodies having point or line contact as shown in Fig. 1.8 is known as higher pairs and they are usually found in gears and cam-follower mechanisms. Figure 1.9 shows body 2 sliding on the fixed horizontal surface of body 1. As shown in this figure, if the motion of body 2 is restricted to the plane of paper, there can be relative translation and rotation about the line or point of contact. Hence, the two degrees of freedom can be represented as pair variables x and θ. Thus, a higher pair that is part of a planar mechanism will have two degrees of freedom. More discussion on planar mechanisms will follow in Sect. 1.5.

Degree of Freedom

The degrees of freedom of a mechanism are given by the number of independent variables needed to completely define the relative movements between all its links. A rigid body in space has three translation and three rotation degrees of freedom. Degrees of freedom of various kinematic pairs are summarized in Table 1.3; they will be naturally less than a rigid body in space.

1.4 Representation of Links

Kinematic Chain

A series of links connected by kinematic pairs is known as a kinematic chain. A closed chain is one in which every link is connected to at least two other links.

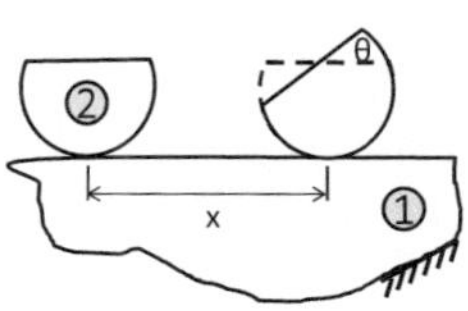

Fig. 1.9 Degrees of freedom of a higher pair

Table 1.3 Degree of freedom of various kinematic pairs

Type	Symbol	DOF	Pair variable
Revolute	R (hinge)	1	θ
Prismatic	P (sliding)	1	x
Screw (helix)	H	1	θ or x
Cylindrical	C	2	θ and x
Spherical	G	3	α, β, and γ
Planar	E	3	x, y, and θ
Higher pair (plane motion)	–	2	x and θ
Pin-in-slot	PS	2	x and θ

Fig. 1.10 Fixed frame

Singular link

Connected to only one other link.

Binary link

Connected to two other links.

Ternary link

Connected to three other links.

1.4.1 Fixed Link

Representation of a fixed link is shown in Fig. 1.10. A straight line with crosshatching is used as a representation for a fixed link and it is generally given a number and all the fixed links in the mechanism are given the same number.

1.4.2 Revolute Pair

A revolute pair is shown in Fig. 1.11 that consists of two links 2 and 3 hinged at a point. Figure 1.11a is the actual construction and the same pair can be represented by a line diagram as shown in Fig. 1.11b by representing the linkages 2 and 3 by straight lines that are joined at a hinge represented by a dark circle. However, it should be noted that the angle between the links is the same in both figures. This type of representation is possible when the connecting links are of similar size.

Representation of a revolute pair between two links which have a significant difference in their size is shown in Fig. 1.12. The actual construction is shown in

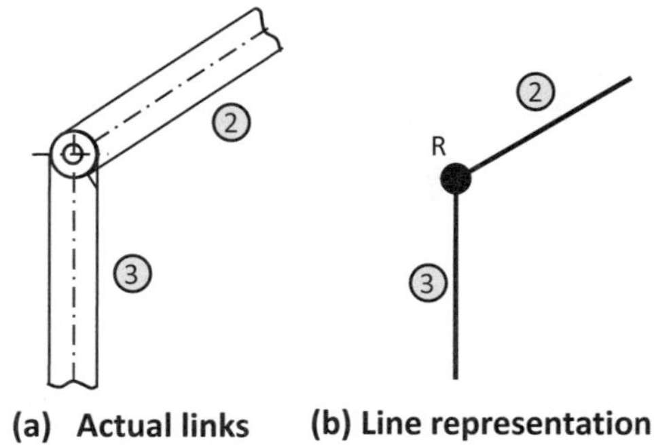

Fig. 1.11 Line representation of a revolute pair (similar-sized links)

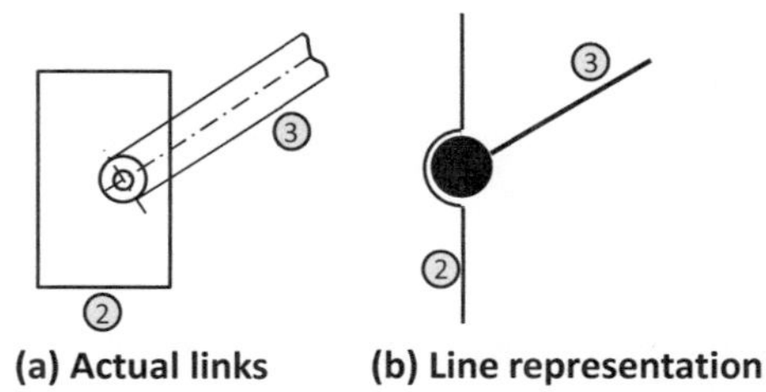

Fig. 1.12 Line representation of a revolute pair (different-sized links)

Fig. 1.12a in which a small link 3 is hinged to a large-sized link 2. A line representation of the same is shown in Fig. 1.12b. The larger link 2 is represented by joining two straight lines to a semi-circle and the smaller link 3 is represented by a straight line with a dark circle at the hinged end that fits inside the semi-circle of link 2.

1.4.3 Prismatic Pair

Figure 1.13 shows the representation of a prismatic pair. A typical prismatic pair construction is shown in Fig. 1.13a that consists of a piston sliding inside a cylinder. The line representation can be either as shown in Fig. 1.13b or c. Both forms are extensively used in representing prismatic pairs.

1.4.4 Pin-in-Slot Pair

A pin-in-slot pair is shown in Fig. 1.14. Figure 1.14a shows the actual construction pair and Fig. 1.14b its schematic representation.

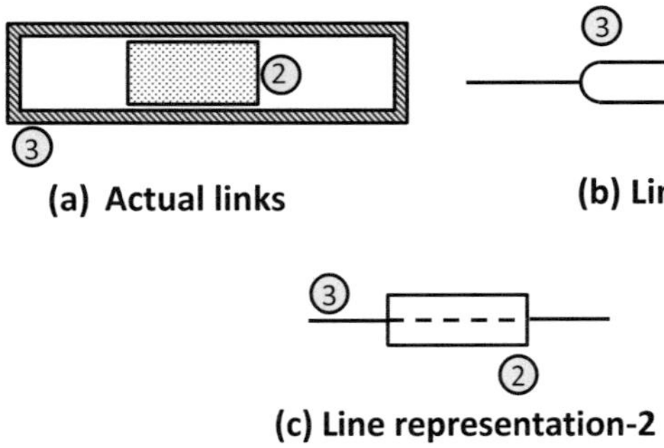

Fig. 1.13 Representation of a prismatic pair

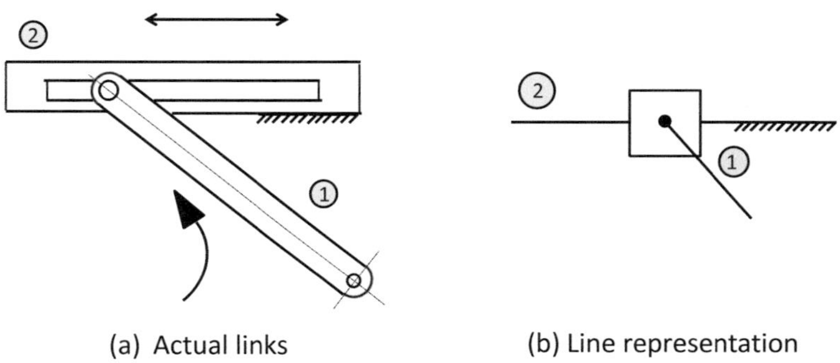

Fig. 1.14 Pin-in-slot pair

1.4.5 Quaternary Link

As shown in Fig. 1.15a, a quaternary is one in which four links (1, 2, 3, 4) are hinged onto a larger link that resembles a plate. A line representation of the same is shown in Fig. 1.15b in which the four links are shown as straight lines that are shown to be hinged at points, shown as small circles, to the larger link.

Example 1.1
Consider the mechanism shown in Fig. 1.16.

(a) Draw a line diagram representation of the mechanism.
(b) Describe how the links are connected.
(c) Identify the type of kinematic pairs in them.
(d) Draw a block diagram, with each block representing a link, and show the type of connection, in addition to clearly marking the fixed link.
Line diagram representation of the mechanism of Fig. 1.16 is shown in Fig. 1.17.
(f) Links 2 and 4 are hinged to the fixed link 1 and link 3 is sliding on link 2 and links 3 and 4 are hinged to each other.

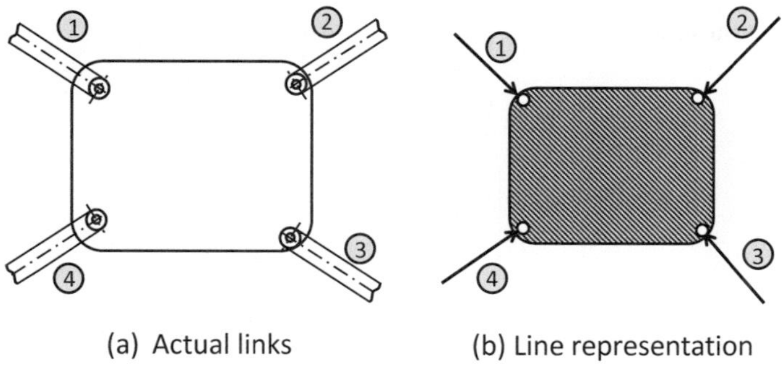

Fig. 1.15 Line representation of a quaternary link

Fig. 1.16 Schematic of Example 1.1

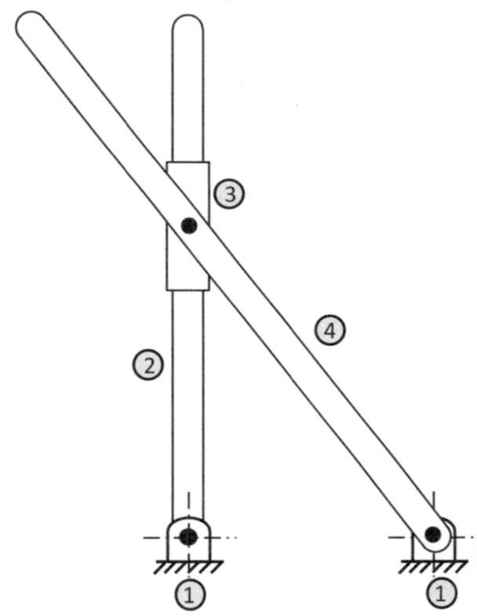

Fig. 1.17 Line diagram of Example 1.1

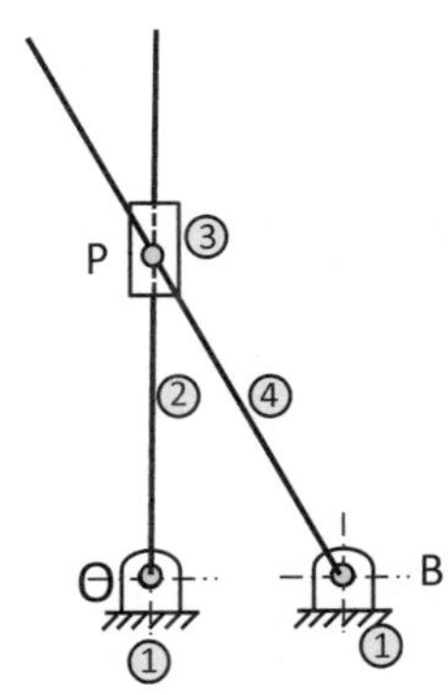

Table 1.4 Kinematic pairs of Example 1.1

Paired links	Type of kinematic pair
1–2	R
2–3	P
3–4	R
4–1	R

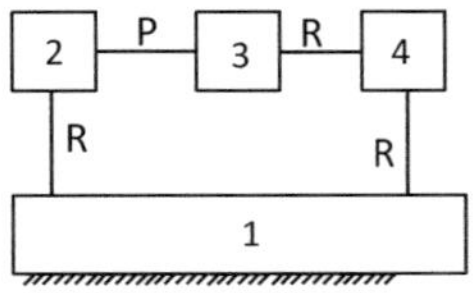

Fig. 1.18 Block diagram of Example 1.1

(c) The kinematic pairing of links is shown in Table 1.4 and the relationship between the links for the four pairs is shown; there are three revolute pairs and one prismatic pair.

(d) The block diagram representation of kinematic pairing of links is shown in Fig. 1.18. Although Table 1.4 shows similar information, a pictorial representation gives a better physical insight into the kinematic pairing between various links.

1.5 Classification of Mechanisms

Mechanisms can be classified as (1) planar, (2) spherical, and (3) spatial.

1.5.1 Planar Mechanisms

A four-bar planar mechanism is shown in Fig. 1.19. This is called a 4R-planar linkage since it has four revolute pairs. There are four links 1, 2, 3, and 4 which are connected by four revolute pairs at O_2, A, B, and O_4. The input motion of link 2 is transferred to the output link 4 through link 3. The input shaft at O_2 and the output shaft at O_4 are parallel. In addition, all the four revolute joints have parallel axes at A, O_2, O_4, and B. There are four revolute joints and the axes of all these four revolute joints are parallel and perpendicular to the plane of motion.

The above four-bar linkage shown in Fig. 1.19 is representative of a planar mechanism. The following are the characteristic features of a planar mechanism:

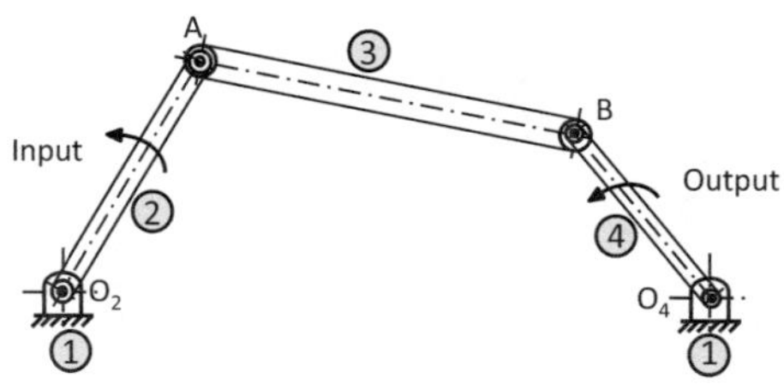

Fig. 1.19 Four-bar linkage

1. All points of the mechanism move in parallel planes.
2. They can be projected onto a single plane called the plane of motion.
3. The axes of the revolute pairs are parallel to each other and perpendicular to the plane of motion.
4. Plane motion reveals the true motion of all points of a mechanism.

Most of the mechanisms discussed in this textbook are planar mechanisms.

In the context of discussing planar mechanisms, it would be relevant to discuss the plane motion of a rigid body as shown in Fig. 1.20. Let XY be the plane of motion and the z-axis perpendicular to the XY plane. Assume the rigid body denoted by ABCD, that was initially parallel to the x- and y-axes, moves to a different position by moving a certain distance along both the coordinate directions in the xy plane and also changing its angular position with respect to the x-axis to θ. Hence, the rigid body can have three independent coordinates Δ_x, Δ_y, and θ and thus have three degrees of freedom. Therefore, a rigid body in space that had six degrees of freedom now has three degrees of freedom when it is constrained to move along a plane. If this rigid body is connected to another rigid body constrained to move along the same xy plane, each of them will have to lose some of their degrees of freedom depending on the type of kinematic pairs. They are shown in Table 1.5.

Now let us discuss some basic definitions. Each rigid body contributing elements to form kinematic pair is referred to as a link. Kinematic chain is defined as a series

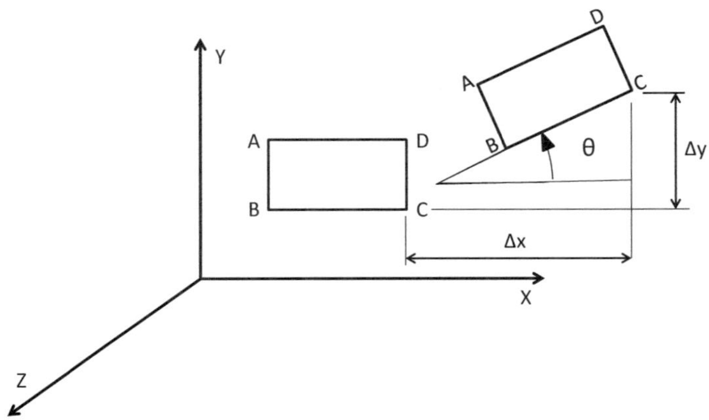

Fig. 1.20 Plane motion of a rigid body

Table 1.5 Kinematic pairs for planar mechanisms

Type	Symbol	DOF	Planar
Revolute	R (hinge)	1	Yes
Prismatic	P (sliding)	1	Yes
Screw (Helix)	S	1	No
Cylindrical	C	2	No
Spherical	G	3	No
Planar	E	3	Yes
Higher pair	H	2	Yes
Pin-in-slot	PS	2	Yes

of links connected by kinematic pairs. Such a kinematic chain is said to be closed, if every link is connected to at least two other links. That means, it may be connected to three, four, or five links, but definitely at least two other links. If a particular link or some links are connected to only one other link, then the chain is said to be an open chain.

1.5.2 Spherical Mechanisms

An example of a spherical mechanism, Hooke's joint, is shown in Fig. 1.21. It consists of two shafts that have a yoke at each end and the yokes are connected by a cross that has a pin at each of the four ends. These pins of the cross have corresponding slots in the yoke. The yokes have revolute pairs with the cross that helps in transmitting power between intersecting shafts. The four rigid bodies are connected by four revolute pairs and all the four revolute pairs intersect at a single point. Thus, the above arrangement of shafts with yokes, linked by the cross, can be used to connect two intersecting shafts that are not parallel.

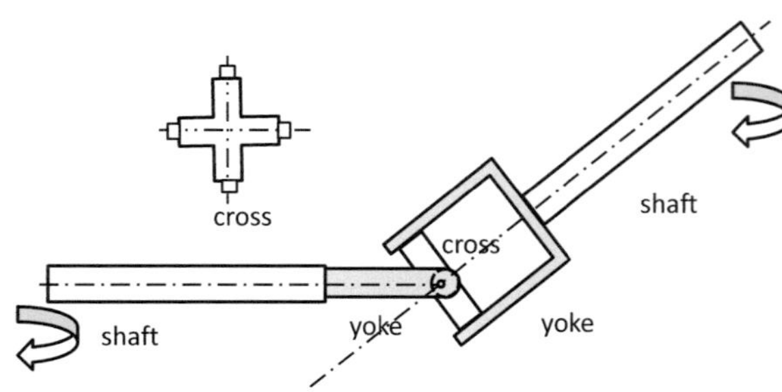

Fig. 1.21 Hooke's joint

1.5.3 *Spatial Mechanisms*

The basic characteristic of a spatial mechanism is that the axis of rotation of the revolute pairs is not parallel. If the links of the planar mechanism of Fig. 1.19 are twisted, the planar mechanism becomes a spatial mechanism and of course, the spatial mechanisms can exist in other forms and the common feature in them is that the axis of rotation of their revolute pairs is not parallel. We shall not discuss spatial mechanisms in this book.

DOF and Number of Inputs

If the degrees of freedom and the number of inputs are the same, then it is known as a constant mechanism A four-bar linkage, dumper truck, etc., are examples of a constant mechanism and one exception to this rule is the differential of an automobile.

Mechanism

A more precise definition of a mechanism is that a closed kinematic chain with one link fixed link is called a mechanism. Hence, mechanisms are derived from their respective kinematic chains. In this context, a four-bar kinematic chain and slider-crank chain are important since they result in a wide variety of mechanisms used in practical applications.

1.6 Commonly Used Mechanisms

1.6.1 *Slider-Crank Mechanism*

A slider-crank mechanism that is used in automobile engines is shown in Fig. 1.22. Figure 1.22a shows the construction of this mechanism that consists of a piston that slides inside a cylinder block. One end of the connecting rod is hinged to the piston and the other end is hinged to the crankshaft. Gas pressure inside the cylinder at appropriate instants of time pushes the piston toward left and this reciprocating motion is converted into to rotary motion of the crankshaft in the counterclockwise direction. A line diagram representation of the slider-crank mechanism is shown in Fig. 1.22b; the fixed link (1) is the frame of the engine, crank (2), connecting rod (3) and piston (4). Since there are three revolute pairs and one prismatic pair, this is known as a 3R-1P mechanism and the type of kinematic pairs are shown in Fig. 1.22b.

1.6.2 *Eccentric Punch*

An eccentric punch used in a punching press, as shown in Fig. 1.23a, has many similarities with a slider-crank mechanism. Sometimes, the kinematic pairs cannot

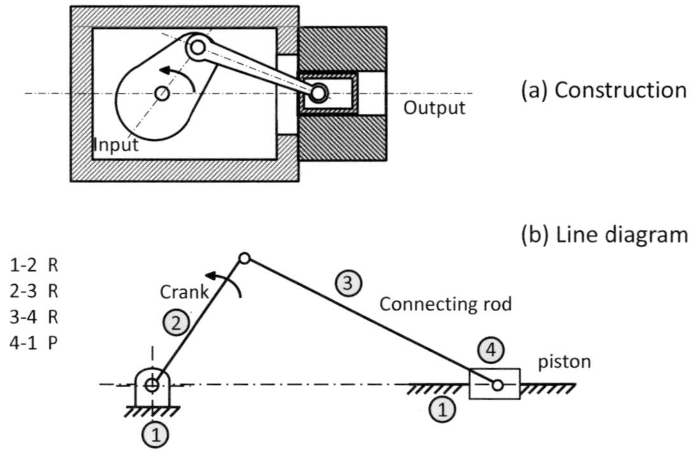

Fig. 1.22 Slider-crank mechanism

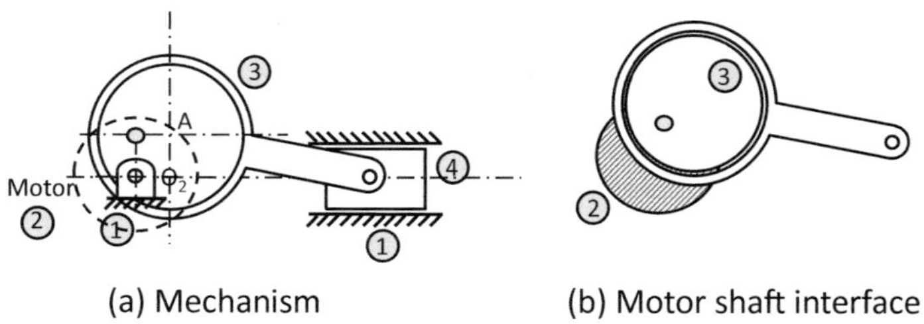

Fig. 1.23 Eccentric punch

be easily identified due to practical considerations and this is such an example. The eccentric is connected to the motor shaft eccentrically and the connecting rod is driven by the eccentric and the punching tool is connected to the connecting rod. The continuous rotation of the motor is thus converted into reciprocating motion of the punching tool. The prismatic pair of the punch tool with the machine frame is easily identifiable and also the revolute pair between the punching tool and the connecting rod and the revolute pair between this eccentric and the crank shaft. However, the kinematic pair between the connecting rod and the eccentric is not easily identifiable.

1.6.3 Four-Bar Mechanism

Figure 1.24 shows the line representation of a four-bar mechanism. The input link is the crank and the output link, the follower. The link that connects the follower and coupler is known as the coupler and is connected to them at the moving hinges. The

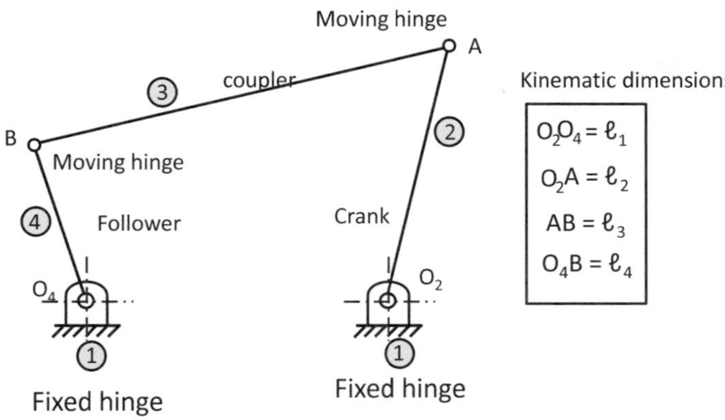

Fig. 1.24 Line representation of a four- bar mechanism

crank and the follower have one of their ends hinged to the fixed frame. As discussed earlier, this is known as a 4R mechanism as there are only revolute pairs.

1.6.4 Equivalence of a Curved Slider

Many times, it would be difficult to figure out the kinematic pairs just by inspection. One such example is shown in Fig. 1.25. Figure 1.25a shows a curved slider in which the piston slides on a curvature of radius ρ. The revolute pairs 1–2, 2–3 and 3–4 can be easily figured out. But the curved slider interface 4–1 cannot be classified as a prismatic pair as it violates the rule of being along a straight line. However, if the curved slider can be converted into an equivalent four-bar linkage as shown in Fig. 1.25b, by replacing the slider with a link of radius ρ, it can be easily recognized as a 4R mechanism and all the analysis can be carried out with respect to this equivalent mechanism.

1.6.5 Equivalence of a Slider-Crank with a Four-Bar Mechanism

Since a four-bar linkage shown in Fig. 1.26a and the slider-crank mechanism shown in Fig. 1.26b are some of the most commonly used mechanisms, it would be of interest to know the equivalence between them. If the link 4 of the four-bar linkage is made extremely long, then the path moved by the point B will be almost a straight line. Then, it becomes equivalent to a slider-crank mechanism. In other words, a slider-crank mechanism can be considered as the derivation from a four-bar linkage

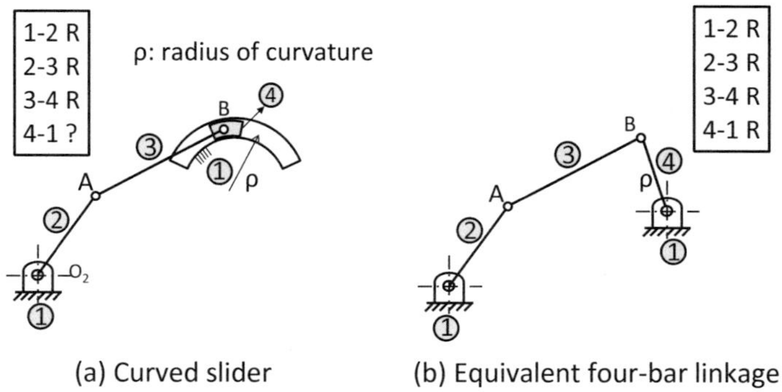

Fig. 1.25 Equivalence of a curved slider

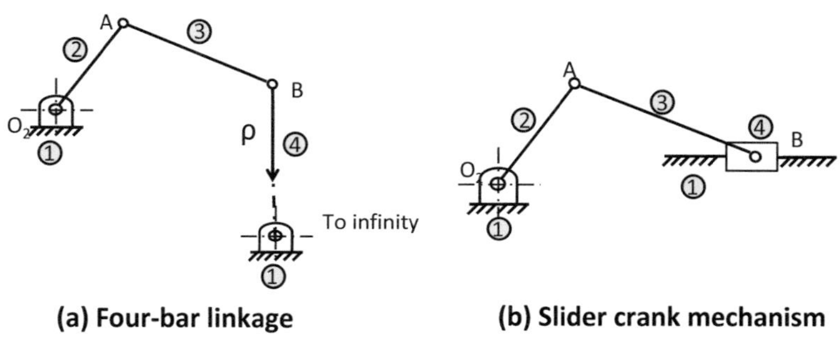

Fig. 1.26 Equivalence of four-bar linkage and slider-crank mechanism

in which the follower link is extremely long. Hence, the method of determining equivalence of a mechanism is many times useful in their kinematic analysis.

1.7 Kinematic Inversion

Fixing different links of the same kinematic chain to produce different, distinct mechanisms is known as kinematic inversion. The relative motion between various links, however, remains unchanged under a kinematic inversion.

1.7.1 Three-Bar Kinematic Chain

Figure 1.27 shows a three-bar kinematic chain in which one of the links is fixed. Since this results in a structure with zero degree of freedom, irrespective of which links is

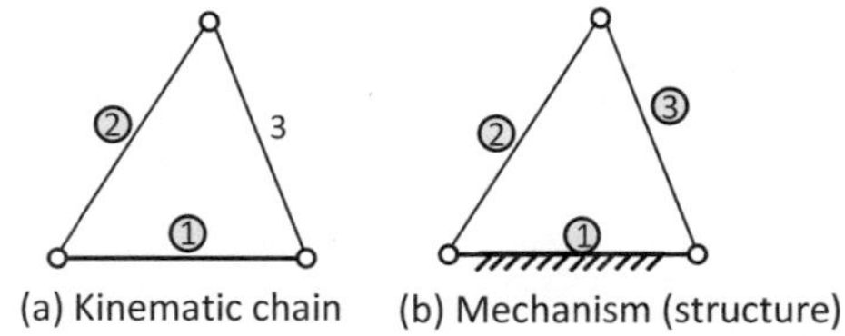

Fig. 1.27 Three-bar kinematic chain and its mechanism

fixed, a three-bar kinematic chain is not of any interest in the study of mechanisms. Hence, we move next to a four-bar kinematic chain.

1.7.2 Four-Bar Kinematic Chain

Figure 1.28a shows a four-bar kinematic chain that consists of four-bars hinged as shown. Since the bars are connected through revolute kinematic pairs, a four-bar chain is also called as a 4R chain. and Fig. 1.28b–e show various inversions of this kinematic chain by changing the fixed links as shown. Grashof's criterion that gives specific inversions of a four-bar mechanism depending on the proportion of their dimensions and which of the links will be a fixed link. Although the relative motion remains unchanged due to kinematic inversion, the type of mechanism that results due to various inversions will be dependent on the dimensions of various links. The following inversions result: double-rocker, crank-rocker, crank-crank and they will be discussed in Sect. 1.10.

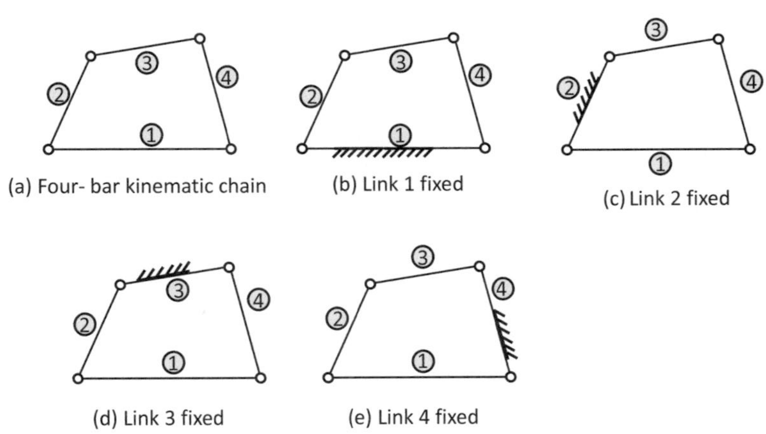

Fig. 1.28 Four-bar kinematic chain and its inversions

1.7.3 Slider-Crank Kinematic Chain

A slider-crank kinematic chain is shown in Fig. 1.29. This is also known as a 3R-1P chain as there are three revolute pairs and one prismatic pair that is the piston. It consists of a crank OA, connecting rod AB and the link OB. They are given their usual names so they can be easily recognized and they many not always play the traditional role known by their names as any of them can be fixed to derive a kinematic inversion. They are shown in Fig. 1.30. Although the piston, crank and connecting rods as they are traditionally known with respect to a slider-crank mechanism, the link 1 is not given any specific name so it can easily assigned different roles in the respective inversions of the slider-crank chain.

Reciprocating Mechanism

We have already discussed the slider-crank mechanism shown in Fig. 1.22, also known as reciprocating mechanism. In order to demonstrate that the same slider-crank kinematic chain of Fig. 1.22 can be used to obtain different inversions, we are discussing again. Now, if link 1 of the slider-crank kinematic chain shown in Fig. 1.29 is fixed, it results in the slider-crank mechanism or the reciprocating mechanism shown in Fig. 1.30a. So, link 1 not only provides the fixed pivot O for the crank but also provides a fixed frame for the piston to oscillate within it.

Oscillating Cylinder Mechanism

If link 3 that was designated as connecting rod in the slider-crank kinematic chain of Fig. 1.29 is fixed, then it results in the oscillating cylinder mechanism shown in Fig. 1.30b. If link 2, designated as crank rotates on pivot A, it tries to push link 1 resulting in its oscillation as the piston slides in it.

Whitworth Quick Return Mechanism

If link 2 that was designated as crank in the slider-crank kinematic chain of Fig. 1.29 is fixed, then it results in the Whitworth quick return mechanism shown in Fig. 1.30c. When link 3 rotates, it moves the piston to move inside link 1 resulting in its oscillations. Link 3 can be slightly extended and connected to tool head 6 through link 5 using revolute pairs; the tool head reciprocates in the horizontal direction whose line of action passes through point O. In one complete rotation of link 3, it cuts the line of action of the tool head at points B_1 and B_2 that correspond to points at which the tool head changes its direction along the horizontal. $\angle B_1AB_2$ is the angle moved by link 3 during the return motion of the tool head and $360^\circ - \angle B_1AB_2$ is the angle

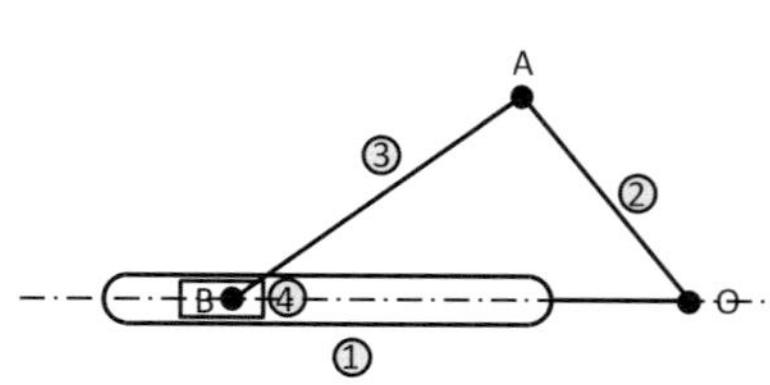

Fig. 1.29 Slider-crank kinematic chain

Fig. 1.30 Inversions of a slider-crank chain

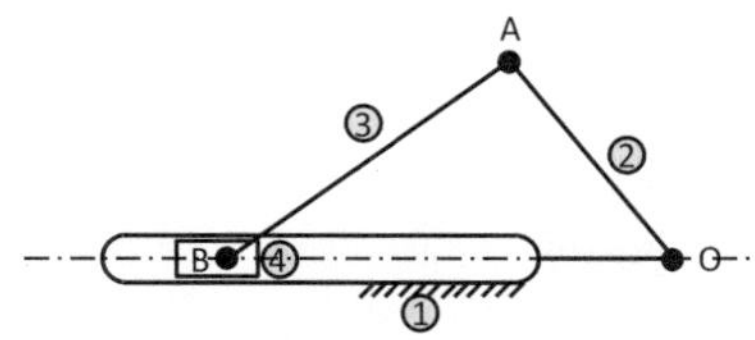

(a) Reciprocating engine mechanism

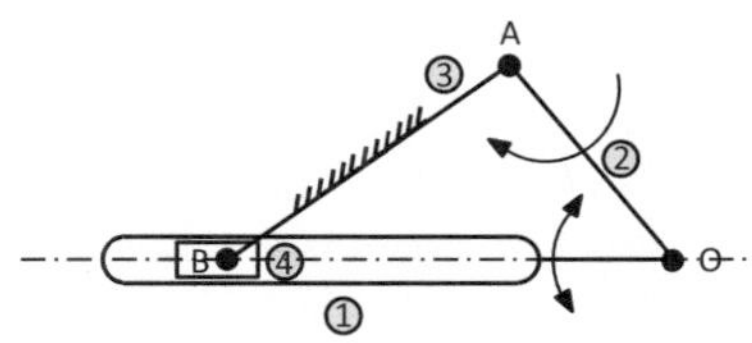

(b) Oscillating cylinder mechanism

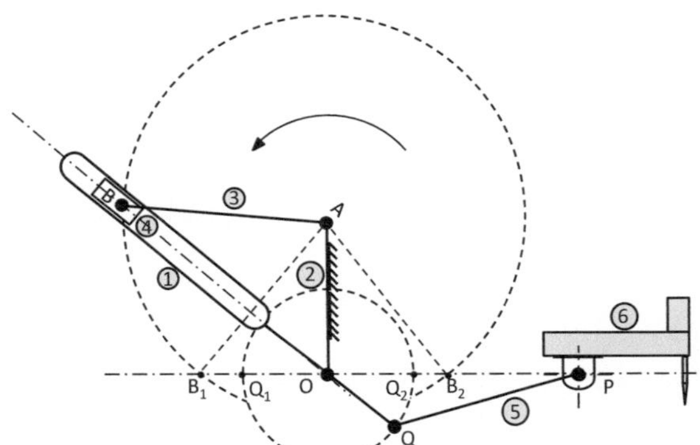

(c) Whitworth quick-return mechanism

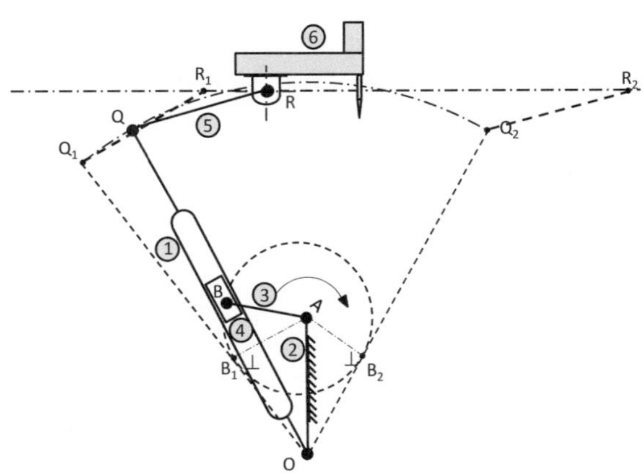

(d) Slotted lever quick return mechanism

moved by the tool head in the forward motion. Hence, the time taken during the forward motion is much more than the reverse motion that would allow the tool to move at a much slower speed to enable an appropriate cutting speed. The reverse motion is much faster as there is no cutting as the tool just grazes the cutting surface. By adjusting the relevant lengths of the links, the above angles can be adjusted for a particular operation.

Slotted-Lever Quick Return Mechanism

In this mechanism, link 2 that was designated as crank in the slider-crank kinematic chain of Fig. 1.29 is fixed, very similar to the Whitworth quick return mechanism. However, the only difference is that link 3 that was designated as connecting rod is much shorter than link 2 that was designated as crank in the slider-crank kinematic chain of Fig. 1.29. In addition, however, there is a difference in the manner in which the tool head is connected to link 1 using link 5. In this case, link 1 is pivoted at O but extended upwards. The tip of link 1, Q, oscillates between the extreme positions Q_1 and Q_2. Pivot point B cuts the extreme positions of link 1 at B_1 and B_2, respectively. $\angle B_1AB_2$ is the angle moved by link 3 during the return motion of the tool head and $360°-\angle B_1AB_2$ is the angle moved by the tool head in the forward motion. Hence, similar to the Whitworth quick return mechanism, the tool head moves much faster in the reverse motion than in the forward motion to account for an appropriate cutting speed.

A choice between the quick return mechanisms will have to be made depending on specific requirements of its application. The time ratio for both the quick return mechanisms is defined as the time taken for the forward stroke to the return stroke. See exercise problem 8.25 later in the text, in which the above mechanisms for the same time ratio, length of stroke, and input angular speed, are required to be compared in terms of velocity, acceleration, force, moments, etc.

Figure 1.31 is a 3R-1P mechanism that is a modification of the reciprocating mechanism shown in Fig. 1.30b in which a piston is sliding on an oscillating link. Crank 2 is hinged to piston 3 at A and the piston slides along the link 4 that is hinged to the fixed link at o_4. The offset e can be changed according to the requirements of a mechanism. By changing the offset to zero and attaching another link to the extension of link 4, the above mechanism can be converted to a slotted-link quick return mechanism as shown in Fig. 1.32, which is same as the slotted-link quick return mechanism obtained in Fig. 1.30d. However, the inversions obtained in Fig. 1.30 from the slider-crank kinematic chain of Fig. 1.29 better demonstrate the principle of obtaining inversions from the same kinematic chain.

The 3R-1P oscillating cylinder mechanism shown in Fig. 1.30b can be slightly modified by connecting the piston (Link 4) to a fixed hinge as shown in Fig. 1.33, resulting in an oscillating cylinder mechanism; this mechanism is practically used in a foot pump as shown in Fig. 1.34. This shows that practical applications, although perform the same relative motion, might look different from the original inversion.

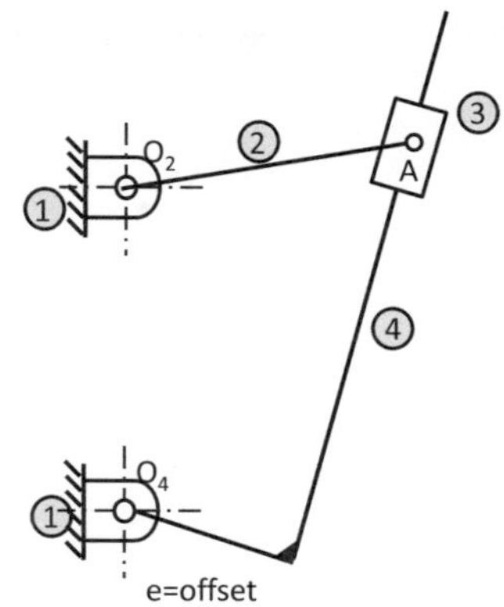

Fig. 1.31 Sliding piston on an oscillating link

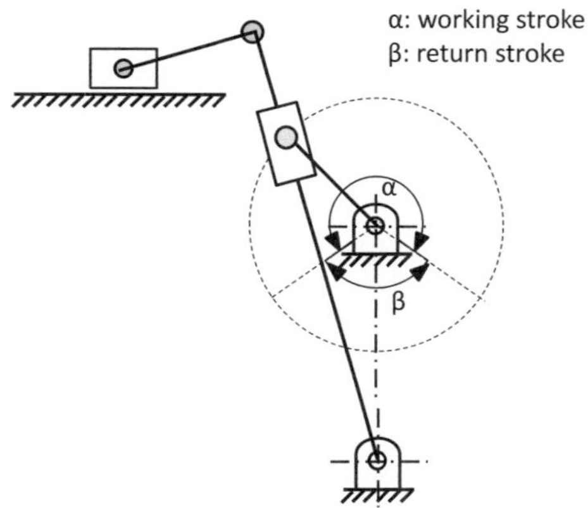

Fig. 1.32 Slotted-link quick return mechanism

Example 1.2

1. Consider the slotted-link quick return mechanism shown in Fig. 1.34a. Draw a block diagram representation of all the links and show their relationship with each other. Clearly mention the type of kinematic pairing between the links and show the fixed links.

The block diagram representation of the mechanism shown in Fig. 1.34a is shown in Fig. 1.34b and they are also summarized in Table 1.6.

1.7.4 R-R-P-P Mechanisms

By converting two revolute pairs of a four-bar kinematic chain into prismatic pairs, we can get an R-R-P-P chain as shown in Fig. 1.35. Figure 1.35b is the block diagram of the kinematic chain and Fig. 1.35a is a typical construction of the kinematic chain. It

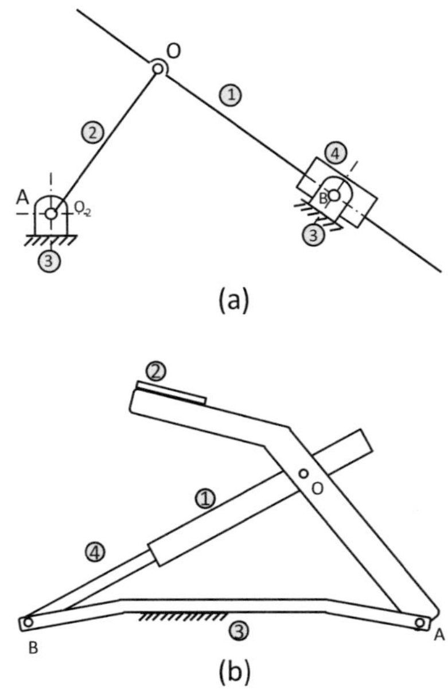

Fig. 1.33 **a** Oscillating cylinder mechanism **b** Foot pump

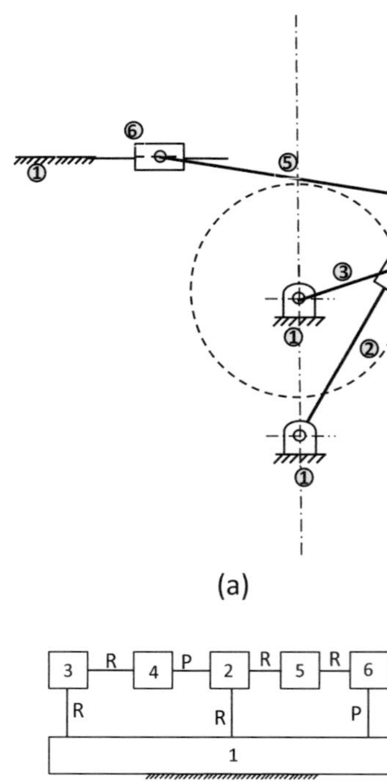

Fig. 1.34 **a** Mechanism of Example 1.2 **b** Block diagram of the mechanism of Example 1.2

Table 1.6 Kinematic pairs of Example 1.2

Link nos	1–2	1–3	1–6	2–4	2–5	4–3	5–6
Type of kinematic pair	R	R	P	P	R	R	R

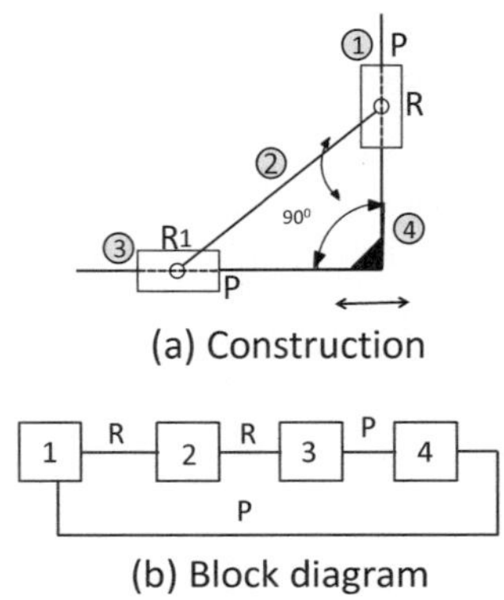

Fig. 1.35 R-R-P-P kinematic chain

consists of two pistons that are, respectively, sliding along the horizontal and vertical directions on frame 4 and the pistons are connected by link 2 using two revolute pairs.

Now we can discuss various inversions of the R-R-P-P chain shown in Fig. 1.35.

Scotch yoke mechanism

By fixing link 1 (piston) of the kinematic chain of Fig. 1.35, a scotch yoke mechanism can be obtained as shown in Fig. 1.36. In this mechanism, piston 3 slides vertically inside link 4 and link 4 slides horizontally in link 1. This mechanism can be used to convert rotary motion into reciprocating motion.

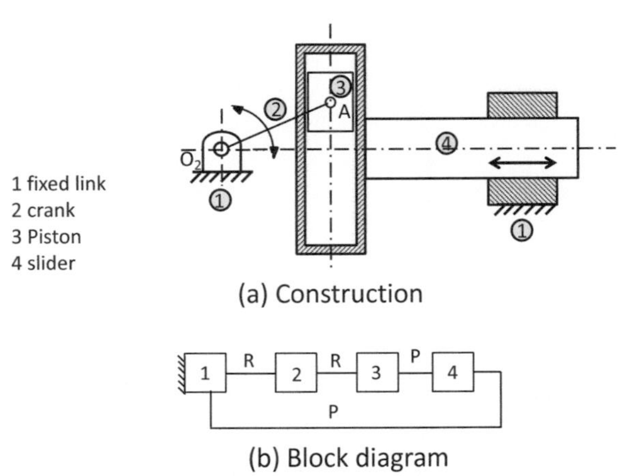

Fig. 1.36 Scotch yoke mechanism

Elliptical trammel

By fixing frame (4) of the kinematic chain of Fig. 1.35, an elliptic trammel can be obtained as shown in Fig. 1.37. The frame has both vertical and horizontal slots for the pistons to traverse. The end of link 2 traces an ellipse. Figure 1.37a shows the block diagram and line representation, and Fig. 1.37b shows the actual construction.

Oldham's coupling

By fixing link 2 of the kinematic chain of Fig. 1.35, an Oldham's coupling can be obtained. Figure 1.38a shows the block diagram and line presentation and Fig. 1.38b shows the actual mechanism. In this mechanism, frame 4 slides inside 3 and 1 and any rotation of link 3 will result in a similar rotation of link 1. Link 2 is the fixed frame of the machine with reference to which both the input and output shafts rotate. This principle is used to connect two parallel shafts with a small eccentricity as shown in the figure. There is sliding friction between 3–4 and 4–1, which drives the shaft ends.

Fig. 1.37 Elliptical trammel

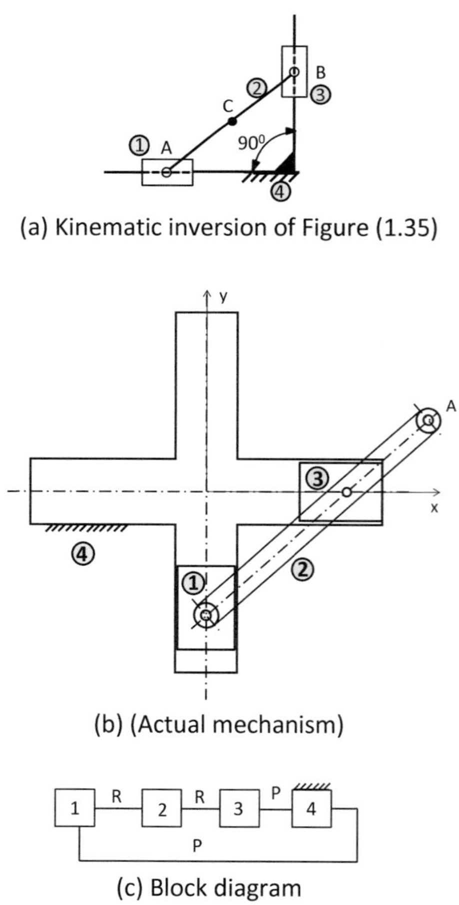

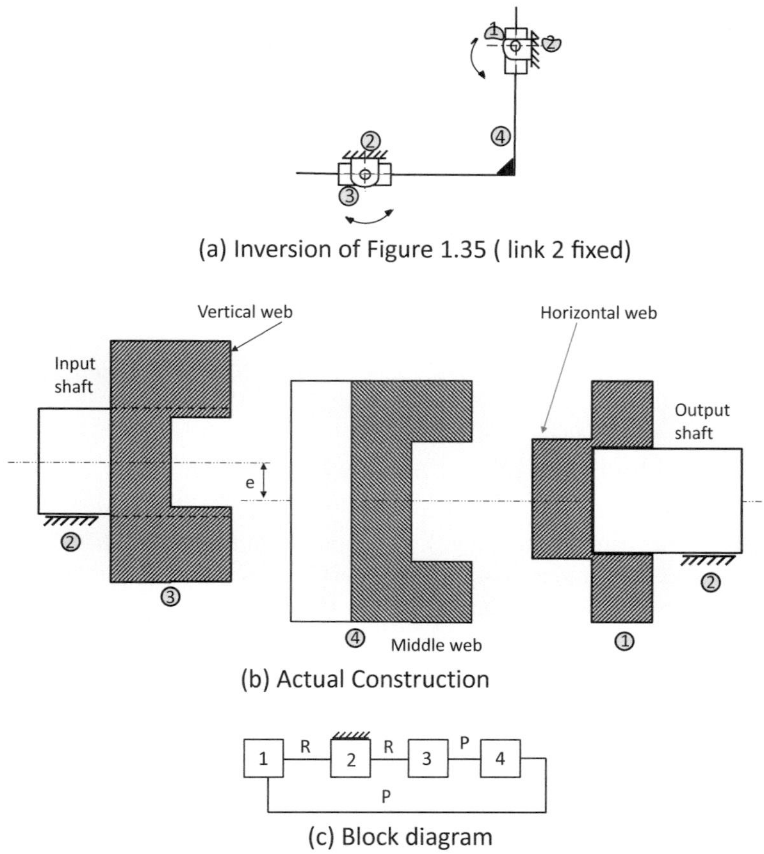

Fig. 1.38 Oldham's coupling

1.7.5 R-P-R-P Mechanisms

In this mechanism, the revolute and prismatic pairs are alternatively placed. Such a mechanism is used in Davi's steering gear as shown in Fig. 1.39; block diagram is shown in Fig. 1.39a and line representation is shown in Fig. 1.39b. The main objective of this mechanism is to move the wheels attached to link 2 by the same angle. A rack and pinion mechanism (not shown) moves link 4 to the left or right depending on the direction in which the wheels are to be turned. Link 4 is hinged to piston 3 that slides on link 2. Hence, 1–2 and 3–4 are revolute pairs, whereas 2–3 and 1–4 are prismatic pairs. The mechanism is symmetrical to the horizontal central line of the vehicle. Hence, symmetric movement of both wheels is possible.

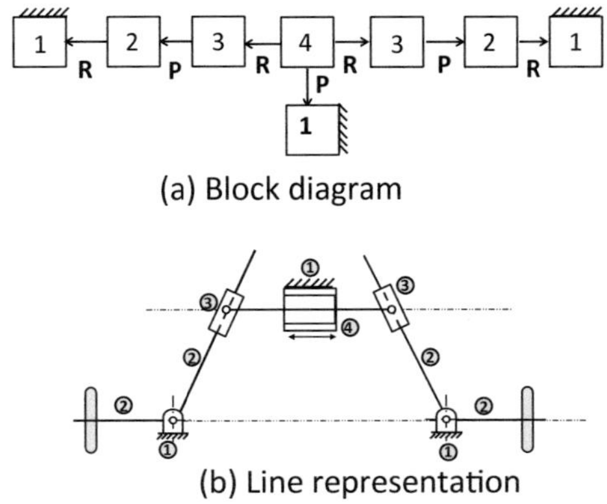

Fig. 1.39 Davi's steering gear

1.8 Equivalence of Higher Pairs as Lower Pairs

Consider a higher pair shown in Fig. 1.40. It consists of two bodies that have a point contact and they are rotating about fixed pivots O_2 and O_4. If the center of curvature of the contacting conjugate surfaces is known, the above higher pair can be, only for the instant shown above, represented as a four-bar mechanism as shown in Fig. 1.41. This is obtained by introducing revolute joints A, B that are connected by links 2, 4, and 3 to fixed pivots O_2 and O_4.

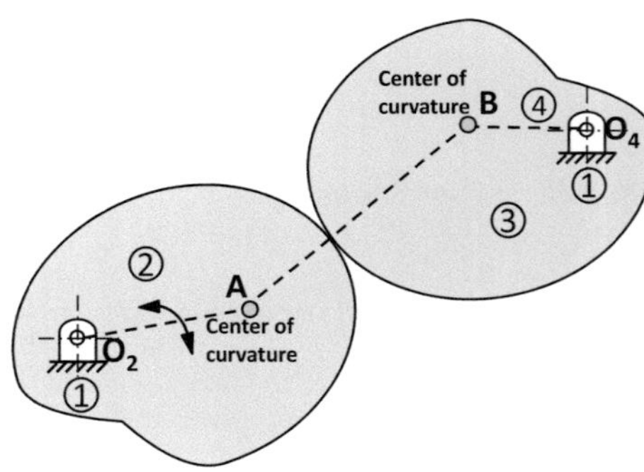

Fig. 1.40 Higher pair

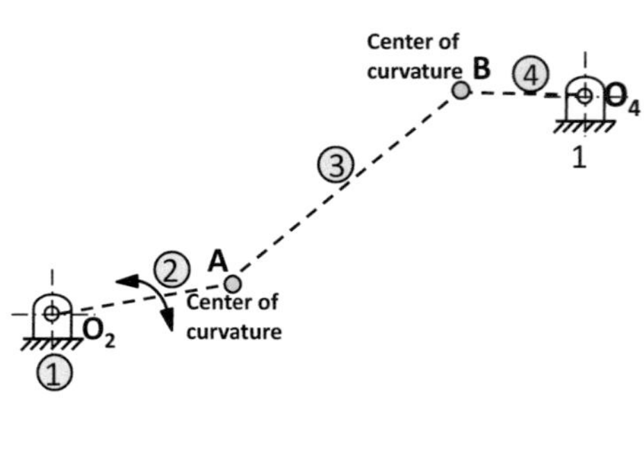

Fig. 1.41 Equivalent four-bar linkage representation of the higher pair shown in Fig. 1.40

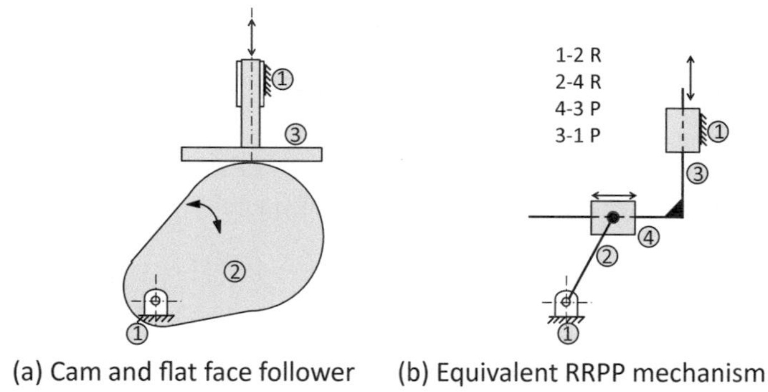

Fig. 1.42 Equivalent mechanism for a flat-face cam-follower

The higher pair of a cam and flat-face follower is shown in Fig. 1.42a and the equivalent RRPP mechanism with lower pairs is shown in Fig. 1.42b. The equivalence is only for the instance shown. The RRPP mechanism consists of a 1 that is fixed and frame 3 slides in it in the vertical direction. There is another cylinder 4 that slides on link 3 in the horizontal direction and link 4 is hinged to crank 2 that is hinged at the other end to a fixed pivot 1. As the crank rotates, frame 3 moves vertically up and down.

1.9 Mobility Analysis

The number of degrees of freedom, also known as mobility, of a mechanism is an important consideration in its practical application. It can be obtained by accounting for the number of links and the manner in which they are connected using kinematic pairs. We shall first start with a simple mechanism that has only revolute pairs and then move on to account for other variations that are possible in mechanisms.

1.9.1 Basic Gruebler's Criterion

An isolated planar link has 3 degrees of freedom. Consider a mechanism with ***n*** such planar links. A fixed link is not counted for the degree of freedom of the mechanism. Hence, the total number of degrees of freedom for all the moving planar links not connected to each other = 3(n − 1). Let j be the total number of lower pairs (revolute or prismatic) that will connect the links. Each lower pair connects two links and hence restricts two degrees of freedom.

Assuming that only two links are hinged at a joint, the total number of degrees of freedom of a mechanism of n links and j kinematic pairs is given by **Grubler's criterion as**

$$F = 3(n - 1) - 2j \tag{1.1}$$

Example 1.3

Test Grubler's criterion for a four-bar 4R linkage is shown in Fig. 1.19.

For a four-bar linkage n = 4, j = 4 (all revolute pairs).

From Eq. (1.1),

$F = 3(n - 1) - 2j = 3(4 - 1) - 2 \times 4 = 1.$

Example 1.4

Test Grubler's criterion for the mechanism is shown in Fig. 1.43a.

Block diagram representation of the mechanism shown in Fig. 1.43a is shown in Fig. 1.43b.

n = 7 (number of links).

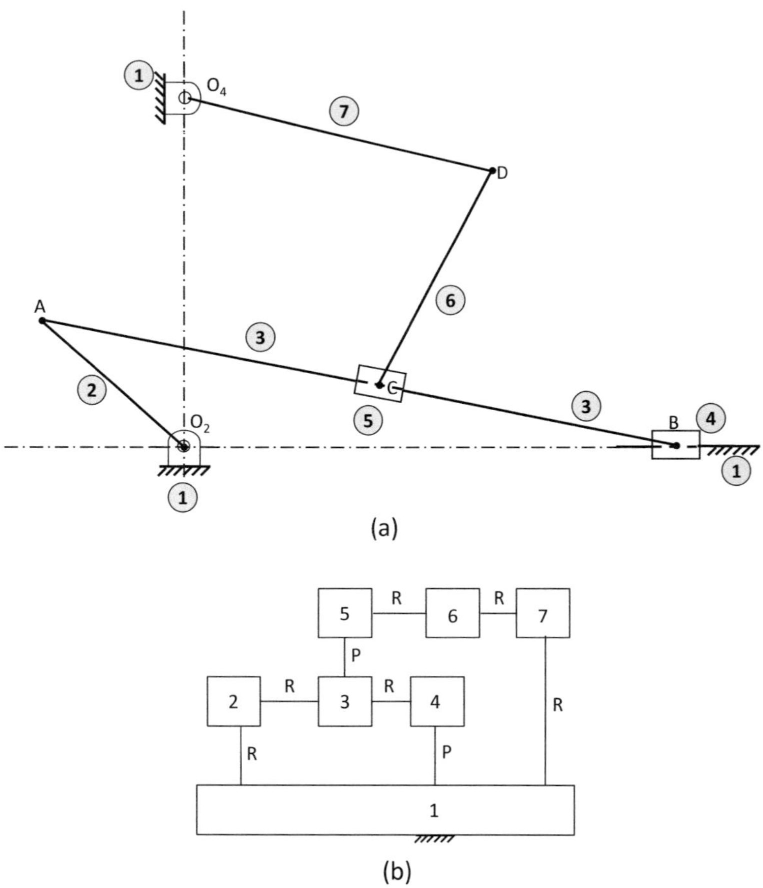

Fig. 1.43 **a** Mechanism of Example 1.4 **b** Block diagram of Example 1.4

J = 8 (number of revolute pairs (6) and prismatic pairs (2)).
From Eq. (1.1),
$F = 3(7 - 1) - 2 \times 8 = 2$ (degrees of freedom).

Example 1.5
Determine the number of degrees of freedom of the slotten-link quick return mechanism of Example 1.2.

Number of binary links $n = 6$.

Number of lower pairs, $j = 7$.

From Grubler's Eq. (1.1), $F = 3(n - 1) - 2j = 3(6 - 1) - 7 \times 2 = 1$. It has one degree of freedom, and since there is only one input through crank O_2A, this is a constant mechanism.

1.9.2 Grubler's Criterion for Higher Pairs and Pin-in-Slot Pairs

Since a higher pair and a pin-in-slot can restrict one degree of freedom, for a mechanism with lower, higher, and pin-in-slot pairs, Grubler's criterion is given by

$$F = 3(n - 1) - 2j - h - k \tag{1.2}$$

F: degrees of freedom, n: number of links, j: equivalent number of hinges, h: number of higher pairs, k: number of pin-in-slot pairs.

1.9.3 Grubler's Criterion for Compound or Higher Order Hinges

The above Grubler's criterion will have to be modified for mechanisms that may be different from those discussed above.

If there are more than two links at a hinge, some modification will have to be made for Grubler's criterion. Figure 1.44a shows that the hinge with links 1, 2, and 3 can be construed as a hinge obtained by joining 1–2 and 2–3 separately and then merging them as shown in Fig. 1.44b. This proves that a hinge with three links can be counted as two hinges. Similarly, a hinge with four links can be counted as equivalent to three hinges and so on.

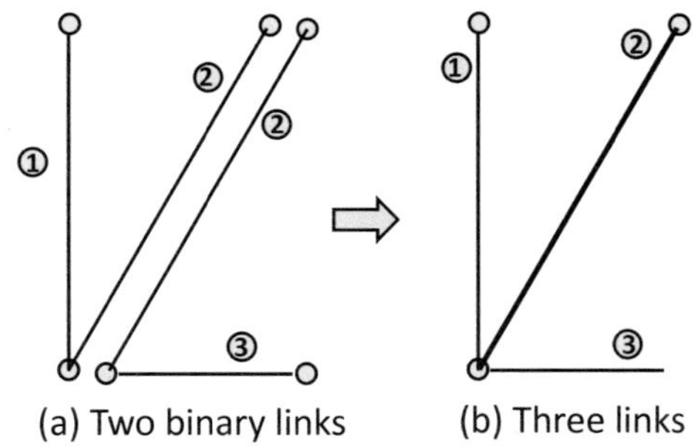

Fig. 1.44 Compound or higher order hinges

Therefore, the number of hinges that should be accounted for a mechanism with compound or higher order hinges is given by

$$j_{eff} = j_1 + 2j_2 + 3j_3 + \cdots \tag{1.3}$$

j_1 number of hinges with two links
j_2 number of hinges with three links
j_3 number of hinges with four links

Grubler's criterion for mechanisms with compound or higher order hinges is given by

$$F = 3(n - 1) - 2j_{eff} - h - k \tag{1.4}$$

F: degrees of freedom, n: number of links, j_{eff}: equivalent number of hinges, h: number of higher pairs, k: number of pin-in-slot pairs.

Example 1.6
Draw the block diagram representation of the links of the mechanism shown in Fig. 1.45 and determine the degrees of freedom of the mechanism.

The block diagram is shown in Fig. 1.46.

Number of links n = 8.

Number of hinges with two links, 6 : 1–2, 1–7, 3–4, 3–5, 4–6, 5–6,

Number of hinges with three links, 2: 2–3–8, 4–7–8.

From Eq. (1.3), effective number of lower pairs $j_{eff} = 6 + 2 \times 2 = 10$.

Using Grubler's criterion of Eq. (1.4),

$F = 3(n - 1) - 2j_{eff} = 3(8 - 1) - 2 \times 10 = 1$.

1.9.4 Redundant Degree of Freedom

Figure 1.47 shows an RPPR mechanism. Link 2 and link 4 are hinged and link 2 acts like a crank and link 4 is the receiver. If a link (link 3 above) can move without

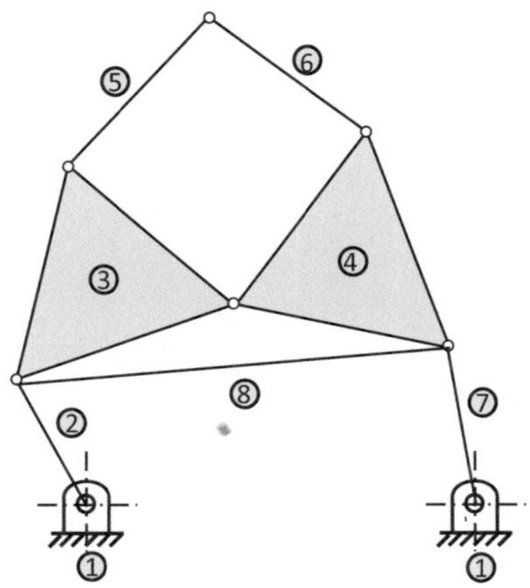

Fig. 1.45 Mechanism of Example 1.6

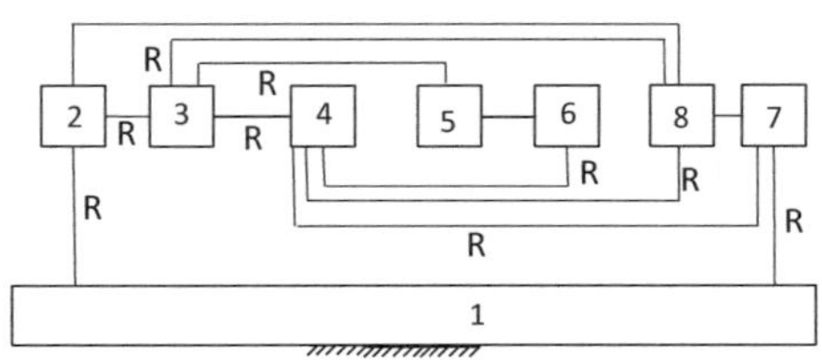

Fig. 1.46 Block diagram representation of Example 1.6

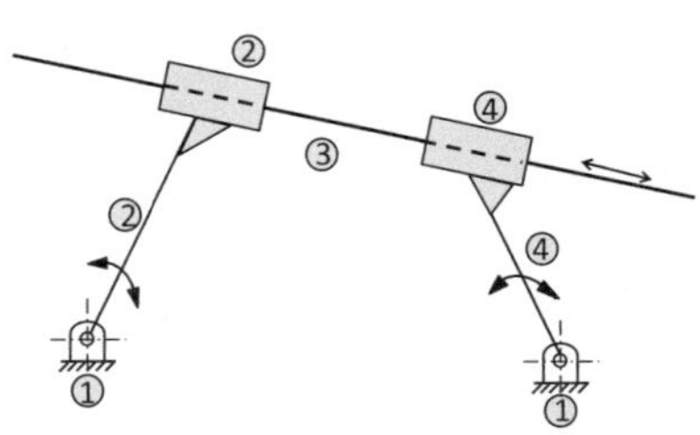

Fig. 1.47 Mechanism with redundant degree of freedom

causing any movement in the rest of the mechanism, then the link is said to have a redundant degree of freedom. Although a direct calculation by using Grubler's criterion of Eq. (1.2) gives a single degree of freedom for the above mechanism ($n = 4$, $j = 4$, $F = 3(n - 1) - 2j - h = 3(4 - 1) - 2 \times 4 = 1$). The mechanism has zero degree of freedom as it gets locked due to the above redundancy. That means one has to subtract the number of redundant degrees of freedom in Grubler's criterion of Eq. (1.4). But the above redundancy can be removed by slightly modifying link 3 as shown in Fig. 1.48, in which case the follower rotates at the same speed as the crank.

Another example of redundant degree of freedom is a cam with roller follower shown in Fig. 1.49. If we apply Grubler's Eq. (1.2).

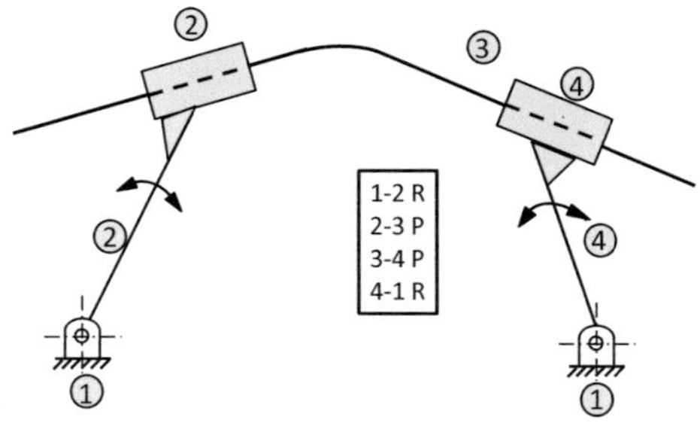

Fig. 1.48 Redundant degree freedom of Fig. 1.47: link 3 is modified

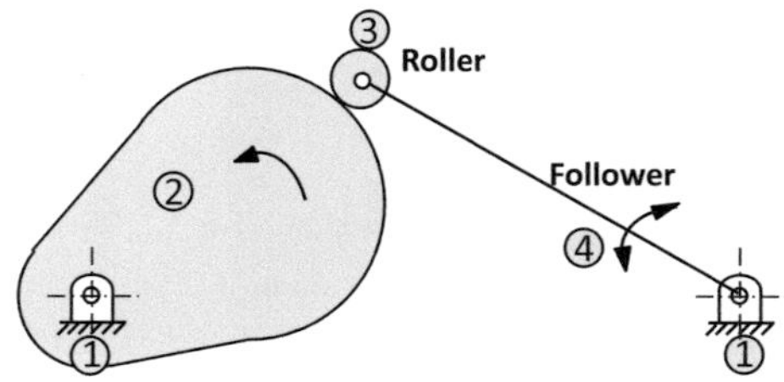

Fig. 1.49 Cam with roller follower

$n = 4, j = 3, h = 1, k = 0$

$F = 3(n - 1) = 2j - h = 3(4 - 1) - 2 \times \ 3 - 1 = 2$

However, the cam with roller follower has only one degree of freedom since rotation of the roller is a redundant degree of freedom. The only output obtained in this system is the oscillation of link 4 depending on the cam profile, irrespective of whether the roller turns or not. Now Grubler's equation can be modified as

$$F_{eff} = 3(n - 1) - 2j - h - k - F_r \tag{1.5}$$

where F_{eff}: effective degrees of freedom, n: number of links, j: equivalent number of lower pairs, h: number of higher pairs, k: number of pin-in-slot pairs, F_r: number of redundant degrees of freedom.

Example 1.7

Draw the block diagram representation of the links of the mechanism shown in Fig. 1.50 and determine the degrees of freedom of the mechanism.

The block diagram of the mechanism shown in Fig. 1.50 is shown in Fig. 1.51. Number of links = 8.

Number of hinges with 2 links, 9: 1–2, 3–4, 4–1, 4–5, 5–6, 6–1, 6–7, 7–8, 8–1.

Number of higher pairs, 1: 2–3.

Redundant degree of freedom $F_r = 1$.

From Grubler's criterion of Eq. (1.5).

$F = 3(8 - 1) - 2 \times 9 - 1 - 1 = 1.$

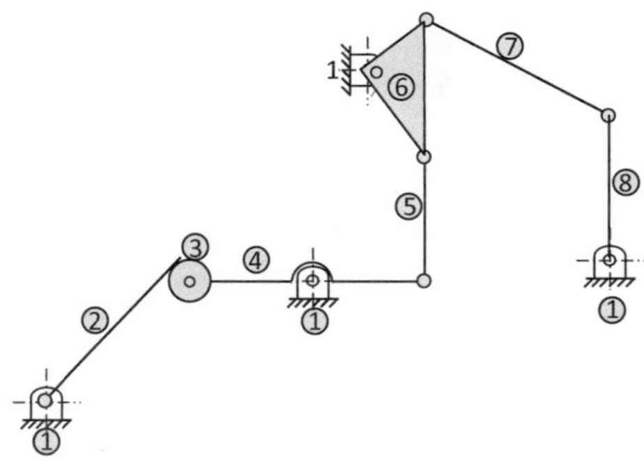

Fig. 1.50 Mechanism

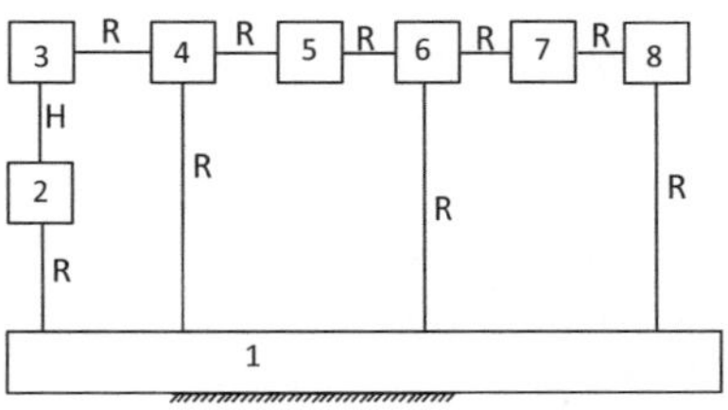

Fig. 1.51 Block diagram

1.9.5 Redundant Kinematic Pairs

Figure 1.52 shows an example of redundant kinematic pairs. In the most commonly encountered kinematic pair of a shaft and bearing shown in Fig. 1.52, we find that the second bearing is a redundant kinematic pair.

Now the modified Grubler's condition accounting for redundancy of kinematic pairs is given by

$$F_{eff} = 3(n-1) - 2(j_{eff} - j_r) - h - k - F_r \tag{1.6}$$

F_{eff}: effective degrees of freedom, n: number of links, j: number of lower pairs, j_r: number of redundant lower pairs, h: number of higher pairs, k: number of pin-in-slot pairs, F_r: number of redundant degrees of freedom.

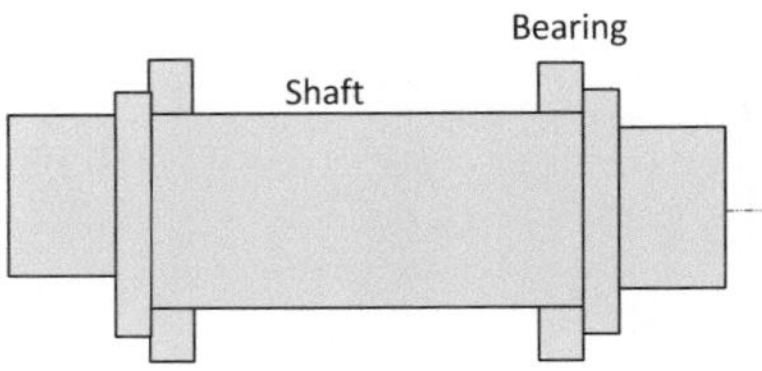

Fig. 1.52 Redundant kinematic pairs of a shaft and bearing

1.9.6 Prismatic Pairs in Different Directions

Figure 1.53 shows prismatic pairs between links 1, 2, and 3 occurring in different directions;1–2 along the horizontal, 2–3 along the inclined, and 3–1 along the vertical directions. Due to these prismatic pairs acting in different directions, there results an additional degree of freedom that has to be accounted for.

If we try to use Gruebler's criterion to both mechanisms that have similar number of links and prismatic joints, $n = 3, j = 3(P), F = 3(3-1) - 2 \times 3 = 0$. This shows that both mechanisms have zero degrees of freedom, which is not true.

Hence, accounting for the additional degree of freedom, the new Grubler's condition becomes

$$F_{eff} = 3(n - 1) - 2(j - j_r) - h - F_r + P_r \tag{1.7}$$

Where F_{eff}: effective degrees of freedom, n: number of links, j: equivalent number of lower pairs, j_r: number of redundant lower pairs, h: number of higher pairs, k: number of pin-in-slot pairs, F_r: number of redundant, degrees of freedom, P_r: number of three-link closed chains having three prismatic pairs in different directions.

1.9.7 Degree of Freedom Due to Kinematic Dimensions

Figure 1.54 shows a four-bar mechanism with a crank and follower of the same length that are always parallel to each other and in addition, the coupler and fixed links are also of the same length and parallel to each other. There is an additional link 5 that is parallel to the coupler hinged to links 2 and 4 that is also of the same length as the coupler. In this configuration, the above mechanism's degree of freedom determined by Grubler's criterion is given by.

$n = 5, j = 6$

$F = 3(5 - 1) - 2 \times 6 = 0$

But the above mechanism has a single degree of freedom similar to that of a four-bar mechanism and the additional link 5 is a redundant link with no effect on the degree of freedom.

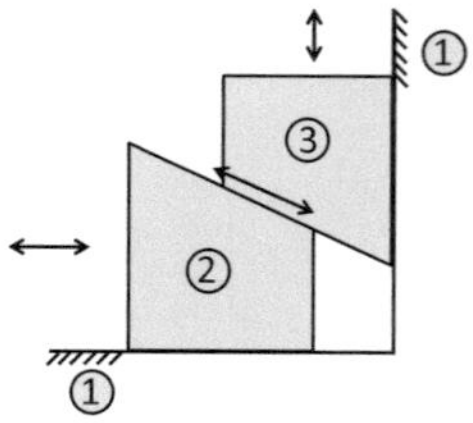

Fig. 1.53 Prismatic pairs in different directions

Fig. 1.54 Four-bar mechanism with an additional equal link parallel to the coupler

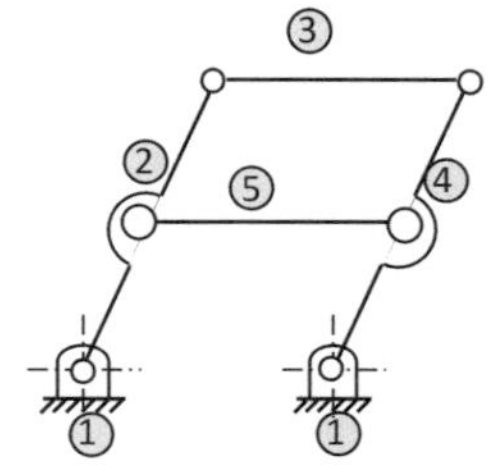

However, if the dimensions of the above mechanism are changed, it may not have any degree of freedom at all. Thus, the mechanism's degree of freedom depends on the dimensions of the links.

1.10 Grashof's Criterion for a Four-Bar Chain

A four-bar kinematic chain with links 1, 2, 3, and 4 is shown in Fig. 1.55. As discussed before, the above kinematic chain becomes a mechanism when one of its links gets fixed. Since a four-bar chain can have different lengths of links, depending on which of the four links gets fixed, it results in different types of mechanisms. The minimum and maximum lengths are also assigned and they could be for any of the linkages, not necessarily in the same order.

Grashof's criterion, which can systematically classify the type of a four-bar mechanism for the above chain is expressed as

$$\ell_{\max} + \ell_{\min} < \ell_1 + \ell_2 \tag{1.8}$$

For a Grashof chain satisfying the above equation, a crank-rocker mechanism is obtained if the shortest link is the crank and the adjacent link is the fixed link. Since there are two possibilities of achieving this, by either fixing the link of length ℓ_1 or ℓ_2, both are shown in Fig. 1.56a, b.

For a Grashof chain satisfying the above Eq. (1.8), a double rocker is obtained if coupler is the shortest link and it undergoes complete rotation. It is shown in Fig. 1.57.

Fig. 1.55 Four-bar kinematic chain

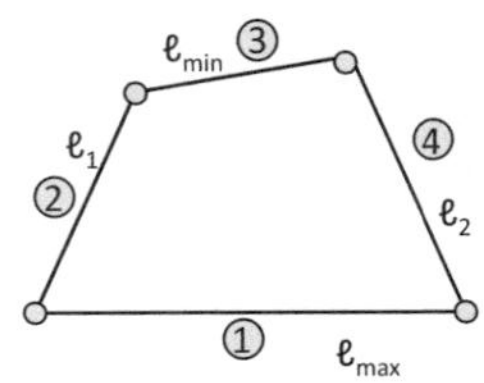

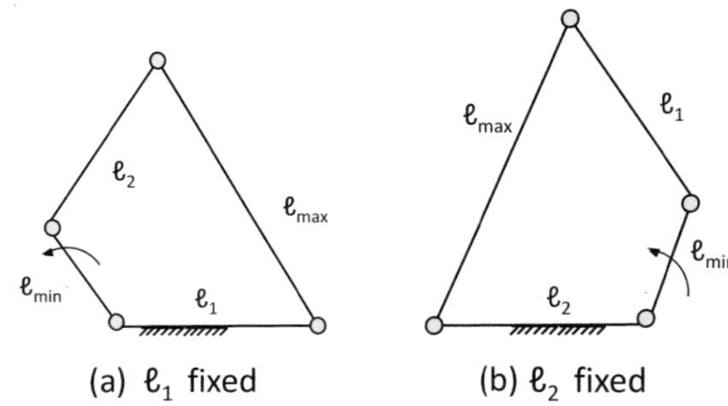

Fig. 1.56 Crank-rocker mechanism

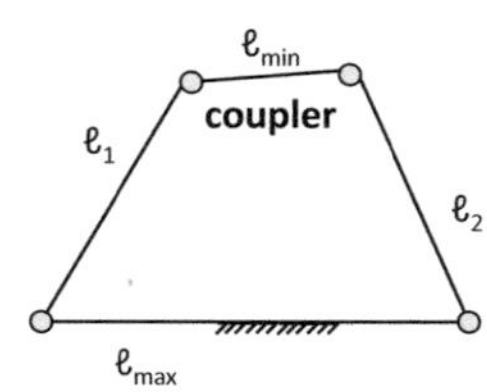

Fig. 1.57 Double rocker

A crank-crank four-bar mechanism is obtained if the shortest link is fixed. It will be discussed in the next example.

Example 1.8

A planar four-bar linkage can be constructed from links of the following lengths: 25, 75, 125, and 125 mm. Obtain various inversions of the mechanism. Checking Grashof's criterion of Eq. (1.8) by these links.

$$\begin{gathered} \ell_{max} + \ell_{min} < \ell_1 + \ell_2 \\ 125 + 25 < 75 + 125 \end{gathered} \tag{1}$$

Hence, Grashof's criterion is satisfied.

Figure 1.58 shows various inversions of the four-bar Grashof's chain.

Non-Grashof mechanism

A non-Grashof link shown in Fig. 1.59 satisfies the following equation:

$$\ell_{max} + \ell_{min} > \ell_1 + \ell_2 \tag{1.9}$$

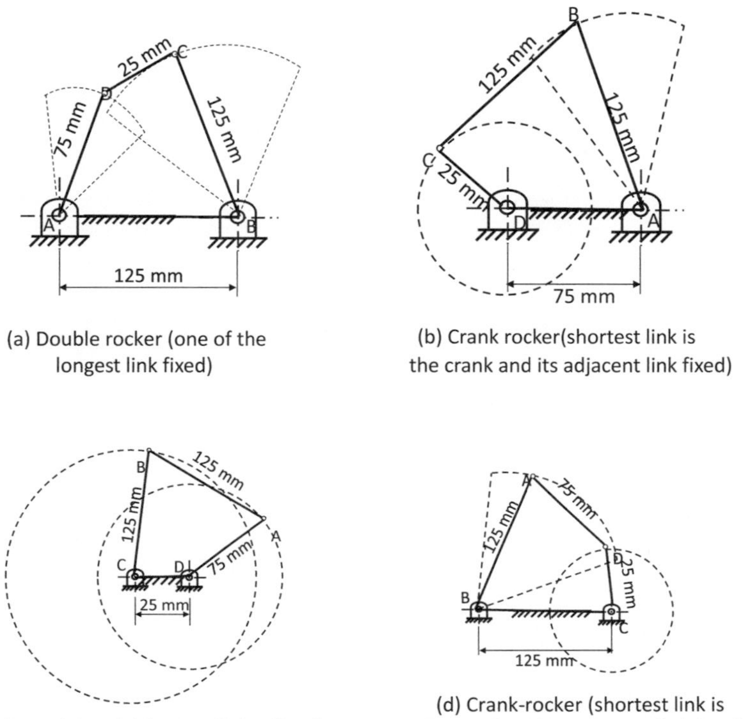

(a) Double rocker (one of the longest link fixed)

(b) Crank rocker(shortest link is the crank and its adjacent link fixed)

(c) Crank-crank (shortest link is fixed)

(d) Crank-rocker (shortest link is the crank and its adjacent link fixed)

Fig. 1.58 Inversions of the four-bar chain of Example 1.8

Fig. 1.59 Non-Grashof link

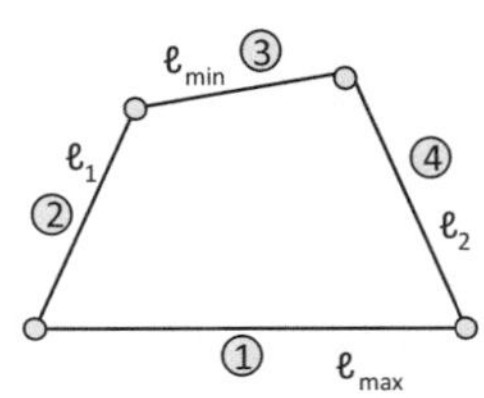

Following are the properties of a non-Grashof linkage

1. All the four inversions are double-rocker mechanisms.
2. The angle of oscillation can be greater than 180°.
3. The rockers can cross the line of the frame.
4. The mechanism can be driven from one configuration to its mirror image configuration.
5. Position of the longest link with respect to the frame decides the type of rocking movement.

Transitional Linkage

If a four-bar linkage shown in Fig. 1.56 satisfies the following equation

$$\ell_{\max} + \ell_{\min} = \ell_1 + \ell_2 \tag{1.10}$$

then it is known as a transitional linkage. As a general case, a transitional linkage behaves like a Grashof's linkage. But sometimes can enter uncertain configurations. All four inversions yield double-crank mechanisms. The above equations can be satisfied with two pairs of equal links; if links of equal length are not adjacent, it will result in parallelogram linkages. However, they can slip into anti-parallelogram configuration.

1.11 Position Analysis

1.11.1 Four-Bar Linkage

Figure 1.60 shows the angular positions of various links of a four-bar mechanism. Let O_2A be the crank and O_4B the follower. The angles between the linkages change as a function of time and it would be necessary to determine these angles for further analysis of the mechanism. The basic rule followed in determining the position of various linkages is the vector closure approach wherein we represent various links as vectors and the sum of the vectors will be equal to zero.

Before we use the vector closure approach, let us try to determine the angles based on trigonometry.

From Fig. 1.60, from triangle O_2O_4A, the distance $y = O_4A$ is given by

$$y^2 = \ell_1^2 + \ell_2^2 - 2\ell_1\ell_2 \cos\theta_2 \tag{1.11}$$

Similarly, from triangle O_4AB, y is given by

$$y^2 = \ell_3^2 + \ell_4^2 - 2\ell_3\ell_4 \cos\gamma \tag{1.12}$$

Equating the above equation, the transmission angle (the angle between the coupler and the follower) can be expressed as

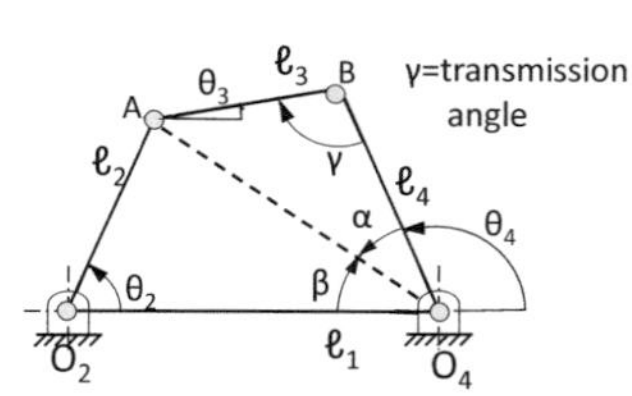

Fig. 1.60 Angular positions of a four-bar linkage

$$\gamma = \cos^{-1}\left\{\frac{\ell_1^2 + \ell_2^2 - \ell_3^2 - \ell_4^2 - 2\ell_1\ell_2\cos\theta_2}{-2\ell_3\ell_4}\right\} \tag{1.13}$$

$$\gamma = \cos^{-1}\left\{\frac{y^2 - \ell_3^2 - \ell_4^2}{-2\ell_3\ell_4}\right\} \tag{1.14}$$

$$\alpha = \cos^{-1}\left\{\frac{y^2 + \ell_4^2 - \ell_3^2}{2y\ell_4}\right\} \tag{1.15}$$

$$\beta = \cos^{-1}\left\{\frac{y^2 + \ell_1^2 - \ell_2^2}{2y\ell_1}\right\} \tag{1.16}$$

$$\theta_4 = 180^\circ - (\alpha + \beta) \tag{1.17}$$

Equation (1.11) and Eqs. (1.14)–(1.16) can be used to compute the positions of the four-bar linkage for various crank angles.

But we shall discuss the vector closure principle for carrying out position analysis that is a general procedure that can be used for any complex mechanism.

The position vectors of the four-bar mechanism can be written as

$$\begin{aligned} &-\ell_1\hat{\mathbf{i}} \\ &\ell_2\cos\theta_2\hat{\mathbf{i}} + \ell_2\sin\theta_2\hat{\mathbf{j}} \\ &\ell_3\cos\theta_3\hat{\mathbf{i}} + \ell_3\sin\theta_3\hat{\mathbf{j}} \\ &\ell_4\cos(180^\circ + \theta_4)\hat{\mathbf{i}} + \ell_4\sin(180^\circ + \theta_4)\hat{\mathbf{j}} \end{aligned} \tag{1.18}$$

Because of vector closure, the x and y components of the above equations should be equal to zero. Assuming θ_2 is known, these result in two non-linear equations in terms of angles θ_3 and θ_4. Therefore, for a given angle of the crank angle, θ_2, θ_3, and θ_4 can be obtained by using Newton–Raphson's method outlined in Appendix.

Equating components along the $\hat{\mathbf{i}}$ direction,

$$f_1(\theta_3, \theta_4) = \begin{pmatrix} -\ell_1 + \ell_2\cos\theta_2 + \ell_3\cos\theta_3 \\ +\ell_4\cos(180^\circ + \theta_4) \end{pmatrix} = 0 \tag{1.19}$$

Equating components along the $\hat{\mathbf{j}}$ direction,

$$\begin{aligned} &f_2(\theta_3, \theta_4) = \ell_2\sin\theta_2 + \ell_3\sin\theta_3 \\ &\ell_4\sin(180^\circ + \theta_4) = 0 \end{aligned} \tag{1.20}$$

$$\begin{bmatrix} \frac{\partial f_1(\theta_3,\theta_4)}{\partial\theta_3} & \frac{\partial f_1(\theta_3,\theta_4)}{\partial\theta_4} \\ \frac{\partial f_2(\theta_3,\theta_4)}{\partial\theta_3} & \frac{\partial f_2(\theta_3,\theta_4)}{\partial\theta_4} \end{bmatrix} \begin{Bmatrix} \Delta\theta_3 \\ \Delta\theta_4 \end{Bmatrix} = -\begin{Bmatrix} f_1(\theta_3, \theta_4) \\ f_2(\theta_3, \theta_4) \end{Bmatrix} \tag{1.21}$$

Using Eqs. (1.19), for a specific crank angle θ_2 and using arbitrary starting values of θ_3 and θ_4, the incremental step is calculated using Eq. (1.21). With the updated values of θ_3 and θ_4, the iteration is continued till convergence is obtained. Thus, the actual values of θ_3 and θ_4 are computed for each crank angle θ_2 and the process is repeated to obtain values of θ_3 and θ_4 for other crank angles.

The above methodology of solving multi-variable equations will be useful in many chapters of this book.

1.11.2 Slider-Crank Mechanism

Figure 1.61 shows a slider-crank mechanism of crank length R and connecting rod length L with an initial position of the crank at angle of θ.

The displacement measured from the pivot point of the crank is given by

$$\begin{aligned} x &= R(1-\cos\theta) + L(1-\cos\phi) \\ &= R(1-\cos\theta) + L\left[1-\sqrt{1-\left(\frac{R}{L}\right)^2\sin^2\theta}\right] \end{aligned} \tag{1.22}$$

The crank and connecting rod lengths and angles are related by

$$R\sin\theta = L\sin\phi \tag{1.23}$$

The second term of Eq. (1.22) can be approximated as

$$\sqrt{1-\left(\frac{R}{L}\right)^2\sin^2\theta} \approx 1-\frac{1}{2}\left(\frac{R}{L}\right)^2\sin^2\theta \tag{1.24}$$

From Eqs. 1.22 and 1.24, the displacement equation for the slider becomes

$$\begin{aligned} x &= R(1-\cos\theta) + L\left[1-1+\frac{1}{2}\left(\frac{R}{L}\right)^2\sin^2\theta\right] \\ &= R(1-\cos\theta) + \frac{R^2}{2L}\sin^2\theta \end{aligned} \tag{1.25}$$

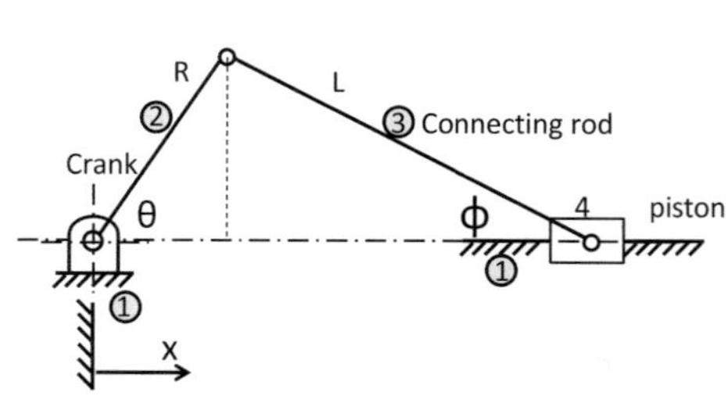

Fig. 1.61 Slider-crank mechanism

The above equation will be used in later chapters to compute approximate velocity and acceleration of the piston.

1.11.3 Scotch Yoke Mechanism

The displacement equation for a scotch yoke mechanism shown in Fig. 1.62 is given by

$$x = r(1 - \cos\theta) \tag{1.26}$$

Example 1.9

Block 2 in Fig. 1.63 is sliding in the horizontal direction and block 4 is sliding along an inclined direction of 15° due to the motion of block 2; both the blocks are connected by link 3 using hinge joints. Solve analytically and determine the position of block 4 by using position vectors. AB = 200 mm.

O in Fig. 1.64 is obtained by extending the inclined and horizontal planes of Fig. 1.63.

OA = AB sin 45°/sin 15° = 546.4 mm

OB = OA cos 15° + AB cos 15° = 669.2 mm

The vector $\vec{OA} = 546.4(\cos 15°\hat{\mathbf{i}} + \sin 15°\hat{\mathbf{j}}) = 527.79\hat{\mathbf{i}} + 141.42\hat{\mathbf{j}}$ mm.

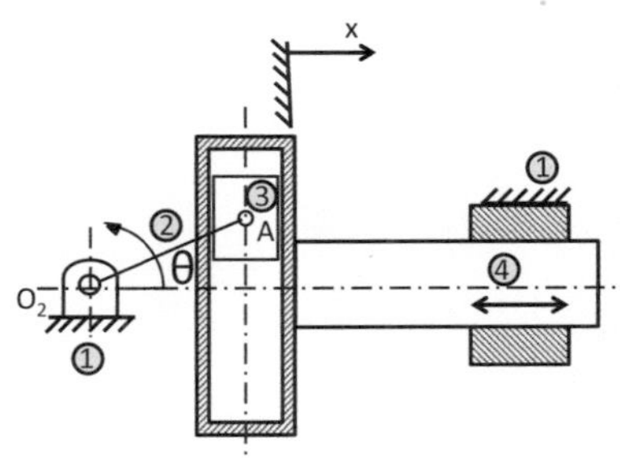

Fig. 1.62 Scotch yoke mechanism

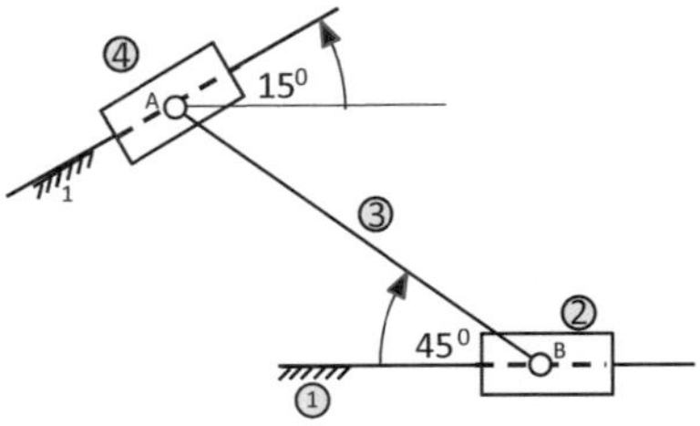

Fig. 1.63 Mechanism of Example 1.9

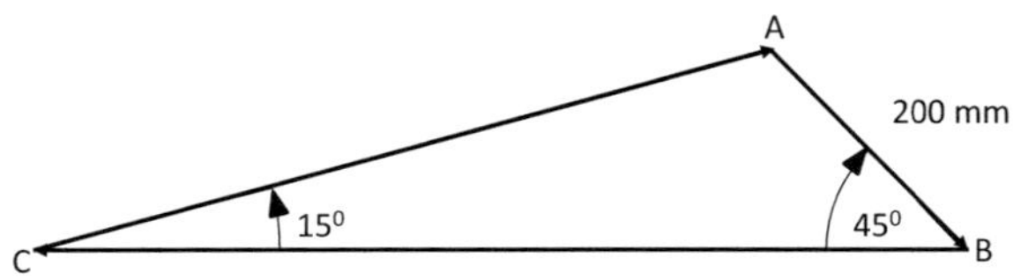

Fig. 1.64 Position vectors of the mechanism of Example 1.9

Example 1.10

Refer to Example 1.2

(a) Determine the position of the sliding block C when $\theta_1 = 135°$ by using the method of vector closure.
(b) Determine the total distance traveled by the sliding block C for each rotation of the crank O_2A.

Following are the various dimensions of the mechanism:

$O_1O_2 = 210$ mm; $O_1B = 400$ mm; $O_2A = 120$ mm; $CD = EF = 350$ mm; $BC = 130$ mm (Fig. 1.65).

The position vectors of Fig. 1.66 are given by

$$\begin{aligned}\mathbf{r}_{O_1A} &= \mathbf{r}_{O_1O_2} + r_{O_2A} \\ &= 210\hat{\mathbf{j}} + 120\left(\cos 135°\hat{\mathbf{i}} + \sin 135°\hat{\mathbf{j}}\right) \\ &= -84.85\hat{\mathbf{i}} + 294.85\hat{\mathbf{j}}\,\text{mm} \end{aligned} \quad (1)$$

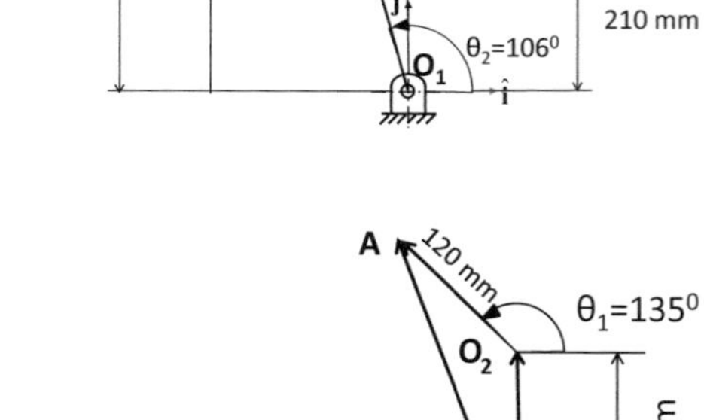

Fig. 1.65 Mechanism of Example 1.10

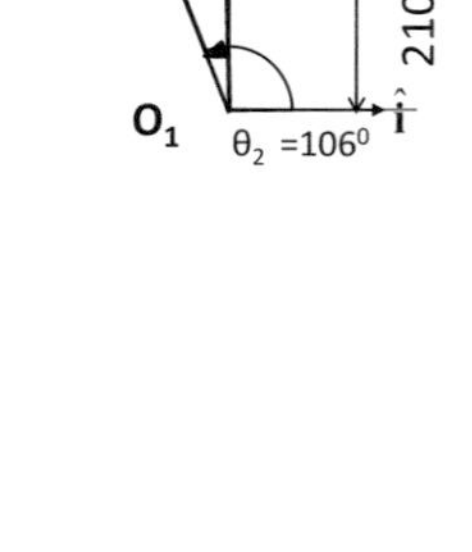

Fig. 1.66 Vector closure of O_1A, O_2A and O_1O_2 of Fig. 1.65

The magnitude of the above vector is given by

$$\left|\mathbf{r}_{O_1A}\right| = \sqrt{84.85^2 + 294.85^2} = 306.82\,\text{mm} \tag{2}$$

From Eqs. (1) and (2), the unit vector along the O_1A direction is given by

$$\hat{\mathbf{e}} = \frac{\mathbf{r}_{O_1A}}{\left|\mathbf{r}_{O_1A}\right|} = \frac{-84.85\hat{\mathbf{i}} + 294.85\hat{\mathbf{j}}}{306.82} = -0.2766\hat{\mathbf{i}} + 0.9610\hat{\mathbf{j}} \tag{3}$$

The angle θ_2 from Eq. (3) is given by

$$\theta_2 = \tan^{-1}\left(\frac{0.9610}{-0.2766}\right) = 106^\circ \tag{4}$$

From Eq. (3), the position vector for O_1B in Fig. 1.67 can be written as

$$\begin{aligned}\mathbf{r}_{O_1B} &= O_1B\hat{\mathbf{e}} = 400\left(-0.2766\hat{\mathbf{i}} + 0.9610\hat{\mathbf{j}}\right)\\ &= -110.62\hat{\mathbf{i}} + 384.40\hat{\mathbf{j}}\ \text{mm}\end{aligned} \tag{5}$$

The position vectors of Fig. 1.67 can be related by

$$\begin{aligned}&\mathbf{r}_{O_1C} = \mathbf{r}_{O_1B} + \mathbf{r}_{BC}\\ &x_{CO}\hat{\mathbf{i}} + 350\hat{\mathbf{j}} = \left(-110.62\hat{\mathbf{i}} + 384.40\hat{\mathbf{j}}\right) + 130\left(\cos\theta_3\hat{\mathbf{i}} + \sin\theta_3\hat{\mathbf{j}}\right)\\ &\qquad = (-110.62 + 130\cos\theta_3)\hat{\mathbf{i}} + (384.40 + 120\sin\theta_3)\hat{\mathbf{j}}\end{aligned} \tag{6}$$

Equating vector components on both sides of Eq. (6)

$$\begin{aligned}\theta_3 &= \sin^{-1}\tfrac{350-384.40}{130} = 195.37^\circ(\text{third quadrant})\\ x_{CO} &= -110.62 + 130\cos 195.37^\circ = -235.97\,\text{mm}\end{aligned} \tag{7}$$

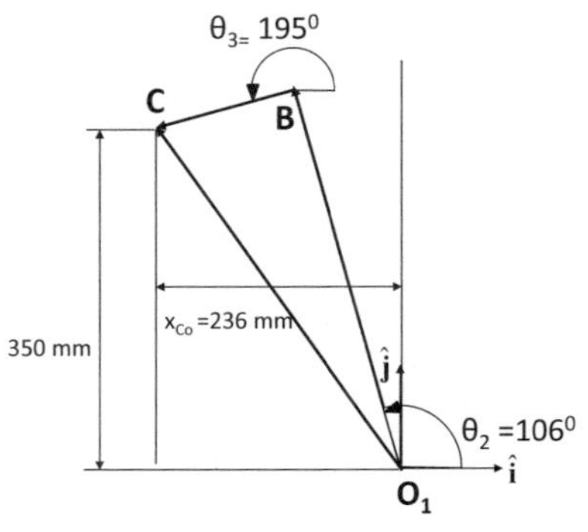

Fig. 1.67 Vector closure of O_1C, BC and O_1B of Fig. 1.65 ($\theta_1 = 135^\circ$)

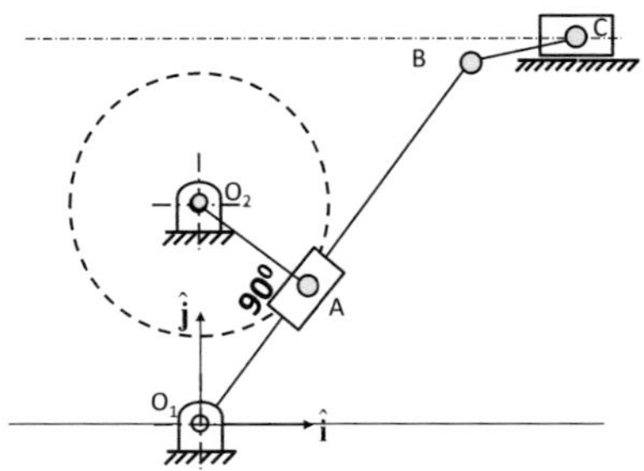

Fig. 1.68 Right extreme position of the mechanism of Example 1.10

Therefore, the position of the slider in the configuration shown in Fig. 1.67, when the crank O_{2A} is making an angle of 135°, is given by

$$\mathbf{r}_{O_1C} = -235.97\hat{\mathbf{i}} + 350\hat{\mathbf{j}}\,\text{mm}$$

(a) **Extreme right position of the piston**

When the piston is at the extreme right position as shown in Fig. 1.68, the link O_1B is tangential to the circle traversed by the crank O_2A and it makes an angle of 90° with it.

Since O_1O_2A is a right-angled triangle (Fig. 1.69), the length of O_1A is given by

$$\begin{aligned} O_1A &= \sqrt{O_1O_2^2 - O_2A^2} \\ &= \sqrt{O_1O_2^2 - O_2A^2} = \sqrt{210^2 - 120^2} = 172.34\ \text{mm} \end{aligned} \tag{8}$$

The angle θ_1 with reference to Fig. 1.69 is given by

$$\theta_4 = \tan^{-1}\frac{O_1A}{O_2A} = \tan^{-1}\frac{172.34}{120} = 55.15° \tag{9}$$

The position vector O_1A is given by

$$\begin{aligned} \mathbf{r}_{O_1A} &= 172.34\left(\cos 55.15°\hat{\mathbf{i}} + \sin 55.15°\hat{\mathbf{j}}\right) \\ &= 98.48\hat{\mathbf{i}} + 141.43\hat{\mathbf{j}}\ \text{mm} \end{aligned} \tag{10}$$

Fig. 1.69 Vector closure of O_1A, O_2A, and O_1O_2 of Fig. 1.68 (right extreme position)

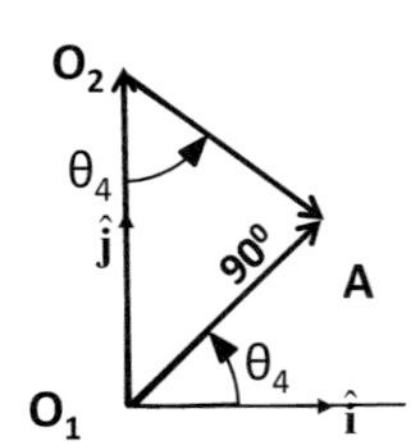

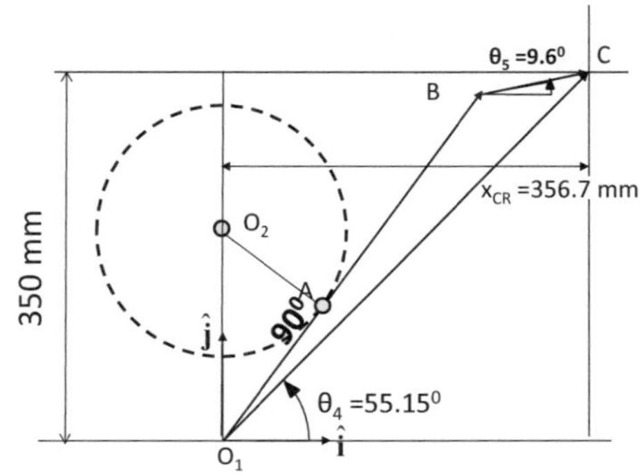

Fig. 1.70 Vector closure of O_1C, BC, BO_1 of Fig. 1.68 (right extreme position)

The unit vector along the O_1A direction is given by

$$\hat{\mathbf{e}} = \frac{\mathbf{r}_{O_1A}}{|\mathbf{r}_{O_1A}|} = \frac{98.48\hat{\mathbf{i}} + 141.43\hat{\mathbf{j}}}{172.34} = 0.5714\hat{\mathbf{i}} + 0.8207\hat{\mathbf{j}} \tag{11}$$

The position vector for O_1B can be written as

$$\begin{aligned}\mathbf{r}_{O_1B} &= O_1B\hat{\mathbf{e}} = 400(0.5714\hat{\mathbf{i}} + 0.8207\hat{\mathbf{j}}) \\ &= 228.56\hat{\mathbf{i}} + 328.28\hat{\mathbf{j}} \text{ mm}\end{aligned} \tag{12}$$

The position vectors of Fig. 1.70 can be related by

$$\begin{aligned}&\mathbf{r}_{O_1C} = \mathbf{r}_{O_1B} + \mathbf{r}_{BC} \\ &\Rightarrow x_{CR}\hat{\mathbf{i}} + 350\hat{\mathbf{j}} = (228.56\hat{\mathbf{i}} + 328.28\hat{\mathbf{j}}) + 130(\cos\theta_5\hat{\mathbf{i}} + \sin\theta_5\hat{\mathbf{j}}) \\ &\qquad = (228.56 + 130\cos\theta_5)\hat{\mathbf{i}} + (328.28 + 130\sin\theta_5)\hat{\mathbf{j}}\end{aligned} \tag{13}$$

Equating vector components on both sides of Eq. (13)

$$\theta_5 = \sin^{-1}\frac{350 - 328.28}{130} = 9.62^\circ \tag{14}$$

$$x_{CR} = 228.56 + 130\cos 9.62^\circ = 356.73 \text{ mm} \tag{15}$$

Therefore, the position of the slider in the configuration at the right extreme position is given by

$$\mathbf{r}_{O_1C} = 356.73\hat{\mathbf{i}} + 350\hat{\mathbf{j}} \text{ mm} \tag{16}$$

(c) **Extreme left position of the piston**

From Eq. (13) and symmetry ($|\theta_6| = |\theta_4| = 55.15^\circ$), the position vector for O_1B (Fig. 1.71) for the left extreme position can be written as

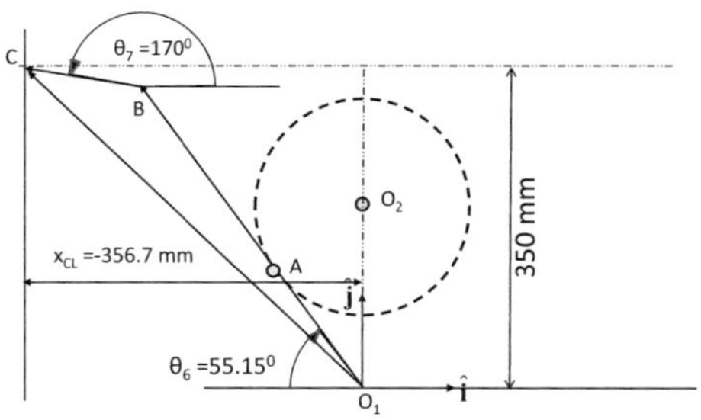

Fig. 1.71 Vector closure of O_1C, BC, O_1B of Fig. 1.68 (left extreme position)

$$\begin{aligned}\mathbf{r}_{O_1B} &= O_1B\hat{\mathbf{e}} = 400\big(-0.5714\hat{\mathbf{i}} + 0.8207\hat{\mathbf{j}}\big)\\ &= -228.57\hat{\mathbf{i}} + 328.26\hat{\mathbf{j}} \text{ mm}\end{aligned} \tag{17}$$

The position vectors of Fig. 1.71 can be related by

$$\begin{aligned}&\mathbf{r}_{O_1C} = \mathbf{r}_{O_1B} + \mathbf{r}_{BC}\\ &\Rightarrow x_c\hat{\mathbf{i}} + 350\hat{\mathbf{j}} = \big(-228.57\hat{\mathbf{i}} + 328.26\hat{\mathbf{j}}\big) + 130\big(\cos\theta_7\hat{\mathbf{i}} + \sin\theta_7\hat{\mathbf{j}}\big)\\ &\qquad = (-228.57 + 130\cos\theta_7)\hat{\mathbf{i}} + (328.26 + 130\sin\theta_7)\hat{\mathbf{j}}\end{aligned} \tag{18}$$

Equating vector components on both sides of Eq. (18)

$$\theta_7 = \sin^{-1}\frac{350 - 328.26}{130} = 170^\circ \tag{19}$$

$$x_{CL} = -228.57 + 130\cos 170^\circ = -356.73 \text{ mm} \tag{20}$$

Therefore, the position of the slider in the configuration at the left extreme position is given by

$$\mathbf{r}_{O_1C} = -356.73\hat{\mathbf{i}} + 350\hat{\mathbf{j}}\,\text{mm} \tag{21}$$

From Eqs. (16) and (21),

Total distance traveled by the slider in revolution of the crank is 2(356.73) = 713 mm.

The position analysis of the above mechanism can be easily extended to determine crank angles in the forward and return stroke that will give the ratio of time taken for the forward to return stroke.

Example 1.11

Consider the drag-link quick-return mechanism shown in Fig. 1.72. Determine the positions of all the linkages when the crank is making an angle of 180° with respect to the horizontal. O_2O_4 = 50 mm, O_2A = 60 mm, AB = 79 mm, BC = O_4B = O_4C = 63 mm, and CD = 115 mm.

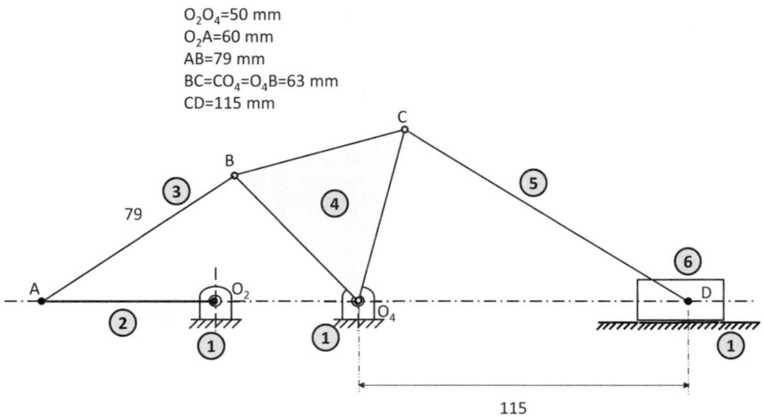

Fig. 1.72 Mechanism of Example 1.11

From Fig. 1.73, the angle φ is given by

$$\phi = \cos^{-1}\frac{O_4A^2 - AB^2 - O_4B^2}{-2 \times AB \times O_4B}$$

$$= \cos^{-1}\frac{110^2 - 79^2 - 63^2}{-2 \times 79 \times 63} = 101^\circ \tag{1}$$

When the law of sines is applied to the figure,

$$\frac{\sin\theta}{AB} = \frac{\sin\phi}{O_4A} = \frac{\sin\beta}{O_4B} \tag{2}$$

From Eq. (2),

$$\theta = \sin^{-1}\left(\frac{\sin\phi \times AB}{O_4A}\right)\sin^{-1} = \left(\frac{\sin 101^\circ \times 79}{110}\right) = 45^\circ \tag{3}$$

And

$$\beta = \sin^{-1}\left(\frac{\sin\phi \times O_4B}{O_4A}\right) = \sin^{-1}\left(\frac{\sin 101^\circ \times 63}{110}\right) = 34^\circ \tag{4}$$

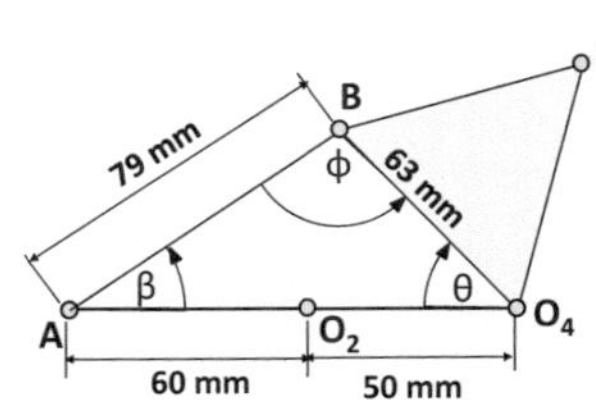

Fig. 1.73 Vector closure of O_4A, AB, and BO_4 of Fig. 1.72

From Fig. 1.74, the angle δ is given by

$$\delta = \cos^{-1}\frac{BC^2 - O_4B^2 - O_4C^2}{-2\times O_4B\times O_4C}$$
$$= \cos^{-1}\frac{63^2 - 63^2 - 63^2}{-2\times 63\times 63} = 60^\circ \quad (5)$$

The angle α is given by

$$\alpha = 180^\circ - \theta - \delta = 180^\circ - 45^\circ - 60^\circ = 75^\circ \quad (6)$$

By applying the law of since, the angle λ.

$$\lambda = \sin^{-1}\frac{O_4C\sin\alpha}{CD} = \sin^{-1}\frac{63\sin 75}{115} = 32^\circ \quad (7)$$

The position vectors of various linkages are as follows:

$$\mathbf{r}_{O_2A} = -O_2A\hat{\mathbf{i}} = -60\hat{\mathbf{i}}\text{ mm} \quad (8)$$

$$\mathbf{r}_{AB} = AB\left(\cos\beta\hat{\mathbf{i}} + \sin\beta\hat{\mathbf{j}}\right) = 79\left(\cos 34^\circ\hat{\mathbf{i}} + \sin 34^\circ\hat{\mathbf{j}}\right)$$
$$= \left(65.5\hat{\mathbf{i}} + 44.18\hat{\mathbf{j}}\right)\text{ mm} \quad (9)$$

$$r_{O_4B} = O_4B\left(-\cos\theta\hat{\mathbf{i}} + \sin\theta\hat{\mathbf{j}}\right) = 63\left(-\cos 45^\circ\hat{\mathbf{i}} + \sin 45^\circ\hat{\mathbf{j}}\right)$$
$$= \left(-44.55\hat{\mathbf{i}} + 44.55\hat{\mathbf{j}}\right)\text{ mm} \quad (10)$$

$$\mathbf{r}_{O_4C} = O_4C\left(\cos\alpha\hat{\mathbf{i}} + \sin\alpha\hat{\mathbf{j}}\right) = 63\left(\cos 75^\circ\hat{\mathbf{i}} + \cos 75^\circ\hat{\mathbf{j}}\right)$$
$$= \left(16.33\hat{\mathbf{i}} + 60.85\hat{\mathbf{j}}\right)\text{ mm} \quad (11)$$

$$\mathbf{r}_{CD} = CD\left(\cos\lambda\hat{\mathbf{i}} - \sin\lambda\hat{\mathbf{j}}\right) = 115\left(\cos 32^\circ\hat{\mathbf{i}} - \sin 32^\circ\hat{\mathbf{j}}\right)$$
$$= \left(97.53\hat{\mathbf{i}} - 60.94\hat{\mathbf{j}}\right)\text{ mm} \quad (12)$$

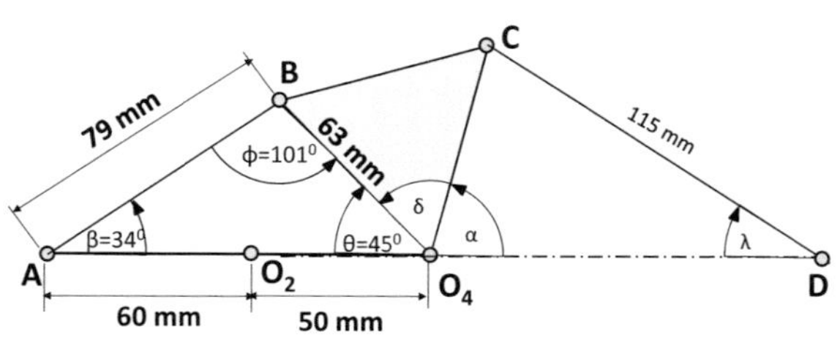

Fig. 1.74 Vector closure of O_4B, BC, and CO_4 and O_4C, CD, and DO_4 of Fig. 1.72

1.12 Conclusions

In this chapter, we have studied various types of mechanisms and their basic characteristics. Many of the mechanisms that were discussed in this chapter will be revisited in the later chapters on kinematics, synthesis, and kinetics. We have also conducted position analysis of mechanisms that is a first step before computing velocity and accelerations of various links and can then conduct their force analysis. Discussing the same mechanisms in different contexts in the later chapter will provide continuity in getting a holistic view of the basics of machines and mechanisms. The block diagram representation of various mechanisms presented in this chapter should be helpful in clearly identifying a mechanism; block diagrams will be provided to most of the exercise problems of this chapter also, and the same mechanisms will be revisited in the later chapters to analyze them from different points of view. The basic objective of this chapter was to give a basic idea of how to understand the working of a mechanism and also possibly to design new mechanisms.

1.13 Exercise Problems

1.1 Consider the mechanism shown in Fig. P1.1a and its block diagram representation in Fig. P1.1b. There are three pistons, $\mathcal{B}_2$ sliding in the vertical slot and the other two, $\mathcal{B}_1$ and , $\mathcal{B}_3$, sliding in the horizontal slot. $\mathcal{B}_3$ and $\mathcal{B}_2$ are connected by a rigid link $\mathcal{B}_5$, pinned, respectively, at A and B. $\mathcal{B}_2$ and $\mathcal{B}_1$ are connected by a rigid link $\mathcal{B}_4$, pinned respectively, at B and C. Determine its mobility.

1.2 Consider the mechanism shown in Fig. P1.2a and its block diagram representation in Fig. P1.2b. Pistons 2 and 6 slide in the horizontal direction; link 3 is pinned to piston 2 and link 5 is pinned to piston 6; links 3, 4, and 5 are pinned at C. The mechanism is driven by link 4 in the counterclockwise direction at the given instant. Determine its mobility.

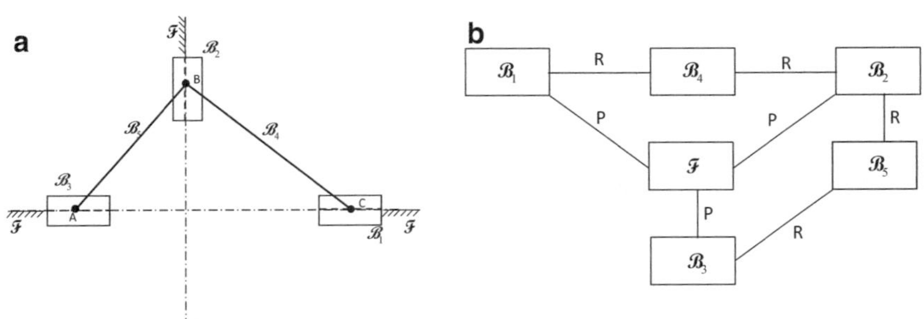

Fig. P1.1 **a** Mechanism of exercise problem 1.1. **b** Block diagram of the mechanism of exercise problem 1.1

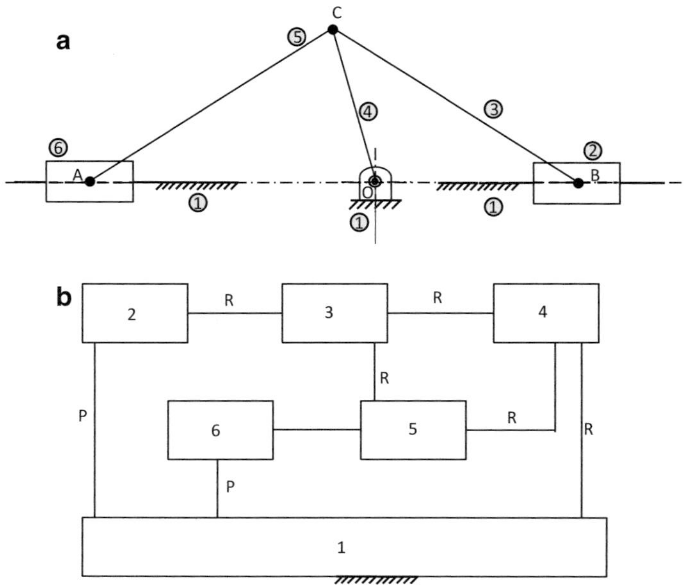

Fig. P1.2 **a** Mechanism of exercise problem 1.2. **b** Block diagram of the mechanism of exercise problem 1.2

1.3 For the mechanism shown in Fig. P1.3a and its block diagram representation shown in Fig. P1.3b, determine its mobility. Links 2 and 3 are slender rods and link 4 is a disc.

1.4 Determine the mobility of the mechanism shown in Fig. P1.4a and its block diagram representation shown in Fig. P1.4b.

1.5 Determine the mobility of the mechanism shown in Fig. P1.5a and its block diagram representation shown in Fig. P1.5b.

1.6 Determine the mobility of the mechanism shown in Fig. P1.6a and its block diagram representation shown in Fig. P1.6b.

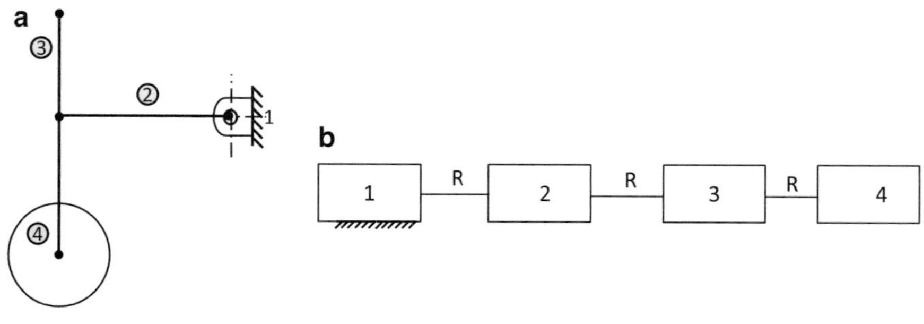

Fig. P1.3 **a** Mechanism of exercise problem 1.3. **b** Block diagram of the mechanism of exercise problem 1.3

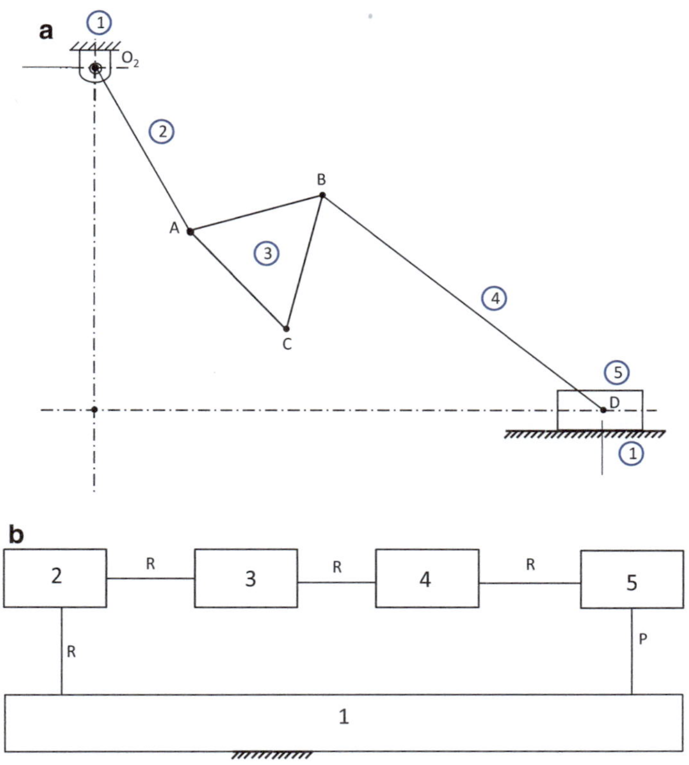

Fig. P1.4 **a** Mechanism of exercise problem 1.4. **b** Block diagram of the mechanism of exercise problem 1.4

1.7 Consider the mechanism shown in Fig. P1.7 in which the link $\mathcal{D}$ rotating in the clockwise direction is pushing link $\mathcal{B}$ by sliding action; its block diagram is shown in Fig. P1.7b. Determine its mobility.

1.8 Determine the mobility of the mechanism shown in Fig. P1.8a and its block diagram representation shown in Fig. P1.8b.

1.9 Determine the mobility of the mechanism shown in Fig. P1.9a and its block diagram representation shown in Fig. P1.9b.

1.10 Determine the mobility of the mechanism shown in Fig. P1.10a and its block diagram representation shown in Fig. P1.10b.

1.11 Determine the mobility of the mechanism shown in Fig. P1.11a and its block diagram representation shown in Fig. P1.11b.

1.12 Determine the mobility of the crank-shaper mechanism shown in Fig. P1.12a and its block diagram representation shown in Fig. P1.12b.

1.13 Determine the mobility of the automobile throttle (accelerator) mechanism shown in Fig. P1.13. Number all the linkages, identify all the kinematic pairs, and draw their block diagram representation.

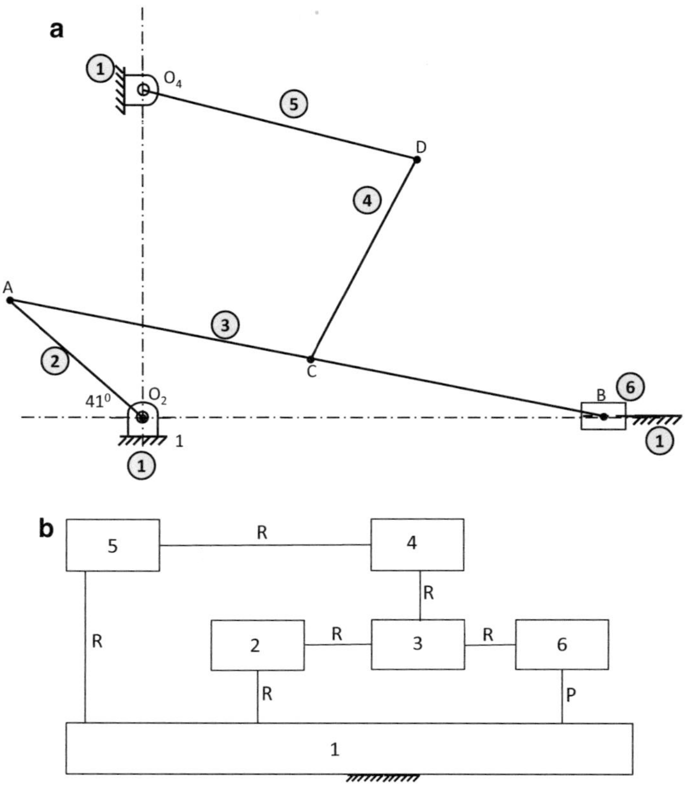

Fig. P1.5 **a** Mechanism of exercise problem 1.5. **b** Block diagram of the mechanism of exercise problem 1.5

1.14 Consider the mechanism shown in Fig. P1.14.

(a) Draw a block diagram representation showing the kinematic pairs.
(b) Determine the mobility.

1.15–1.26 Determine the mobility of the mechanism and draw the corresponding mechanism, whose block diagrams are shown in the following Figs. P1.15, P1.16, P1.17, P1.18, P1.19, P1.20, P1.21, P1.22, P1.23, P1.24, P1.25, P1.26.

1.27 Determine the time ratio of the crank-shaper quick-return mechanism shown in Fig. P1.27.

1.28 Determine the time ratio of the quick-return mechanism shown in Fig. P1.28.

1.29 Determine the position vectors of various links of the mechanism shown in Fig. P1.29 using the vector closure equation. See Fig. P1.1b for its block diagram representation.

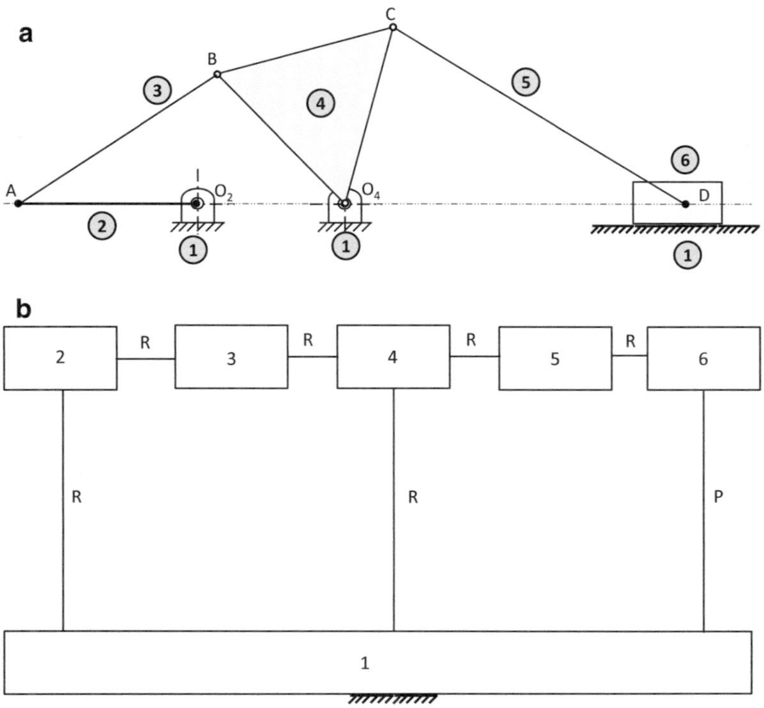

Fig. P1.6 **a** Mechanism of exercise problem 1.6. **b** Block diagram of the mechanism of exercise problem 1.6

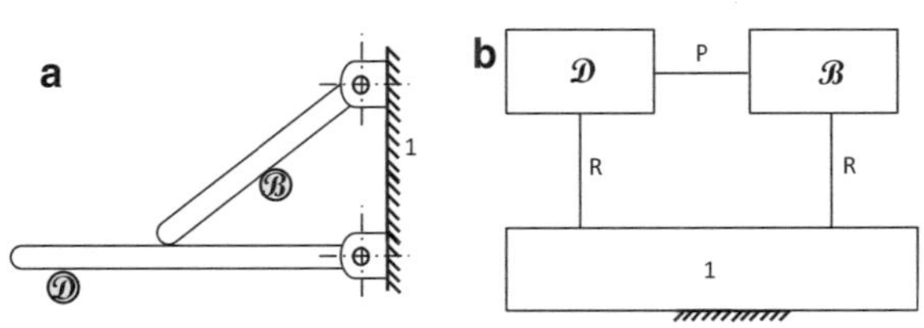

Fig. P1.7 **a** Mechanism of exercise problem 1.7. **b** Block diagram of the mechanism of exercise problem 1.7

1.30 Determine the position vectors of various links using the vector closure equation of the mechanism shown in Fig. P1.30.

1.31 Determine the position vectors of various links of the mechanism shown in Fig. P1.31 using the vector closure equation.

1.32 Determine the position vectors of various links of the mechanism shown in Fig. P1.32 using the vector closure equation. See Fig. P1.4b for its block diagram representation.

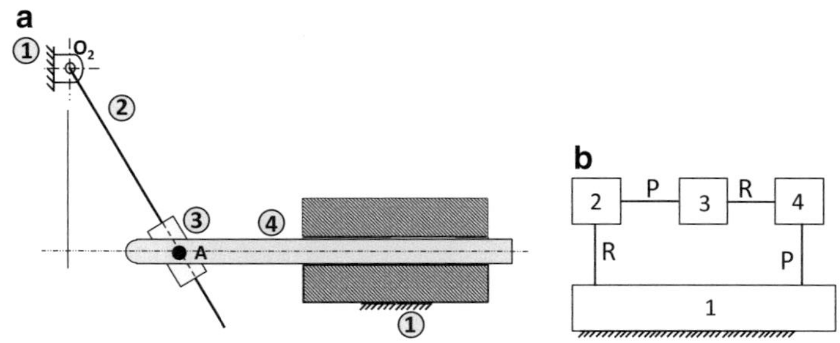

Fig. P1.8 **a** Mechanism of exercise problem 1.8. **b** Block diagram of the mechanism of exercise problem 1.8

Fig. P1.9 **a** Mechanism of exercise problem 1.9. **b** Block diagram of the mechanism of exercise problem 1.9

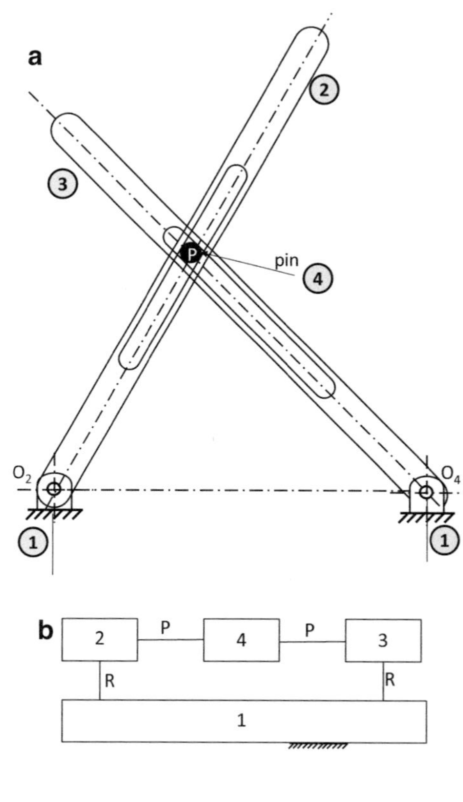

1.33 Determine the position vectors of various links of the mechanism shown in Fig. P1.33 using the vector closure equation. See Fig. P1.5b for the block diagram.

1.34 Determine the position vectors of various links of the mechanism shown in Fig. P1.34 using the vector closure equation, when link 2 (crank) makes an angle of 45° in the counterclockwise direction with respect to the horizontal. See Fig. P1.6b for the block diagram.

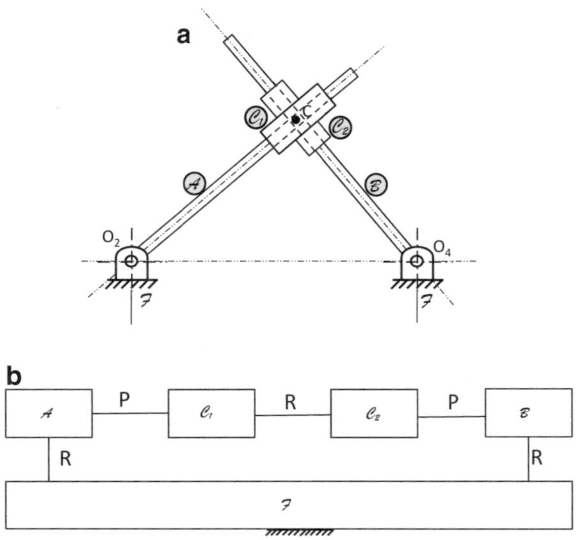

Fig. P1.10 **a** Mechanism of exercise problem 1.10. **b** Block diagram of the mechanism of exercise problem 1.10

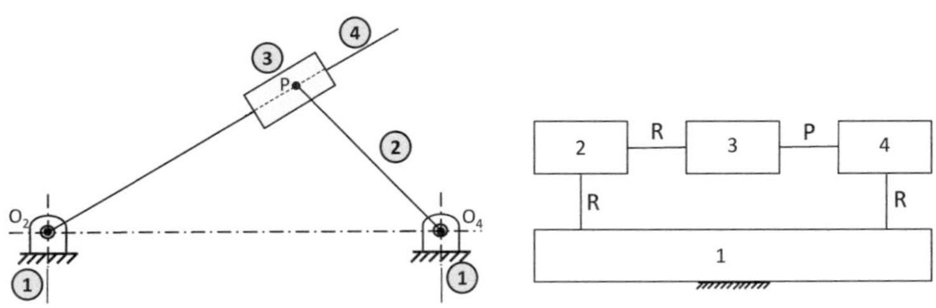

Fig. P1.11 **a** Mechanism of exercise problem 1.11. **b** Block diagram of the mechanism of exercise problem 1.11

1.35 Determine the position vectors of various links of the mechanism shown in Fig. P1.35 using the vector closure equation.
1.36 Determine the position vectors of various links of the mechanism shown in Fig. P1.36 using the vector closure equation. See Fig. P1.8b for its block diagram representation.
1.37 Determine the position vectors of various links of the mechanism shown in Fig. P1.37 using the vector closure equation. See Fig. P1.9b for its block diagram representation.
1.38 Determine the position vectors of various links of the mechanism shown in Fig. P1.38 using the vector closure equation. See Fig. P1.10b for its block diagram representation.

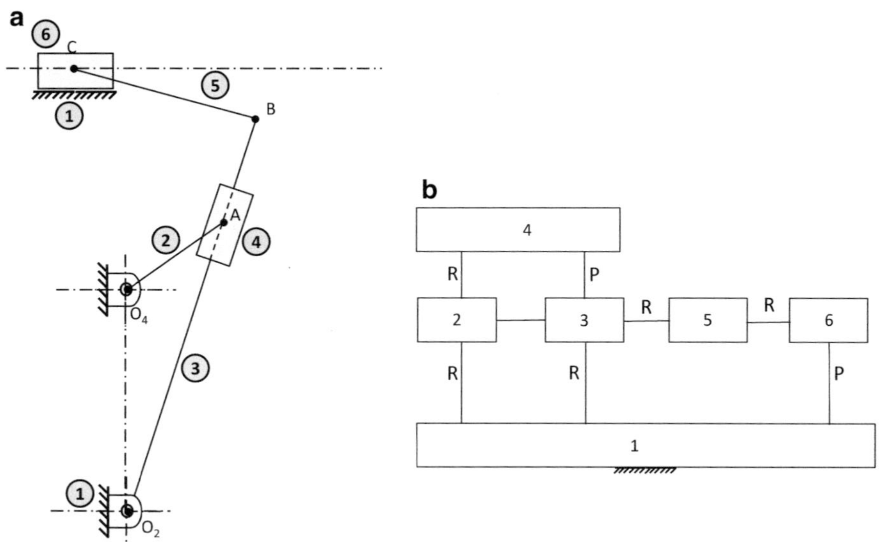

Fig. P1.12 **a** Mechanism of exercise problem 1.12. **b** Block diagram of the mechanism of exercise problem 1.12

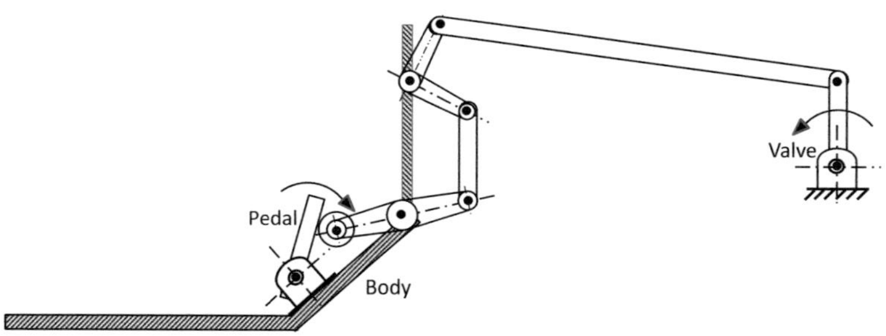

Fig. P1.13 Mechanism of exercise problem 1.13

Fig. P1.14 Mechanism of exercise problem 1.14

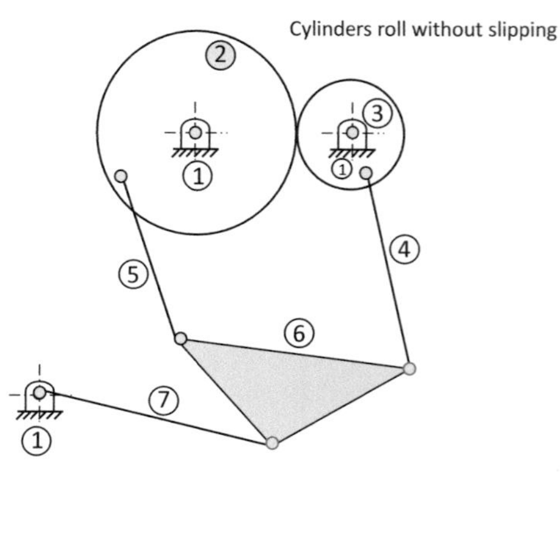

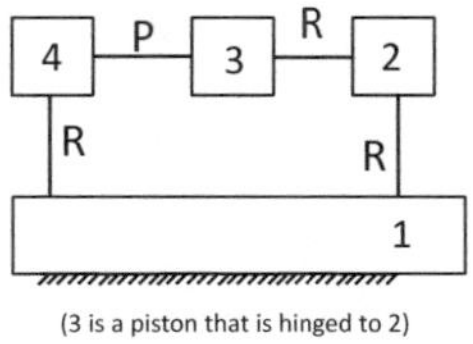

Fig. P1.15 Block diagram of the mechanism of exercise problem 1.15

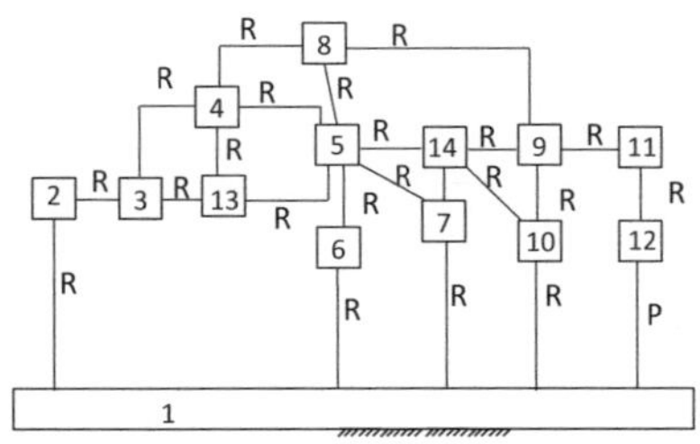

Fig. P1.16 Block diagram of the mechanism of exercise problem 1.16

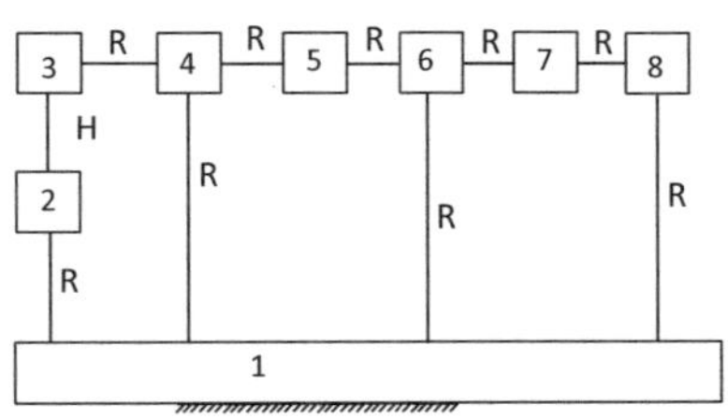

Fig. P1.17 Block diagram of the mechanism of exercise problem 1.17

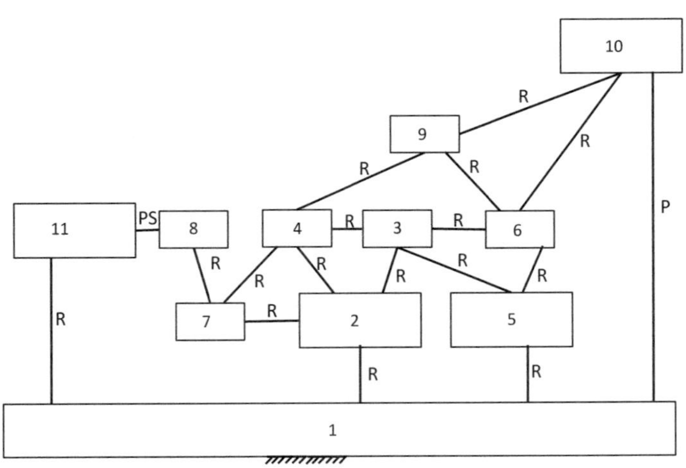

Fig. P1.18 Block diagram of the mechanism of exercise problem 1.18

Fig. P1.19 Block diagram of the mechanism of exercise problem 1.19

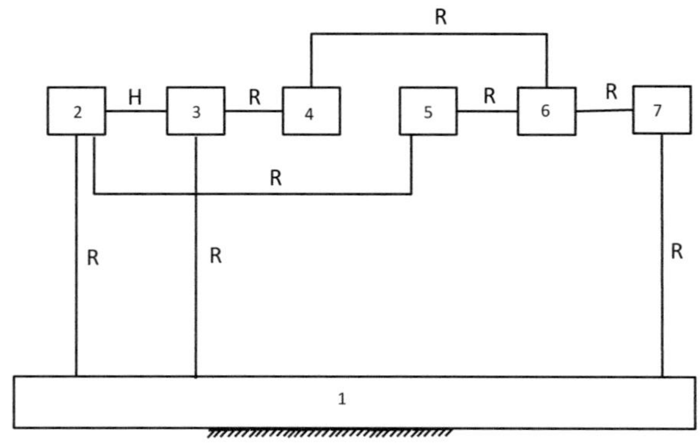

Fig. P1.20 Block diagram of the mechanism of exercise problem 1.20

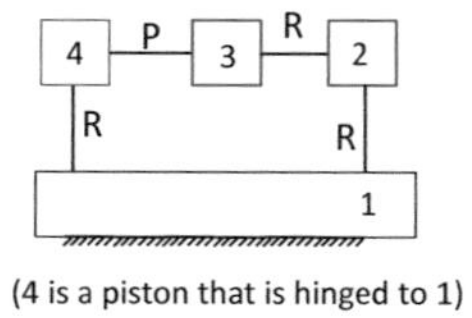

Fig. P1.21 Block diagram of the mechanism of exercise problem 1.21

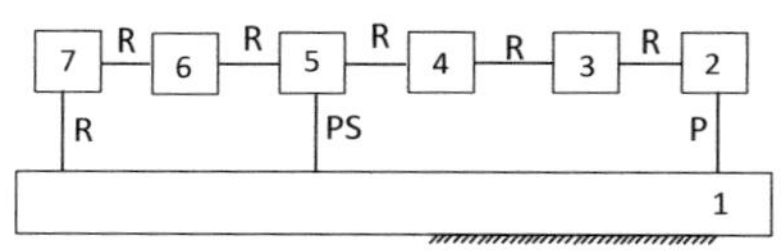

Fig. P1.22 Block diagram of the mechanism of exercise problem 1.22

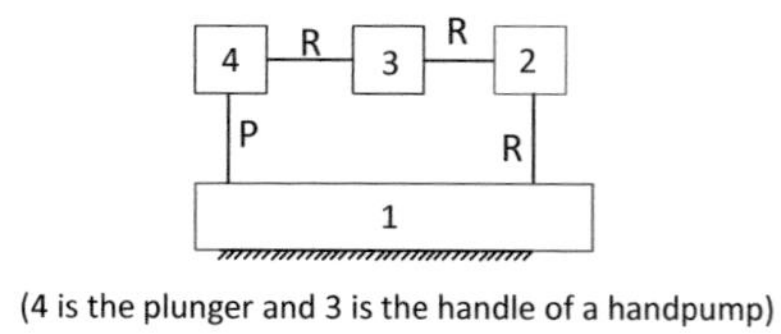

Fig. P1.23 Block diagram of the mechanism of exercise problem 1.23

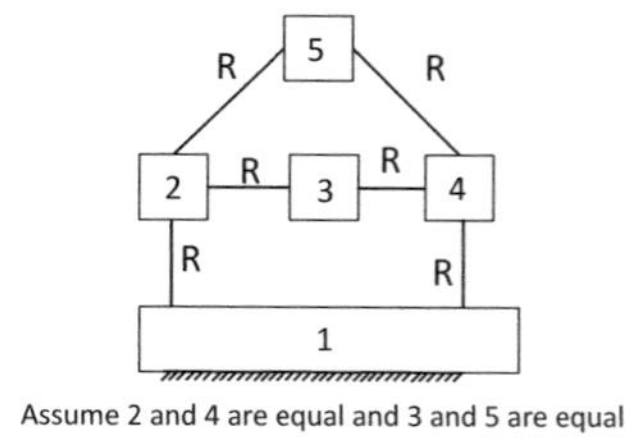

Fig. P1.24 Block diagram of the mechanism of exercise problem 1.24

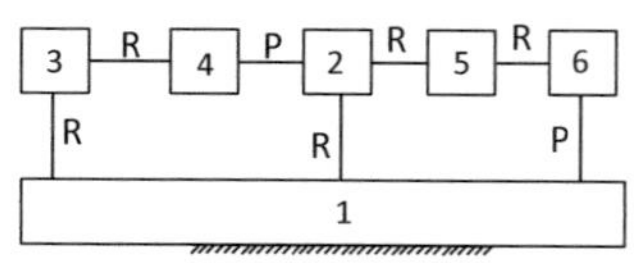

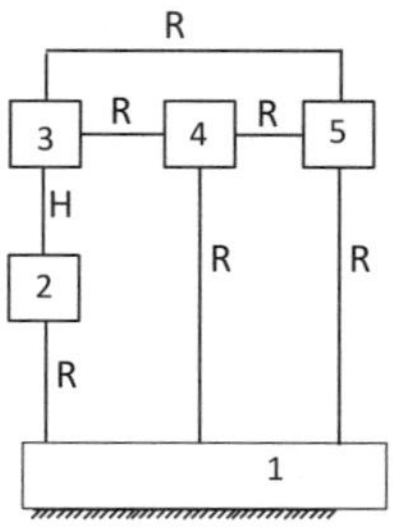

Fig. P1.25 Block diagram of the mechanism of exercise problem 1.25

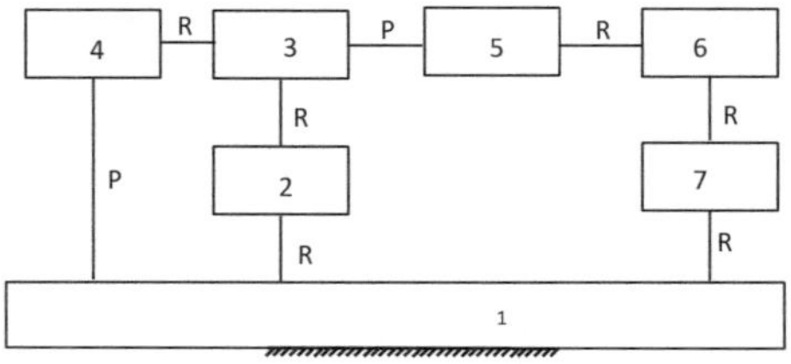

Fig. P1.26 Block diagram of the mechanism of exercise problem 1.26

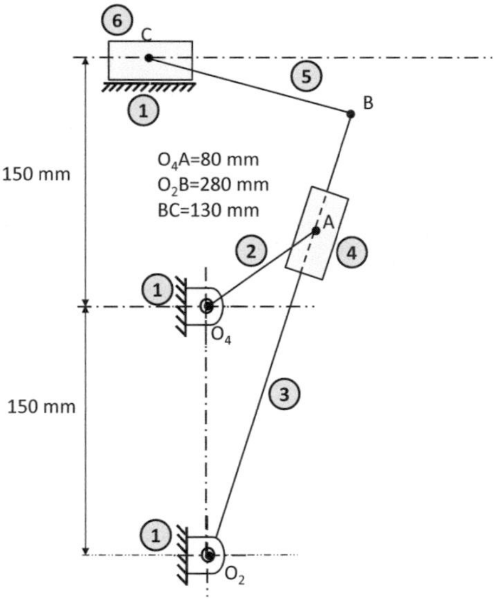

Fig. P1.27 Mechanism of exercise problem 1.27

1.39 Determine the position vectors of various links of the mechanism shown in Fig. P1.39 using the vector closure equation. See Fig. P1.11b for its block diagram representation.

1.40 Determine the position vectors of various links of the mechanism shown in Fig. P1.40 using the vector closure equation. See Fig. P1.12b for its block diagram.

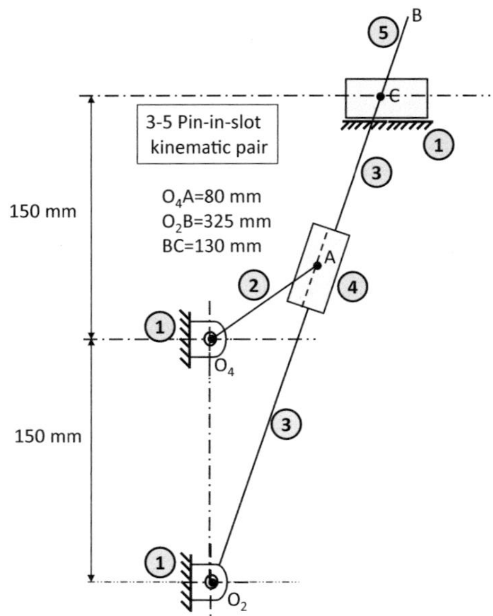

Fig. P1.28 Mechanism of exercise problem 1.28

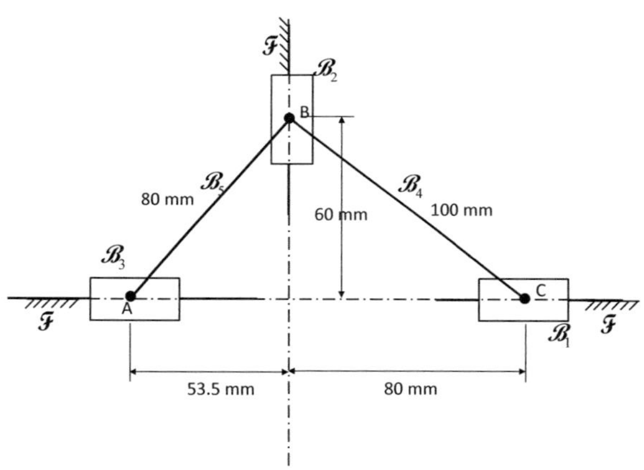

Fig. P1.29 Mechanism of exercise problem 1.29

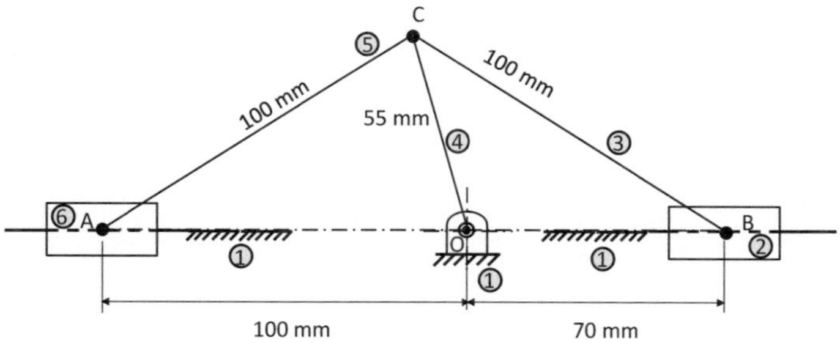

Fig. P1.30 Mechanism of exercise problem 1.30

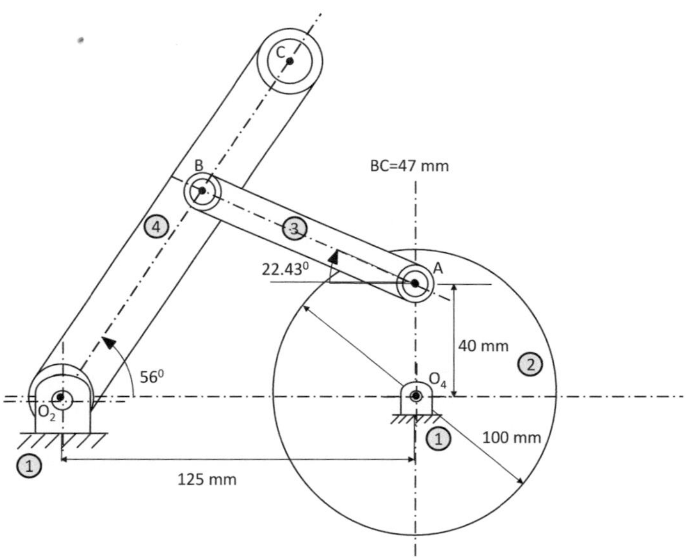

Fig. P1.31 Mechanism of exercise problem 1.31

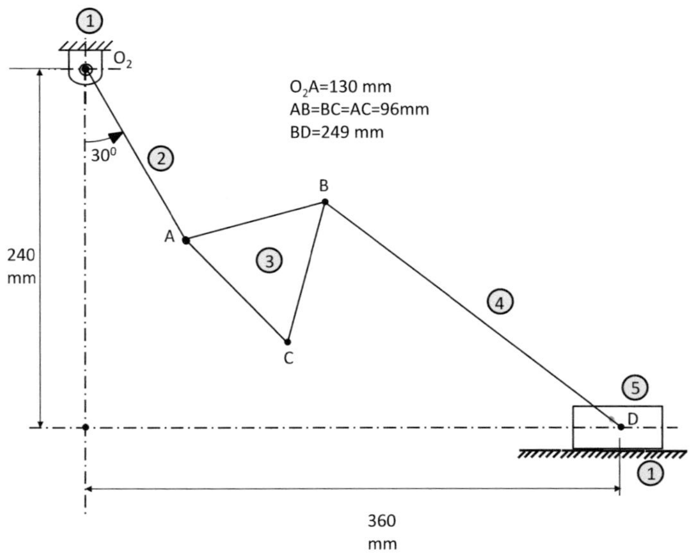

Fig. P1.32 Mechanism of exercise problem 1.32

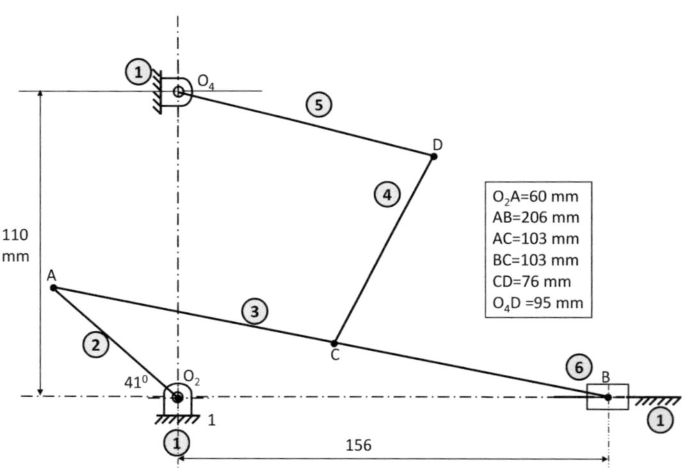

Fig. P1.33 Mechanism of exercise problem 1.33

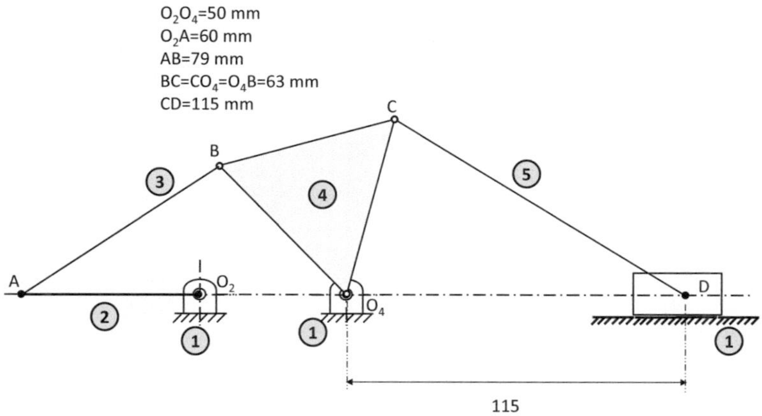

Fig. P1.34 Mechanism of exercise problem 1.34

Fig. P1.35 Mechanism of exercise problem 1.35

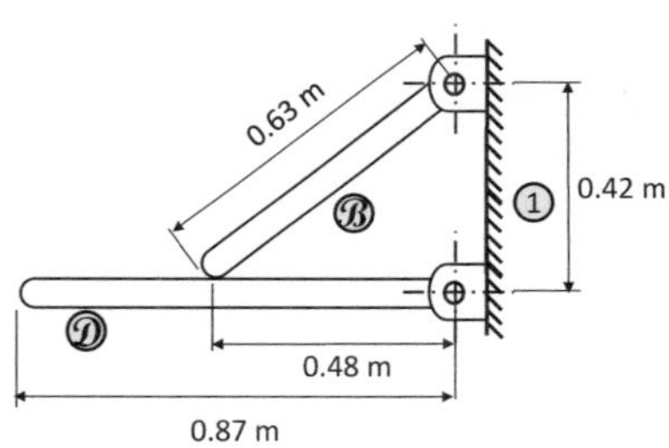

Fig. P1.36 Mechanism of exercise problem 1.36

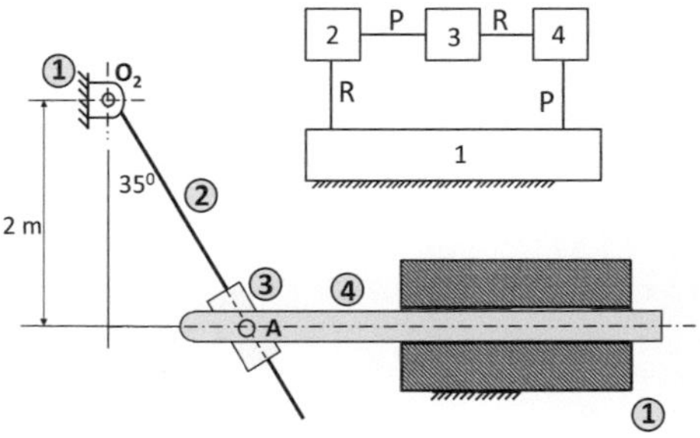

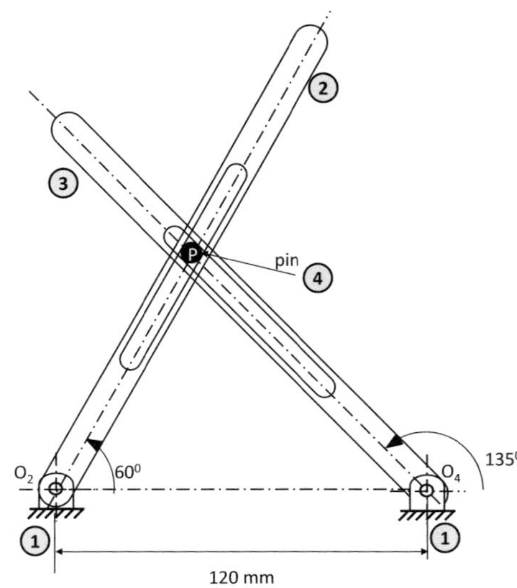

Fig. P1.37 Mechanism of exercise problem 1.37

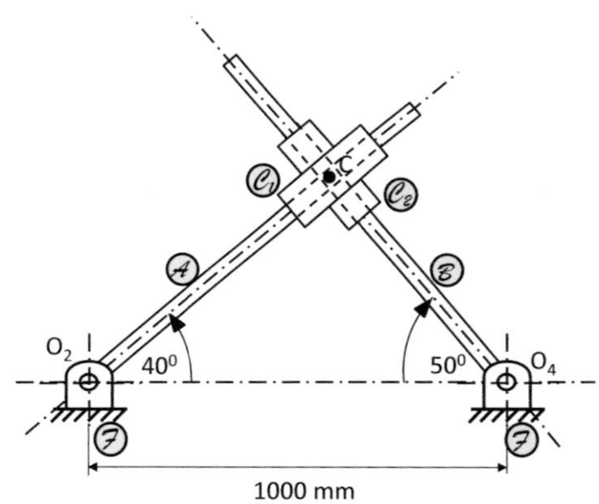

Fig. P1.38 Mechanism of exercise problem 1.38

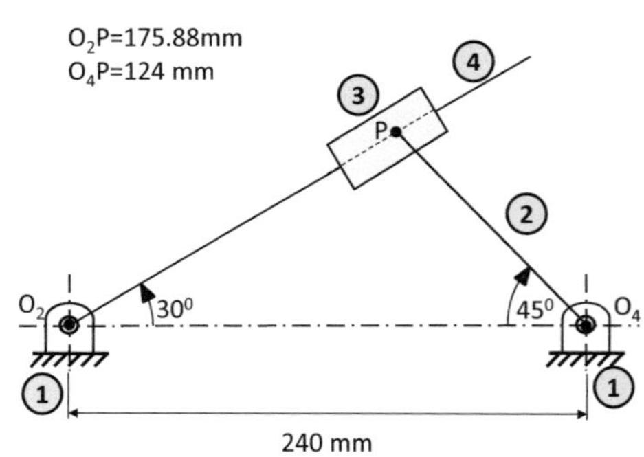

Fig. P1.39 Mechanism of exercise problem 1.39

Fig. P1.40 .

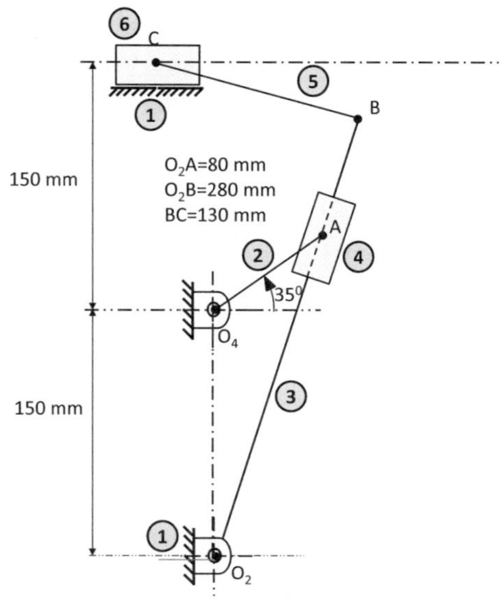

Appendix

Newton-Raphson's method for solving a set of multivariable functions

Consider a multivariable function $f_i(x_1, x_2, ...x_n),\ \ i = 1\, to\, n$. Extend the Newton-Raphson's method that is applied to a single variable function to numerically solve a multivariable function.

For a single variable function f(x) = 0, Newton-Raphson's method gives

$$x_{n+1} = x_n - f(x_n)/f'(x_n) \tag{1.27}$$

From Eq. (1.27), the incremental change of the independent variable can thus be obtained as

$$\Delta x \ = x_{n+1} - x_n = -f(x_n)/f'(x_n) \tag{1.28}$$

Equation (1.28) can be written as

$$f'(x_n)\Delta x \ = \ -f(x_n) \tag{1.29}$$

Equation (1.29) can now be extended to a multivariable function $f_i(x_1, x_2, ...x_n),\ \ i = 1\, to\, n$ as

$$\sum_{i=1}^{n} \frac{\partial f_j(x_1, x_2, ...x_n)}{\partial x_i} \Delta x_i = -f_i(x_1, x_2, ...x_n) \tag{1.30}$$

Now partial derivatives will be involved since the function has many independent variables. Hence, by starting from a set of trial values, the incremental change of these values can be obtained from Eq. (1.30). Eq. (1.30) can be successively applied till convergence is obtained.

For a multivariable function of two variables, Eq. (1.30) can be written as

$$\begin{bmatrix} \frac{\partial f_1(x_1,x_2)}{\partial x_1} & \frac{\partial f_1(x_1,x_2)}{\partial x_2} \\ \frac{\partial f_2(x_1,x_2)}{\partial x_1} & \frac{\partial f_2(x_1,x_2)}{\partial x_2} \end{bmatrix} \begin{Bmatrix} \Delta x_1 \\ \Delta x_2 \end{Bmatrix} = - \begin{Bmatrix} f_1(x_1, x_2) \\ f_2(x_1, x_2) \end{Bmatrix} \tag{1.31}$$

Equation (1.31) can be written as

$$\begin{Bmatrix} \Delta x_1 \\ \Delta x_2 \end{Bmatrix} = - \begin{bmatrix} \frac{\partial f_1(x_1,x_2)}{\partial x_1} & \frac{\partial f_1(x_1,x_2)}{\partial x_2} \\ \frac{\partial f_2(x_1,x_2)}{\partial x_1} & \frac{\partial f_2(x_1,x_2)}{\partial x_2} \end{bmatrix}^{-1} \begin{Bmatrix} f_1(x_1, x_2) \\ f_2(x_1, x_2) \end{Bmatrix} \tag{1.32}$$

If the value of the functions and their partial derivatives are evaluated at a certain trial value, the incremental change for a better estimate of the solution can be obtained from Eq. (1.32). It should be noted that there should be as many equations as that of the number of variables.

Chapter 2
Fundamentals of Kinematics

2.1 Introduction

Now that we have learnt the basics of configuring a mechanism in the previous chapter, in this chapter, we shall study velocity and acceleration relationship between points of a rigid body in motion. In order to compute velocity and acceleration of a point in motion, we need to consider its position as a function of time; subsequent derivatives will give velocity and acceleration. Hence, using information about the position vector of a point, its velocity and acceleration can be, respectively, computed by successive differentiation. A similar process can be extended to obtaining velocity and acceleration relationships between two points of a rigid body in terms of its angular velocity and angular acceleration or linear velocity and linear acceleration, as the case may be.

To study velocity and acceleration relationships, we deal only with the consideration of space and time, and this branch of study is known as kinematics. The branch that deals with forces and motions of a rigid body is known as kinetics, which will be discussed later. Since force-motion relationships require kinematic considerations, kinematics is studied first.

2.2 Coordinate Systems

In Chap. 1, we have studied the vector closure equations in the Cartesian coordinate system. Since studying motion relationships needs a broader understanding of various coordinate systems and derivatives of position vectors expressed in them, we shall discuss then in detail as follows:

D. N. Manik, *Fundamentals of Mechanisms and Machines*,
https://doi.org/10.1007/978-981-97-1810-8_2

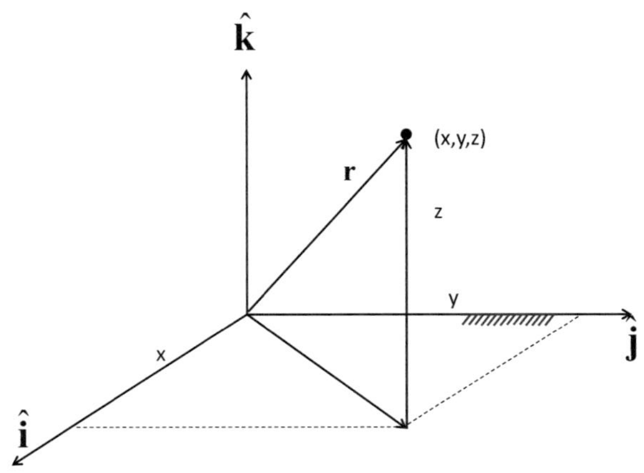

Fig. 2.1 Cartesian coordinate system

2.2.1 Cartesian Coordinates

A Cartesian coordinate system is the simplest of all the coordinate systems and a three-dimensional Cartesian coordinate system is shown in Fig. 2.1. It consists of three mutually perpendicular axes, respectively, defined by unit vectors $\hat{\mathbf{i}}$, $\hat{\mathbf{j}}$, and $\hat{\mathbf{k}}$. The position vector of a point in three-dimensional space in a Cartesian coordinate system is given by

$$\mathbf{r} = x\hat{\mathbf{i}} + y\hat{\mathbf{j}} + z\hat{\mathbf{k}} \tag{2.1}$$

where x, y, and z are coordinates in the corresponding coordinate directions.

2.2.2 Polar Coordinates

A polar coordinate system, drawn with reference to a planar Cartesian coordinate system, is shown in Fig. 2.2. Consider a point P moving along a curve. A unit vector $\hat{\mathbf{e}}_r$ along the direction of a line drawn from the origin to point P and a unit vector $\hat{\mathbf{e}}_\theta$ perpendicular to $\hat{\mathbf{e}}_r$ in the direction of its counterclockwise rotation are as shown. The line drawn from the origin of the Cartesian coordinate system, OP, defines the length of the position vector that makes an angle θ with respect to the horizontal direction.

The position vector of point P in the polar coordinate system is given by

$$\mathbf{r}_{OP} = r\hat{\mathbf{e}}_r \tag{2.2}$$

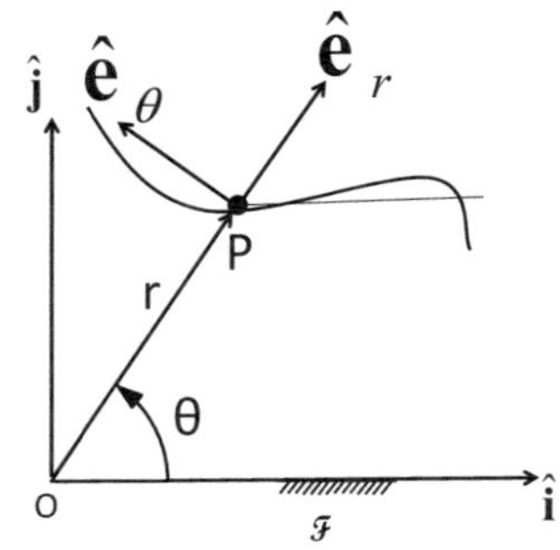

Fig. 2.2 Polar coordinate system

The unit vector $\hat{\mathbf{e}}_r$ along the position vector in the Cartesian coordinate system is defined by

$$\hat{\mathbf{e}}_r = \cos\theta\hat{\mathbf{i}} + \sin\theta\hat{\mathbf{j}} \tag{2.3}$$

And the unit vector $\hat{\mathbf{e}}_\theta$ perpendicular to the position vector in the Cartesian coordinate system is defined by

$$\hat{\mathbf{e}}_\theta = -\sin\theta\hat{\mathbf{i}} + \cos\theta\hat{\mathbf{j}} \tag{2.4}$$

2.2.3 *Spherical Coordinates*

A spherical coordinate system defined in three-dimensional space is shown in Fig. 2.3. It is defined by three parameters: radial distance r from the origin, polar angle θ measured with respect to the $\hat{\mathbf{k}}$ direction(zenith direction), and the azimuth angle ϕ that is the angle subtended by the projection of the position vector on the reference plane of $\hat{\mathbf{i}}\hat{\mathbf{j}}$ vectors and the $\hat{\mathbf{i}}$ vector.

The relationship between the Cartesian (x, y, x) and spherical system (r, θ, ϕ) coordinates can be expressed as follows:

$$\begin{aligned}
&r = \sqrt{x^2 + y^2 + z^2} \\
&\theta = \cos^{-1}\frac{z}{\sqrt{x^2+y^2+z^2}} = \cos^{-1}\frac{z}{r}\ , \ 0 \le \theta \le \pi \\
&\phi = \tan^{-1}\frac{y}{x}, \qquad -\pi \le \phi \le \pi
\end{aligned} \tag{2.5}$$

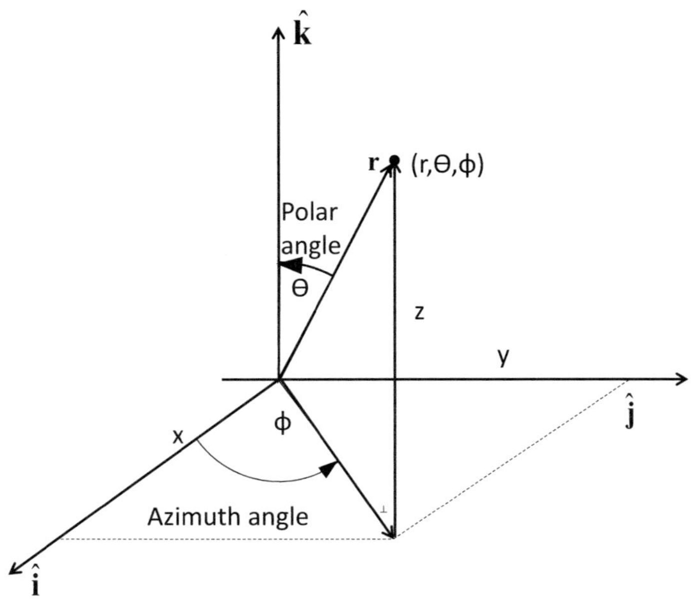

Fig. 2.3 Spherical coordinate system

The unit vectors of the spherical coordinate system and the Cartesian coordinate system are related as follows:

$$\begin{Bmatrix} \hat{\mathbf{e}}_r \\ \hat{\mathbf{e}}_\theta \\ \hat{\mathbf{e}}_\phi \end{Bmatrix} = \begin{bmatrix} \cos\phi\sin\theta & \sin\phi\sin\theta & \cos\theta \\ \cos\phi\cos\theta & \sin\phi\cos\theta & -\sin\theta \\ -\sin\phi & \cos\phi & 0 \end{bmatrix} \begin{Bmatrix} \hat{\mathbf{i}} \\ \hat{\mathbf{j}} \\ \hat{\mathbf{k}} \end{Bmatrix} \tag{2.6}$$

The Cartesian coordinates of the spherical coordinate system are related as follows:

$$\begin{aligned} x &= r\sin\theta\cos\phi \\ y &= r\sin\theta\sin\phi \\ z &= r\cos\theta \end{aligned} \tag{2.7}$$

The unit vectors of a spherical coordinate system are shown in Fig. 2.4. The unit vector $\hat{\mathbf{e}}_r$ is in the direction of the line joining the origin and the point defined by the spherical coordinates; the unit vector $\hat{\mathbf{e}}_\theta$ is perpendicular to the position vector; the unit vector $\hat{\mathbf{e}}_\phi$ is perpendicular to the projection of the position vector on the reference plane of $\hat{\mathbf{i}}\hat{\mathbf{j}}$ vectors, r′. The unit vectors $\hat{\mathbf{e}}_r$, $\hat{\mathbf{e}}_\theta$ and $\hat{\mathbf{e}}_\phi$ are mutually perpendicular to each other.

Fig. 2.4 Unit vectors of a spherical coordinate system

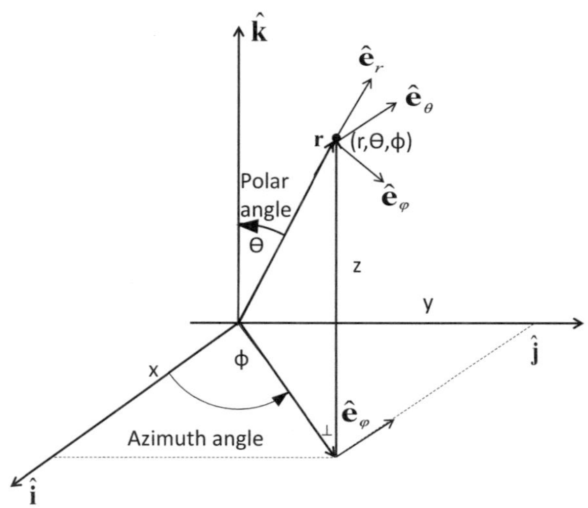

The position vector of a point in the spherical coordinate system is given by

$$\mathbf{r} = r\hat{\mathbf{e}}_r \tag{2.8}$$

2.2.4 Cylindrical Coordinate System

A cylindrical coordinate system shown in Fig. 2.5 is a three-dimensional coordinate system that can be considered as a special case of the spherical coordinate system; the axis of one of the coordinate directions of a cylindrical coordinate system is always in the direction of the Cartesian coordinate unit vector $\hat{\mathbf{k}}$. It can also be construed as an extension of the polar coordinates in the $\hat{\mathbf{k}}$ direction. This coordinate system is useful in cases where there is an axisymmetry. The unit vector $\hat{\mathbf{e}}_r$ is in the direction of the radial component in the reference plane formed by unit vectors $\hat{\mathbf{i}}\hat{\mathbf{j}}$ and $\hat{\mathbf{e}}_\theta$ is perpendicular to the unit vector $\hat{\mathbf{e}}_r$ and the unit vector $\hat{\mathbf{k}}$ is always along the same direction in both systems.

The position vector of a point in the cylindrical coordinate system is given by

$$\mathbf{r} = r\hat{\mathbf{e}}_r + z\hat{\mathbf{k}} \tag{2.9}$$

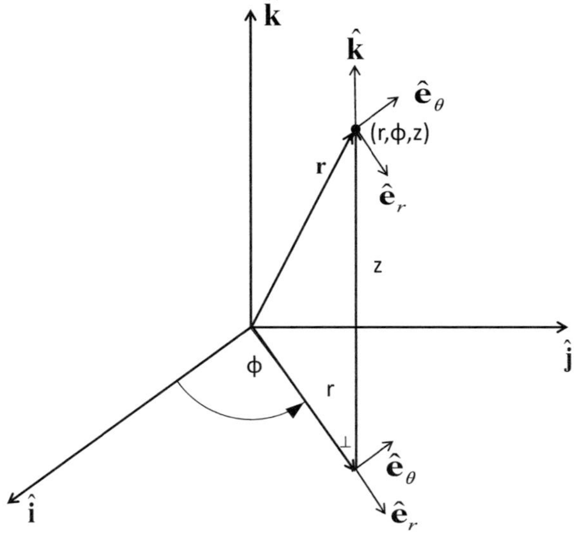

Fig. 2.5 Cylindrical coordinate system

From Eq. (2.7) and Fig. 2.5, the unit vectors of the cylindrical coordinate system and the Cartesian coordinate system are related by

$$\begin{Bmatrix} \hat{\mathbf{e}}_r \\ \hat{\mathbf{e}}_\phi \\ \hat{\mathbf{k}} \end{Bmatrix} = \begin{bmatrix} \cos\phi & \sin\phi & 0 \\ -\sin\phi & \cos\phi & 0 \\ 0 & 0 & 1 \end{bmatrix} \begin{Bmatrix} \hat{\mathbf{i}} \\ \hat{\mathbf{j}} \\ \hat{\mathbf{k}} \end{Bmatrix} \tag{2.10}$$

2.3 Reference Frames and Vector Derivatives

As we are going to generally use the vector approach throughout the book, taking derivatives of vectors will be required. There is nothing to fear about taking derivatives of a vector; since a vector has both magnitude and direction, we need to account for changes that occur as a function of time for both magnitude and direction (represented by the unit vectors).

The derivative of a vector can be obtained by using the product rule of differentiation in which we consider the product of magnitude and direction of each element of the vector. Let us consider a vector **A**; bold characters are used to define vectors and matrices in this book. The derivative of this vector is given by

$$\frac{d\mathbf{A}}{dt} = \lim_{\Delta t \to 0}\left[\frac{\mathbf{A}(t+\Delta t) - \mathbf{A}(t)}{\Delta t}\right] \tag{2.11}$$

Since the vector has both magnitude and direction, the time derivative of both will have to be accounted. If the vector can be written as the product of its magnitude and a unit vector that defines its direction, the derivative of this vector will contain time derivatives of the magnitude as well as that of the unit vector. We shall see later in this chapter how the time derivative of unit vectors can be systematically computed in various coordinate systems.

Consider a point P moving along a locus, as shown in Fig. 2.6, in the fixed three-dimensional space defined by a Cartesian coordinate system, by unit vectors $\hat{\mathbf{i}}$, $\hat{\mathbf{j}}$, and $\hat{\mathbf{k}}$. Let a moving Cartesian coordinate system, defined by unit vectors $\hat{\mathbf{e}}_1$, $\hat{\mathbf{e}}_2$, and $\hat{\mathbf{e}}_3$ with their origin at P move along with it. The relationship between the moving coordinates and fixed coordinates is not defined at the moment but will take definite relationships later. Consider the following vector defined in a moving coordinate system as shown in Fig. 2.6:

$$\mathbf{a} = a_1\hat{\mathbf{e}}_1 + a_2\hat{\mathbf{e}}_2 + a_3\hat{\mathbf{e}}_3 \tag{2.12}$$

where a_1, a_2, and a_3 are scalar magnitudes of the vector $\mathbf{a}$ defined in the moving coordinate system. Now the main objective is to obtain derivatives of a vector that itself is referenced to a moving frame of reference. When we take the derivative of a vector, we must clearly state the coordinate system in which the derivative is being taken. The fixed coordinate system represented by the Cartesian coordinate system is generally the coordinate system with reference to which the derivatives are taken in this book as they represent "absolute" derivatives.

The derivative of the vector defined by Eq. (2.12) in the fixed coordinate system is given by

$$\begin{aligned}\frac{{}^{\mathcal{F}}d\mathbf{a}}{dt} &= \frac{da_1}{dt}\hat{\mathbf{e}}_1 + \frac{da_2}{dt}\hat{\mathbf{e}}_2 + \frac{da_3}{dt}\hat{\mathbf{e}}_3 \\ &\quad + a_1\frac{{}^{\mathcal{F}}d\hat{\mathbf{e}}_1}{dt} + a_2\frac{{}^{\mathcal{F}}d\hat{\mathbf{e}}_2}{dt} + a_3\frac{{}^{\mathcal{F}}d\hat{\mathbf{e}}_3}{dt}\end{aligned} \tag{2.13}$$

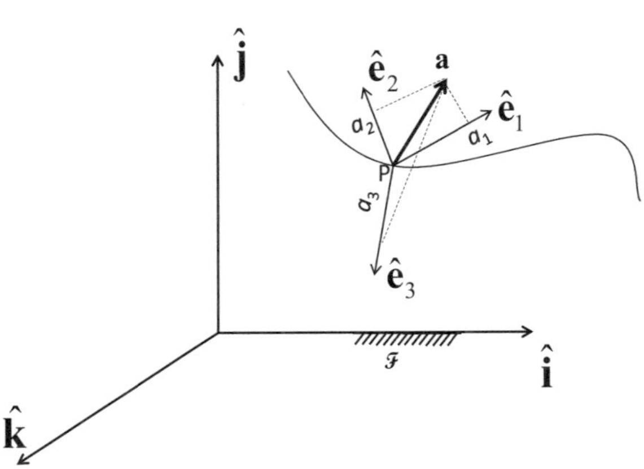

Fig. 2.6 A vector defined in terms of a moving coordinate system

In Eq. (2.13), the derivative of the scalars is straightforward as usual. But the derivative of the unit vectors with respect to the fixed frame of reference needs some discussion. If the relationship between the unit vectors $\hat{\mathbf{e}}_1$, $\hat{\mathbf{e}}_2$, $\hat{\mathbf{e}}_3$, and $\hat{\mathbf{i}}$, $\hat{\mathbf{j}}$, $\hat{\mathbf{k}}$ are known, then the derivatives $\frac{^{\mathcal{F}}d\hat{\mathbf{e}}_1}{dt}$, $\frac{^{\mathcal{F}}d\hat{\mathbf{e}}_2}{dt}$, $\frac{^{\mathcal{F}}d\hat{\mathbf{e}}_3}{dt}$ can be obtained with respect to the fixed frame of reference. These aspects will become much clearer as we move to specific examples. But, at this point of time, you need to understand that derivative of a vector has terms corresponding to changes in both magnitude and direction.

If the vector $\mathbf{a}$ is referenced to unit vectors $\hat{\mathbf{e}}_1$, $\hat{\mathbf{e}}_2$, $\hat{\mathbf{e}}_3$ that do not change with respect to time, the derivative of this vector becomes

$$\frac{^{\mathcal{F}}d\mathbf{a}}{dt} = \frac{da_1}{dt}\hat{\mathbf{e}}_1 + \frac{da_2}{dt}\hat{\mathbf{e}}_2 + \frac{da_3}{dt}\hat{\mathbf{e}}_3 \tag{2.14}$$

Sometimes, we are lucky to be in the above situation leading to Eq. (2.14). But, as a general case, Eq. (2.13) is encountered in many practical situations.

2.4 Velocity and Acceleration of a Point

2.4.1 Cartesian Coordinate System

Consider a point P defined in a Cartesian coordinate system as shown in Fig. 2.7. In addition, the coordinate system is fixed in space. That makes it easier for computing the velocity and acceleration of point P as it would require computation of the derivatives of the scalar components only.

The position of vector of point P in the above coordinate system is given by

$$\mathbf{r}_{OP} = x\hat{\mathbf{i}} + y\hat{\mathbf{j}} + z\hat{\mathbf{k}} \tag{2.15}$$

Fig. 2.7 A point P defined in a Cartesian coordinate system

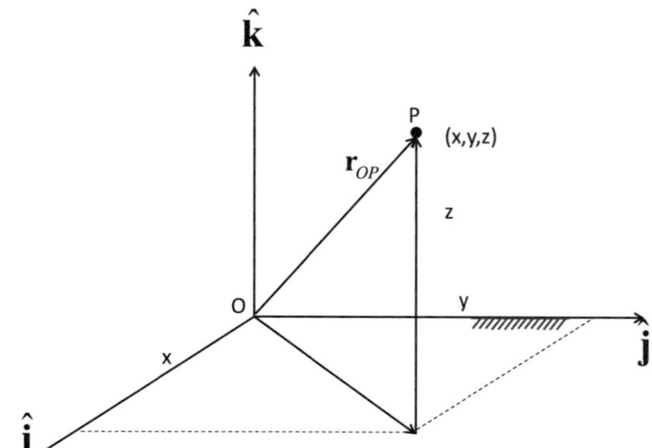

The velocity of point P is given by

$$\mathbf{v}_P = \dot{x}\hat{\mathbf{i}} + \dot{y}\hat{\mathbf{j}} + \dot{z}\hat{\mathbf{k}} \tag{2.16}$$

The acceleration of point P is given by

$$\mathbf{a}_P = \ddot{x}\hat{\mathbf{i}} + \ddot{y}\hat{\mathbf{j}} + \ddot{z}\hat{\mathbf{k}} \tag{2.17}$$

Example 2.1

Consider a point P that moves with respect to a fixed Cartesian coordinate system such that its components in coordinate directions vary with respect to time as follows: $x = 3t^3 + 5t^2 + 3$ cm; $y = 7t^2 + 5t + 2$ cm; $z = 10t + 5$ cm. (a) Determine the general equations for velocity and acceleration of point P. (b) Determine position, velocity, and acceleration of point P at t = 1 s. (c) What is the distance of point P at t = 1 s from its initial position at t = 0. (d) Direction of position, velocity, and acceleration at t = 1 s.

The position vector of point P at any time instant is given by

$$\mathbf{r}_{OP} = \left(3t^3 + 5t^2 + 3\right)\hat{\mathbf{i}} + \left(7t^2 + 5t + 2\right)\hat{\mathbf{j}} + (10t + 5)\hat{\mathbf{k}}\ \text{cm} \tag{1}$$

(a) From Eq. (2.16), the velocity of point P at any time instant is given by

$$\mathbf{v}_P = \left(9t^2 + 10t\right)\hat{\mathbf{i}} + (14t + 5)\hat{\mathbf{j}} + (10)\hat{\mathbf{k}}\ \text{cm/s} \tag{2}$$

From Eq. (2.17), the acceleration of point P at any time instant is given by

$$\mathbf{a}_P = (18t + 10)\hat{\mathbf{i}} + (14)\hat{\mathbf{j}} + (0)\hat{\mathbf{k}}\ \text{cm/s}^2 \tag{3}$$

(b) From Eq. (1), the position of point P at t = 1 s is given by

$$\begin{aligned}\mathbf{r}_{OP} &= (3 + 5 + 3)\hat{\mathbf{i}} + (7 + 5 + 2)\hat{\mathbf{j}} + (10 + 5)\hat{\mathbf{k}} \\ &= (11)\hat{\mathbf{i}} + (14)\hat{\mathbf{j}} + (15)\hat{\mathbf{k}}\ \text{cm}\end{aligned} \tag{4}$$

From Eq. (2), the velocity of point P at t = 1 s is given by

$$\begin{aligned}\mathbf{v}_P &= (9 + 10)\hat{\mathbf{i}} + (14 + 5)\hat{\mathbf{j}} + (10)\hat{\mathbf{k}}\ \text{cm/s} \\ &= (19)\hat{\mathbf{i}} + (19)\hat{\mathbf{j}} + (10)\hat{\mathbf{k}}\ \text{cm/s}\end{aligned} \tag{5}$$

From Eq. (3), the acceleration of point P at t = 1 is given by

$$\begin{aligned}\mathbf{a}_P &= (18+10)\hat{\mathbf{i}} + (14)\hat{\mathbf{j}} + (0)\hat{\mathbf{k}} \\ &= (28)\hat{\mathbf{i}} + (14)\hat{\mathbf{j}} + (0)\hat{\mathbf{k}}\ \text{cm/s}^2\end{aligned} \tag{6}$$

(c) From Eq. (1), the position vector at t = 0 is given by

$$\mathbf{r}_{OP}(t=0) = 3\hat{\mathbf{i}} + 2\hat{\mathbf{j}} + 5\hat{\mathbf{k}}\ \text{cm} \tag{7}$$

From Eqs. (4) and (7), the distance between the position of the point at t = 0 and 1 s is given by

$$d = \sqrt{(11-3)^2 + (14-2)^2 + (5-15)^2} = 17.54\ \text{cm} \tag{8}$$

(d) From Eq. (4), the direction of the position vector at t = 1 s is given by

$$\hat{\mathbf{e}}_r = \frac{(11)\hat{\mathbf{i}} + (14)\hat{\mathbf{j}} + (15)\hat{\mathbf{k}}}{\sqrt{11^2+14^2+15^2}} = 0.47\hat{\mathbf{i}} + 0.60\hat{\mathbf{j}} + 0.64\hat{\mathbf{k}} \tag{9}$$

From Eq. (5), the direction of velocity of point P at t = 1 s is given by

$$\hat{\mathbf{e}}_v = \frac{(19)\hat{\mathbf{i}} + (19)\hat{\mathbf{j}} + (10)\hat{\mathbf{k}}}{\sqrt{19^2+19^2+10^2}} = 0.66\hat{\mathbf{i}} + 0.66\hat{\mathbf{j}} + 0.35\hat{\mathbf{k}} \tag{10}$$

From Eq. (6), the direction of acceleration of point P at t = 1 s is given by

$$\hat{\mathbf{e}}_a = \frac{(28)\hat{\mathbf{i}} + (14)\hat{\mathbf{j}} + (0)\hat{\mathbf{k}}}{\sqrt{28^2+14^2+0^2}} = 0.89\hat{\mathbf{i}} + 0.45\hat{\mathbf{j}} \tag{11}$$

2.4.2 Polar Coordinates

Motion of a particle represented by point P in polar coordinates is shown in Fig. 2.8 and the motion is restricted to the plane of paper. The motion of this particle is also traced with respect to a Cartesian coordinate system that is fixed. The unit vector $\hat{\mathbf{e}}_r$ is in the direction of position vector $\mathbf{r}_{OP}$ and the unit vector $\hat{\mathbf{e}}_\theta$ is perpendicular to $\hat{\mathbf{e}}_r$ and in the direction of positive value of θ in the counterclockwise direction. The above unit vectors have been defined earlier in Sect. 2.2.2 and are reproduced again as follows:

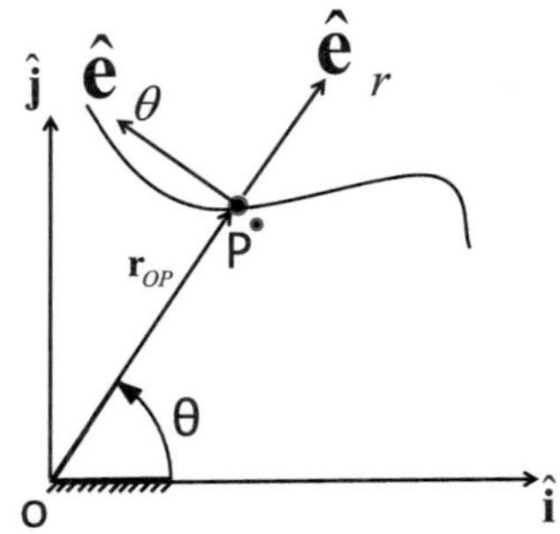

Fig. 2.8 Motion of a particle in polar coordinates

Velocity

The position vector of a point P in the polar coordinate system defined by Fig. 2.8 is given by

$$\mathbf{r}_{OP} = r\hat{\mathbf{e}}_r \tag{2.18}$$

The unit vector $\hat{\mathbf{e}}_r$ along the position vector in the Cartesian coordinate system is defined by

$$\hat{\mathbf{e}}_r = \cos\theta\hat{\mathbf{i}} + \sin\theta\hat{\mathbf{j}} \tag{2.19}$$

And the unit vector $\hat{\mathbf{e}}_\theta$ perpendicular to the position vector in the Cartesian coordinate system is defined by

$$\hat{\mathbf{e}}_\theta = -\sin\theta\hat{\mathbf{i}} + \cos\theta\hat{\mathbf{j}} \tag{2.20}$$

Differentiating Eq. (2.18) with respect to time, the velocity of point P is given by

$$\dot{\mathbf{r}}_{OP} = \mathbf{v}_P = \dot{r}\hat{\mathbf{e}}_r + r\dot{\hat{\mathbf{e}}}_r \tag{2.21}$$

From Eq. (2.19), the time derivative of the unit vector $\hat{\mathbf{e}}_r$ is given by

$$\dot{\hat{\mathbf{e}}}_r = \left(-\sin\theta\hat{\mathbf{i}} + \cos\theta\hat{\mathbf{j}}\right)\dot{\theta} = \dot{\theta}\hat{\mathbf{e}}_\theta \tag{2.22}$$

From Eqs. (2.21) and (2.22), the velocity of point P in polar coordinates is given by

$$\mathbf{v}_P = \dot{r}\hat{\mathbf{e}}_r + r\dot{\theta}\hat{\mathbf{e}}_\theta \tag{2.23}$$

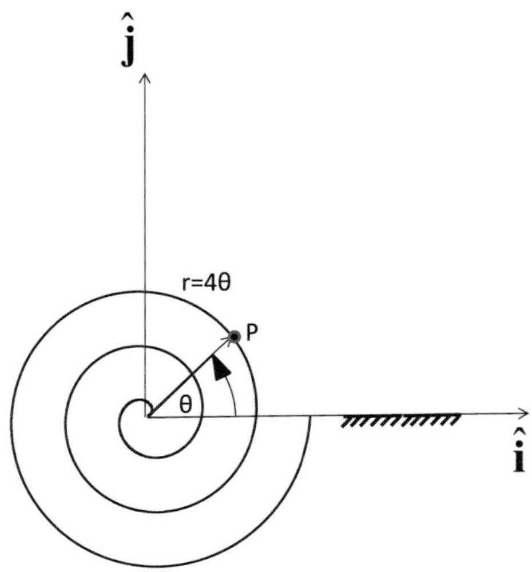

Fig. 2.9 A point moving around the spiral of Archimedes

Example 2.2
Consider a point P moving around the spiral of Archimedes as shown in Fig. 2.9, according to r = 4θ. If it is moving at a constant speed of 2 m/s, determine the radial, tangential, and Cartesian components of velocity at θ = 165°.

The radius and angle are related by

$$r = 4\theta \tag{1}$$

The time derivative of Eq. (1) is given by

$$\dot{r} = 4\dot{\theta} \tag{2}$$

From Eqs. (1), (2), and (2.23), the velocity of point P is given by

$$\begin{aligned} \mathbf{v}_P &= \dot{r}\hat{\mathbf{e}}_r + r\dot{\theta}\hat{\mathbf{e}}_\theta \\ &= 4\dot{\theta}\hat{\mathbf{e}}_r + 4\theta\dot{\theta}\hat{\mathbf{e}}_\theta \end{aligned} \tag{3}$$

From Eq. (3), the speed of point P is given by

$$|\mathbf{v}_P| = \sqrt{\left(4\dot{\theta}\right)^2 + \left(4\theta\dot{\theta}\right)^2} \tag{4}$$

From Eq. (4), the time derivative of θ is given by

$$\dot{\theta} = \frac{|\mathbf{v}_P|}{4\sqrt{1+\theta^2}} \tag{5}$$

At the given speed of point P and angular position of 165°, Eq. (5) becomes

$$\dot{\theta} = \frac{|\mathbf{v}_P|}{4\sqrt{1+\theta^2}} = \frac{2}{4\sqrt{1+\left(165\frac{\pi}{180}\right)^2}} = 0.1640\,\text{rad/s} \tag{6}$$

From Eqs. (3) and (6), the velocity of point P in polar coordinates at the angular position of 165° and the given speed is given by

$$\begin{aligned}\mathbf{v}_P &= 4\dot{\theta}\hat{\mathbf{e}}_r + 4\theta\dot{\theta}\hat{\mathbf{e}}_\theta \\ &= 4\times 0.1640\hat{\mathbf{e}}_r + 4\times\frac{165\times\pi}{180}\times 0.1640\hat{\mathbf{e}}_\theta \\ &= 0.66\hat{\mathbf{e}}_r + 1.9\hat{\mathbf{e}}_\theta\ \text{m/s}\end{aligned} \tag{7}$$

Using Eqs. (2.19) and (2.20) in Eq. (7), the velocity of point P in Cartesian coordinates is given by

$$\begin{aligned}\mathbf{v}_P &= 0.66\hat{\mathbf{e}}_r + 1.9\hat{\mathbf{e}}_\theta\ \text{m/s} \\ &= 0.66(\cos\theta\hat{\mathbf{i}} + \sin\theta\hat{\mathbf{j}}) + 1.9(-\sin\theta\hat{\mathbf{i}} + \cos\theta\hat{\mathbf{j}}) \\ &= (0.66\cos\theta - 1.9\sin\theta)\hat{\mathbf{i}} + (0.66\sin\theta + 1.9\cos\theta)\hat{\mathbf{j}} \\ &= \left(0.66\cos\frac{165\times\pi}{180} - 1.9\sin\frac{165\times\pi}{180}\right)\hat{\mathbf{i}} + \left(0.66\sin\frac{165\times\pi}{180} + 1.9\cos\frac{165\times\pi}{180}\right)\hat{\mathbf{j}} \\ &= -1.12\hat{\mathbf{i}} - 1.65\hat{\mathbf{j}}\ \text{m/s}\end{aligned} \tag{8}$$

Acceleration of a Point in Polar Coordinates

Differentiating Eq. (2.23) with respect to time, the acceleration of point P is given by

$$\mathbf{a}_P = \ddot{r}\hat{\mathbf{e}}_r + \dot{r}\dot{\hat{\mathbf{e}}}_r + \dot{r}\dot{\theta}\hat{\mathbf{e}}_\theta + r\ddot{\theta}\hat{\mathbf{e}}_\theta + r\dot{\theta}\dot{\hat{\mathbf{e}}}_\theta \tag{2.24}$$

The time derivatives of the unit vectors of polar coordinates are given by

$$\dot{\hat{\mathbf{e}}}_r = \dot{\theta}\hat{\mathbf{e}}_\theta;\ \dot{\hat{\mathbf{e}}}_\theta = -\dot{\theta}\hat{\mathbf{e}}_r \tag{2.25}$$

From Eqs. (2.24) and (2.25), the acceleration of point P in polar coordinates is given by

$$\begin{aligned}\mathbf{a}_P &= \ddot{r}\hat{\mathbf{e}}_r + \dot{r}\dot{\theta}\hat{\mathbf{e}}_\theta + \dot{r}\dot{\theta}\hat{\mathbf{e}}_\theta + r\ddot{\theta}\hat{\mathbf{e}}_\theta - r\dot{\theta}^2\hat{\mathbf{e}}_r \\ &= \left(\ddot{r} - r\dot{\theta}^2\right)\hat{\mathbf{e}}_r + \left(2\dot{r}\dot{\theta} + r\ddot{\theta}\right)\hat{\mathbf{e}}_\theta\end{aligned} \tag{2.26}$$

Example 2.3

Consider a point moving around a closed curve with the length of its position vector varying as a function of its angular position given by $r = 5(1 + \cos\theta)$. If the angle changes at a constant angular speed of 10 rad/s, determine the velocity and acceleration of the point at an angular position of 240°.

Since the angular speed is constant, $\ddot{\theta} = 0$.

The radial coordinate of the point is given by

$$r = 5(1 + \cos\theta)\text{cm} \tag{1}$$

The time derivative of Eq. (1) is given by

$$\dot{r} = -5\dot{\theta}\sin\theta \tag{2}$$

The time derivative of Eq. (2) is given by

$$\ddot{r} = -5\left(\ddot{\theta}\sin\theta + \dot{\theta}^2\cos\theta\right) = -5\left(\dot{\theta}^2\cos\theta\right) \tag{3}$$

From Eqs. (1), (2), and (2.23), the velocity of the point at any instant is given by

$$\begin{aligned}\mathbf{v}_P &= \dot{r}\hat{\mathbf{e}}_r + r\dot{\theta}\hat{\mathbf{e}}_\theta \\ &= -5\dot{\theta}\sin\theta\hat{\mathbf{e}}_r + 5(1 + \cos\theta)\dot{\theta}\hat{\mathbf{e}}_\theta\end{aligned} \tag{4}$$

The velocity of the point at an angular position of 240° is given by

$$\begin{aligned}\mathbf{v}_P &= -5\dot{\theta}\sin\theta\hat{\mathbf{e}}_r + 5(1 + \cos\theta)\dot{\theta}\hat{\mathbf{e}}_\theta \\ &= -5 \times 10 \times \sin 240\hat{\mathbf{e}}_r + 5 \times 10(1 + \cos 240)\hat{\mathbf{e}}_\theta \\ &= 43.3\hat{\mathbf{e}}_r + 25\hat{\mathbf{e}}_\theta \ \text{cm/s}\end{aligned} \tag{5}$$

From Eqs. (1), (2), (3), and (2.26), the acceleration of the point at any instant is given by

$$\begin{aligned}\mathbf{a}_P &= \left(\ddot{r} - r\dot{\theta}^2\right)\hat{\mathbf{e}}_r + \left(2\dot{r}\dot{\theta} + r\ddot{\theta}\right)\hat{\mathbf{e}}_\theta \\ &= \left[-5\left(\dot{\theta}^2\cos\theta\right) - 5(1 + \cos\theta)\dot{\theta}^2\right]\hat{\mathbf{e}}_r + \left[-10\dot{\theta}^2\sin\theta\right]\hat{\mathbf{e}}_\theta \\ &= \left[-10\left(\dot{\theta}^2\cos\theta\right) - 5\dot{\theta}^2\right]\hat{\mathbf{e}}_r + \left[-10\dot{\theta}^2\sin\theta\right]\hat{\mathbf{e}}_\theta\end{aligned} \tag{6}$$

Acceleration of the point and angular position of 240° is given by

$$\begin{aligned}\mathbf{a}_P &= \left[-10\left(\dot{\theta}^2\cos\theta\right) - 5\dot{\theta}^2\right]\hat{\mathbf{e}}_r + \left[-10\dot{\theta}^2\sin\theta\right]\hat{\mathbf{e}}_\theta \\ &= \left[-10\left(10^2\cos 240\right) - 5 \times 10^2\right]\hat{\mathbf{e}}_r + \left[-10 \times 10^2\sin 240\right]\hat{\mathbf{e}}_\theta \\ &= 866\hat{\mathbf{e}}_\theta \ \text{cm/s}^2\end{aligned} \tag{7}$$

2.4.3 Tangential and Normal Components

Consider a point P moving along a locus on a plane surface as shown in Fig. 2.10. Its distance as measured along the locus starts at a reference point s = 0. The position vectors of point P at s and s + Δs are as shown. If the position vector of P was known as a function of time, its velocity could have been easily obtained by differentiating with time. But the position vector is now a function of arc length and its distance from the reference position of arc length is a function of time. Therefore, velocity of point P can be written as

$$\mathbf{v}_P = \frac{d\mathbf{r}_{OP}[s(t)]}{ds}\frac{ds}{dt} = \dot{s}\frac{d\mathbf{r}_{OP}[s(t)]}{ds} \tag{2.27}$$

Variation of the position vector with respect to the arc length can be written as

$$\begin{aligned}\frac{d\mathbf{r}_{OP}[s(t)]}{ds} &= \lim_{\Delta s\to 0}\frac{\mathbf{r}_{OP}(s+\Delta s)-\mathbf{r}_{OP}(s)}{\Delta s}\\ &= \lim_{\Delta s\to 0}\frac{\Delta\mathbf{r}_{OP}}{\Delta s} = \hat{\mathbf{e}}_t\end{aligned} \tag{2.28}$$

Equation 2.28 defines a unit vector in the tangential direction. Hence, from Eqs. (2.27) and (2.28), velocity of point P defines a unit vector in the tangential direction to the locus at any instant of time.

Velocity of point P can be defined in terms of the tangential unit vector as

$$\mathbf{v}_P = \dot{s}\hat{\mathbf{e}}_t \tag{2.29}$$

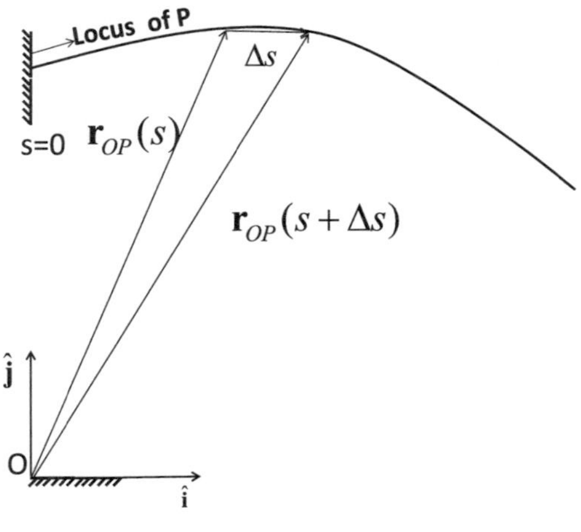

Fig. 2.10 Position vector of a point moving along a locus

Fig. 2.11 Center of curvature, tangential, and normal vector at a point on a locus

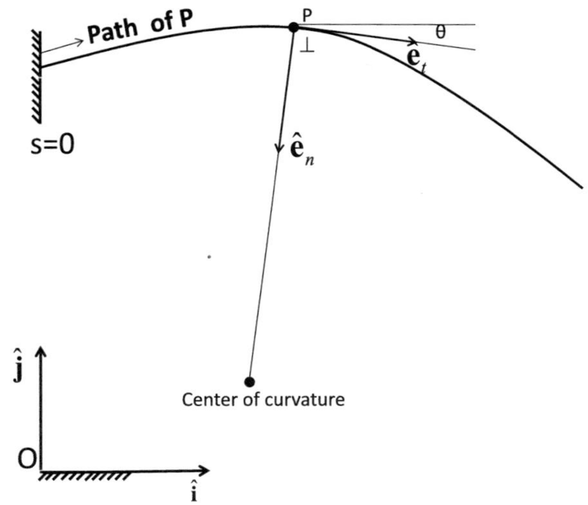

Acceleration of point P can be defined as

$$\mathbf{a}_P = \ddot{s}\hat{\mathbf{e}}_t + \dot{s}^2\frac{d\hat{\mathbf{e}}_t}{ds} = \ddot{s}\hat{\mathbf{e}}_t + \dot{s}^2\frac{d\hat{\mathbf{e}}_t}{d\theta}\frac{d\theta}{ds} \tag{2.30}$$

To further evaluate terms of Eq. (2.30), consider a modification of Fig. 2.10 shown in Fig. 2.11 below:

A normal vector $\hat{\mathbf{e}}_n$ can be defined as shown in Fig. 2.11 that is perpendicular to the tangential vector $\hat{\mathbf{e}}_t$ in a direction towards the center of curvature of the locus at point P.

The tangential vector can be defined in terms of the global Cartesian coordinate system as

$$\hat{\mathbf{e}}_t = \cos\theta\hat{\mathbf{i}} - \sin\theta\hat{\mathbf{j}} \tag{2.31}$$

And the normal vector is given by

$$\hat{\mathbf{e}}_n = -\sin\theta\hat{\mathbf{i}} - \cos\theta\hat{\mathbf{j}} \tag{2.32}$$

From Eqs. (2.31) and (2.32),

$$\frac{d\hat{\mathbf{e}}_t}{d\theta} = \hat{\mathbf{e}}_n \tag{2.33}$$

For a convex upward locus shown in Fig. 2.11, the following equations can be written as

$$\frac{d\theta}{ds} = \frac{1}{\rho} \tag{2.34}$$

From Eqs. 2.30–2.34, acceleration of point P in terms of normal and tangential coordinates can be written as

$$\mathbf{a}_P = \ddot{s}\hat{\mathbf{e}}_t + \frac{\dot{s}^2}{\rho}\hat{\mathbf{e}}_n \tag{2.35}$$

where ρ is the radius of curvature.

Even for a concave downward locus also, the above equation remains the same except the direction of the normal vector as shown in Fig. 2.12.

The cross-product of velocity and acceleration in tangential and normal coordinates expressed in Eqs. (2.29) and (2.30) yields the following information:

$$\mathbf{v}_P \times \mathbf{a}_P = \begin{vmatrix} \hat{\mathbf{e}}_t & \hat{\mathbf{e}}_n & \hat{\mathbf{e}}_b \\ \dot{s} & 0 & 0 \\ \ddot{s} & \frac{\dot{s}^2}{\rho} & 0 \end{vmatrix} = \frac{\dot{s}^3}{\rho}\hat{\mathbf{e}}_b \tag{2.36}$$

where $\hat{\mathbf{e}}_b$ is a unit vector perpendicular to both the unit vectors $\hat{\mathbf{e}}_t$ and $\hat{\mathbf{e}}_n$. Hence, the magnitude of the above cross-product is $|\mathbf{v}_P \times \mathbf{a}_P| = \frac{\dot{s}^3}{\rho}$. Therefore, even if the velocity and acceleration of a point moving along a locus are in the Cartesian coordinates, their cross-product can give information about the curvature of the locus at the given instant.

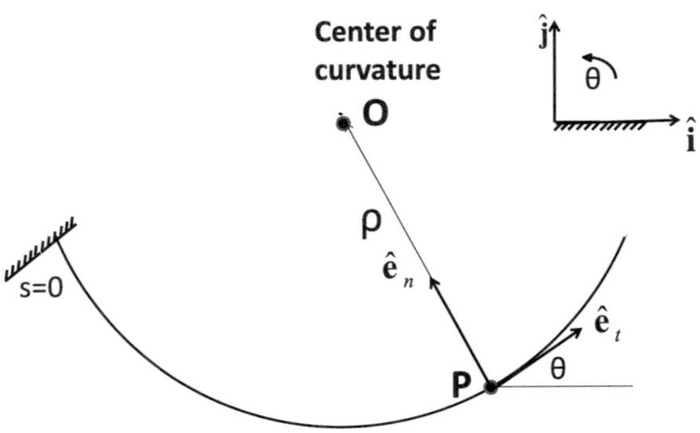

Fig. 2.12 Normal and tangential coordinates for a concave downward curve

Example 2.4

The position vector of a point P is given by $\mathbf{r} = 2t\hat{\mathbf{i}} + t^2\hat{\mathbf{j}}$ m. At t = 1 s, determine the tangential and normal components of velocity, acceleration, and radius of curvature of the locus and unit vectors.

Position vector of point P as a function of time is given by

$$\mathbf{r} = 2t\hat{\mathbf{i}} + t^2\hat{\mathbf{j}} \text{ m} \tag{1}$$

Velocity vector of point P as a function of time is given by

$$\mathbf{v} = 2\hat{\mathbf{i}} + 2t\hat{\mathbf{j}} \text{ m/s} \tag{2}$$

Acceleration vector of point P as a function of time is given by

$$\mathbf{a} = 2\hat{\mathbf{j}} \text{ m/s}^2 \tag{3}$$

From Eq. (2), velocity of point P at t = 1 s is given by

$$\mathbf{v}(t = 1) = 2\hat{\mathbf{i}} + 2\hat{\mathbf{j}} \text{ m/s} \tag{4}$$

From Eq. (3), acceleration of point P is given by

$$\mathbf{a} = 2\hat{\mathbf{j}} \text{ m/s}^2 \tag{5}$$

Magnitude of the velocity of point P at t = 1 s is given by

$$|\mathbf{v}(t = 1)| = \dot{s} = \sqrt{2^2 + 2^2} = \sqrt{8} \text{ m/s} \tag{6}$$

The tangential unit vector that gives the direction of the velocity is given by

$$\hat{\mathbf{e}}_t = \frac{\mathbf{v}(t = 1)}{\dot{s}} = \frac{2\hat{\mathbf{i}} + 2\hat{\mathbf{j}}}{\sqrt{8}} = 0.71\hat{\mathbf{i}} + 0.71\hat{\mathbf{j}} \tag{7}$$

From Eqs. (4) and (5), the cross-product of the velocity and acceleration vectors at t = 1 s is given by

$$|\mathbf{v}_P \times \mathbf{a}_P| = \begin{vmatrix} \hat{\mathbf{i}} & \hat{\mathbf{j}} & \hat{\mathbf{k}} \\ 2 & 2 & 0 \\ 0 & 2 & 0 \end{vmatrix} = 4 = \frac{\dot{s}^3}{\rho} \tag{8}$$

From Eqs. (6) and (8), curvature of the locus of point P at t = 1 s is given by

$$\frac{1}{\rho} = \frac{4}{\dot{s}^3} = \frac{4}{8^{\frac{3}{2}}} = 0.18\, m^{-1} \tag{9}$$

From Eq. (2.35), magnitude of the tangential acceleration is given by

$$\ddot{s} = \sqrt{|\mathbf{a}_P|^2 - \left(\frac{\dot{s}^2}{\rho}\right)^2} = \sqrt{2^2 - (8 \times 0.1768)^2} = 1.41\ \text{m/s}^2 \qquad (10)$$

Direction of the normal acceleration in Cartesian coordinates is given by

$$\hat{\mathbf{e}}_n = \frac{\mathbf{a}_P - \ddot{s}\hat{\mathbf{e}}_t}{\frac{\dot{s}^2}{\rho}} = \frac{2\hat{\mathbf{j}} - 1.41\left(0.707\hat{\mathbf{i}} + 0.707\hat{\mathbf{j}}\right)}{8 \times 0.1768} - 0.707\hat{\mathbf{i}} + 0.707\hat{\mathbf{j}} \qquad (11)$$

2.5 Derivative of a Vector Referenced to a Moving Frame of Reference

If a vector is defined with respect to a moving coordinate system that is fixed on a moving body, it is possible to obtain its derivative in terms of the derivative of the scalar components of the vector and the angular velocity of the moving body. We shall first obtain the derivative of a two-dimensional vector on a moving body performing plane motion. Then the derivative for a general three-dimensional vector can be easily obtained.

Consider a fixed planar Cartesian coordinate system of unit vectors $\hat{\mathbf{i}}$ and $\hat{\mathbf{j}}$ shown in Fig. 2.13. A body $\mathcal{B}$ performing plane motion is rotating at an angular speed of ω rad/s; unit vectors of the moving Cartesian coordinate system of unit vectors $\hat{\mathbf{e}}_i$ and $\hat{\mathbf{e}}_j$ are fixed on the moving body and the unit vector $\hat{\mathbf{e}}_i$ makes an angle θ with unit vector $\hat{\mathbf{i}}$ of the fixed coordinate system at the given instant.

Vector $\mathbf{a}$ is defined with respect to the moving coordinate system and has scalar components a_x and a_y as shown and is given by

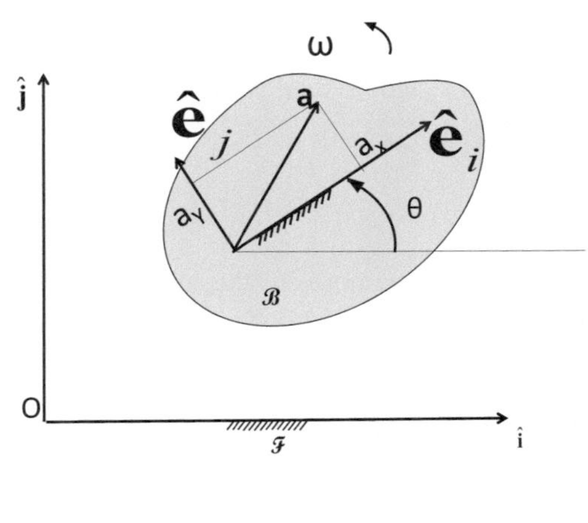

Fig. 2.13 Vector on a moving body in plane motion

$$\mathbf{a} = a_x \hat{\mathbf{e}}_i + a_y \hat{\mathbf{e}}_j \tag{2.37}$$

The unit vectors of the moving coordinates with respect to the fixed coordinates are, respectively, defined as follows:

$$\hat{\mathbf{e}}_i = \cos\theta \hat{\mathbf{i}} + \sin\theta \hat{\mathbf{j}} \tag{2.38}$$

$$\hat{\mathbf{e}}_j = -\sin\theta \hat{\mathbf{i}} + \cos\theta \hat{\mathbf{j}} \tag{2.39}$$

The derivatives of the unit vectors defined in Eqs. (2.38) and (2.39) are, respectively, given by

$$\frac{d\hat{\mathbf{e}}_i}{dt} = \left(-\sin\theta \hat{\mathbf{i}} + \cos\theta \hat{\mathbf{j}}\right)\dot{\theta} = \dot{\theta}\hat{\mathbf{e}}_j \tag{2.40}$$

$$\frac{d\hat{\mathbf{e}}_j}{dt} = -\left(\cos\theta \hat{\mathbf{i}} + \sin\theta \hat{\mathbf{j}}\right)\dot{\theta} = -\dot{\theta}\hat{\mathbf{e}}_i \tag{2.41}$$

From Eq. (2.37) through Eq. (2.41), the time derivative of **a** is given by

$$\begin{aligned}
\dot{\mathbf{a}} &= \dot{a}_x \hat{\mathbf{e}}_i + \dot{a}_y \hat{\mathbf{e}}_j + a_x \dot{\hat{\mathbf{e}}}_i + a_y \dot{\hat{\mathbf{e}}}_j \\
&= \dot{a}_x \hat{\mathbf{e}}_i + \dot{a}_y \hat{\mathbf{e}}_j + a_x \dot{\theta} \hat{\mathbf{e}}_j - a_y \dot{\theta} \hat{\mathbf{e}}_i \\
&= \dot{a}_x \hat{\mathbf{e}}_i + \dot{a}_y \hat{\mathbf{e}}_j + \dot{\theta}\left(-a_y \hat{\mathbf{e}}_i + a_x \hat{\mathbf{e}}_j\right) \\
&= {}^{\mathcal{B}}\dot{\mathbf{a}} + \dot{\theta}\hat{\mathbf{k}} \times \mathbf{a}
\end{aligned} \tag{2.42}$$

since

$$\hat{\mathbf{k}} \times \mathbf{a} = \begin{vmatrix} \hat{\mathbf{e}}_i & \hat{\mathbf{e}}_j & \hat{\mathbf{k}} \\ 0 & 0 & 1 \\ a_x & a_y & 0 \end{vmatrix} = -a_y \hat{\mathbf{e}}_i + a_x \hat{\mathbf{e}}_j \tag{2.43}$$

And that the time derivative of the scalars is basically the derivative of the vector in the moving coordinate system on which it is located, the time derivative of vector **a** is given by

$$\dot{\mathbf{a}} = {}^{\mathcal{B}}\dot{\mathbf{a}} + \dot{\theta}\hat{\mathbf{k}} \times (\mathbf{a}) \tag{2.44}$$

In the above equation, the unit vector $\hat{\mathbf{k}}$ refers to the direction of the axis of rotation of the moving body. Hence, $\omega\hat{\mathbf{k}}$ is the angular velocity vector of the rigid body in plane motion. If the moving body is performing three-dimensional motion defined by the angular velocity vector $\boldsymbol{\omega} = \omega\hat{\mathbf{r}}$, with its axis of rotation $\hat{\mathbf{r}}$ being a three-dimensional vector, the derivative of a general three-dimensional vector on a rigid body moving in the three-dimensional space can be written as

$$\dot{\mathbf{a}} = {}^{\mathcal{B}}\dot{\mathbf{a}} + \boldsymbol{\omega} \times \mathbf{a} \tag{2.45}$$

Hence, the derivative of a vector defined in a moving coordinate system is therefore a sum of the derivative of the scalars of the vector and the cross-product of the angular velocity of the body on which the moving coordinate system is located with the vector. Equation (2.45) is a very useful equation that is used throughout the book that can compute the derivative of a vector defined with respect to a moving coordinate system.

2.6 Velocity Relations for a Rigid Body in Plane Motion

Consider a rigid body in plane motion as shown in Fig. 2.14. The rigid body is assumed to be rotating counterclockwise at an angular velocity of ω rad/s. Consider any two points P and Q on the rigid body that are referenced to the fixed frame of reference. The line joining the points P and Q makes an angle θ with respect to the horizontal at the instant shown.

To determine the velocity relationship between the two points P and Q on the rigid body, the relevant position vectors can be redrawn as shown in Fig. 2.15

$$\mathbf{r}_{OQ} = \mathbf{r}_{OP} + \mathbf{r}_{PQ} \tag{2.46}$$

$\mathbf{r}_{OQ}$ and $\mathbf{r}_{OP}$ are absolute position vectors referenced to the fixed coordinate system, whereas vector $\mathbf{r}_{PQ}$ is a vector on the moving rigid body. Hence, although the magnitude of $\mathbf{r}_{PQ}$ remains the same because of rigid body considerations, its direction keeps changing with respect to time. This has to be accounted for while taking the time derivative Eq. (2.46).

Time derivative Eq. (2.46) results in

$$\dot{\mathbf{r}}_{OQ} = \dot{\mathbf{r}}_{OP} + \dot{\mathbf{r}}_{PQ} \tag{2.47}$$

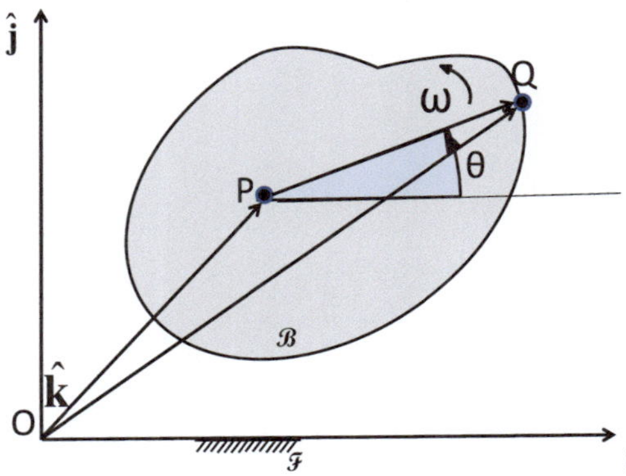

Fig. 2.14 Rigid body in plane motion

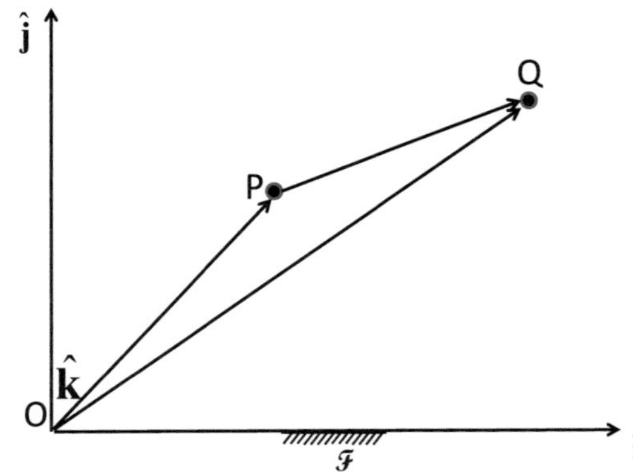

Fig. 2.15 Position vectors of points P and Q of Fig. 2.14

Expressing in terms of velocities of the respective points and the relative velocities between them, Eq. (2.47) becomes

$$\mathbf{v}_Q = \mathbf{v}_P + \mathbf{v}_{Q/P} \tag{2.48}$$

$\mathbf{v}_Q$ and $\mathbf{v}_P$ are, respectively, the absolute velocities of the points Q and P, and $\mathbf{v}_{Q/P}$ is the relative velocity of point P with respect to Q. Equation (2.48) is graphically shown in Fig. 2.16.

For a rigid body performing plane motion, either magnitude or direction of some of the components of Fig. 2.16 will be known through which the velocity diagram can be completed. It is important to note that $\mathbf{v}_{Q/P}$ is perpendicular to the direction of PQ as it is the cross-product of the angular velocity vector and the position vector, which will be proved later in Eq. (2.53). Although we are going to use the vector approach throughout the textbook for determining various parameters of dynamic motion, the graphical approach will also be presented as a complimentary alternative to get better physical insight.

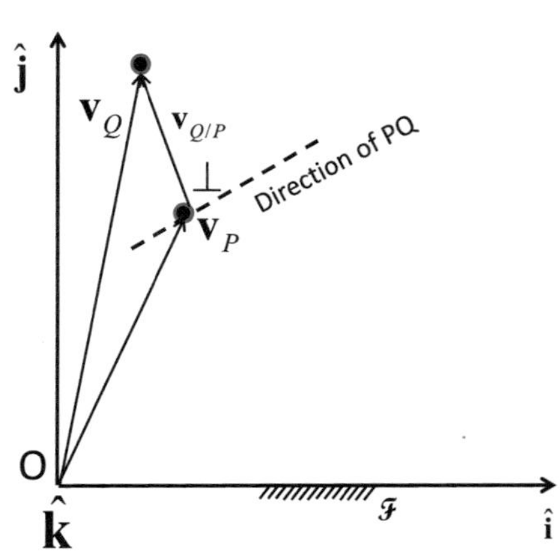

Fig. 2.16 Velocity diagram of points P and Q of Fig. 2.14

2.6.1 Vector Approach

Now we shall return to the vector approach for determining the velocity relationship between points P and Q. We have to focus determining the relative velocity $\mathbf{v}_{Q/P}$. Referring to Fig. 2.15, position vector $\mathbf{r}_{PQ}$ can be written in terms of its length and a unit vector along its length. Now we have to define the new local coordinates axes, which are aligned with the global directions, but their origin at point P. Since the directions of both local and global coordinates are the same, as we are only interested in computing velocity, there is no reason for defining local coordinates for points that are on the moving body. Eventually, we are going to deal with the kinetic and kinematic analysis of mechanisms that have various links. Hence, it would be sufficient to work with one set of global coordinates, instead of defining local coordinates for every rigid body or its points. The position vector is thus given by

$$\mathbf{r}_{PQ} = r_{PQ}\hat{\mathbf{u}} \tag{2.49}$$

As shown in Fig. 2.17, the unit vector along PQ is given by

$$\hat{\mathbf{u}} = \cos\theta\hat{\mathbf{i}} + \sin\theta\hat{\mathbf{j}} \tag{2.50}$$

Differentiating the unit vector of Eq. (2.50)

$$\begin{aligned}\hat{\mathbf{u}} &= \left(-sin\theta\hat{\mathbf{i}} + \cos\theta\hat{\mathbf{j}}\right)\dot{\theta} \\ &= \left(-sin\theta\hat{\mathbf{i}} + \cos\theta\hat{\mathbf{j}}\right)\omega\end{aligned} \tag{2.51}$$

From Eqs. (2.49) and (2.50), relative velocity $\mathbf{v}_{PQ}$ is given by

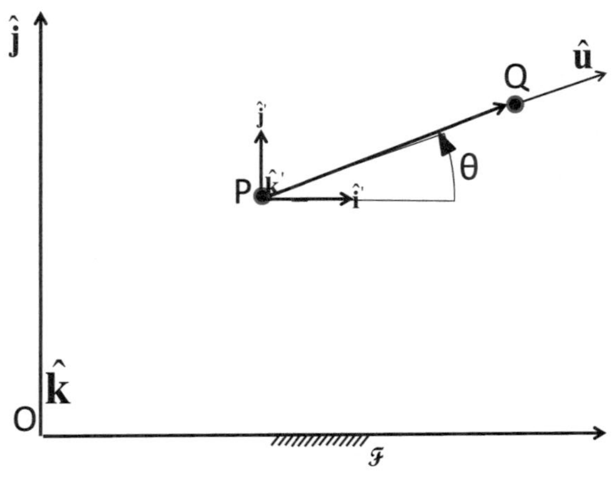

Fig. 2.17 Unit vector along PQ of Fig. 2.14

$$\mathbf{v}_{Q/P} = r_{PQ}\left(-sin\theta\hat{\mathbf{i}} + \cos\theta\hat{\mathbf{j}}\right)\omega \tag{2.52}$$

If we take the dot product of Eqs. (2.50) and (2.52), by noting that $\hat{\mathbf{i}}.\hat{\mathbf{j}} = 0, \hat{\mathbf{i}}.\hat{\mathbf{i}} = 1$ and $\hat{\mathbf{j}}.\hat{\mathbf{j}} = 1$,

$$\mathbf{v}_{Q/P}.\hat{\mathbf{u}} = r_{PQ}\omega\left(-sin\theta\hat{\mathbf{i}} + \cos\theta\hat{\mathbf{j}}\right).\left(\cos\theta\hat{\mathbf{i}} + \sin\theta\hat{\mathbf{j}}\right) = 0 \tag{2.53}$$

Equation(2.53) demonstrates that the relative velocity $\mathbf{v}_{PQ}$ and the unit vector along PQ are perpendicular to each other. This will be an important factor in solving the velocity diagram using graphical method.

Since we would like to use the vector $\mathbf{r}_{PQ}$ in our calculations, we can modify Eq. (2.53) using the concept of cross-product of vectors. Since we know that the cross-product of two unit vectors perpendicular to each other is unity, we can always express the direction of relative velocity $\mathbf{v}_{Q/P}$ in terms of cross-product of the unit vector $\hat{\mathbf{k}}$ with the unit vector $\hat{\mathbf{u}}$. But before that, we have to prove that $\hat{\mathbf{k}} \times \hat{\mathbf{u}}$ gives the direction of the relative velocity $\mathbf{v}_{Q/P}$.

Cross-product of the unit vector $\hat{\mathbf{k}}$ with the unit vector $\hat{\mathbf{u}}$ is given by

$$\hat{\mathbf{k}} \times \hat{\mathbf{u}} = \hat{\mathbf{k}} \times \left(\cos\theta\hat{\mathbf{i}} + \sin\theta\hat{\mathbf{j}}\right) = \left(-sin\theta\hat{\mathbf{i}} + \cos\theta\hat{\mathbf{j}}\right) \tag{2.54}$$

From Eqs. (2.52) and (2.54), the relative velocity $\mathbf{v}_{Q/P}$ is given by

$$\mathbf{v}_{Q/P} = r_{PQ}\omega\hat{\mathbf{k}} \times \left(\cos\theta\hat{\mathbf{i}} + \sin\theta\hat{\mathbf{j}}\right) = \omega\hat{\mathbf{k}} \times \mathbf{r}_{PQ} \tag{2.55}$$

Equation (2.55) represents the cross-product between the angular velocity vector and the position vector of PQ. Note that the angular velocity vector is given by the product of the magnitude of angular velocity and the direction of the axis of rotation. The axis of rotation is $\hat{\mathbf{k}}$ for the plane motion we have assumed in this case. But the above Eq. (2.55) can be extended to a general three-dimensional motion with the relevant direction of the axis of rotation and position vector PQ.

From Eqs. (2.48) and (2.55), the velocity relationship between two points of a rigid body is given by

$$\begin{aligned}\mathbf{v}_Q &= \mathbf{v}_P + \omega\hat{\mathbf{k}} \times \mathbf{r}_{PQ} \\ &= \mathbf{v}_P + \mathbf{v}_{Q/P}\end{aligned} \tag{2.56}$$

Since PQ are points on the moving body, we could have directly used Eq. (2.44) to obtain the time derivative of $\mathbf{r}_{PQ}\mathbf{r}_{PQ}$

$$\dot{\mathbf{r}}_{PQ} = {}^{\mathcal{B}}\dot{r}_{PQ}\hat{\mathbf{u}} + \omega\hat{\mathbf{k}} \times \mathbf{r}_{PQ} \tag{2.57}$$

Since P and Q are points on the rigid body, the distance between them remains the same. Hence, ${}^{\mathcal{B}}\dot{r}_{PQ} = 0$. Equation (2.57) then becomes

$$\dot{\mathbf{r}}_{PQ} = \mathbf{v}_{Q/P} = \omega\hat{\mathbf{k}} \times \mathbf{r}_{PQ} \tag{2.58}$$

Hence, Eq. (2.58) is same as Eq. (2.50). Since we also wanted to prove that the relative velocity between two points of a rigid body is perpendicular to the line joining them, an important point in velocity analysis, we had to go through the derivations from Eqs. (2.49)–(2.58). However, we need Eq. (2.44) many times later in the book.

2.6.2 Graphical Approach

Figure 2.16 is reproduced in Fig. 2.18. Points P and Q in many cases are either fixed pivots or common points between linkages connected by revolute pairs. In many cases, either one of the velocities or its direction is known. We can use the above velocity triangle to determine the unknown components of velocity from known components by completing the velocity triangle. This is demonstrated in the following examples.

Example 2.5
Consider the line diagram of a reciprocating mechanism shown in Fig. 2.19a and the block diagram representation of its kinematic pairs shown in Fig. 2.19b. It consists of a crank (2) hinged to the fixed link (1) at O and to the connecting rod (3) at B. The connecting rod is hinged to the reciprocating piston (4) at A and the reciprocating mass is allowed to slide along the horizontal direction on the fixed link (1). The kinematic pairs of the above mechanism have already been discussed earlier. Point B is a common point between the crank and connecting and point A is a common point between the piston and the connecting rod. If the crank is rotating counterclockwise at an angular speed of 15 rad/s, determine the velocity of the piston at this instant of time.

Fig. 2.18 Graphical representation of velocity information of Fig. 2.14

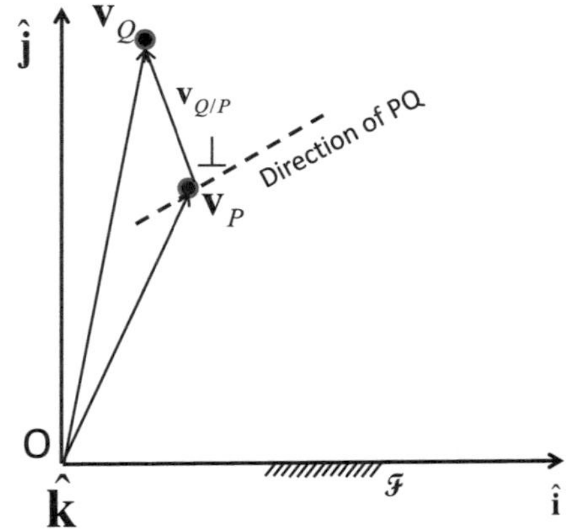

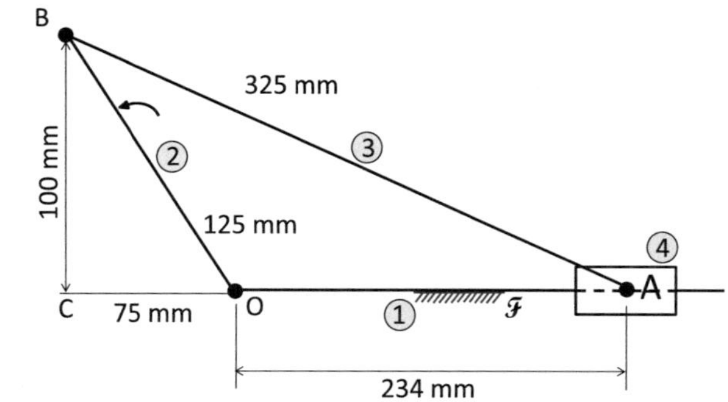

Fig. 2.19 Line diagram of a reciprocating mechanism of Example 2.5

Since we know the angular velocity, ω_2, of the crank, we should start the velocity analysis with link 2 as shown in Fig. 2.20. From Eq. (2.56), the velocity relationship between the points B and O can be written. Points O and B are obvious points on the crank since we know that the velocity at point O at which the crank is hinged is zero and point B is a common point between the crank and the connecting rod. Hence, the following equation can be written as

$$\begin{aligned} \mathbf{v}_B &= \underbrace{\mathbf{v}_O}_{=0} + \omega_2 \hat{\mathbf{k}} \times \mathbf{r}_{OB} \\ &= 0 + 15\hat{\mathbf{k}} \times \left(-75\hat{\mathbf{i}} + 100\hat{\mathbf{j}}\right) \\ &= -1500\hat{\mathbf{i}} - 1125\hat{\mathbf{j}} \text{mm/s} \end{aligned} \tag{1}$$

Magnitude of the velocity of B is given by

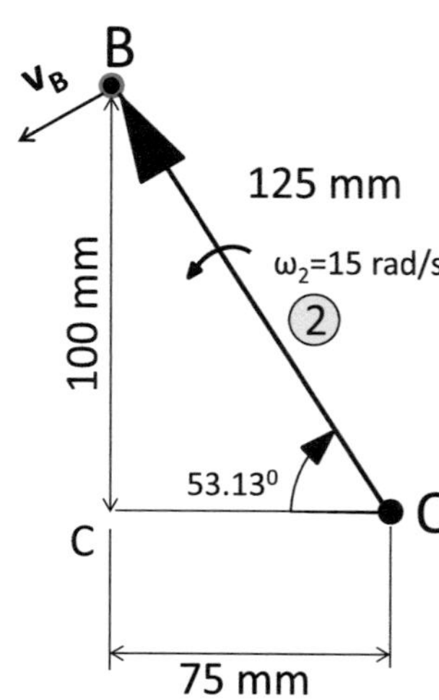

Fig. 2.20 Velocity relationships between various points of the crank of Example 2.5

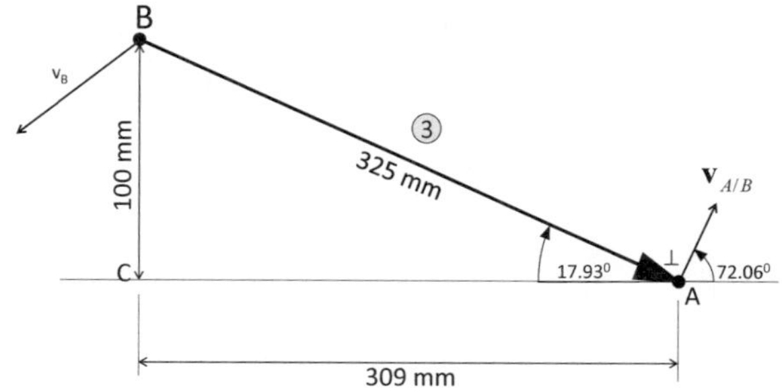

Fig. 2.21 Connecting rod position and velocity vectors of Example 2.5

$$|\mathbf{v}_B| = \sqrt{1500^2 + 1125^2} = 1875\,\text{mm/s} \tag{2}$$

As per our earlier discussion, the velocity of point B, v_B, is perpendicular to OB. However, that information is not required here to compute the velocity of B. But that will be useful in the graphical analysis.

Now we can move to the next link, the connecting rod. Since B is a common point between the crank and the connecting rod, the velocity of point B on the connecting rod is same as computed from Eq. (1). In addition, point A on the piston is constrained to be moved only in the horizontal direction. Using this information, we can compute the angular velocity, ω_3, of the connecting rod and velocity of the piston (Fig. 2.21).

From Eqs. (2.48) and (2.55), relating the velocities of points A and B on the connecting rod,

$$\begin{aligned}\mathbf{v}_A &= \mathbf{v}_B + \omega_3\hat{\mathbf{k}} \times \mathbf{r}_{BA} = -1500\hat{\mathbf{i}} - 1125\hat{\mathbf{j}} + \omega_3\hat{\mathbf{k}} \times \left(309\hat{\mathbf{i}} - 100\hat{\mathbf{j}}\right)\\ &= (-1500 + 100\omega_3)\hat{\mathbf{i}} + (-1125 + 309\omega_3)\hat{\mathbf{j}}\ \text{mm/s}\end{aligned} \tag{3}$$

There is no vertical component for the velocity of A. Hence, its $\hat{\mathbf{j}}$ component must be equal to zero. Therefore, angular velocity of the connecting rod, ω_3, can be obtained by equating the $\hat{\mathbf{j}}$ component of velocity of A in Eq. (3) to zero. It is given by

$$\omega_3 = \frac{1125}{309} = 3.64\,\text{rad/s (counterclockwise)} \tag{4}$$

Using the angular velocity of the connecting rod from Eq. (4), noting that it has only horizontal direction, velocity of the piston can be obtained from Eq. (3) as

$$\begin{aligned}\mathbf{v}_A &= (-1500 + 100\omega_3)\hat{\mathbf{i}} = (-1500 - 100 \times 3.64)\hat{\mathbf{i}} \\ &= -1136\,\hat{\mathbf{i}}\ \text{mm/s} \\ &\text{(Towards left along the horizontal direction)}\end{aligned} \quad (5)$$

Now that we have all the kinematics-related information, we can compute the relative velocity $\mathbf{v}_{A/B}$ as follows:

$$\begin{aligned}\mathbf{v}_{A/B} &= \omega_3\hat{\mathbf{k}} \times \mathbf{r}_{BA} = 3.64\hat{\mathbf{k}} \times \left(309\hat{\mathbf{i}} - 100\hat{\mathbf{j}}\right) \\ &= 3641\hat{\mathbf{i}} + 1125\hat{\mathbf{j}} = 1182.4\,\text{mm/s}\angle 72.06^\circ\end{aligned} \quad (6)$$

Equation(6) will be useful in checking the results of the graphical method.

Graphical method

(1) The crank makes an angle of 53.13° ($\tan^{-1}\left(\frac{100}{75}\right)$ as shown in Fig. 2.22a. Velocity of point B is perpendicular to OB and its magnitude is given by

$$v_B = OB \times \omega_2 = 125 \times 15 = 1875\,\text{mm/s}$$

(2) Draw a line at an angle of 53.13° representing the direction of the crank OB. Along any point along OB, draw a vector representing the velocity of B at an angle of (180 – 53.13 + 90) =216.86° as shown in Fig. 2.22b. As you know, this velocity vector must be perpendicular to the direction of OB. Point O is not only used as the hinge position of the crank but also used as the absolute point of velocity in all the velocity diagrams.
(3) The connecting rod makes an angle of 17.93° as shown in Fig. 2.22c and hence, relative velocity $\mathbf{v}_{A/B}$ is perpendicular to the connecting rod at point A. The direction of $\mathbf{v}_{A/B}$ is in the direction of a line inclined at an angle of (180 – 90 – 17.93)= 72.06° as shown in the figure.
(4) Continue with velocity diagram of Fig. 2.22c and draw the direction of AB inclined at 17.93°, the direction of $\mathbf{v}_{A/B}$ perpendicular to AB, and the direction of the velocity of A along the horizontal direction as shown in Fig. 2.22d. The direction of $\mathbf{v}_{A/B}$ meets the horizontal line at *a*.
(5) Refer to Fig. 2.22e. Point *a* at which the direction of the velocity of A and the direction of the relative velocity $\mathbf{v}_{A/B}$ meet gives the actual magnitudes of each of them. From Eq. (2.56), the relevant velocity equation is $\mathbf{v}_A = \mathbf{v}_B + \mathbf{v}_{A/B}$. Hence, the velocity vector $\mathbf{v}_{A/B}$ should start from the tip of the velocity vector $\mathbf{v}_B$ and move in the direction of $\mathbf{v}_{A/B}$ at 72.06°. the velocity vector $\mathbf{v}_A$ should start from the absolute (zero) starting point of the velocity vector of $\mathbf{v}_B$. Hence, the velocity triangle is complete as shown in Fig. 2.22e. Measuring the velocity diagram, $|\mathbf{v}_A| = 1108\,\text{mm/s}$ and is towards left and $\left|\mathbf{v}_{A/B}\right| = 1159.4\,\text{mm/s}$ in the direction of 72.06° with respect to the horizontal.

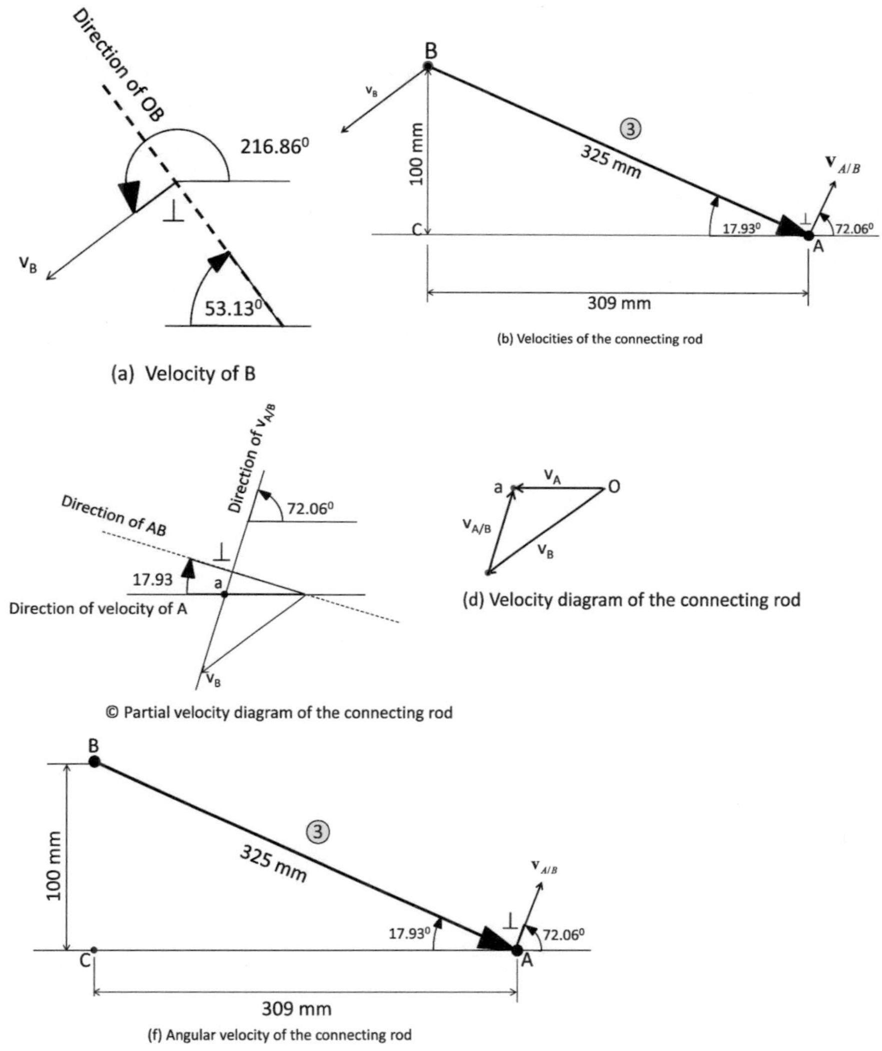

Fig. 2.22 Sequence of velocity diagrams of Example 2.5

(6) Refer to Fig. 2.22f showing a line diagram of the connecting rod. Now that we know the direction of $\mathbf{v}_{A/B}$, it is clear that the connecting rod must be rotating in the counterclockwise direction and its magnitude is given by $\frac{|\mathbf{v}_{A/B}|}{AB} = 3.56$ rad/s.

We may not be able to present the same sequence of diagrams for all the graphical methods presented in this book. A detailed sequence was presented here to show the connection between vector and graphical methods.

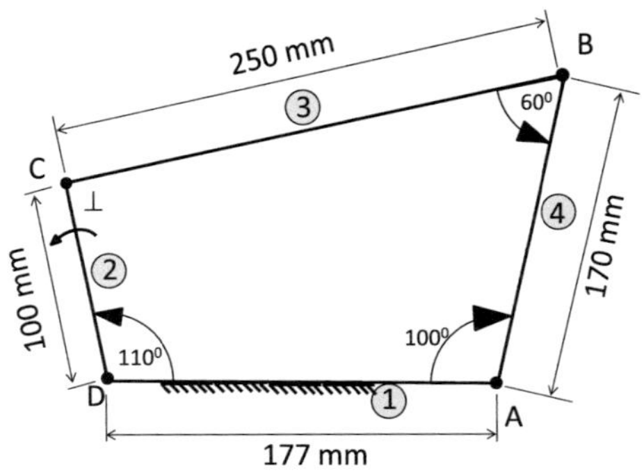

Fig. 2.23 Four-bar mechanism of Example 2.6

Example 2.6

Consider the four-bar mechanism shown in Fig. 2.23. If the angular velocity of crank (link 2), ω_2, is 5 rad/s (counterclockwise), determine the angular velocities of the floating link (link 3), ω_3 and the follower (link 4), ω_4.

Figure 2.24 shows a line diagram of the crank (link 2). Position vector $\mathbf{r}_{DC}$ is given by

$$\begin{aligned}\mathbf{r}_{DC} &= 100\cos 110\hat{\mathbf{i}} + 100\sin 110\hat{\mathbf{j}} \\ &= -34.20\hat{\mathbf{i}} + 93.96\hat{\mathbf{j}}\ \text{mm}\end{aligned} \tag{1}$$

From Eq. 2.56, velocity of point C, $\mathbf{v}_C$, is given by

$$\begin{aligned}\mathbf{v}_C &= \mathbf{v}_D + \omega_2\hat{\mathbf{k}} \times \mathbf{r}_{DC} \\ &= 5\hat{\mathbf{k}} \times \left(-34.20\hat{\mathbf{i}} + 93.96\hat{\mathbf{j}}\right) \\ &= -469.84\hat{\mathbf{i}} - 171.01\hat{\mathbf{j}}\ \text{mm/s}\end{aligned} \tag{2}$$

Consider the line diagram of link 3 shown in Fig. 2.25. Position vector $\mathbf{r}_{CB}$ is given by

$$\begin{aligned}\mathbf{r}_{CB} &= 250\cos 20\hat{\mathbf{i}} + 250\sin 20\hat{\mathbf{j}} \\ &= 234.92\hat{\mathbf{i}} + 85.5\hat{\mathbf{j}}\ \text{mm}\end{aligned} \tag{3}$$

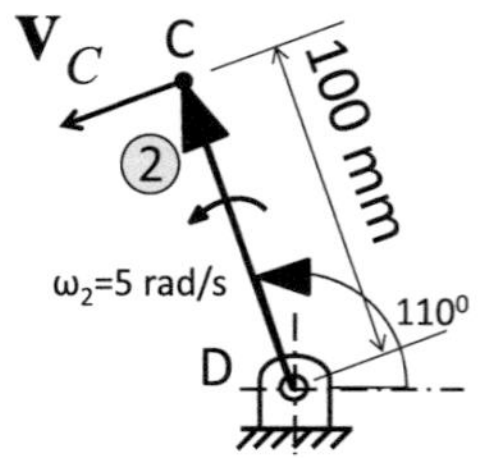

Fig. 2.24 Line diagram of the crank (link 2) of Example 2.6

Fig. 2.25 Line diagram of the coupler (link 3) of Example 2.6

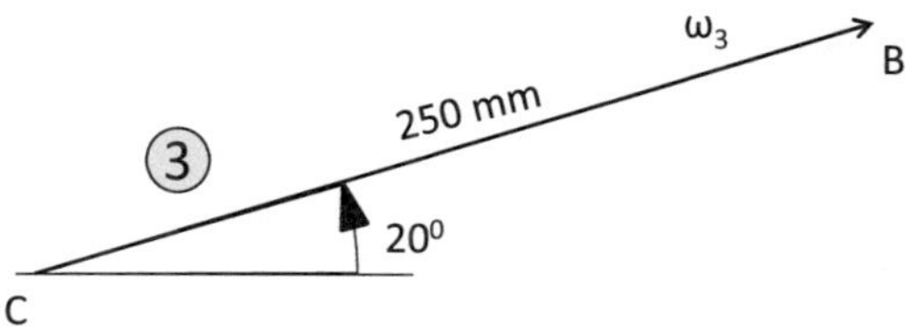

Velocity of point B, $\mathbf{v}_B$, is given by

$$\begin{aligned}\mathbf{v}_B &= \mathbf{v}_C + \omega_3\hat{\mathbf{k}} \times \mathbf{r}_{CB} \\ &= -469.84\hat{\mathbf{i}} - 171.01\hat{\mathbf{j}} \; + \omega_3\hat{\mathbf{k}} \times \left(234.92\hat{\mathbf{i}} + 85.5\hat{\mathbf{j}}\right) \\ &= (-469.84 - 85.5\omega_3)\hat{\mathbf{i}} + (-171.01 + 234.92\omega_3)\hat{\mathbf{j}} \text{ mm/s} \qquad (4)\end{aligned}$$

Consider the line diagram of link 4 (follower) shown in Fig. 2.26. Position vector $\mathbf{r}_{AB}$ is given by

$$\begin{aligned}\mathbf{r}_{AB} &= 177\cos 80\hat{\mathbf{i}} + 177\sin 80\hat{\mathbf{j}} \\ &= 29.52\hat{\mathbf{i}} + 167.41\hat{\mathbf{j}} \qquad (5)\end{aligned}$$

Velocity of point B, $\mathbf{v}_B$, is given by

$$\begin{aligned}\mathbf{v}_B &= \mathbf{v}_A + \omega_4\hat{\mathbf{k}} \times \mathbf{r}_{AB} \\ &= \omega_4\hat{\mathbf{k}} \times \left(29.52\hat{\mathbf{i}} + 167.41\hat{\mathbf{j}}\right) \\ &= -167.41\omega_4\hat{\mathbf{i}} + 29.52\omega_4\hat{\mathbf{j}} \qquad (6)\end{aligned}$$

From Eqs. (4) and (6),

$$\begin{bmatrix} -85.5 & +167.41 \\ 234.92 & -29.52 \end{bmatrix} \begin{Bmatrix} \omega_3 \\ \omega_4 \end{Bmatrix} = \begin{Bmatrix} 469.84 \\ 171.01 \end{Bmatrix} \qquad (7)$$

Fig. 2.26 Line diagram of the follower (link 4) of Example 2.6

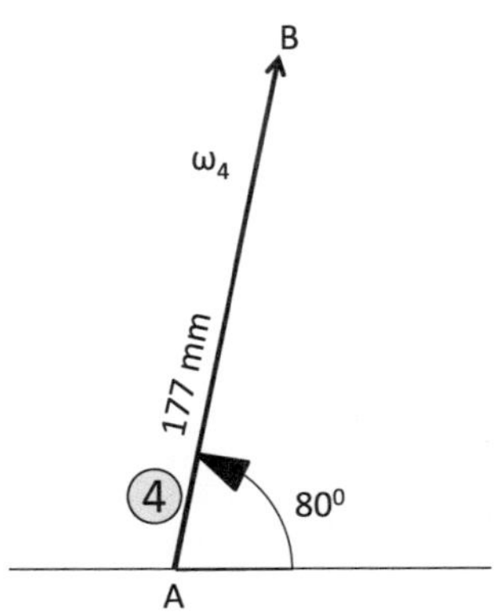

Solving the system of linear Eq. (7), $\omega_3 = 1.1547$rad/s (counterclockwise) and $\omega_4 = 3.3963$ rad/s (counterclockwise).

Graphical Method

The following steps are to be followed:

(a) Refer to Fig. 2.27a. Draw a line at an angle of 110° representing the direction of CD. Velocity of the point C, $\mathbf{v}_C = \omega_2 CD = 5 \times 100 = 500\,\text{mm/s}$. Since $\mathbf{v}_C$ is perpendicular to CD, draw the velocity vector perpendicular to CD of magnitude 500 mm/s. Point "o" represents the origin of the velocity diagram.

(b) Refer to Fig. 2.27b. Draw a line representing BC at an angle of 20°. Since BC is perpendicular to CD, BC is in line with the direction of velocity $\mathbf{v}_C$. From the tip of velocity vector $\mathbf{v}_C$, draw a line perpendicular to BC that represents the direction of $\mathbf{v}_{B/C}$.

(c) Refer to Fig. 2.27c. Draw a line AB at an angle of 80°. Draw the direction of the velocity vector $\mathbf{v}_B$, perpendicular to AB, to meet the direction of $\mathbf{v}_{B/C}$ at point a.

(d) Refer to Fig. 2.27d. Complete the velocity triangle. Magnitude of $|\mathbf{v}_{B/C}| =$ 288.68 mm/s and the magnitude of $|\mathbf{v}_B| = 577.35$ mm/s. Angular velocity

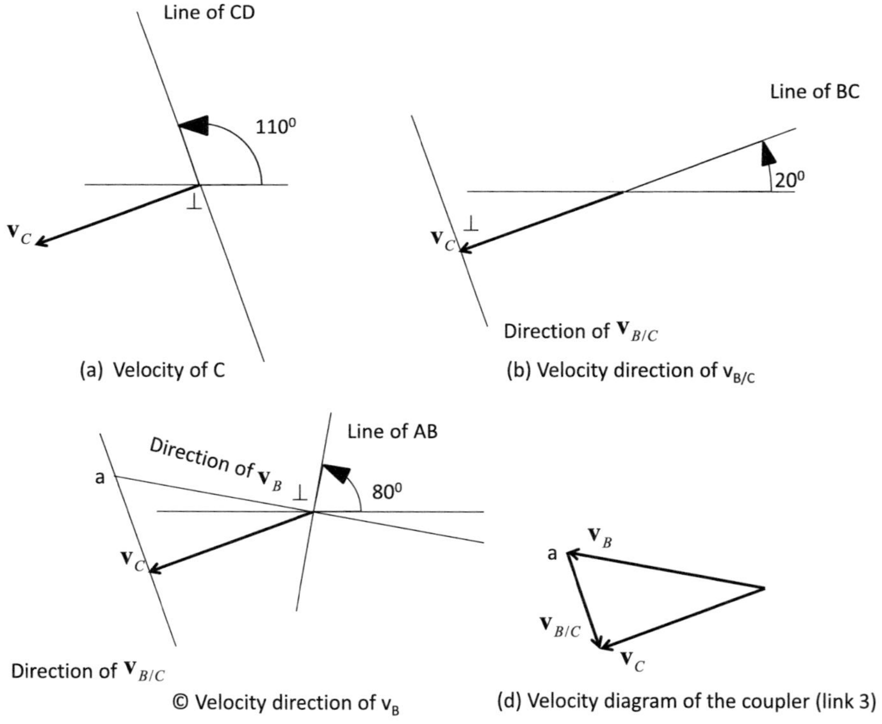

Fig. 2.27 Graphical method of solving the velocity of a four-bar mechanism of Example 2.6

$\omega_3 = \frac{|\mathbf{v}_{B/C}|}{BC} = \frac{288.68}{250} = 1.1547$ rad/s; angular velocity $\omega_2 = \frac{|\mathbf{v}_B|}{AB} = \frac{577.35}{170} =$ 3.3963 rad/s.

2.7 Instantaneous Center

For any rotating body, it is always possible to find an instantaneous center of zero velocity about which it is rotating at any given instant; in addition, it is possible to find an instantaneous center that may have a velocity common to two rotating bodies about which they are rotating. Both the concepts are useful in the kinematic analysis of rigid bodies.

First, let us discuss the instantaneous center of zero velocity. Consider points P and Q on a rigid body in plane motion shown in Fig. 2.28a that is rotating with an angular velocity ω. From Eq. 2.56, the velocities of points P and Q are related by

$$\begin{aligned}\mathbf{v}_Q &= \mathbf{v}_P + \omega\hat{\mathbf{k}} \times \mathbf{r}_{PQ} \\ &= \mathbf{v}_P + \mathbf{v}_{Q/P}\end{aligned} \tag{2.59}$$

We can always find a point P, either on the body or an imaginary extension moving with it, that has zero velocity. Such a point is known as the instantaneous center of zero velocity that is only valid at a particular instant of time of position of the body.

The concept of instantaneous center of zero velocity with respect to a single body with respect to a fixed frame of reference is illustrated in Fig. 2.28a. A point I that is either on the actual body or an imaginary extension of the body and moving with it is assumed to be the instantaneous center of zero velocity. Since point I has zero velocity, velocity of point Q is $\mathbf{v}_Q = \omega\hat{\mathbf{k}} \times \mathbf{r}_{IQ}$ and is perpendicular to the line joining IQ. Hence, if we know the directions of velocities of two points of a rigid body in motion, the instantaneous center of zero velocity can be obtained, by drawing lines perpendicular to the absolute velocities. Once the instantaneous center of zero velocity is located, the unknown parameters at the instant like magnitude of one of the velocities and angular velocity of the body can be computed.

The above concept of instantaneous center of zero velocity can be further extended to instantaneous centers of velocity that can relate two rigid bodies in motion to a common point that may or may not have a zero velocity. For example, Fig. 2.28b shows a four-bar linkage that has six instantaneous centers of velocity: (1) P_{12} is the point about which link 2 rotates about the hinge in the fixed reference 1, (2) P_{23} is the point about which link 3 rotates with respect to link 2, (3) P_{34} is the point about which link 4 rotates about link 3, (4) P_{14} is the point about which link 4 rotates about a hinge in the fixed reference 1, (5) P_{24} is the common point about which links 2 and 4 rotate, and (6) P_{31} is the point about which link 3 rotates about the fixed frame of reference 1, which is nothing but the instantaneous center of zero velocity discussed earlier and can be useful in determining the angular velocity of the coupler link that is a floating linkage.

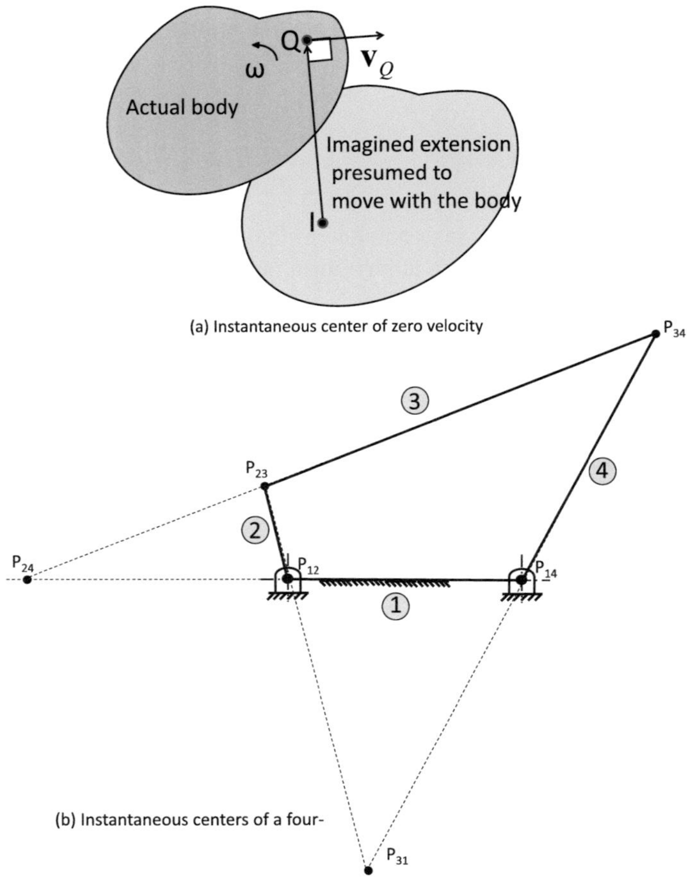

Fig. 2.28 **a** Instantaneous center of zero velocity **b** Instantaneous centers of a four-bar linkage

It is easy to locate instantaneous centers P_{12}, P_{14}, P_{23}, and P_{34}; they are the hinge points of the given links. As per Kennedy's theorem, P_{24}, P_{12}, and P_{14} should be along the same line and P_{24}, P_{23}, and P_{34} should be along the same line, which would help locate P_{24} by extending $P_{12}P_{14}$ and $P_{12}P_{14}$ to meet at P_{24}. Since the direction of velocities of P_{23} and P_{34} are, respectively, perpendicular to links 2 and 4, the directions of these velocities are known, by extending links 2 and 4 to meet at P_{31} will give the instantaneous center of zero velocity that can be used for determining the angular velocity of the coupler, 3.

The instantaneous centers P_{12}, P_{14}, and P_{31} have zero velocity and the instantaneous centers P_{24}, P_{23}, and P_{34} have in general have non-zero velocities. The instantaneous center P_{24} is not that much relevant for the present chapter, but is useful in making useful conclusions regarding the angular velocities of rigid bodies that are part of dynamic systems such as gears, cams, etc. For the above four-bar linkage, the

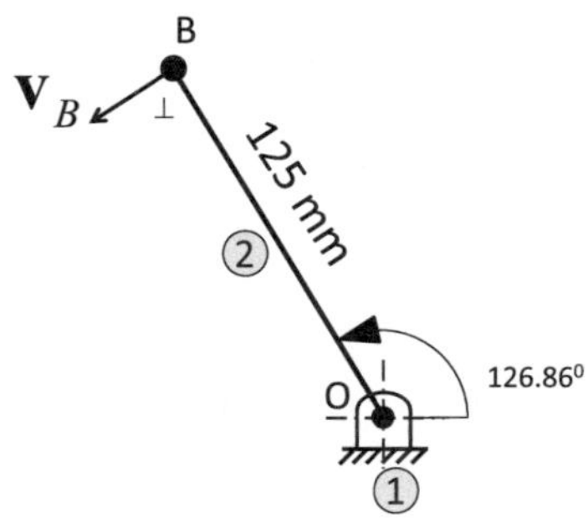

Fig. 2.29 Velocity of point B of Example 2.7

ratio of the angular velocities of links 2 and 4 is given by the ratio of $\frac{\omega_2}{\omega_4} = \frac{P_{24}P_{12}}{P_{24}P_{14}}$ (See Shigley for further details).

Example 2.7
Consider Example 2.5. Determine the angular velocity of the connecting rod and speed of the piston by using the method of instantaneous center of zero velocity.

Velocity of point B is given by

$$|\mathbf{v}_B| = OB \times \omega_1 = 125 \times 15 = 1875\,\text{mm/s} \tag{1}$$

And it is perpendicular to OB as shown in Fig. 2.29.

A line diagram of the connecting rod (link 3) with known velocities (magnitude and/or direction) is shown in Fig. 2.30. Magnitude and direction of $\mathbf{v}_B$ is known from velocity analysis of the crank that is perpendicular to OB and the direction of $\mathbf{v}_A$ is along the horizontal. Now, by drawing perpendicular lines to the direction of velocities $\mathbf{v}_A$ and $\mathbf{v}_B$, the instantaneous center of zero velocity can be determined as shown in Fig. 2.31.

Referring to Fig. 2.31, from similar triangles BCO and OAI,

$$OI = \frac{OA \times BO}{CO} = \frac{125 \times 234}{75} = 390\,\text{mm} \tag{2}$$

Hence, $AI = \sqrt{390^2 - 234^2} = 312$ mm, BI = 390 + 125 = 515 mm.

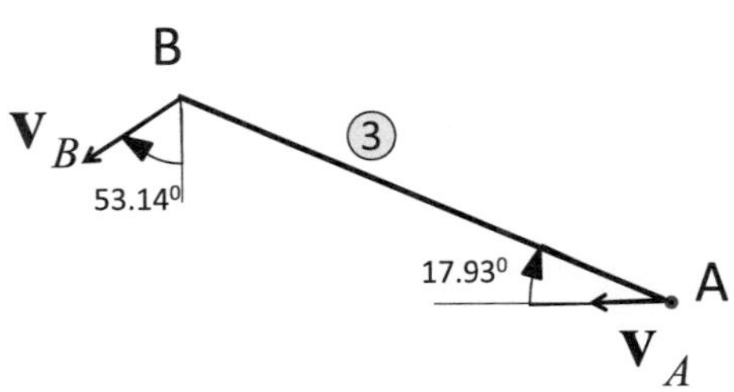

Fig. 2.30 Line diagram of the connecting rod with known velocities (magnitude and/or directions) of Example 2.7

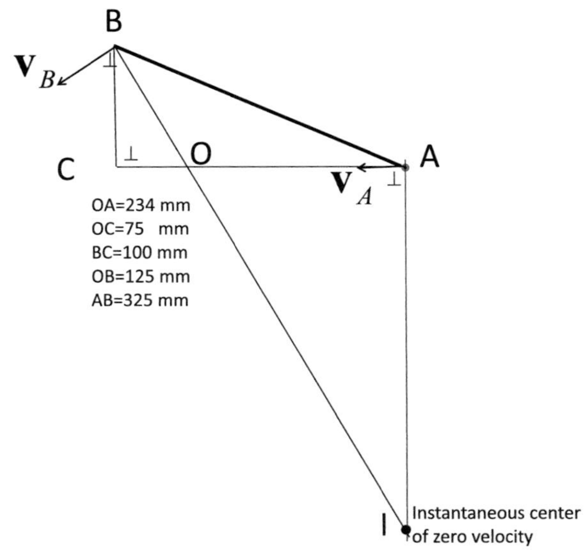

Fig. 2.31 Locating the instantaneous center of zero velocity for the connecting rod of Example 2.7

The angular velocity of link 3 is given by

$$\omega_3 = \frac{|\mathbf{v}_B|}{BI} = \frac{1875}{515} = 3.64\,\text{rad/s} \tag{3}$$

Magnitude of the velocity of point A is given by

$$|\mathbf{v}_A| = \omega_3 AI = 3.64 \times 312 = 1136\,\text{mm/s} \tag{4}$$

Example 2.8

Consider Example 2.6. Determine the angular velocities of the coupler and follower using the method of instantaneous centers.

Figure 2.32 shows the instantaneous center of the coupler, which can be obtained as follows:

Draw the outline of the four-bar mechanism. Since the angular velocity of the crank (link 2) is known, magnitude of velocity of the point C, $\mathbf{v}_C = 500$ mm/s and it is perpendicular to the crank. Magnitude of the velocity of point B, $\mathbf{v}_B$, is not known but its direction is perpendicular to the follower. Therefore, by extending the crank and follower, the instantaneous center I can be obtained.

BCI is a right-angled triangle. Hence, if we can create another similar right-angled triangle within this triangle, location of the instantaneous center, I, can be obtained. Draw a line from A perpendicular to CI that meets CI at point F. $\angle DAF = 20°$; AF = AD cos 20 = 166.33 mm; DF = AD sin 20 = 60.54 mm.

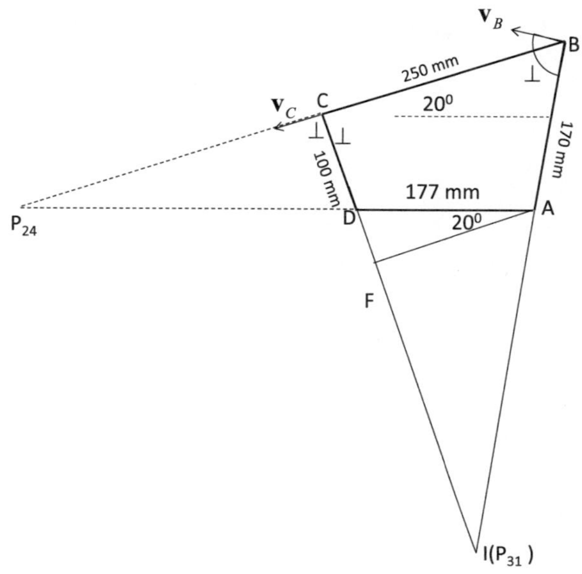

Fig. 2.32 Instantaneous center of the coupler of Example 2.8

From similar triangles BCI and AFI,

$$\begin{aligned}\frac{BC}{AF} &= \frac{CI}{FI}\\ &= \frac{CD + DF + FI}{FI}\end{aligned} \quad (1)$$

From Eq. (1),

$$\begin{aligned}FI &= \frac{CD + DF}{\frac{BC}{AF} - 1}\\ &= \frac{100 + 60.54}{\frac{250}{166.33} - 1} = 319.1120\,\text{mm}\end{aligned} \quad (2)$$

From similar triangles BCI and AFI,

$$\begin{aligned}\frac{BC}{AF} &= \frac{BI}{AI}\\ &= \frac{AB + AI}{AI}\end{aligned} \quad (3)$$

From Eq. (3),

$$\begin{aligned} AI &= \frac{AB}{\frac{BC}{AF} - 1} \\ &= \frac{170}{\frac{250}{166.33} - 1} = 337.92\,\text{mm} \end{aligned} \tag{4}$$

Distance CI is given by

$$\begin{aligned} CI &= CD + DF + FI \\ &= 100 + 60.54 + 319.11 = 479.65\ \text{mm} \end{aligned} \tag{5}$$

Distance BI is given by

$$\begin{aligned} BI &= AB + AI \\ &= 170 + 337.92 = 507.92\,\text{mm} \end{aligned} \tag{6}$$

Angular velocity of the coupler (link 3) is given by

$$\omega_3 = \frac{|\mathbf{v}_C|}{CI} = \frac{500}{479.65} = 1.04\,\text{rad/s} \tag{7}$$

Magnitude of the velocity of point B is given by

$$|\mathbf{v}_B| = \omega_3 \times BI = 1.04 \times 507.92 = 529.47\,\text{mm/s} \tag{8}$$

Since point B is common between the coupler and the follower, angular velocity of point B is given by

$$\omega_4 = \frac{|\mathbf{v}_B|}{AB} = \frac{529.47}{170} = 3.1145\ \text{rad/s} \tag{9}$$

2.8 Acceleration Relations for a Rigid Body

Consider a rigid body in plane motion as shown in Fig. 2.33. The rigid body is assumed to be rotating counterclockwise at an angular velocity of ω rad/s and an angular acceleration of α rad/s^2 with $\hat{\mathbf{k}}$ as the axis of rotation. Consider any two points P and Q on the rigid body that are referenced to a fixed frame of reference. The line joining points P and Q makes an angle θ with respect to the horizontal at the instant shown.

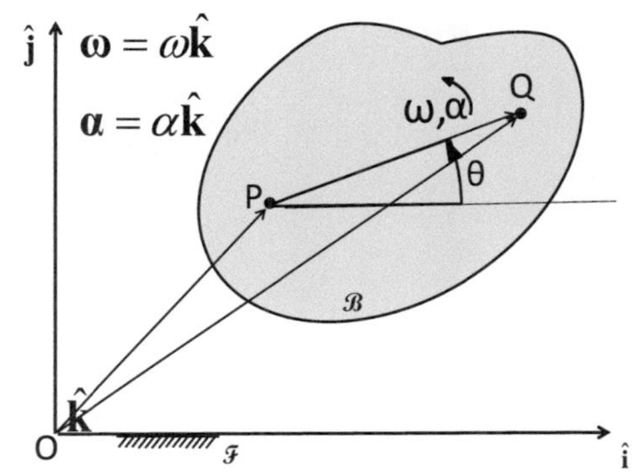

Fig. 2.33 Rigid body in plane motion rotating with an angular velocity and angular acceleration

2.8.1 Vector Approach

Consider Eq. 2.56, the velocity relationship between two points of a rigid body. Differentiating Eq. 2.56 and using Eq. 2.44, the acceleration relationship between P and Q is given by

$$\begin{aligned} \mathbf{a}_Q &= \mathbf{a}_P + \ddot{\theta}\hat{\mathbf{k}} \times \mathbf{r}_{PQ} + \dot{\theta}\hat{\mathbf{k}} \times \dot{\mathbf{r}}_{PQ} \\ &= \mathbf{a}_P + \alpha\hat{\mathbf{k}} \times \mathbf{r}_{PQ} + \omega\hat{\mathbf{k}} \times \dot{\mathbf{r}}_{PQ} \end{aligned} \tag{2.60}$$

Since $\mathbf{v}_\mathrm{P}$ and $\mathbf{v}_\mathrm{Q}$ are absolute velocities, their time derivatives will correspondingly give absolute accelerations, $\mathbf{a}_\mathrm{P}$ and $\mathbf{a}_\mathrm{Q}$. In addition, the time derivative of angular velocity gives angular acceleration.

We have earlier proved that $\dot{\mathbf{r}}_{PQ}$ is relative velocity $\mathbf{v}_{Q/P}$ between points Q and P, given by

$$\dot{\mathbf{r}}_{PQ} = \mathbf{v}_{Q/P} = \omega\hat{\mathbf{k}} \times \mathbf{r}_{PQ} \tag{2.61}$$

From Eqs. (2.60) and (2.61), the acceleration relationship between points P and Q becomes

$$\mathbf{a}_Q = \mathbf{a}_P + \alpha\hat{\mathbf{k}} \times \mathbf{r}_{PQ} + \omega\hat{\mathbf{k}} \times \omega\hat{\mathbf{k}} \times \mathbf{r}_{PQ} \tag{2.62}$$

The third term on the right-hand side of Eq. 2.62 can be simplified as follows. From Eq. 2.54,

$$\hat{\mathbf{k}} \times \mathbf{r}_{PQ} = r_{PQ}\left(-sin\theta\hat{\mathbf{i}} + \cos\theta\hat{\mathbf{j}}\right) \tag{2.63}$$

$$\begin{aligned} \hat{\mathbf{k}} \times \left(\hat{\mathbf{k}} \times \mathbf{r}_{PQ}\right) &= r_{PQ}\hat{\mathbf{k}} \times \left(-sin\theta\hat{\mathbf{i}} + \cos\theta\hat{\mathbf{j}}\right) \\ &= -r_{PQ}\left(\cos\theta\hat{\mathbf{i}} + \sin\theta\hat{\mathbf{j}}\right) = -\mathbf{r}_{PQ} \end{aligned} \tag{2.64}$$

From Eqs. (2.62) and (2.64), the acceleration relationship becomes

$$\mathbf{a}_Q = \mathbf{a}_P + \alpha\hat{\mathbf{k}} \times \mathbf{r}_{PQ} - \omega^2\mathbf{r}_{PQ} \tag{2.65}$$

2.8.2 Graphical Approach

Equation (2.65) can be written as

$$\mathbf{a}_Q = \mathbf{a}_P + \underbrace{\alpha\hat{\mathbf{k}} \times \mathbf{r}_{PQ} - \omega^2\mathbf{r}_{PQ}}_{\mathbf{a}_{Q/P}} \tag{2.66}$$

The acceleration vectors are thus related by the following equation:

$$\mathbf{a}_Q = \mathbf{a}_P + \mathbf{a}_{Q/P} \tag{2.67}$$

From the above Eqs. (2.66) and (2.67), the relative acceleration point Q with respect to P, $\mathbf{a}_{Q/P}$, has two components as follows:

$$\begin{aligned}\mathbf{a}_{Q/P} &= \alpha\hat{\mathbf{k}} \times \mathbf{r}_{PQ} \quad - \quad \omega^2\mathbf{r}_{PQ} \\ &= \underbrace{\mathbf{a}^t_{Q/P}}_{\text{(tangential)}} \quad + \quad \underbrace{\mathbf{a}^r_{Q/P}}_{\text{(radial)}}\end{aligned} \tag{2.68}$$

As we have already proved earlier that vector $\hat{\mathbf{k}} \times \mathbf{r}_{PQ}$ is perpendicular to vector $\mathbf{r}_{PQ}$, tangential component of the relative acceleration between Q and P, $\mathbf{a}^t_{Q/P} = \alpha\hat{\mathbf{k}} \times \mathbf{r}_{PQ}$. The other component of the relative acceleration is in the negative direction of vector $\mathbf{r}_{PQ}$. Hence, it is the radial acceleration and if point P is a fixed point, it can be called centripetal acceleration.

From Eqs. 2.67 and 2.68, the acceleration relationship between points P and Q can be represented as shown in Fig. 2.34. Using this figure, a graphical approach to acceleration analysis can be used.

Example 2.9

Consider Example 2.5. Determine angular acceleration of the connecting rod and acceleration of the piston if the crank is rotating at a uniform angular velocity $\omega_2 =$ 15 rad/s(CCW), that is angular acceleration, $\alpha_2 = 0$.

Vector Approach

From Fig. 2.35 and Eq. 2.65, the acceleration of point B with respect to the fixed hinge point O can be written as

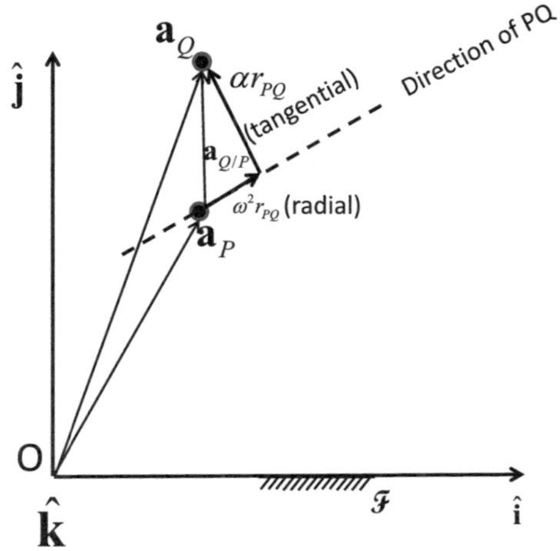

Fig. 2.34 Graphical representation of acceleration relationship between points P and Q of Fig. 2.33

$$\mathbf{a}_B = \underset{=0}{\mathbf{a}_O} + \underset{=0}{\alpha_2}\hat{\mathbf{k}} \times \mathbf{r}_{OB} - \omega_2^2 \mathbf{r}_{OB} \tag{1}$$

Since the hinge point O is fixed and the crank has zero angular acceleration (since it is rotating at uniform angular velocity), the first two terms of Eq. (1) are equal to zero. Thus, acceleration of point B is given by

$$\begin{aligned}\mathbf{a}_B &= -\omega_2^2 \mathbf{r}_{OB} = -15^2\left(-75\hat{\mathbf{i}} + 100\hat{\mathbf{j}}\right) \\ &= 16875\hat{\mathbf{i}} - 22500\hat{\mathbf{j}} \ \text{m/s}^2\end{aligned} \tag{2}$$

The direction of acceleration of point B is given by

$$\hat{\mathbf{u}}_{a_B} = \frac{16875\hat{\mathbf{i}} - 22500\hat{\mathbf{j}}}{\sqrt{16875^2 + 22500^2}} = 0.6\hat{\mathbf{i}} - 0.8\hat{\mathbf{j}} = \angle 306.86° \tag{3}$$

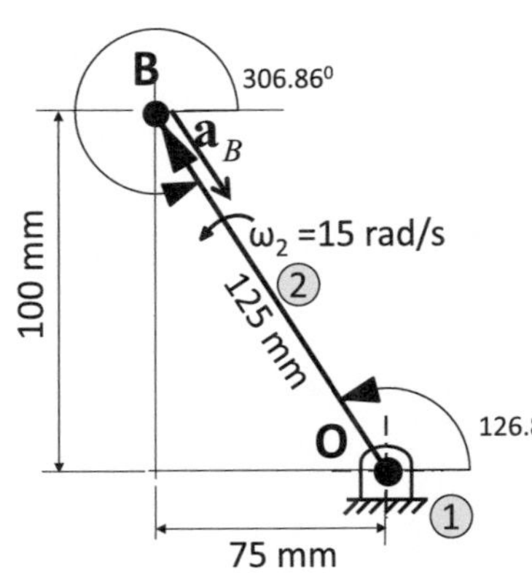

Fig. 2.35 Crank of Example 2.9

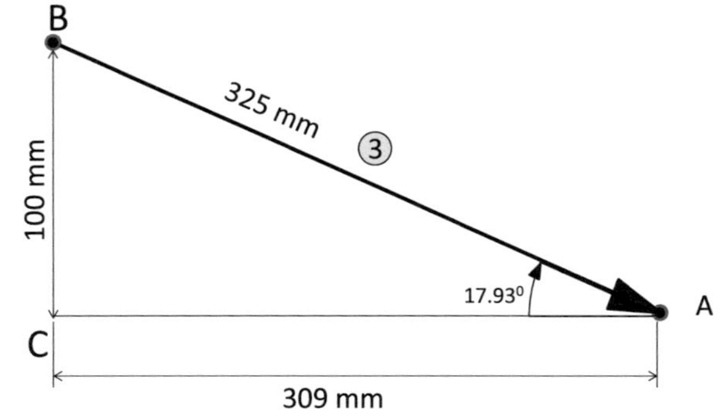

Fig. 2.36 Connecting rod of Example 2.9

and its magnitude is given by

$$|\mathbf{a}_B| = \omega_2^2 r_{OB} = 15^2 \times 75 = 28125 \ \text{m/s}^2 \tag{4}$$

Magnitude of acceleration of point B can also be obtained from the acceleration vector of Eq. (2) by taking the square root of the sum of squares.

From Eq. (2.35) and Fig. (2.36), acceleration of point A can be written as

$$\begin{aligned}\mathbf{a}_A &= \mathbf{a}_B + \alpha_3\hat{\mathbf{k}} \times \mathbf{r}_{BA} - \omega_3^2\mathbf{r}_{BA} \\ &= 16875\hat{\mathbf{i}} - 22500\hat{\mathbf{j}} + \alpha_3\left(\hat{\mathbf{k}} \times \left(309\hat{\mathbf{i}} - 100\hat{\mathbf{j}}\right)\right) - 3.6408^2\left(309\hat{\mathbf{i}} - 100\hat{\mathbf{j}}\right)\end{aligned} \tag{5}$$

Equation (5) can be simplified as

$$\mathbf{a}_A = \left(16875 + 100\alpha_3 - 309 \times 3.6408^2\right)\hat{\mathbf{i}} + \underbrace{\left(-22500 + 309\alpha_3 + 100 \times 3.6408^2\right)}_{=0}\hat{\mathbf{j}} \tag{6}$$

Since the piston is constrained to move along the horizontal direction, acceleration of point A will not have any component in the vertical direction. Hence, equating the vertical component of Eq. (6) to zero, angular acceleration of the connecting rod is given by

$$\alpha_3 = \frac{22500 - 100 \times 3.6408^2}{309} = 68.5257 \ \text{rad/s}^2 \tag{7}$$

Using the connecting rod angular acceleration and the horizontal component of Eq. (6), acceleration of point A is given by

$$\begin{aligned}\mathbf{a}_A &= \left(16875 + 100\alpha_3 - 309 \times 3.6408^2\right)\hat{\mathbf{i}} \\ &= \left(16875 + 100 \times 68.5257 - 309 \times 3.6408^2\right)\hat{\mathbf{i}} = 19632\hat{\mathbf{i}} \ \text{mm/s}^2\end{aligned} \tag{8}$$

Since acceleration of point A is opposite to the direction of velocity, at this instant the piston is decelerating.

Graphical Method

(a) We shall first start with the crank as its angular velocity and angular acceleration are known. Since the crank is rotating at a uniform angular velocity, it has only radial acceleration along the crank towards hinge point O. First draw a line inclined at an angle of 126.86° representing crank direction OB as shown in Fig. 2.37a. Let O also be the origin (absolute zero) of the acceleration diagram. Mark of the radial acceleration of magnitude $|\mathbf{a}_B| = 28125\ \text{m/s}^2$ opposite to the direction of OB as shown that represents acceleration vector $\mathbf{a}_B$.

(b) Since we now know acceleration of point B, we can proceed to compute the relative acceleration of point A with respect to B, $\mathbf{a}_{A/B}$. Since the radial component of $\mathbf{a}_{A/B}$ will be along the connecting rod and the tangential component perpendicular to it, draw a line inclined at an angle of 17.93° representing the direction of line BA as shown in Fig. 2.37b.

(c) Since the angular velocity of the connecting rod is already known, we can compute the radial component magnitude of acceleration of $\mathbf{a}_{A/B}$, $\left|\mathbf{a}^r_{A/B}\right| = \omega_3^2\ r_{BA} = 4305\ \text{mm/s}^2$ and its direction will be opposite to the direction of BA. From the tip of acceleration vector $\mathbf{a}_B$, mark off 4305 mm/s² parallel to BA in

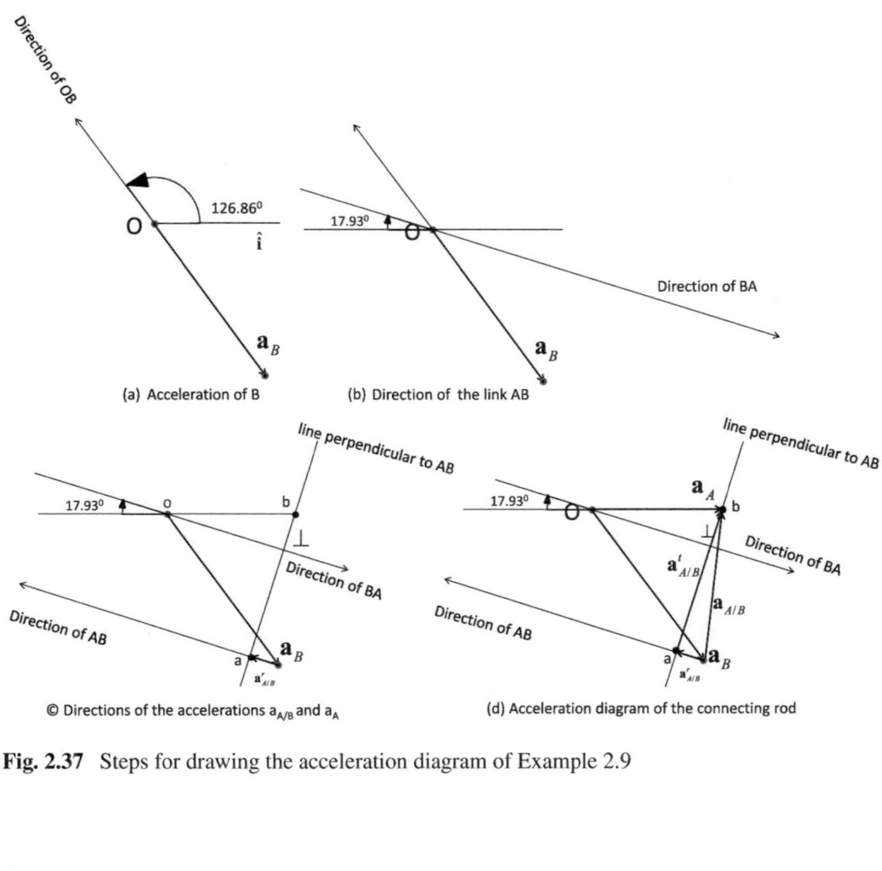

Fig. 2.37 Steps for drawing the acceleration diagram of Example 2.9

the opposite direction as shown in Fig. 2.37c. Mark point a that is the tip of $\mathbf{a}^r_{A/B}$.

(d) Angular acceleration of the connecting rod, hence the tangential component of $\mathbf{a}_{A/B}$ cannot be computed. However, direction of the tangential component that is perpendicular to AB is known. Draw a line from the tip of the acceleration vector $\mathbf{a}^r_{A/B}$ that is a as shown in Fig. 2.37c perpendicular to AB to meet the horizontal line from the origin of the acceleration diagram at point b. Since the acceleration of point A is along the horizontal direction, Ob represents its acceleration.

(e) From Eq. (2.68), the relevant equation to complete the acceleration triangle is $\mathbf{a}_{A/B} = \mathbf{a}^t_{A/B} + \mathbf{a}^r_{A/B}$. As shown in Fig. 2.37d, the acceleration vectors $\mathbf{a}_A$ and $\mathbf{a}^t_{A/B}$ are drawn. Magnitudes of $|\mathbf{a}_A| = 19{,}632$ mm/s^2 and $\left|\mathbf{a}^t_{A/B}\right| = 22{,}256$ mm/s^2 are marked off from the graph. Angular acceleration of the connecting rod, $\alpha_3 = \left|\mathbf{a}^t_{A/B}\right|/\text{AB} = 22{,}256/325 = 68.53$ rad/s^2.

Example 2.10

Consider Example 2.6. Determine the angular acceleration of the coupler (link 3) and follower (link 4) if the crank is rotating with a constant angular velocity of 5 rad/s (CCW).

Refer to Fig. 2.38a. From Eq. (2.65), acceleration of the point C, $\mathbf{a}_C$, is given by

$$\begin{aligned}\mathbf{a}_C &= \mathbf{a}_D + \alpha_2\hat{\mathbf{k}} \times \mathbf{r}_{DC} - \omega_2^2\mathbf{r}_{DC} \\ &= -5^2\left(-34.20\hat{\mathbf{i}} + 93.96\hat{\mathbf{j}}\ \right) \\ &= 855.1\hat{\mathbf{i}} - 2349.2\hat{\mathbf{j}}\ \text{mm/s}^2\end{aligned} \tag{1}$$

Refer to Fig. 2.38b. From Eq. (2.65), acceleration of point B, is given by

$$\begin{aligned}\mathbf{a}_B &= \mathbf{a}_C + \alpha_3\hat{\mathbf{k}} \times \mathbf{r}_{CB} - \omega_3^2\mathbf{r}_{CB} \\ &= 855.1\hat{\mathbf{i}} - 2349.2\hat{\mathbf{j}}\ + \alpha_3\hat{\mathbf{k}} \times \left(234.92\hat{\mathbf{i}} + 85.5\hat{\mathbf{j}}\right) - 1.1547^2\left(234.92\hat{\mathbf{i}} + 85.5\hat{\mathbf{j}}\right) \\ &= (541.82 - 85.51\alpha_3)\hat{\mathbf{i}} + (-2463.2 + 234.92\alpha_3)\hat{\mathbf{j}}\ \text{mm/s}^2\end{aligned} \tag{2}$$

Refer to Fig. 2.38c.

From Eq. (2.65), acceleration of point B is also given by

$$\begin{aligned}\mathbf{a}_B &= \mathbf{a}_A + \alpha_4\hat{\mathbf{k}} \times \mathbf{r}_{AB} - \omega_4^2\mathbf{r}_{AB} \\ &= \alpha_4\hat{\mathbf{k}} \times \left(29.52\hat{\mathbf{i}} + 167.41\hat{\mathbf{j}}\right) - 3.40^2\left(29.52\hat{\mathbf{i}} + 167.41\hat{\mathbf{j}}\right) \\ &= (-340.49 - 167.42\alpha_4)\hat{\mathbf{i}}\ + (-1931 + 29.52\alpha_4)\hat{\mathbf{j}}\ \frac{mm}{s^2}\end{aligned} \tag{3}$$

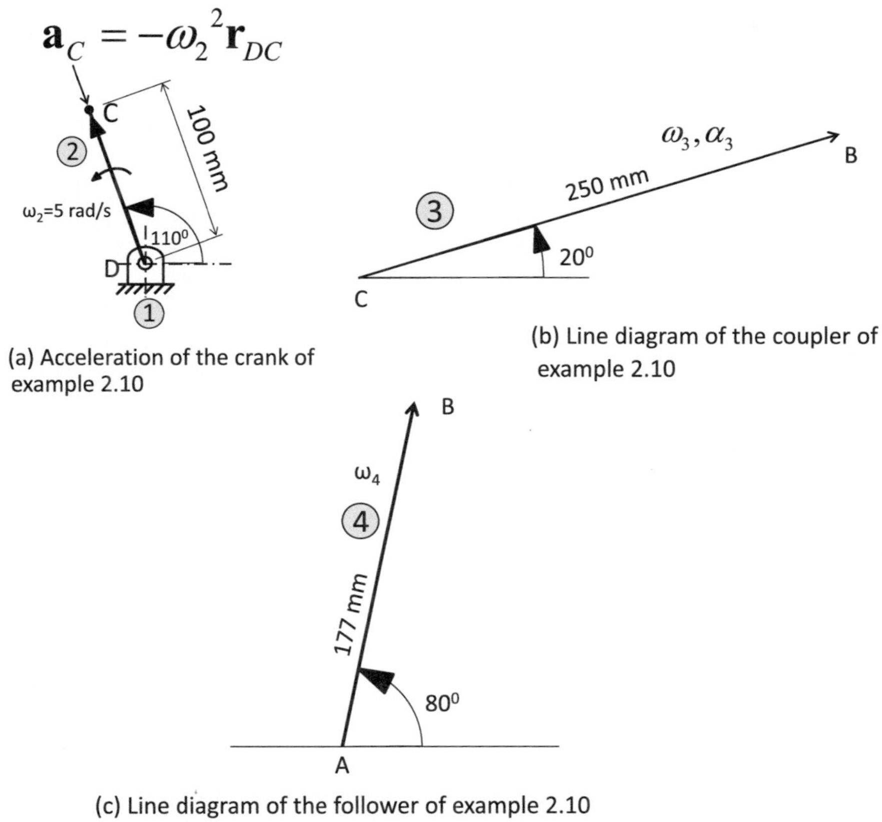

Fig. 2.38 **a** Acceleration of the crank of Example 2.10 **b** Line diagram of the coupler of Example 2.10 **c** Line diagram of the follower of Example 2.10

From Equations (2) and (3),

$$\begin{bmatrix} -85.51 & 167.42 \\ 234.92 & -29.52 \end{bmatrix} \begin{Bmatrix} \alpha_3 \\ \alpha_4 \end{Bmatrix} = \begin{Bmatrix} -882.31 \\ 532.2 \end{Bmatrix} \tag{4}$$

Solving Eq. (4), $\alpha_3 = 1.71$ rad/s^2, $\alpha_4 = -4.40$ rad/s^2.

Solving the above problem using graphical representation is given as Exercise Problem 2.41.

2.9 Rolling

2.9.1 *Rolling on a Flat Surface*

Consider a wheel rolling on a flat surface without slipping as shown in Fig. 2.39. That means the friction present at the interface ensures that the velocity of the contacting point instantly goes to zero. Hence, the contact point can be considered as the instantaneous center of zero velocity. Let r be the wheel radius and ω its angular velocity.

Velocity

From Eq. 2.56, velocity of the wheel center mass is given by

$$\mathbf{v}_C = \omega\hat{\mathbf{k}} \times \mathbf{r}_{IC} = r\omega\hat{\mathbf{k}} \times \hat{\mathbf{j}} = -r\omega\hat{\mathbf{i}} \tag{2.69}$$

Acceleration

Differentiating Eq. (2.69) with respect to time, acceleration of the mass center of the rolling wheel is given by

$$\mathbf{a}_C = -r\alpha\hat{\mathbf{i}} \tag{2.70}$$

Although the instantaneous center has zero velocity, it has acceleration. It can be obtained by relating accelerations of the mass center and instant center of zero velocity. Hence, applying Eq. (2.65) between the above two points (C and I), the acceleration of the point I is given by

$$\begin{aligned} \mathbf{a}_I &= \mathbf{a}_C + \ddot{\theta}\hat{\mathbf{k}} \times \mathbf{r}_{CI} - \dot{\theta}^2\mathbf{r}_{CI} \\ &= \mathbf{a}_C + \alpha\hat{\mathbf{k}} \times \mathbf{r}_{CI} - \omega^2\mathbf{r}_{CI} \\ &= -r\alpha\hat{\mathbf{i}} + r\alpha\hat{\mathbf{i}} - r\omega^2\hat{\mathbf{j}} \\ &= -r\omega^2\hat{\mathbf{j}} \end{aligned} \tag{2.71}$$

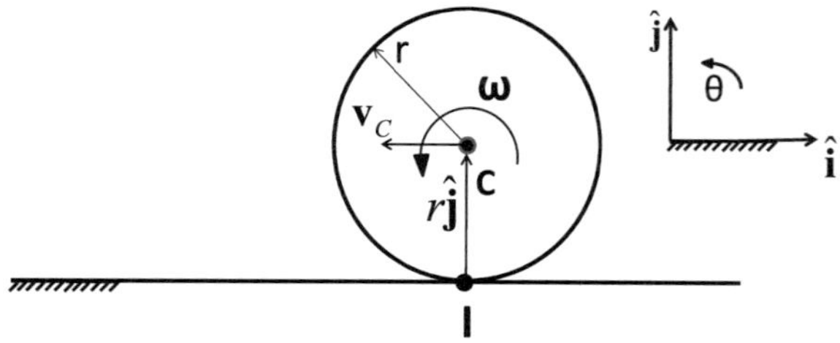

Fig. 2.39 A disc rolling on a flat surface

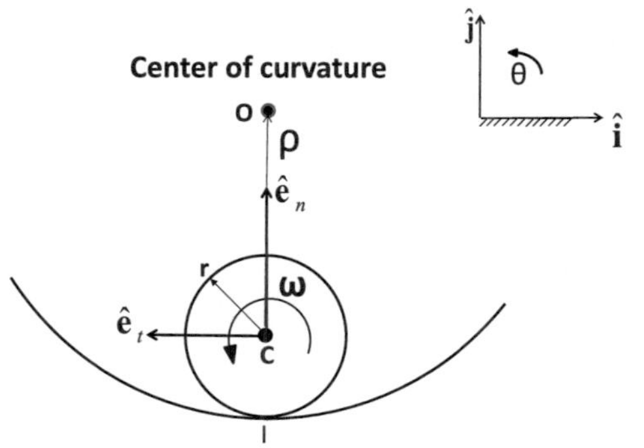

Fig. 2.40 Disc rolling on a curved surface

2.9.2 *Rolling on a Curved Surface*

A disc rolling on a curved surface that is assumed to be non-slipping is shown in Fig. 2.40. The velocity of the mass center is $\mathbf{v}_C = -r\omega\hat{\mathbf{e}}_t$, the tangential acceleration is $\mathbf{a}_C^t = -r\alpha\hat{\mathbf{e}}_t$, and the normal acceleration $a_C^n = \frac{(r\omega)^2}{\rho}\hat{\mathbf{e}}_n$. The above discussions will be useful later.

2.10 Velocity of a Point with Respect to a Fixed and Moving Coordinate System

For kinematic and kinetic analysis of many mechanisms, it may become necessary to define a coordinate system that is located on a moving body. Hence, there is a need to establish relations that relate the absolute motion with the moving body. First, we shall discuss the velocity relations and then followed by acceleration relations, when moving coordinate systems are referenced.

Consider a point P moving on a plane surface, along a certain locus, as shown in Fig. 2.41. Let this point P be referred to a fixed coordinate system $\mathcal{F}$ (unit vectors $\hat{\mathbf{i}}$, $\hat{\mathbf{j}}$) and to a moving coordinate system that is fixed on a moving body $\mathcal{B}$ (unit vectors $\hat{\mathbf{e}}_i$,$\hat{\mathbf{e}}_j$) that is rotating with an angular velocity ω. We can always imagine an extension of the moving body that moves along with it. In this case, the extension shown in light shade is made large enough to cover point P under consideration. In addition, we can also imagine a point P′ that is coincident with point P but fixed on the above extension of the body $\mathcal{B}$.

2.10.1 *Vector Approach*

The position vector of point P can be defined as

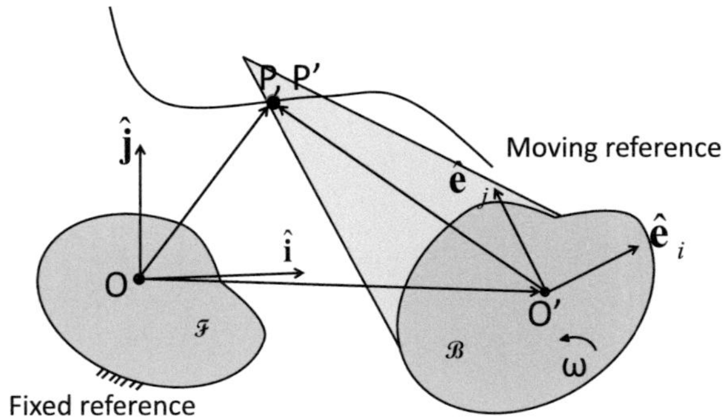

Fig. 2.41 Position vector referenced to a fixed and moving reference

$$\mathbf{r}_{OP} = \mathbf{r}_{OO'} + \mathbf{r}_{O'P} \tag{2.72}$$

Taking the time derivative of Eq. (2.72) on both sides,

$$\dot{\mathbf{r}}_{OP} = \dot{\mathbf{r}}_{OO'} + \dot{\mathbf{r}}_{O'P} \tag{2.73}$$

$\dot{\mathbf{r}}_{OP}$ defines the time derivative of the position vector of point P with respect to a fixed frame of reference and hence defines the absolute velocity of point P. Similarly, $\dot{\mathbf{r}}_{OO'}$ defines the absolute velocity of the point O′ that is the origin of the coordinate system located on the moving body $\mathcal{B}$. Then we are left with $\dot{\mathbf{r}}_{O'P}$ that is the time derivative of point P with respect to a moving coordinate system located on $\mathcal{B}$; this time derivative can be computed using Eq. (2.44).

From Eqs. (2.44) and (2.73),

$$\mathbf{v}_{P/\mathcal{F}} = \mathbf{v}_{O'/\mathcal{F}} + {}^{\mathcal{B}}\dot{\mathbf{r}}_{O'P} + \dot{\theta}\hat{\mathbf{k}} \times \mathbf{r}_{O'P} \tag{2.74}$$

Since velocities referred to the fixed frame of reference $\mathcal{F}$ are absolute velocities, we can drop their reference to $\mathcal{F}$. Then Eq. (2.74) becomes

$$\mathbf{v}_P = \mathbf{v}_{O'} + {}^{\mathcal{B}}\dot{\mathbf{r}}_{O'P} + \dot{\theta}\hat{\mathbf{k}} \times \mathbf{r}_{O'P} \tag{2.75}$$

Since $\mathbf{r}_{O'P} \equiv \mathbf{r}_{O'P'}$, changing O'P to O'P' of the third term of RHS, Eq. (2.75) becomes

$$\mathbf{v}_P = \mathbf{v}_{O'} + \overbrace{{}^{\mathcal{B}}\dot{\mathbf{r}}_{O'P}}^{\mathbf{v}_{P/\mathcal{B}}} + \overbrace{\dot{\theta}\hat{\mathbf{k}} \times \mathbf{r}_{O'P'}}^{\mathbf{v}_{P'/\mathcal{B}}} \tag{2.76}$$

${}^{\mathcal{B}}\dot{\mathbf{r}}_{O'P}$ is the relative velocity of point P with respect to the moving body $\mathcal{B}$. $\dot{\theta}\hat{\mathbf{k}} \times \mathbf{r}_{O'P'}$ is the relative velocity of point P′ with respect to the moving body $\mathcal{B}$. From these considerations, Eq. (2.76) can be written as

$$\mathbf{v}_P = \mathbf{v}_{O'} + \mathbf{v}_{P/\mathcal{B}} + \mathbf{v}_{P'/\mathcal{B}} \tag{2.77}$$

2.10.2 Graphical Approach

Equation (2.77) is graphically shown in Fig. (2.42).

Note that although P and P′ are coincident points, their velocities with respect to the moving body $\mathcal{B}$ are not the same; P is an independent point moving along a locus, whereas P’ is a point on the extension of $\mathcal{B}$.

As the location of the point O′ on the moving body is arbitrary, further simplification of Eq. (2.77) will result if it is moved to point P′. Then Eq. (2.77) becomes

$$\mathbf{v}_P = \mathbf{v}_{P'} + \mathbf{v}_{P/\mathcal{B}} \tag{2.78}$$

Equation (2.78) can be expressed in words as follows:

Absolute velocity of a point P can be expressed as a sum of its velocity with respect to a moving reference and absolute velocity of a point on the moving reference (P′) that is either on it or its extended portion and coincident to it (P). Equation (2.78) is graphically presented in Fig. 2.43.

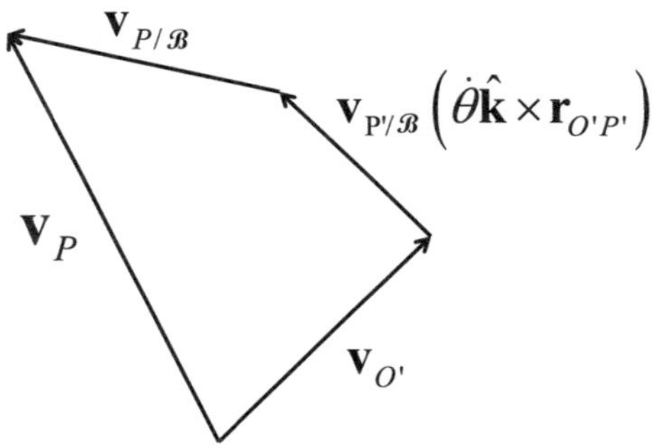

Fig. 2.42 Velocity of a point P expressed in terms of a moving coordinate system

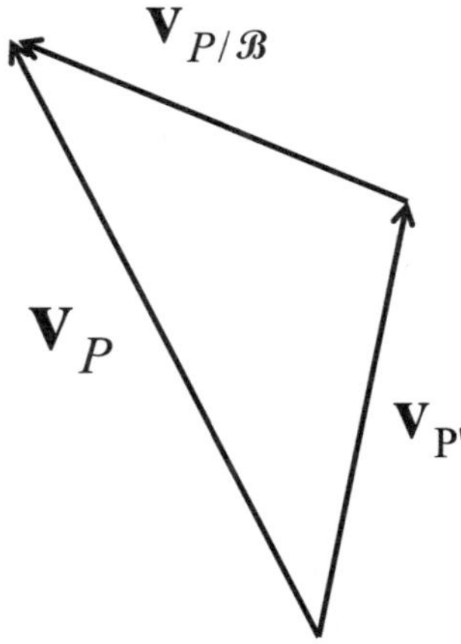

Fig. 2.43 Velocity triangle of Eq. (2.78)

Example 2.11

Consider the mechanism shown in Fig. 2.44a and the corresponding block diagram representation of its kinematic pairs in Fig. 2.44b. Link 2 is hinged at O of the fixed link 1 and it also slides inside piston 3. Piston 3 is hinged to link 4 at P. If link 4 is rotating at an angular speed of 2 rad/s in the counterclockwise direction, determine the angular speed of link 2.

In these types of mechanisms in which a piston moves inside an oscillating link, a moving coordinate system will have to be used to account for the change in the relative velocity and acceleration direction, and the respective magnitudes

$$BP = 2000\,\text{mm}$$

$$OB = 1000\,\text{mm}$$

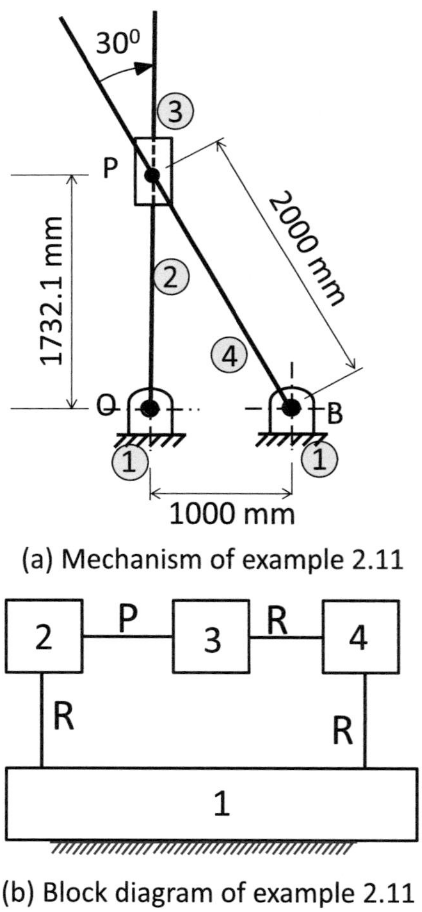

Fig. 2.44 **a** Mechanism of Example 2.11 **b** Block diagram of Example 2.11

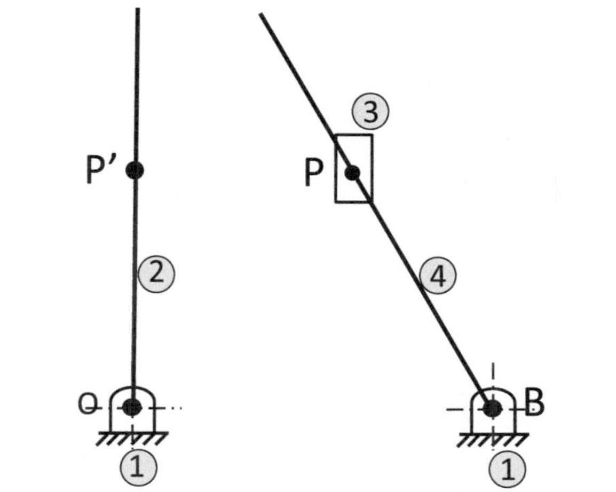

Fig. 2.45 **a** Link 2 of Example 2.1 **b** Links 3 and 4 of Example 2.11

$$\text{OP} = \sqrt{BP^2 - OB^2} = \sqrt{2000^2 - 1000^2} = 1732.1\,\text{mm}$$

The mechanism shown in Fig. 2.44a is disassembled into two parts as shown in Fig. 2.45a, b for better clarity. Point P′ that is coincident with the hinge point P between links 3 and 4 is located on link 2. It does not matter whether you locate P′ on link 2 and P on link 3 or vice versa. As long as consistent equations are written, it would not make any difference. In fact, you can try the other options of P on link 2 and P′ on link 3 and verify that you get the same answer.

Position vector $\mathbf{r}_{BP}$ is given by

$$\mathbf{r}_{BP} = (-OB\hat{\mathbf{i}} + OP\hat{\mathbf{j}}) = \left(-1000\hat{\mathbf{i}} + 1732.1\hat{\mathbf{j}}\right)\text{mm} \tag{1}$$

Velocity of point P is given by

$$\begin{aligned}\mathbf{v}_P &= \mathbf{v}_B + \omega_4\hat{\mathbf{k}} \times \mathbf{r}_{BP}\\ &= 2\hat{\mathbf{k}} \times \left(-1000\hat{\mathbf{i}} + 1732.1\hat{\mathbf{j}}\right) = -\left(3464.2\hat{\mathbf{i}} + 2000\hat{\mathbf{j}}\right)\text{mm/s}\end{aligned} \tag{2}$$

In the reciprocating mechanism presented in Examples 1 and 2, the common points between the links remained the same throughout the motion of all the links. Hence, we are able to use the velocity or acceleration computed on the common points of linkages. However, in the present case, point P at which link 3 and link 4 are hinged will not have a common point on link 2 at all positions since P has a relative velocity with link 2.

Hence, from Eq. (2.78), velocity of point P can also be written as

$$\mathbf{v}_P = \mathbf{v}_{P/2} + \mathbf{v}_{P'} \tag{3}$$

Since the piston is constrained to slide in link 2, the direction of the relative velocity of point P with respect to link 2, $\mathbf{v}_{P/2}$, is always along the length of link 2. Hence, in the given configuration, $\mathbf{v}_{P/2}$ is in the vertical direction. Since point P' is on link 2, Eq. (3) becomes

$$\begin{aligned}\mathbf{v}_P &= v_{P/2}\hat{\mathbf{j}} + \mathbf{v}_O + \omega_2\hat{\mathbf{k}} \times \mathbf{r}_{OP'} \\ &= -1732.1\omega_2\hat{\mathbf{i}} + v_{P/2}\hat{\mathbf{j}} \text{ mm/s}\end{aligned} \tag{4}$$

From Eqs. (2) and (4) by comparing coefficients along the coordinate directions, $v_{P/2} = -2000$ mm/s (vertically downwards) $\omega_2 = 3464.2/1732.1 = 2$ rad/s (counterclockwise)

Graphical Method

(a) Draw a line inclined at 60° to represent the direction of link 4 as shown in Fig. 2.46a.
(b) Magnitude of velocity of point P, $v_P = BP \times \omega_4 = 2000 \times 2 = 4000$ mm/s. Since the angular velocity is counterclockwise, it is in a direction perpendicular to the direction of the link, as shown in Fig. 2.46b. Mark 4000 mm/s along a direction perpendicular to BP to represent the velocity of point P.
(c) Since the piston slides on link 4, the relative velocity of point P with respect to link 2, $v_{P/2}$ will always be along the length of link 2. Since link 2 is vertical at this instant of time, $v_{P/2}$ will be in the vertical direction. Choose P′ as the coincident point of P, but that is located on link 2. The absolute velocity of P′, $v_P{}'$, will be in the horizontal direction (perpendicular to the direction of link 2). Both the above directions are drawn as shown in Fig. 2.46c and they meet at point a.
(d) Draw the vectors $\mathbf{v}_{P'}$ and $\mathbf{v}_{P/2}$ as shown in Fig. 2.46d. By direct measurement to scale, $|\mathbf{v}_{P'}| = 3464.2$ mm/s and $|\mathbf{v}_{P/2}| = 2000$ mm/s. Angular velocity of link 2, $\omega_2 = \frac{|\mathbf{v}_{P'}|}{OP} = \frac{3464.2}{1732.1} = 2$ rad/s. It is in the counterclockwise direction based on the direction of its velocity vector. Similarly, the relative velocity $\mathbf{v}_{P/2}$ is in the downward direction.

Example 2.12

Consider the quick-return mechanism as shown in Fig. 2.47 in which link 2 is the driver that has a uniform angular velocity of 5 rad/s (counterclockwise.) Determine the velocity of point C at the instant shown (Fig. 2.48).

Based on vector closure, the following equation has to be satisfied

$$\mathbf{r}_{O_4O_2} + \mathbf{r}_{O_2A} + \mathbf{r}_{AB} + \mathbf{r}_{BO_4} = 0 \tag{1}$$

Fig. 2.46 Steps for the graphical method to solve Example 2.11

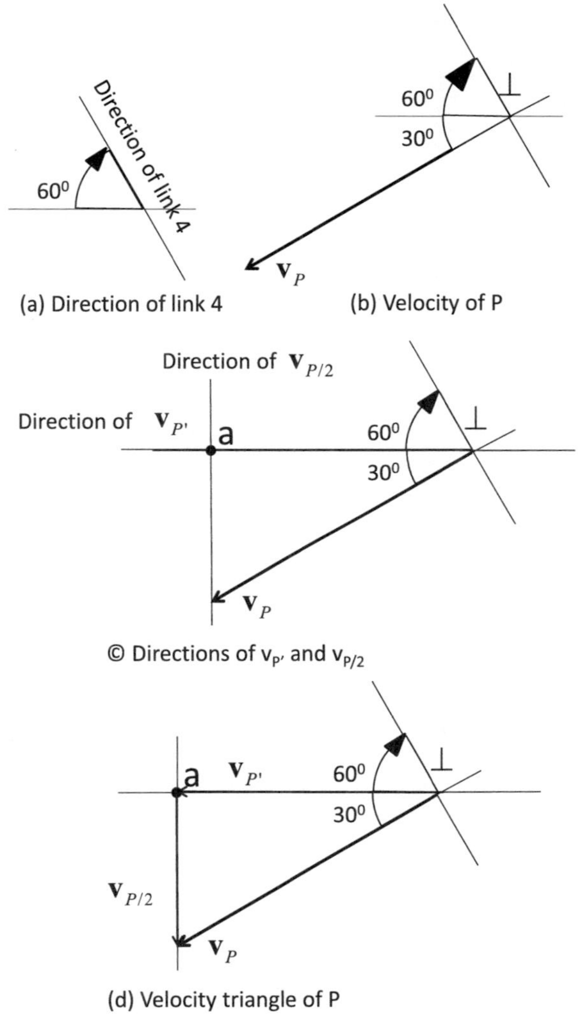

Hence $\mathbf{r}_{AB}$ is given by

$$
\begin{aligned}
\mathbf{r}_{AB} &= -\mathbf{r}_{O_4O_2} - \mathbf{r}_{O_2A} - \mathbf{r}_{BO_4} \\
&= -100\hat{\mathbf{j}} - 40\left(-\cos 60°\hat{\mathbf{i}} + \sin 60°\hat{\mathbf{j}}\right) - 120\left(-\cos 70°\hat{\mathbf{i}} - \sin 70°\hat{\mathbf{j}}\right) \\
&= 61.04\hat{\mathbf{i}} - 21.88\hat{\mathbf{j}}\,\text{mm}\ (\text{AB} = 64.84\ \text{mm})
\end{aligned}
\quad (2)
$$

Position vector $\mathbf{r}_{O_2A}$ of the crank is given by

$$
\begin{aligned}
\mathbf{r}_{O_2A} &= 40\left(-\cos 60°\hat{\mathbf{i}} + \sin 60°\hat{\mathbf{j}}\right)\text{mm} \\
&= \left(-20\hat{\mathbf{i}} + 34.64\hat{\mathbf{j}}\right)\text{mm}
\end{aligned}
\quad (3)
$$

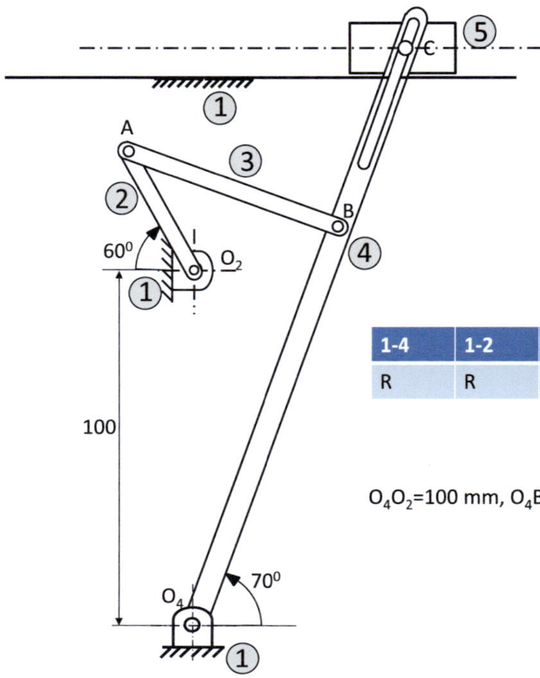

1-4	1-2	2-3	3-4	4-5	5-1
R	R	R	R	PS	P

O_4O_2=100 mm, O_4B=120 mm, BC=55 mm, O_2A=40 mm

Fig. 2.47 Mechanism of Example 2.12

$$\begin{aligned}\mathbf{r}_{O_4B} &= \left[120\left(\cos 70°\hat{\mathbf{i}} + \sin 70°\hat{\mathbf{j}}\right)\right] \\ &= 41.04\hat{\mathbf{i}} + 112.76\hat{\mathbf{j}} \text{ mm}\end{aligned} \tag{4}$$

The velocity of point A is given by

$$\begin{aligned}\mathbf{v}_A &= \mathbf{v}_{O_2} + \omega_2\hat{\mathbf{k}} \times \mathbf{r}_{O_2A} \\ &= 5\hat{\mathbf{k}} \times \left(-20\hat{\mathbf{i}} + 34.64\hat{\mathbf{j}}\right) \\ &= -173.2\hat{\mathbf{i}} - 100\hat{\mathbf{j}} \quad \text{mm/s}\end{aligned} \tag{5}$$

The velocity of point B in terms of ω_3 is given by

$$\begin{aligned}\mathbf{v}_B &= \mathbf{v}_A + \omega_3\hat{\mathbf{k}} \times \mathbf{r}_{AB} \\ &= -173.2\hat{\mathbf{i}} - 100\hat{\mathbf{j}} \ + \omega_3\hat{\mathbf{k}} \times \left(61.04\hat{\mathbf{i}} - 21.88\hat{\mathbf{j}}\right) \\ &= (-173.2 + 21.88\omega_3)\hat{\mathbf{i}} + (-100 + 61.04\omega_3)\hat{\mathbf{j}}\end{aligned} \tag{6}$$

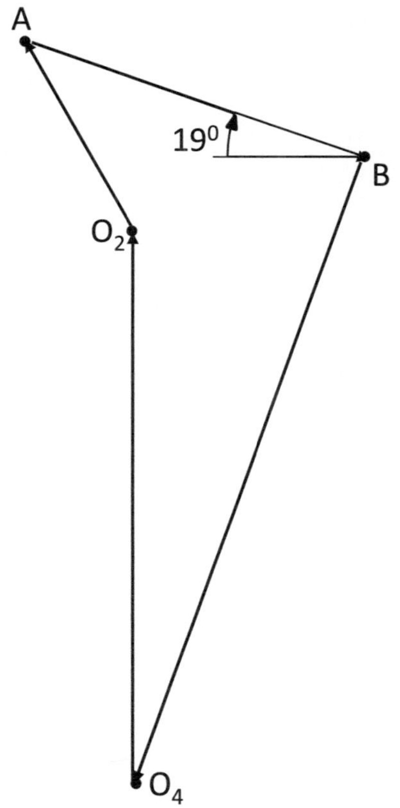

Fig. 2.48 Position analysis of Example 2.12

The velocity of B can also be written in terms of ω_4 as

$$\begin{aligned}\mathbf{v}_B &= \mathbf{v}_{O_4} + \omega_4\hat{\mathbf{k}} \times \mathbf{r}_{O_4B} \\ &= \omega_4\hat{\mathbf{k}} \times \left[\left(41.04\hat{\mathbf{i}} + 112.76\hat{\mathbf{j}}\right)\right] \\ &= -112.76\omega_4\hat{\mathbf{i}} + 41.04\omega_4\hat{\mathbf{j}}\,\text{mm/s}\end{aligned} \tag{7}$$

Equating coefficients of the velocity of B in both coordinate directions

$$\begin{bmatrix} 21.88 & 112.76 \\ 61.04 & -41.04 \end{bmatrix}\begin{Bmatrix} \omega_3 \\ \omega_4 \end{Bmatrix} = \begin{Bmatrix} 173.2 \\ 100 \end{Bmatrix} \tag{8}$$

Solving the above linear equations
$\omega_3 = 2.36$ rad/s, $\omega_4 = 1.08$ rad/s

Position vector $\mathbf{r}_{O_4C}$ is given by

$$\begin{aligned}\mathbf{r}_{O_4C} &= (O_4B + BC)\left(\cos 70°\hat{\mathbf{i}} + \sin 70°\hat{\mathbf{j}}\right) \\ &= \left(59.85\hat{\mathbf{i}} + 164.45\hat{\mathbf{j}}\right)\text{mm}\end{aligned} \tag{9}$$

The unit vector along Position vector $\mathbf{r}_{O_4C}$ is given by

$$\hat{\mathbf{e}} = \frac{\mathbf{r}_{O_4C}}{\left|\mathbf{r}_{O_4C}\right|} = \frac{\left(59.85\hat{\mathbf{i}} + 164.45\hat{\mathbf{j}}\right)}{175} = 0.34\hat{\mathbf{i}} + 0.94\hat{\mathbf{j}} \tag{10}$$

Let C' be a point on link 4 that is coincident with the point C on link 5. Its velocity is given by

$$\begin{aligned}\mathbf{v}_{C'} &= \mathbf{v}_{O_4} + \omega_4\hat{\mathbf{k}} \times \mathbf{r}_{O_4C} \\ &= 1.08\hat{\mathbf{k}} \times \left(59.85\hat{\mathbf{i}} + 164.45\hat{\mathbf{j}}\right) \\ &= -177.61\hat{\mathbf{i}} + 64.64\hat{\mathbf{j}}\ \text{mm/s}\end{aligned} \tag{11}$$

The velocity of C can also be expressed as

$$\begin{aligned}\mathbf{v}_C &= \mathbf{v}_{C/4} + \mathbf{v}_{C'} \\ &= v_{C/4}\hat{\mathbf{e}} + \mathbf{v}_{C'} \\ &= v_{C/4}\left(0.34\hat{\mathbf{i}} + 0.94\hat{\mathbf{j}}\right) - 177.61\hat{\mathbf{i}} + 64.64\hat{\mathbf{j}} \\ &= \left(0.34v_{C/4} - 177.61\right)\hat{\mathbf{i}} + \left(0.94v_{C/4} + 64.64\right)\hat{\mathbf{j}}\end{aligned} \tag{12}$$

Since C is constrained to move along the horizontal direction, it has no vertical component of velocity

$$\left(0.94v_{C/4} + 64.64\right) = 0 \tag{13}$$

Hence

$$v_{C/4} = -\frac{64.64}{0.94} = -68.76\,\text{mm/s} \tag{14}$$

From Eqs. (2.16) and (2.17), the absolute velocity of C is given by

$$\begin{aligned}\mathbf{v}_C &= \left(0.34v_{C/4} - 177.61\right)\hat{\mathbf{i}} \\ &= (0.34 \times -68.76 - 177.61)\hat{\mathbf{i}} \\ &= -201\hat{\mathbf{i}}\,\text{mm/s}\end{aligned} \tag{15}$$

Graphical Method

For determining the velocity of C by using the graphical method, following steps are followed:

(1) Begin with link 2 that is the crank since its angular velocity is known. Velocity of point A is given by $O_2A \times \omega_2 = 40 \times 5 = 200$ mm/s. Draw a line representing the direction of O_2A at an angle of 60° as shown in Fig. 2.49a. With O as the starting point for the velocity of point A, draw v_A perpendicular to O_2A of magnitude 200 mm/s.

(2) Draw O_4B through the point O at an angle of 70° representing the direction of O_2B as shown in Fig. 2.49b. Draw a line through the point O and perpendicular to O_2B representing the direction of the velocity of point B.

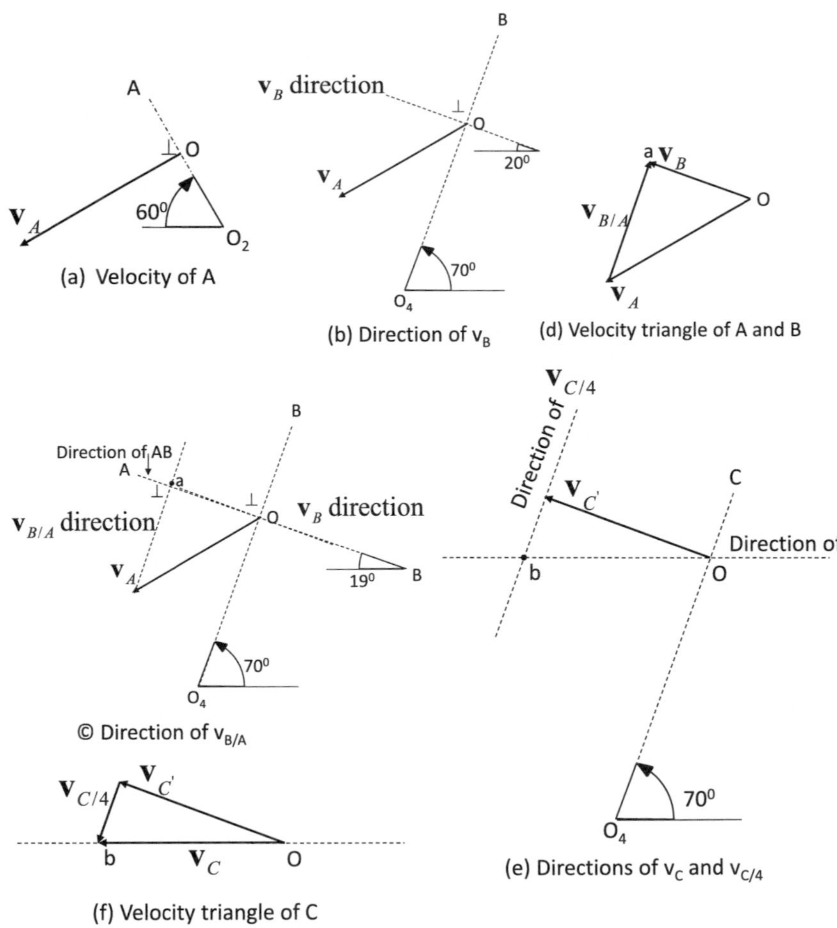

Fig. 2.49 Velocity diagrams of Example 2.12

(3) Draw a line AB at an angle of 19° as shown in Fig. 2.49c. Since the relative velocity $\mathbf{v}_{B/A}$ should start from the tip of the velocity vector of point A, draw a line from the tip of this vector perpendicular to AB representing the direction of $\mathbf{v}_{B/A}$. The directions of $\mathbf{v}_{B/A}$ and $\mathbf{v}_B$ meet at point a. Coincidentally for this problem, the directions of AB and velocity of B differ only by 1°. Hence, it is difficult to clearly distinguish between them.

(4) Redraw the figure of the previous step to represent only the velocity vectors as shown in Fig. 2.49d. By directly measuring to the scale, $|\mathbf{v}_B| = 129.3$ mm/s and $|\mathbf{v}_{B/A}| = 153.2$ mm/s. The angular velocity of link 3, $\omega_3 = \frac{|\mathbf{v}_{B/A}|}{AB} = \frac{153.2}{64.8} = 2.36$ rad/s; angular velocity of link 4, $\omega_4 = \frac{|\mathbf{v}_B|}{O_4B} = \frac{129.3}{120} = 1.08$ rad/s. The signs for angular velocities are obtained by visual inspection of the velocities with respect to the reference points, either fixed or moving.

(5) Consider a point C′ on link 4 that is coincident with the point C on link 5. Velocity of the point C′ is given by $O_2C \times \omega_4 = (120 + 55) \times 1.08 = 189$ mm/s. Draw a line O_4C representing the direction of O_4C. Mark a point O representing the starting point of absolute velocity vectors as shown in Fig. 2.49e. Draw the velocity of vector of C′ of magnitude 189 mm/s and perpendicular to O_4C. Draw a line through the tip of the velocity vector of C′ in the direction parallel to O_4C that represents the direction of the relative velocity $\mathbf{v}_{C/4}$. Draw a horizontal line from O representing the direction of velocity of C. The directions of $\mathbf{v}_{C/4}$ and $\mathbf{v}_C$ meet at point b.

(6) Redraw the figure from the previous step showing only the velocity vectors as shown in Fig. 2.49f. The velocity of point C is 201 mm/s towards left. $|\mathbf{v}_{C/4}| = 68.63$ mm/s along link 4 towards O_4.

2.11 Acceleration Relations in Different Frames of Reference, Coriolis Acceleration

Consider a point P moving on the plane of paper along a certain locus as shown in Fig. 2.50. Let this point P be referred to a fixed coordinate system $\mathcal{F}$ and to a moving coordinate system that is fixed on a moving body $\mathcal{B}$ that is rotating with an angular velocity ω and an angular acceleration α. We can always imagine an extension of the moving body that moves along with it. In this case the extension shown in light shade is made large enough to cover point P under consideration. In addition, we can also imagine a point P’ that is coincident with point P but fixed on the above extension of the body $\mathcal{B}$.

2.11.1 Vector Approach

We have already derived the velocity relationship in Eq. (2.77) for different frames of reference. It is once again reproduced below.

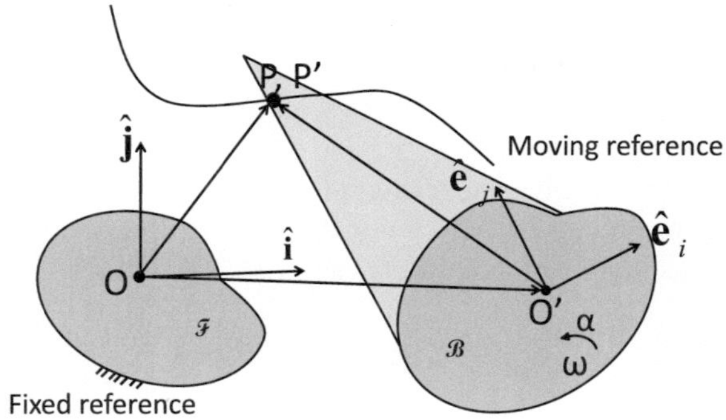

Fig. 2.50 Acceleration of a point referenced to a fixed and moving coordinate system

$$\mathbf{v}_P = \mathbf{v}_{O'} + \mathbf{v}_{P/\mathcal{B}} + \mathbf{v}_{P'/\mathcal{B}} \tag{2.79}$$

From Eqs. (2.61), (2.79) becomes

$$\mathbf{v}_P = \mathbf{v}_{O'} + \mathbf{v}_{P/\mathcal{B}} + \dot{\theta}\hat{\mathbf{k}} \times \mathbf{r}_{O'P} \tag{2.80}$$

Differentiating Eq. (2.80) with respect to time

$$\dot{\mathbf{v}}_P = \dot{\mathbf{v}}_{O'} + \dot{\mathbf{v}}_{P/\mathcal{B}} + \ddot{\theta}\hat{\mathbf{k}} \times \mathbf{r}_{O'P} + \dot{\theta}\hat{\mathbf{k}} \times \dot{\mathbf{r}}_{O'P} \tag{2.81}$$

$\mathbf{v}_P$ and $\mathbf{v}_{O'}$ are absolute velocities and hence their differentiation with respect to time will give the corresponding absolute accelerations. Whereas $\mathbf{v}_{P/\mathcal{B}}$ is referenced to a moving coordinate system on the body $\mathcal{B}$ and hence the corresponding acceleration can be obtained by using Eq. (2.44). [Noting that from Eq. (2.44) $\dot{\mathbf{v}}_{P/B} = \mathbf{a}_{P/B} + \dot{\theta}\hat{\mathbf{k}} \times \mathbf{v}_{P/B}$]

$$\mathbf{a}_P = \mathbf{a}_{O'} + \mathbf{a}_{P/\mathcal{B}} + \dot{\theta}\hat{\mathbf{k}} \times \mathbf{v}_{P/B} + \ddot{\theta}\hat{\mathbf{k}} \times \mathbf{r}_{O'P} + \dot{\theta}\hat{\mathbf{k}} \times \dot{\mathbf{r}}_{O'P} \tag{2.82}$$

From Eq. (2.44) $\dot{\mathbf{r}}_{O'P}$ can be obtained as

$$\dot{\mathbf{r}}_{O'P} = \mathbf{v}_{P/B} + \dot{\theta}\hat{\mathbf{k}} \times \mathbf{r}_{O'P} \tag{2.83}$$

Substituting Eq. 2.83 in Eq. 2.82

$$\mathbf{a}_P = \mathbf{a}_{O'} + \mathbf{a}_{P/\mathcal{B}} + \dot{\theta}\hat{\mathbf{k}} \times \mathbf{v}_{P/B} + \ddot{\theta}\hat{\mathbf{k}} \times \mathbf{r}_{O'P} + \dot{\theta}\hat{\mathbf{k}} \times \left(\mathbf{v}_{P/B} + \dot{\theta}\hat{\mathbf{k}} \times \mathbf{r}_{O'P}\right) \tag{2.84}$$

From Eq. (2.64), $\hat{\mathbf{k}} \times \left(\hat{\mathbf{k}} \times \mathbf{r}_{PQ}\right) = -\mathbf{r}_{PQ}$. Hence Eq. (2.84) becomes

$$\mathbf{a}_P = \mathbf{a}_{O'} + \mathbf{a}_{P/\mathcal{B}} + \underbrace{\ddot{\theta}\hat{\mathbf{k}} \times \mathbf{r}_{O'P}}_{\text{tangential}} - \underbrace{\dot{\theta}^2\mathbf{r}_{O'P}}_{\text{radial}} + \underbrace{2\dot{\theta}\hat{\mathbf{k}} \times \mathbf{v}_{P/B}}_{\text{Coriolis component}} \tag{2.85}$$

When the acceleration of a point is referenced to a fixed and moving coordinate system, in addition to the relative acceleration, relative tangential and radial acceleration, there is an additional component known as Coriolis component as noted in the above equation.

Since the choice of origin O' of the moving coordinate system is arbitrary, it can as well coincide with point P′ that is an imaginary extension of the moving body coinciding with point P. Then, Eq. (2.85) becomes

$$\mathbf{a}_P = \mathbf{a}_{P'} + \mathbf{a}_{P/\mathcal{B}} + \underbrace{2\dot{\theta}\hat{\mathbf{k}} \times \mathbf{v}_{P/\mathcal{B}}}_{\text{Coriolis component}} \tag{2.86}$$

2.11.2 Graphical Method

Equation (2.86) is presented in Fig. 2.51

Example 2.13
Consider the mechanism shown in Fig. 2.52a and its block diagram representation of its kinematic pairs. We have already analyzed this mechanism for velocity in Example 2.11. For the same angular velocity of link 4, $\omega_4 = 2$ rad/s, as in Example 2.11 compute the acceleration of link 2 if the angular acceleration of link 4, $\alpha_4 = 2$ rad/s^2.

From Example 2.11, $\mathbf{v}_{P/2} = -2000\,\mathbf{j}$ mm/s and $\omega_2 = 2$ rad/s.

The mechanism shown in Fig. 2.52a is disassembled into two parts as shown Fig. 2.53a, b for better clarity. Point P′ that is coincident with the hinge point P between links 3 and 4 is located on link 2.

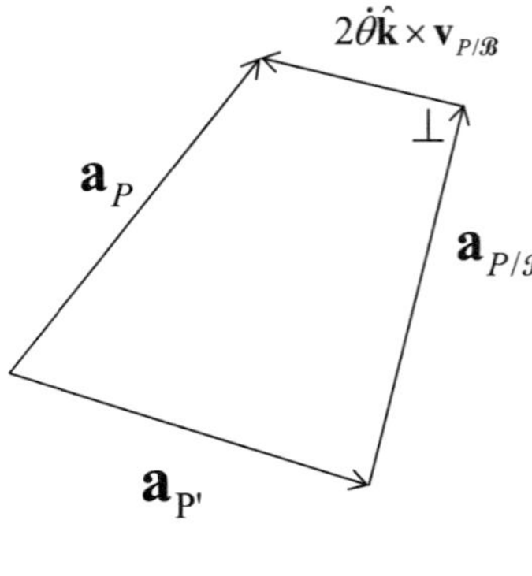

Fig. 2.51 Acceleration of a point P referenced to a moving coordinate system

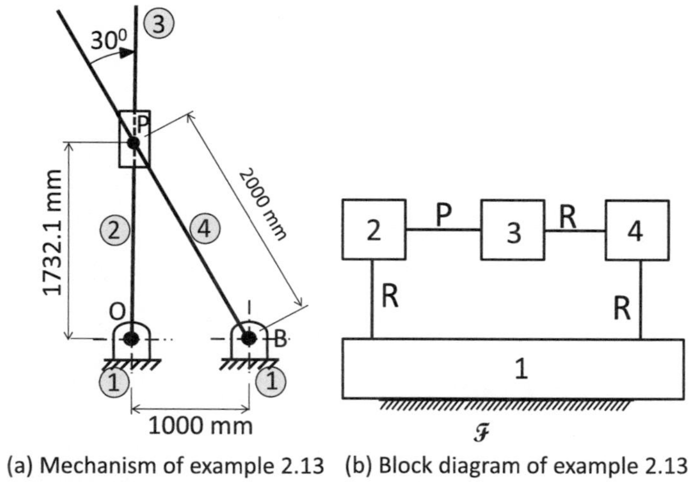

Fig. 2.52 **a** Mechanism of Example 2.3 **b** Block diagram of Example 2.13

Fig. 2.53 **a** Link 2 of Example 2.13 **b** Link 3 and 4 Example 2.13

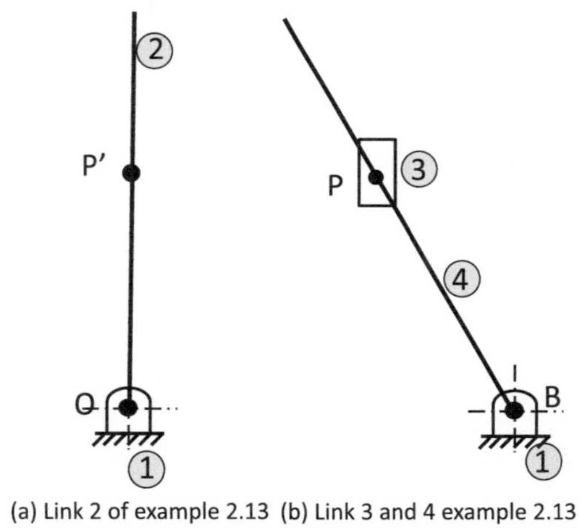

Acceleration of point P is given by

$$\begin{aligned}\mathbf{a}_P &= \mathbf{a}_B - \omega_4^2\mathbf{r}_{BP} + \alpha_4\hat{\mathbf{k}} \times \mathbf{r}_{BP}\\ &= -2^2\left(-1000\hat{\mathbf{i}} + 1732.1\hat{\mathbf{j}}\right) + 2\hat{\mathbf{k}} \times \left(-1000\hat{\mathbf{i}} + 1732.1\hat{\mathbf{j}}\right)\\ &= (4000 - 3464.2)\hat{\mathbf{i}} + (-6928.4 - 2000)\hat{\mathbf{j}}\\ &= 536\hat{\mathbf{i}} - 8928.4\hat{\mathbf{j}}\ \ \text{mm/s}^2 \end{aligned} \tag{1}$$

The acceleration of point P′ that is coincident with point P but located on link 2 is given by

$$\begin{aligned}\mathbf{a}_{P'} &= \mathbf{a}_O + \alpha_2\hat{\mathbf{k}} \times \mathbf{r}_{OP'} - \omega_2^2\mathbf{r}_{OP'} \\ &= \alpha_2\hat{\mathbf{k}} \times 1732.1\hat{\mathbf{j}} - 2^2 \times 1732.1\hat{\mathbf{j}} \end{aligned} \tag{2}$$

Since point P has a relative acceleration with respect to link 2, from Eq. (2.86) the acceleration P can also be written as

$$\begin{aligned}\mathbf{a}_P &= \mathbf{a}_{P/2} + \mathbf{a}_{P'} + 2\omega_2\hat{\mathbf{k}} \times \mathbf{v}_{P/B} \\ &= a_{P/2}\hat{\mathbf{j}} - 1732.1\alpha_2\hat{\mathbf{i}} - 6928.4\hat{\mathbf{j}} + 2 \times 2\hat{\mathbf{k}} \times \left(-2000\,\hat{\mathbf{j}}\right) \\ &= = (-1732.1\alpha_2 + 8000)\hat{\mathbf{i}} + \left(a_{P/2} - 6928.4\right)\hat{\mathbf{j}} \end{aligned} \tag{3}$$

Equating Eqs. (1) and (3), $\alpha_2 = 4.31$ rad/s^2 and $a_{P/2} = -2000$ mm/s^2.

Graphical Method

(1) Since we have complete information regarding the angular velocity and acceleration of link 4, we can start with link 4. Draw a line at an angle of 60° (the angle made by link 4 with respect to the horizontal) as shown in Fig. 2.54a.

(2) Compute the magnitude of radial acceleration of point P on link 4, $\left|\mathbf{a}_P^r\right| = \omega_4^2 BP = 2^2 \times 2000 = 8000\,\text{mm/s}^2$. Mark this acceleration opposite to the direction of BP as shown in Fig. 2.54b.

(3) Compute the magnitude of tangential acceleration of point P on link 4, $\left|\mathbf{a}_P^t\right| = \alpha_4 BP = 2 \times 2000 = 4000\,\text{mm/s}^2$. Mark this acceleration perpendicular to $\mathbf{a}_P^r$ as shown in Fig. 2.54c.

(4) Acceleration of point P, $\mathbf{a}_P$, is the vector sum of its radial and tangential accelerations, which obtained as shown in Fig. 2.54d.

(5) We know the angular velocity of link 2 from which the radial acceleration of point P′ (coincident point of P but on link 2). The radial acceleration of point P′, $\mathbf{a}_{P'}^r = \omega_2^2 OP = 2^2 \times 1732.1 = 6928.2$ mm/s^2, which is downwards. Mark this acceleration as shown in Fig. 2.54e.

(6) We know the angular velocity of link 2 and the relative velocity of point P with respect to link 2, $\mathbf{v}_{P/2}$. Hence, we can compute the magnitude of the Coriolis component of acceleration, $|\mathbf{a}_C| = \left|2\omega_2\hat{\mathbf{k}} \times \mathbf{v}_{P/B}\right| = 2 \times 2 \times 2000 = 8000\,\text{mm/s}^2$. Since the Coriolis component, $\mathbf{a}_C$, is perpendicular to the direction of relative motion of point P with respect to link 2, mark the above acceleration towards the tip of the acceleration vector $\mathbf{a}_P$ in the horizontal direction as shown in Fig. 2.54f.

(7) The direction of the relative acceleration between point P and link 2, $\mathbf{a}_{P/2}$, is in the vertical direction and the direction of the tangential acceleration of P′, $\mathbf{a}_{P'}^t$, is in the horizontal direction towards left. The above loci meet at point a, as shown in Fig. 2.54g.

(8) Join the acceleration vectors as shown in Fig. 2.54h to get $\mathbf{a}_{P'}^t$ and $\mathbf{a}_{P/2}$. From direct measurement, $\left|\mathbf{a}_{P'}^t\right| = 7464.1$ mm/s^2 and $\mathbf{a}_{P/2} = 2000$ mm/s^2 in the downward direction. Angular acceleration of link 2, $\alpha_2 = 7464.1/1732.1 = 4.31$ rad/s^2 in the counterclockwise direction.

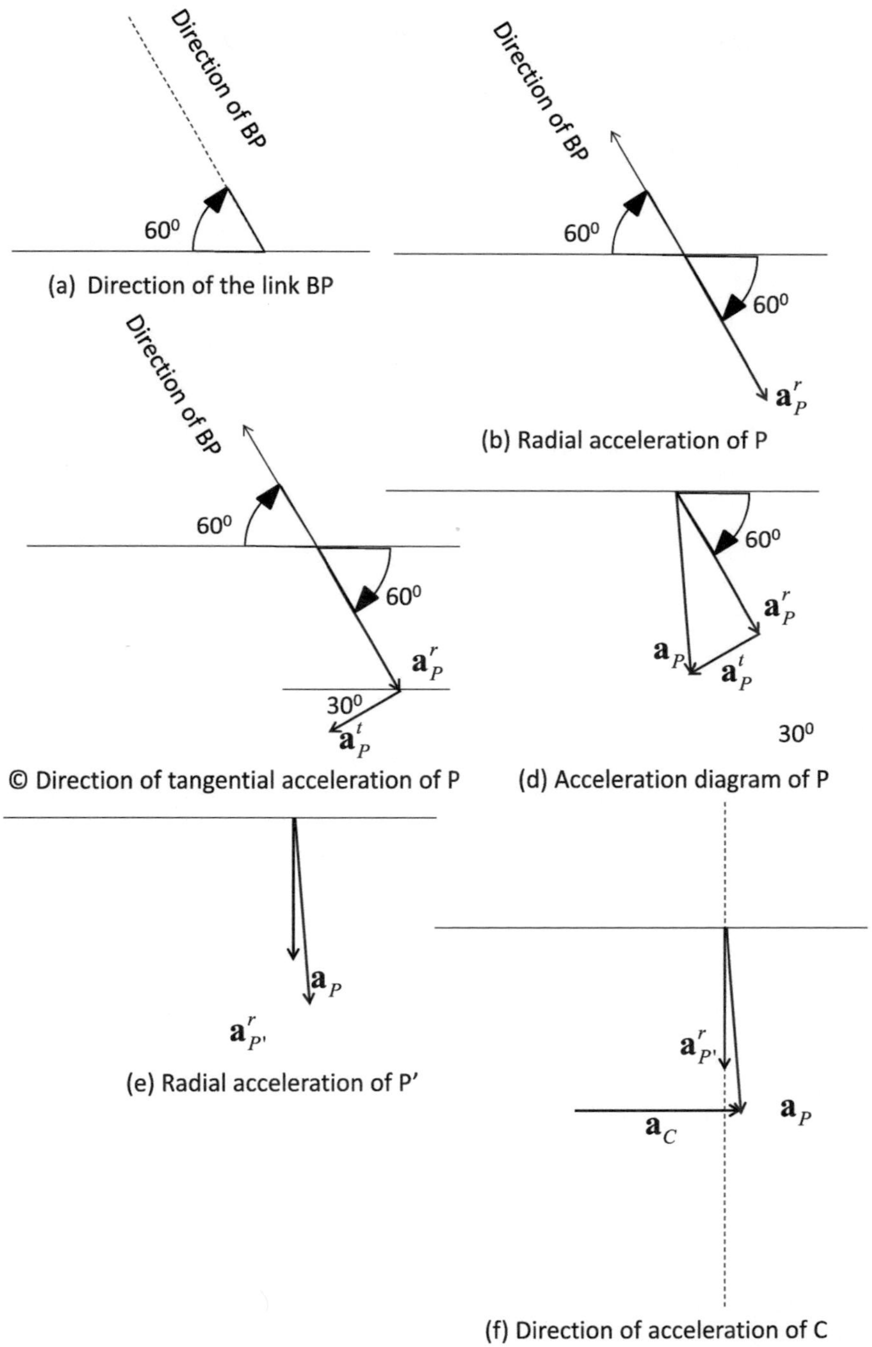

Fig. 2.54 Acceleration diagrams of Example 2.13

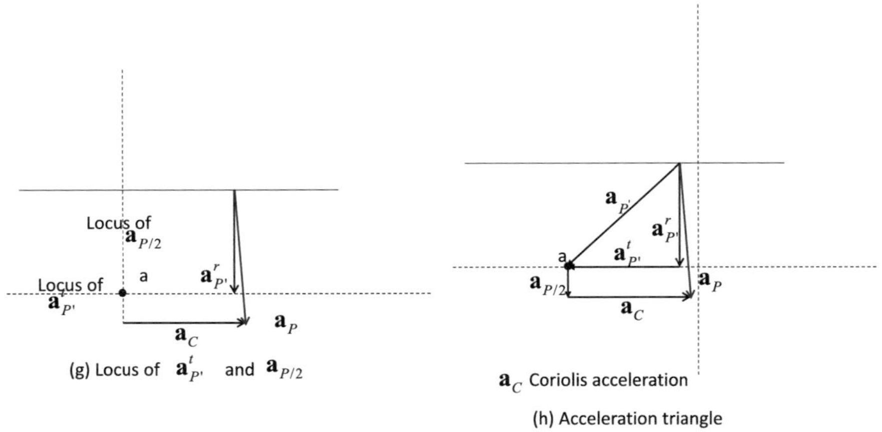

Fig. 2.54 (continued)

Example 2.14

Consider the mechanism given in Example 2.12. Determine the acceleration of the point C.

The acceleration of point A is given by

$$\begin{aligned}\mathbf{a}_A &= \mathbf{a}_{O_2} + \alpha_2\hat{\mathbf{k}} \times \mathbf{r}_{O_2A} - \omega_2^2\mathbf{r}_{O_2A} \\ &= -5^2(-20\hat{\mathbf{i}} + 34.64\hat{\mathbf{j}}) \\ &= 500\hat{\mathbf{i}} - 866\hat{\mathbf{j}}\ \text{mm/s}^2\end{aligned} \tag{1}$$

The acceleration of point B is given by

$$\begin{aligned}\mathbf{a}_B &= \mathbf{a}_A + \alpha_3\hat{\mathbf{k}} \times \mathbf{r}_{AB} - \omega_3^2\mathbf{r}_{AB} \\ &= 500\hat{\mathbf{i}} - 866\hat{\mathbf{j}}\ + \alpha_3\hat{\mathbf{k}} \times (61.04\hat{\mathbf{i}} - 21.88\hat{\mathbf{j}}) - 2.36^2(61.04\hat{\mathbf{i}} - 21.88\hat{\mathbf{j}}) \\ &= (160.03 + 21.88\alpha_3)\hat{\mathbf{i}} + (-744.13 + 61.04\alpha_3)\hat{\mathbf{j}}\ \text{mm/s}^2\end{aligned} \tag{2}$$

The acceleration of B can also be written as

$$\begin{aligned}\mathbf{a}_B &= \mathbf{a}_{O_4} + \alpha_4\hat{\mathbf{k}} \times \mathbf{r}_{O_4B} - \omega_4^2\mathbf{r}_{O_4B} \\ &= \alpha_4\hat{\mathbf{k}} \times (41.04\hat{\mathbf{i}} + 112.76\hat{\mathbf{j}}) - 1.08^2(41.04\hat{\mathbf{i}} + 112.76\hat{\mathbf{j}}) \\ &= (-112.76\alpha_4 - 47.87)\hat{\mathbf{i}} + (41.04\alpha_4 - 131.5)\hat{\mathbf{j}}\end{aligned} \tag{3}$$

From Eqs. (2) and (3)

$$\begin{bmatrix} 21.88 & 112.76 \\ 61.04 & -41.04 \end{bmatrix}\begin{Bmatrix} \alpha_3 \\ \alpha_4 \end{Bmatrix} = \begin{Bmatrix} -160.03 - 47.87 \\ 744.13 - 131.5 \end{Bmatrix} \tag{4}$$

Solving the linear equations of (4),

$$\alpha_3 = 7.78\text{rad/s}^2$$
$$\alpha_4 = -3.35\ \text{rad/s}^2$$

The acceleration of C′ is given by

$$\begin{aligned}\mathbf{a}_{C'} &= \mathbf{a}_{O_4} + \alpha_4\hat{\mathbf{k}} \times \mathbf{r}_{O_4C} - \omega_4^2\mathbf{r}_{O_4C} \\ &= -3.35\hat{\mathbf{k}} \times \left(59.85\hat{\mathbf{i}} + 164.44\hat{\mathbf{j}}\right) - 1.08^2\left(59.85\hat{\mathbf{i}} + 164.44\hat{\mathbf{j}}\right) \\ &= 481.06\hat{\mathbf{i}} - 392.3\hat{\mathbf{j}}\ \text{mm/s}^2\end{aligned} \tag{5}$$

From Eq. (2.86), the acceleration of C is given by (by using $\hat{\mathbf{e}}$ from Example 2.12 for the direction of $\mathbf{a}_{C/4}$)

$$\begin{aligned}\mathbf{a}_C &= \mathbf{a}_{C'} + \mathbf{a}_{C/4} + 2\omega_4\hat{\mathbf{k}} \times \mathbf{v}_{C/4} \\ &= 481.06\hat{\mathbf{i}} - 392.3\hat{\mathbf{j}} + \mathrm{a}_{C/4}\left(0.34\hat{\mathbf{i}} + 0.94\hat{\mathbf{j}}\right) - 2 \times 68.76 \times 1.08 \times \hat{\mathbf{k}} \times \left(0.34\hat{\mathbf{i}} + 0.94\hat{\mathbf{j}}\right) \\ &= \left(0.34\mathrm{a}_{C/4} + 620.67\right)\hat{\mathbf{i}} + \left(0.94\mathrm{a}_{C/4} - 442.8\right)\hat{\mathbf{j}}\end{aligned} \tag{6}$$

Since C is constrained to move along the horizontal direction, it has no vertical component of acceleration. Hence, by equating the vertical component of Eq. (6) to zero

$$\mathrm{a}_{C/4} = \frac{442.8}{0.94} = 471.06\ \text{mm/s}^2 \tag{7}$$

From Eqs. (6) and (7), acceleration of C is given by

$$\begin{aligned}\mathbf{a}_C &= (0.34 \times 471.06 + 620.67)\hat{\mathbf{i}} \\ &= 780.8\hat{\mathbf{i}}\ \text{mm/s}^2\end{aligned} \tag{8}$$

Graphical method

The following steps will have to be followed for carrying out the acceleration analysis of the given mechanism

(1) Draw a line at an angle of 60° representing the direction of O_2 A as shown in Fig. 2.55a. Since link 2 is rotating at a uniform angular velocity of 5 rad/s, point A on the link has only radial acceleration given by $|\mathbf{a}_A| = -O_2A \times \omega_2^2 = -40 \times 5^2 = -1000\ \text{mm/s}^2$. Draw the vector $\mathbf{a}_A$ starting at O along O_2A towards O_2.

(2) Draw a line O_4B at an angle of 70° through the point O as shown in Fig. 2.55b. Since we know the angular velocity of link 4, the radial acceleration of point B is given by $\left|\mathbf{a}_B^r\right| = -O_4B \times \omega_4^2 = -120 \times 1.08^2 = -139.34\ \text{mm/s}^2$ Draw this vector $\mathbf{a}_B^r$ along O_4B towards O_4.

(3) Draw a line inclined at an angle of 19° to represent AB in the same Fig. 2.55c. Since we know the angular velocity of link 3, the relative radial acceleration of

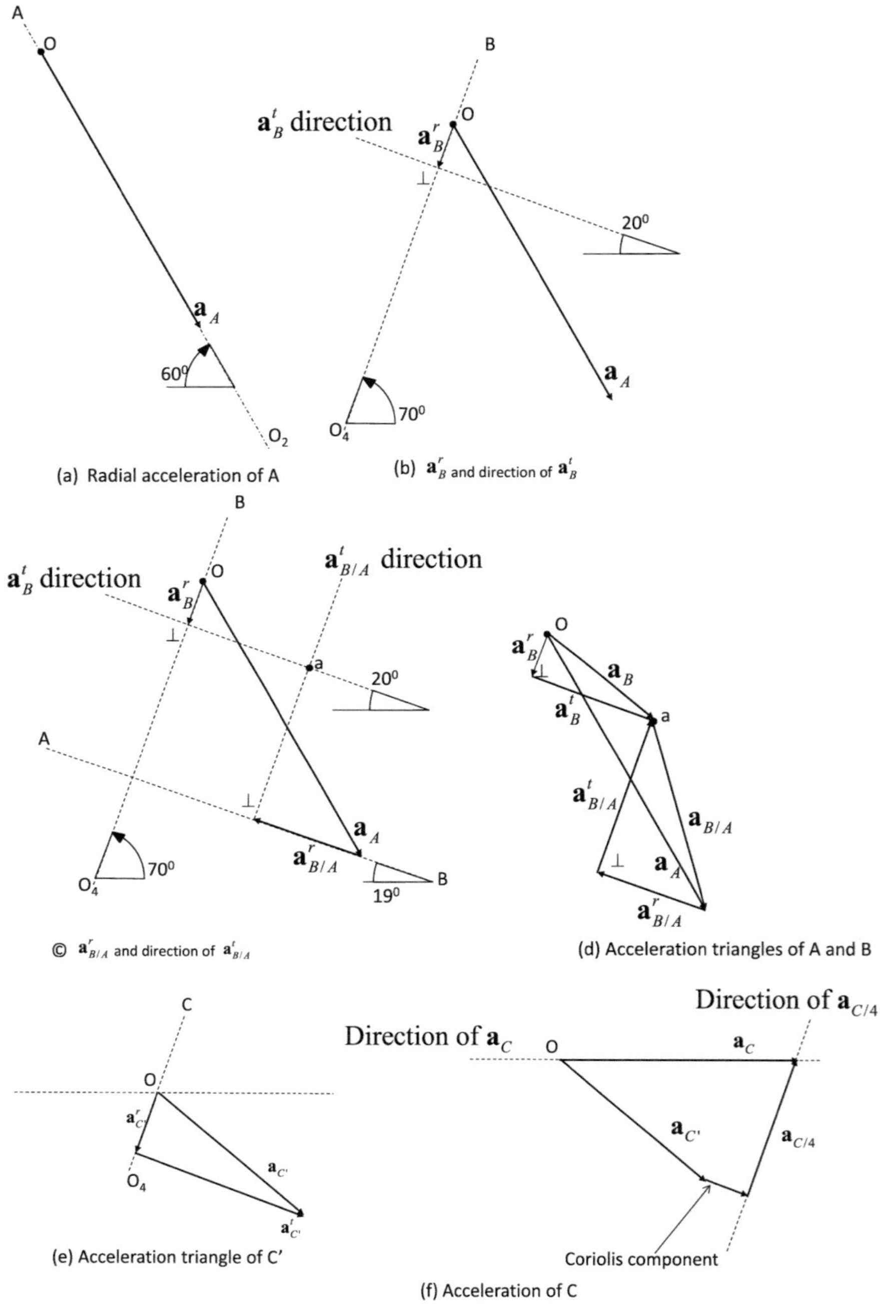

Fig. 2.55 Acceleration diagrams of Example 2.14

point B with respect to A is given by $\left|\mathbf{a}^r_{B/A}\right| = -AB \times \omega_3^2 = -64.84 \times 2.36^2 = -361.13$ mm/s^2 Draw this vector $\mathbf{a}^r_{B/A}$ along AB towards A.

(4) The angular accelerations of links 3 and 4 are not known, and tangential accelerations $\mathbf{a}^t_B$ and $\mathbf{a}^t_{B/A}$ are not known but their directions are known; they are, respectively, perpendicular to the corresponding radial accelerations. By drawing vectors of tangential accelerations from the respective tips of radial accelerations, they meet at point *a* as shown in Fig. 2.55c.

(5) Now the acceleration triangle of accelerations $\mathbf{a}_A$, $\mathbf{a}_B$ and $\mathbf{a}_{B/A}$ can be completed as shown in Fig. 2.55d. From this figure, $\left|\mathbf{a}^t_B\right| = 400\,\text{mm/s}^2$ and $\left|\mathbf{a}^t_{B/A}\right| = 505\,\text{mm/s}^2$

Angular acceleration of link 4 is given by $\alpha_4 = \frac{\left|\mathbf{a}^t_B\right|}{O_4B} = \frac{400}{120} = -3.34\,\text{rad/s}^2$ (clockwise).

Angular acceleration of link 3 is given by $\alpha_3 = \frac{\left|\mathbf{a}^t_{B/A}\right|}{AB} = \frac{505}{64.84} = 7.79\,\text{rad/s}^2$ (counterclockwise).

(6) Radial acceleration of C′, which is a coincident point with C but on link 4, is given by $\left|\mathbf{a}^r_{C'}\right| = -O_4C \times \omega_4^2 = -(120 + 55) \times 1.08^2 = -204.12$ mm/s^2 and tangential acceleration of C′ is given by $\left|\mathbf{a}^t_{C'}\right| = \left|O_4C \times \alpha_4\right| = (120 + 55) \times 3.34 = 584.5$ mm/s^2. $\mathbf{a}^r_{C'}$ is along O_4C towards O_4 and $\mathbf{a}^t_{C'}$ is perpendicular to it. Draw the vector $\mathbf{a}^r_{C'}$ from the starting point O and $\mathbf{a}^t_{C'}$ perpendicular to it as shown in Fig. 2.55e and draw the resultant acceleration vector $\mathbf{a}_{C'}$ as shown.

(7) The Coriolis component of acceleration is perpendicular to O_4C and hence perpendicular to it and its magnitude is given by $\left|2\omega_4\mathbf{v}_{C/4}\right| = 2\times1.08\times68.63 = 148.2$ mm/s^2. Draw this vector from the tip of vector $\mathbf{a}_{C'}$ in the direction of $\mathbf{a}^t_{C'}$ as shown in Fig. 2.55f and draw the directions of vectors $\mathbf{a}_C$(horizontal direction) and $\mathbf{a}_{C/4}$ (along O_4C) to meet at point b. From the diagram, $\left|\mathbf{a}_C\right| = 780\,\text{mm/s}^2$ (towards right) and $\left|\mathbf{a}_{C/4}\right| = 470\,\text{mm/s}^2$ (along O_4B).

2.12 Conclusions

In this chapter, we have discussed in detail how a complete kinematic analysis of any mechanism can be carried out, both by using vector and graphical approaches. It is not always the case that details of a mechanism are completely known. Many times, expected motion parameters will be given and then one has to figure out details of the mechanism that can deliver such motion parameters; this will be discussed in Chap. 7 on synthesis of mechanisms. Furthermore, kinetic analysis of mechanisms will be presented in Chap. 8 that requires information presented in this chapter.

2.13 Exercise Problems

2.1 A missile shown in Fig. P2.1 is traveling at a constant speed of 800 kmph (kilometers per hour). It is tracked by a radar that can find its distance r, angle θ, and rate of change of θ. If the missile is currently at an altitude of 12,000 m, making an angle of 60° with respect to the radar, what is the rate of change of θ sensed by the radar at this instant. Neglect the radar dimensions with respect to the altitude.

2.2 A point P moves tracing a parabola as shown in Fig. P2.2 and $x(t) = 0.01t^3$ meters.

(a) Find the acceleration of P when t = 5 s.
(b) Determine the curvature of the parabola for part (a).
(c) Determine the acceleration of point P in polar coordinates for part (a).
(d) Determine the acceleration of point P in tangential and normal coordinates for part (a).

Fig. P2.1 Radar tracking a missile

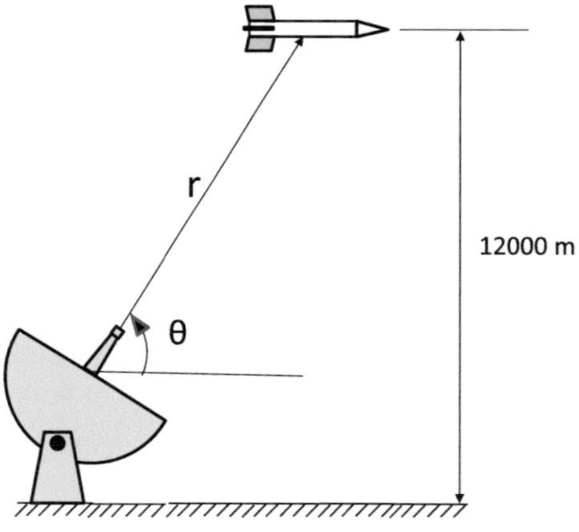

Fig. P2.2 A point moving along a parabola

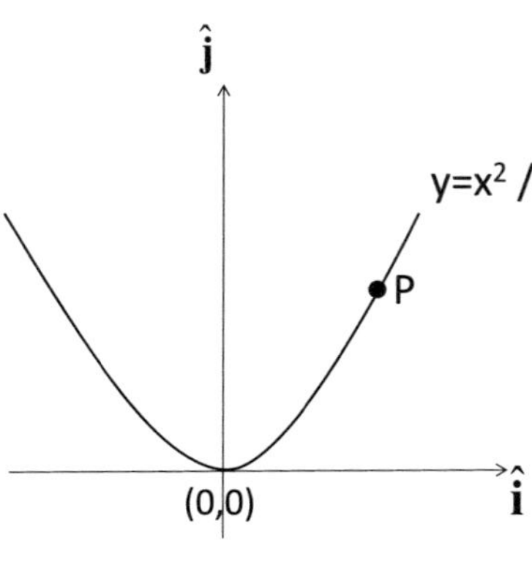

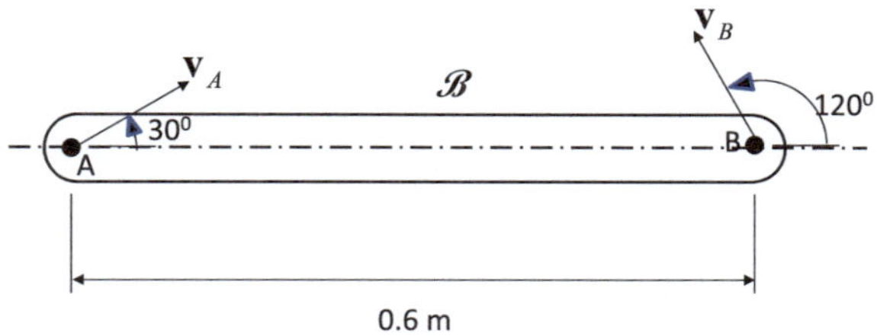

Fig. P2.3 Velocity of two points on a rotating rigid link

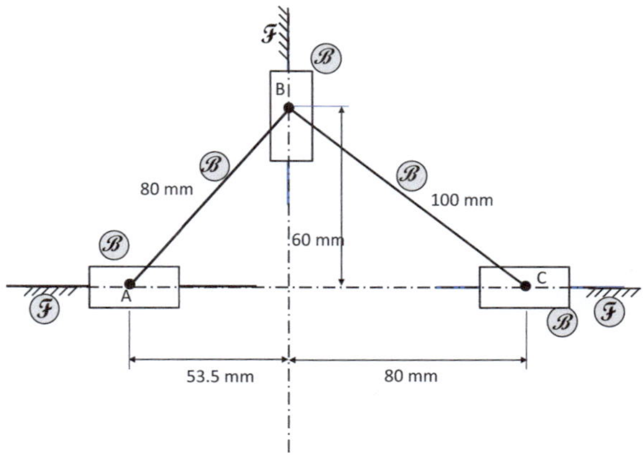

Fig. P2.4 Mechanism of exercise problem 2.4

2.3 Consider a body $\mathcal{B}$ that has points A and B as shown in Fig. P2.3. If the magnitudes of the velocities shown at these points are:

$|v_A| = 0.3$ m/s and $|v_B| = 0.6$ m/s and directions as shown, determine the angular velocity of $\mathcal{B}$ and the velocity of the midpoint between A and B using the vectorial approach.

2.4 Consider the mechanism shown in Fig. P2.4. (a) If $\mathcal{B}_1$ moves towards the right at a velocity of 0.2 m/s, determine the velocities of $\mathcal{B}_2$ and $\mathcal{B}_3$. (b) Determine the angular velocities of $\mathcal{B}_4$ and $\mathcal{B}_5$. See Fig. P1.1b for the mechanism's block diagram representation.

2.5 Consider the mechanism shown in Fig. P2.5. The mechanism is driven by link 4 in the counterclockwise direction at the given instant. See Fig. P1.2b for the mechanism's block diagram representation. If link 4 has an angular velocity of 15 rad/s, determine the velocities of the pistons and the angular velocities of links 3 and 5.

2.6 For the mechanism shown in Fig. P2.6, determine the velocity of P on link 4. The various angular velocities of the links are:$\omega_2 = 3$ rad/s (CCW), $\omega_3 = 2$ rad/s (CCW), and $\omega_4 = -1$ rad/s (CW). See Fig. P1.3b for the mechanism's block diagram representation.

2.7 A four-bar mechanism is shown in Fig. P2.7. The disk (link 2), mounted at its center $O_{4,}$ acts like a crank. Link 3 that acts as a coupler connects to the crank at A and to the follower at B in the position shown and is hinged at O_2. If link

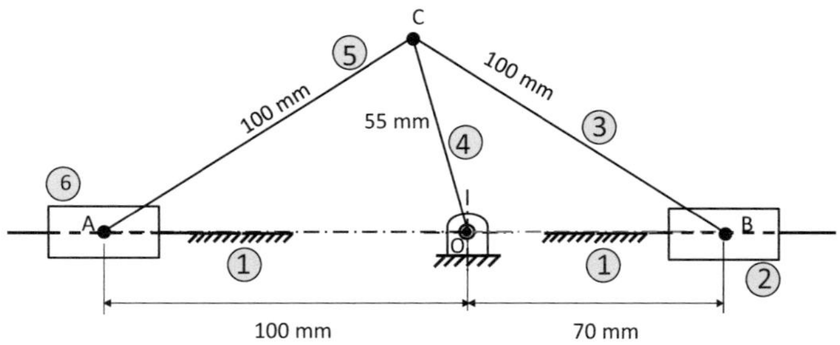

Fig. P2.5 Mechanism of exercise problem 2.5

Fig. P2.6 Mechanism of exercise problem 2.6

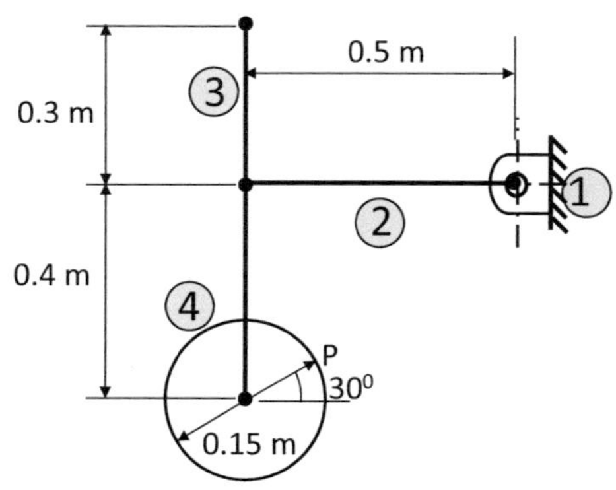

2 is uniformly rotating at 4000 rpm in the clockwise direction, determine the angular velocities of the coupler and follower and the velocity of point C. O_2B = 88 mm, AB = 82 mm, BC = 54 mm.

2.8 Consider the mechanism shown in Fig. P2.8. If the angular velocity of link 2 is 5 rad/s in the counterclockwise direction, determine the angular velocities of links 3 and 4. In addition, determine the velocity of the center of gravity of link 3. See Fig. P1.4b for the mechanism's block diagram representation.

2.9 Consider the mechanism shown in Fig. P2.9. If $\omega_2 = -1$ rad/s (clockwise), determine the angular velocities of links 3, 4, and 5 and the linear velocity of link 6. See Fig. P1.5b for the mechanism's block diagram representation.

2.10 Determine the velocity of slider (link 6) in the given position of the drag link mechanism shown in Fig. P2.10, if the crank (link 2) is rotating clockwise at 15 rad/s.

2.11 Consider a body B that has points A and B as shown in Fig. P2.11. If the magnitudes of the velocities shown at these points are: $|v_A| = 0.3$ m/s and $|v_B| = 0.6$ m/s and the directions as shown, determine the angular velocity of $\mathcal{B}$ and the velocity of the midpoint between A and B using the method of instantaneous centers.

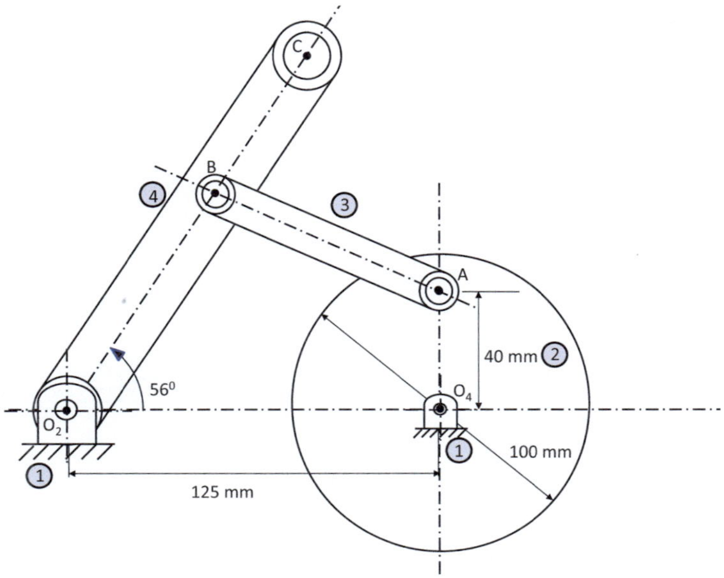

Fig. P2.7 Mechanism of exercise problem 2.7

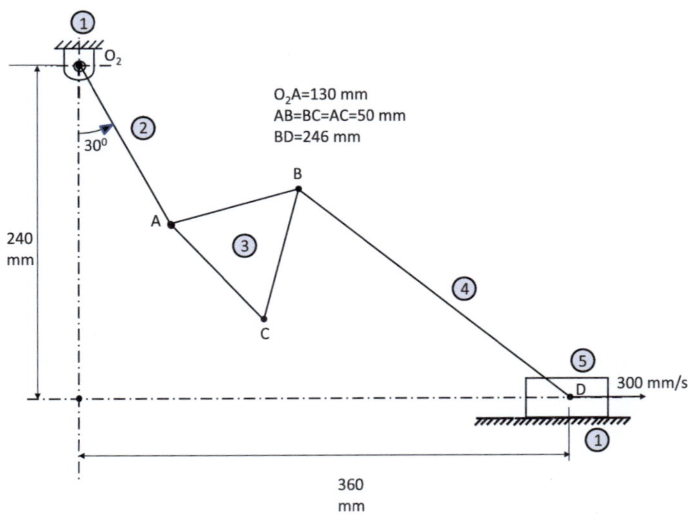

Fig. P2.8 Mechanism of exercise problem 2.8

2.12 Consider the mechanism shown in Fig. P2.12 that was shown earlier in Exercise Problem 2.4a If $\mathcal{B}_1$ moves towards right at a velocity of 0.2 m/s, using the method of instantaneous centers, determine (a) the velocities of $\mathcal{B}_2$ and $\mathcal{B}_3$ and (b) the angular velocities of $\mathcal{B}_4$ and $\mathcal{B}_5$.

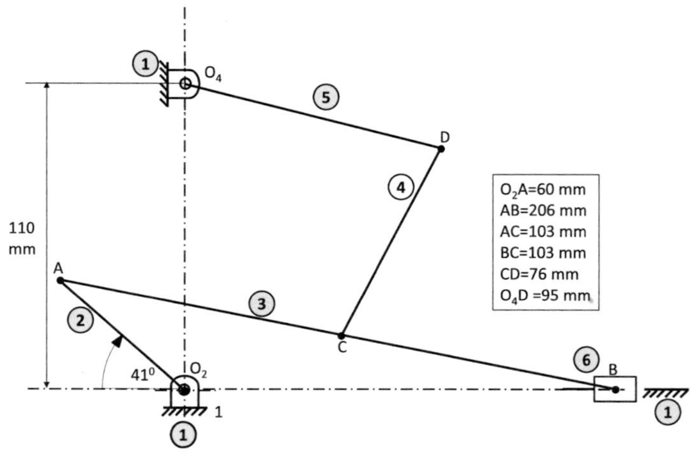

Fig. P2.9 Mechanism of exercise problem 2.9

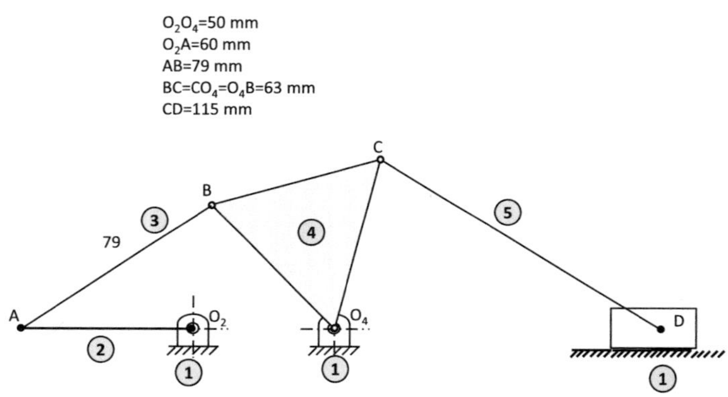

Fig. P2.10 Mechanism of exercise problem 2.10

Fig. P2.11 Velocity of two points on a rotating rigid link

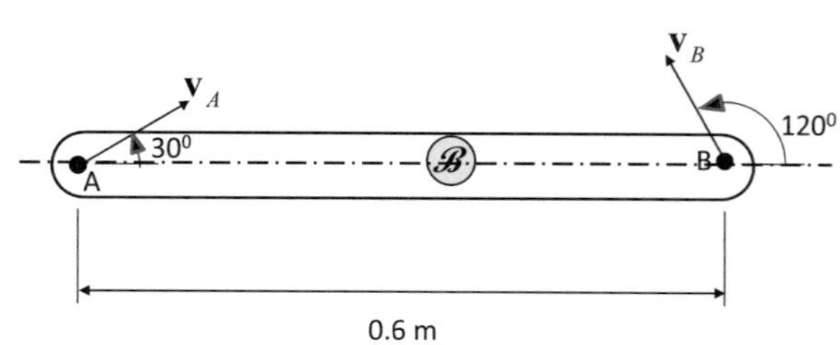

2.13 Consider the mechanism shown in Fig. P2.13 that was earlier shown in Exercise Problem 2.5. The mechanism is driven by link 4 in the counterclockwise direction at the given instant. If link 4 has an angular velocity of 15 rad/s, using

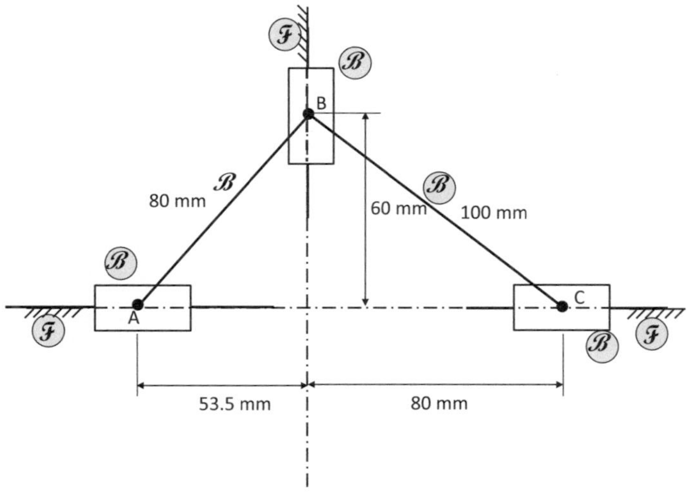

Fig. P2.12 Mechanism of exercise problem 2.12

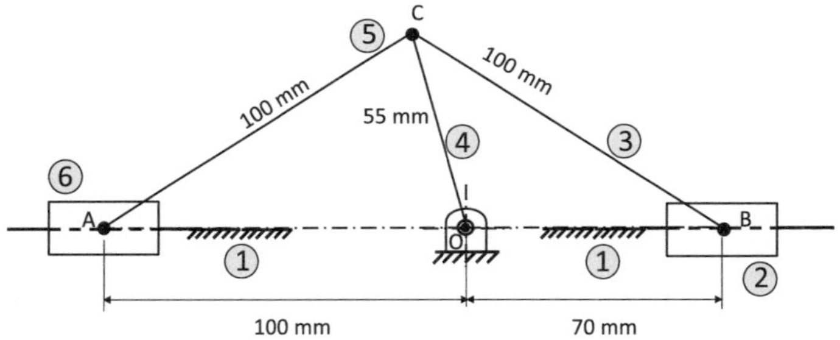

Fig. P2.13 Mechanism of exercise problem 2.13

the method instantaneous centers, determine the velocities of the pistons and the angular velocities of links 3 and 5.

2.14 For the mechanism shown in Fig. P2.14 that was earlier shown in Exercise Problem 2.6, determine the velocity of P on link 4 using the method of instantaneous centers. The various angular velocities of the links are $\omega_2 = 3$ rad/s (CCW), $\omega_3 = 2$ rad/s (CCW) and $\omega_4 = -1$ rad/s (CW).

2.15 A four-bar mechanism is shown in Fig. P2.15 that was earlier shown in Exercise Problem 2.7. If link 2 is uniformly rotating at 4000 rpm in the clockwise direction, using the method of instantaneous centers, determine the angular velocities of the coupler and follower using the method of instantaneous centers and the velocity of point C. $O_2B = 88$ mm, $AB = 82$ mm, $BC = 54$ mm.

2.16 Consider the mechanism shown in Fig. P2.16 that was earlier shown in Exercise Problem 2.8. If the angular velocity of link 2 is 5 rad/s in the counterclockwise direction, using the method of instantaneous centers, determine the angular

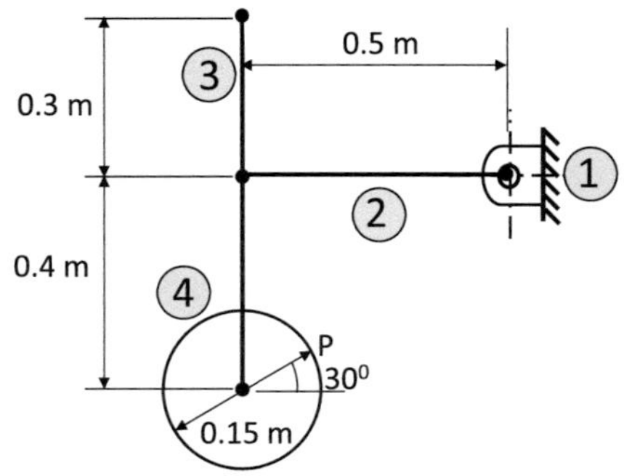

Fig. P2.14 Mechanism of exercise problem 2.14

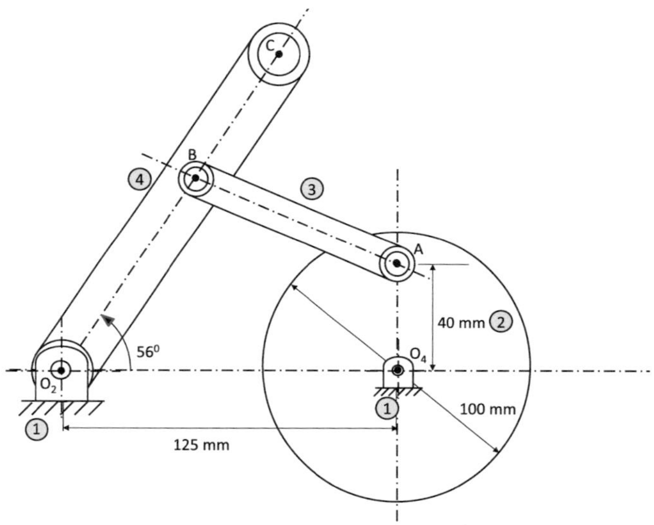

Fig. P2.15 Mechanism of exercise problem 2.15

velocities of links 3 and 4. In addition, determine the velocity of the center of gravity of link 3.

2.17 Consider the mechanism shown in Fig. P2.17 that was shown earlier in Exercise Problem 2.9. If $\omega_2 = -1$ rad/s (clockwise), using the method of instantaneous centers, determine the angular velocities of links 3, 4, and 5 and the linear velocity of link 6.

2.18 Determine the velocity of slider (link 6) in the given position of the drag link mechanism shown in Fig. P2.18 that was earlier shown in Exercise Problem 2.10, if the crank (link 2) is rotating clockwise at 150 rad/s, using the instantaneous center method.

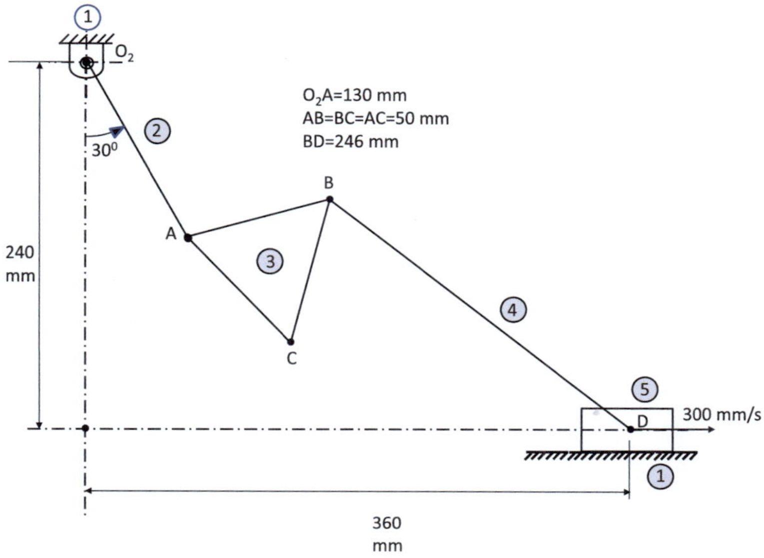

Fig. P2.16 Mechanism of exercise problem 2.16

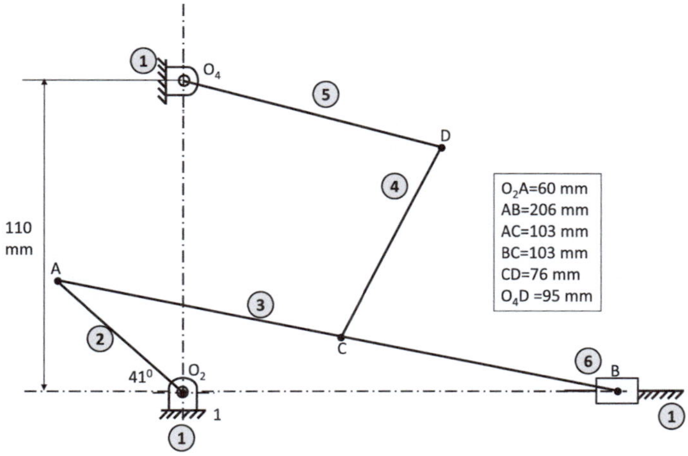

Fig. P2.17 Mechanism of exercise problem 2.17

2.19 Consider the part of a mechanism shown in Fig. P2.19a and its block diagram shown in Fig. P2.1b . If the angular velocity of link 2 is 2 rad/s and the relative velocity of link 3 with respect to link 2 is 50 mm/s in the upward direction determine the total velocity of a point P on link 3; link 3 slides on link 2.

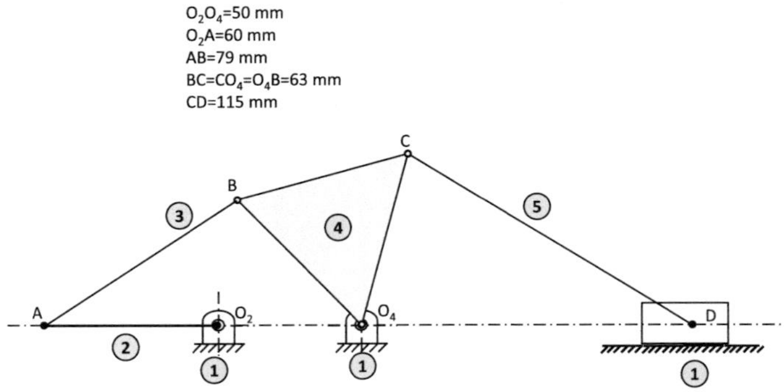

Fig. P2.18 Mechanism of exercise problem 2.18

Fig. P2.19 **a** A slider moving up a rotating rigid link **b** Block diagram

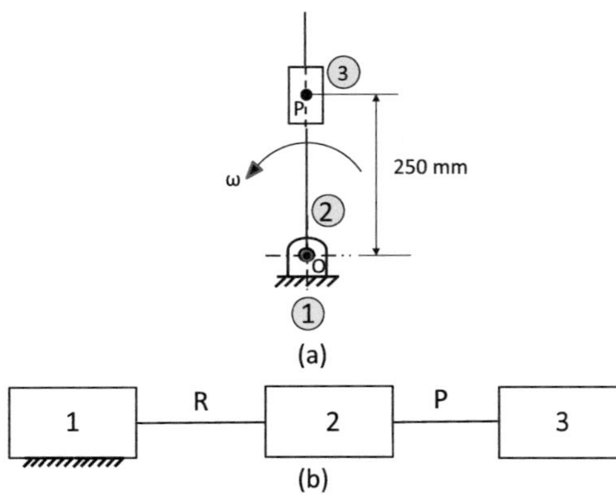

Fig. P2.20 Mechanism of exercise problem 2.20

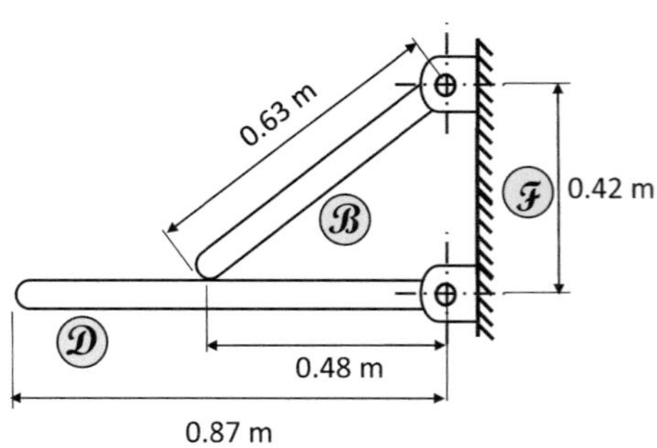

2.20 Consider the mechanism shown in Fig. P2.20 in which the link $\mathcal{D}$ rotating at 2 rad/s in the clockwise direction is pushing link $\mathcal{B}$ by sliding action. Determine the angular velocity of the link $\mathcal{B}$ and the relative sliding velocity of the contact points.

2.21 In the mechanism shown in Fig. P2.21, point A moves towards left at a constant velocity of 75 mm/s. Determine the angular velocity of link 2 and the relative velocity of point A with respect to link 2 See Fig. P1.8b for the mechanism's block diagram representation.

2.22 Consider the mechanism shown in Fig. P2.22. If the slotted links are driven independently with the following angular velocities, determine the absolute velocity of point P on link 4 and its relative velocities with the respective slotted links: $\omega_2 = 20$ rad/s (CCW), $\omega_3 = 25$ rad/s (CCW). See Fig. P1.9b for its block diagram representation.

2.23 The mechanism shown in Fig. P2.23 consists of two slender rods, A and B, respectively, hinged at O_2 and O_4. Two collars, C_1 and C_2, hinged at C, respectively, slide along A and B. If the rod A is rotating in the counterclockwise direction with an angular velocity of 0.5 rad/s and B is rotating in the clockwise direction with an angular velocity of 0.7 rad/s, determine the relative velocities of the collars with respect to the two rods. See Fig. P1.10b for the mechanism's

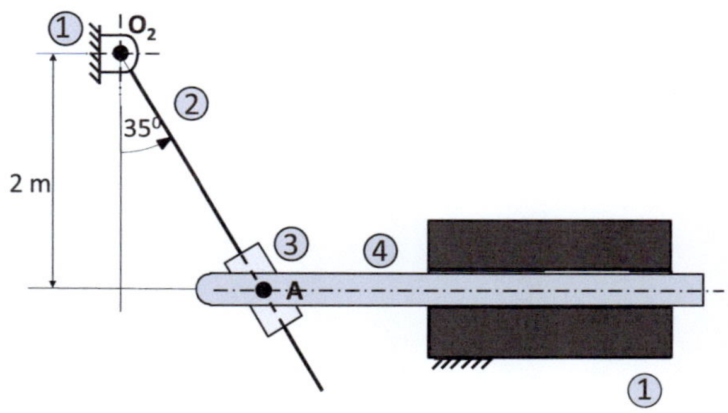

Fig. P2.21 Mechanism of exercise problem 2.21

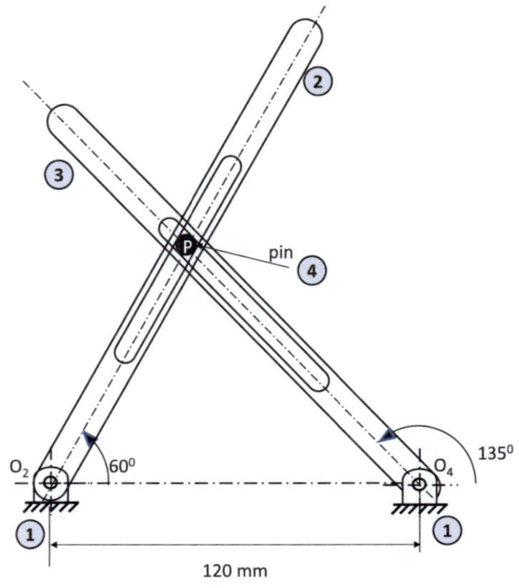

Fig. P2.22 Mechanism of exercise problem 2.22

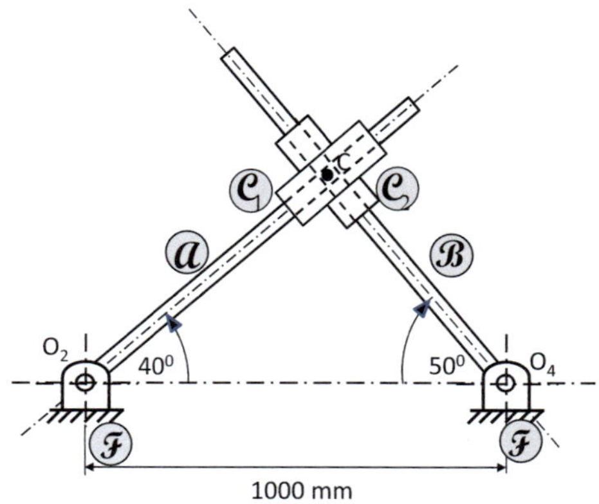

Fig. P2.23 Mechanism of exercise problem 2.23

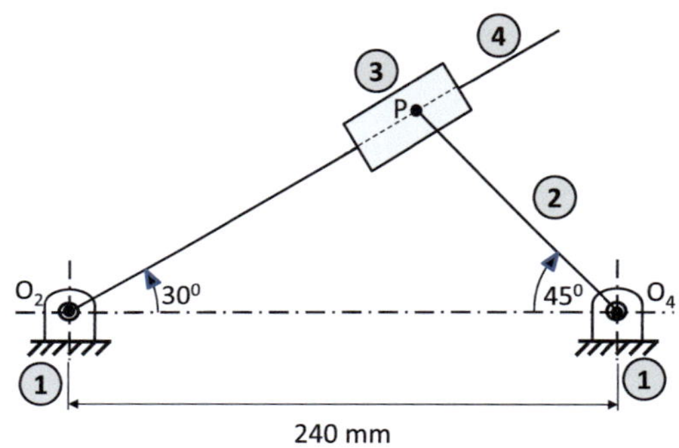

Fig. P2.24 Mechanism of exercise problem 2.24

block diagram representation. What is the difference between the mechanisms presented in Exercise Problems 2.22 and 2.23?

2.24 Consider the mechanism shown in Fig. P2.24. Link 3 (piston) slides on link 4 and pinned to link 2 (crank) at P. If the crank is rotating at angular velocity of 15 rad/s in the clockwise direction, determine the angular velocity of link 4 and the relative velocity of the piston (link 3) with respect to link 4. See Fig. P1.11b for the mechanism's block diagram representation.

2.25 Consider the quick return mechanism shown in Fig. P2.25. If link 2 (crank) is rotating at an angular velocity $\omega_2 = 15$ rad/s (CCW), determine the angular velocities of links 3 and 5, and the linear velocity of link 6. In addition, determine the relative velocity of link 4 with respect to link 3. See Fig. P1.12b for the mechanism's block diagram representation.

2.26 Consider a Body $\mathcal{B}$ that Has Point A and B as Shown in Fig. P2.26. If the magnitudes of the acceleration Shown at these points are: $|\mathbf{a}_A| = 0.3$ m/s^2 and $|\mathbf{a}_B| = 0.6$ m/s^2, determine the angular velocity and angular acceleration of $\mathcal{B}$ and the angular acceleration of the midpoint between A and B.

2.27 Consider the mechanism shown in Fig. P2.27 that was earlier discussed in Exercise Problems 2.4 and 2.12. If $\mathcal{B}_1$ moves towards right with a velocity of 0.2 m/s and an acceleration of of 0.5 m/s^2 determine (a) the velocities

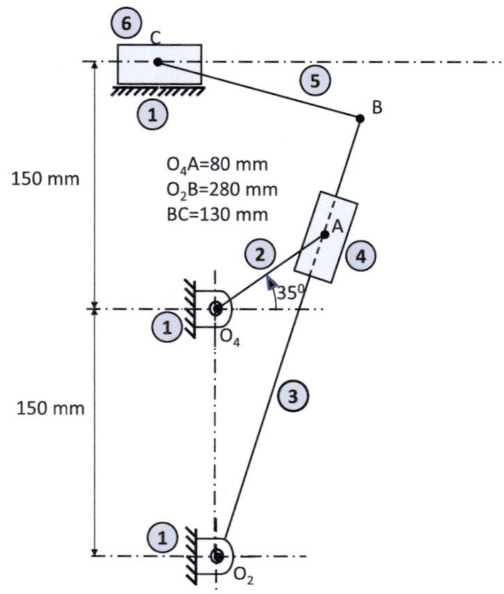

Fig. P2.25 Mechanism of exercise problem 2.25

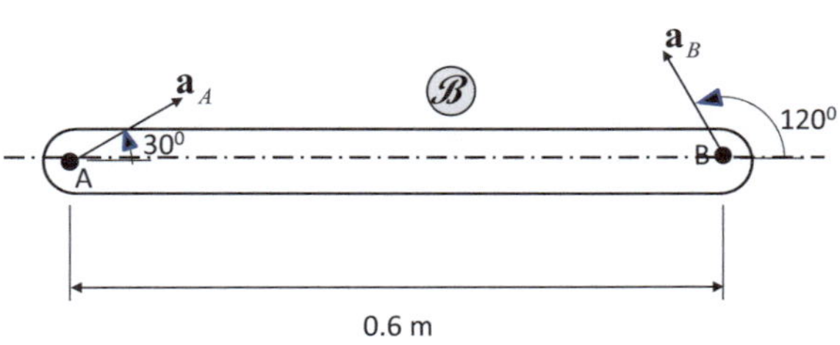

Fig. P2.26 Acceleration of two points on a rotating rigid link

and accelerations of $\mathscr{B}_2$ and $\mathscr{B}_3$ and (b) the angular velocities and angular accelerations of $\mathscr{B}_4$ and $\mathscr{B}_5$. See Fig. P1.1b for its block diagram representation.

2.28 Consider the mechanism shown in Fig. P2.28 that was earlier discussed in Exercise Problems 2.5 and 2.13. If link 4 has an angular velocity of 15 rad/s, determine the accelerations of the pistons and the angular accelerations of links 3 and 5.

2.29 The mechanism shown in Fig. P2.29 that was earlier discussed in Exercise Problems 2.6 and 2.14. Determine the acceleration of P on link 4. The various angular velocities of the links are: $\omega_2 = 3$ rad/s (CCW), $\omega_3 = 2$ rad/s (CCW) and $\omega_4 = -1$ rad/s (CW)

2.30 A four-bar mechanism is shown in Fig. P2.30 that was earlier discussed in Exercise Problems 2.7 and 2.15. If link 2 is uniformly rotating at 4000 rpm in the clockwise direction, determine the angular accelerations of of the coupler and follower and the acceleration of point of point C. $O_2B = 88$ mm, AB = 82 mm, BC = 54 mm.

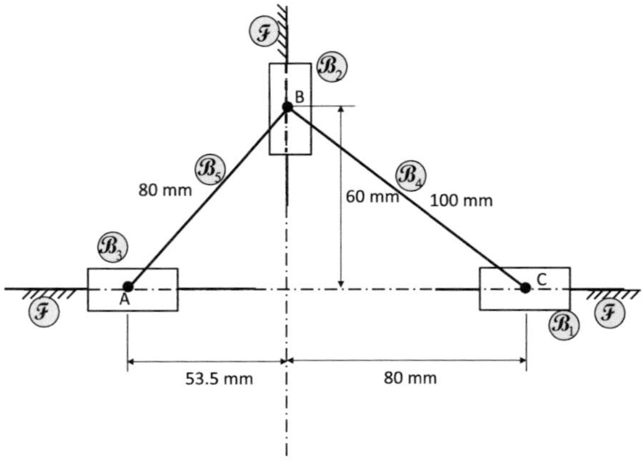

Fig. P2.27 Mechanism of exercise problem 2.27

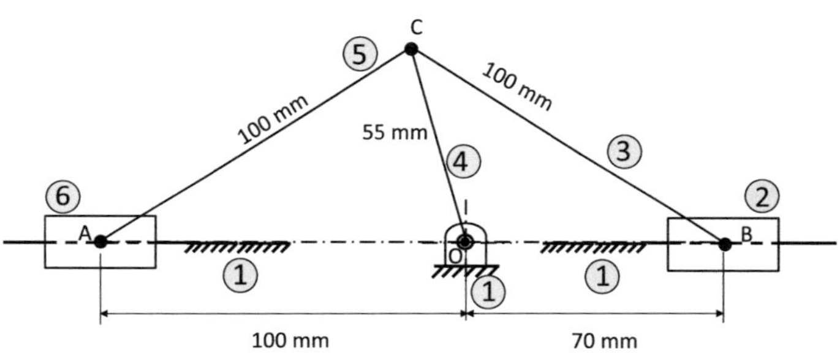

Fig. P2.28 Mechanism of exercise problem 2.28

Fig. P2.29 Mechanism of exercise problem 2.29

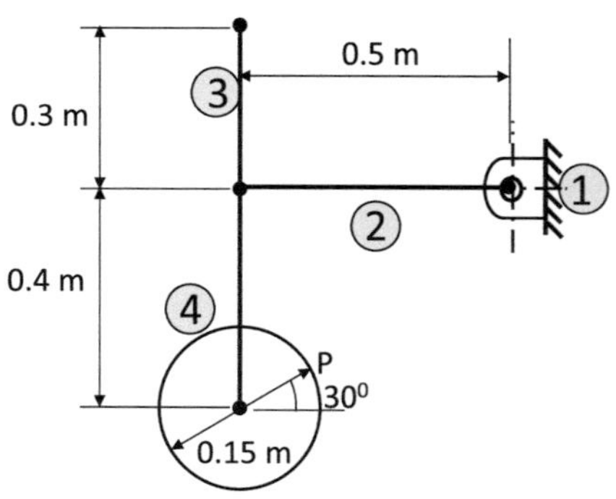

2.31 Consider the mechanism shown in Fig. P2.31 that was earlier discussed in Exercise Problems 2.8 and 2.16. If the angular velocity of link 2 is 5 rad/s in the counterclockwise direction, determine the angular accelerations of links 3 and 4. In addition, determine the acceleration of the center of gravity of link 3 and linear acceleration of link 5.

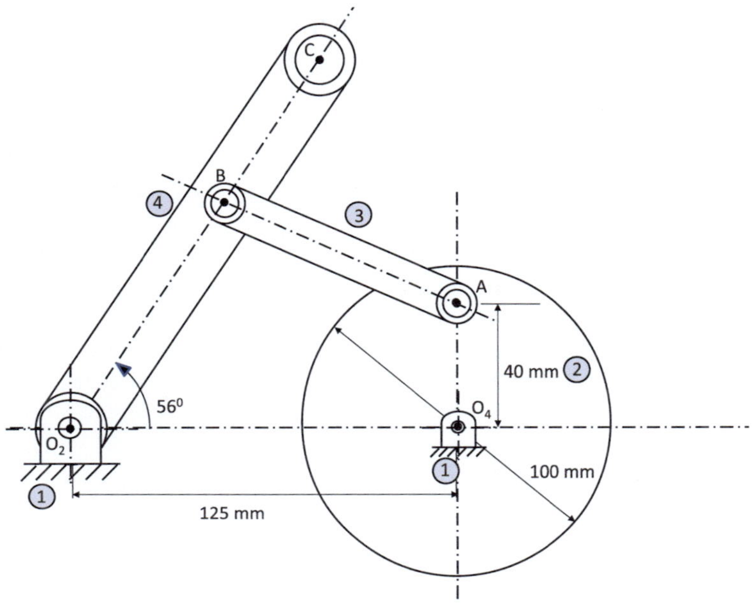

Fig. P2.30 Mechanism of exercise problem 2.30

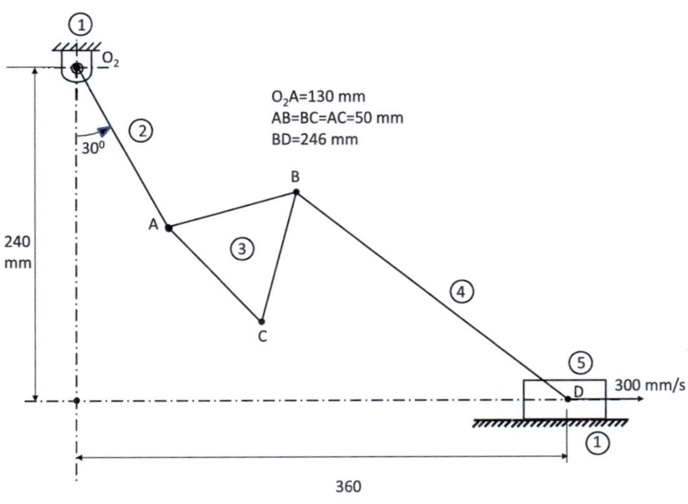

Fig. P2.31 Mechanism of exercise problem 2.31

2.32 Consider the mechanism shown in Fig. P2.32 that was earlier discussed in Exercise Problems 2.9 and 2.17. If $\omega_2 = -1$ rad/s (clockwise), determine the angular accelerations of links 3, 4, and 5 and the linear velocity of link 6.

2.33 Determine the acceleration of slider (link 6) in the given position of the drag link mechanism shown in Fig. P2.33 that was earlier discussed in Exercise Problems 2.10 and 2.18. if the crank (link 2) is rotating clockwise at 150 rad/s.

2.34 Consider the part of a mechanism shown in Fig. P2.34a and its block diagram shown in Fig. P2.19b. If the angular velocity of link 2 is 2 rad/s (CCW) and the angular acceleration is 10 rad/s^2 (CCW); the relative velocity of link 3 with respect to link 2 is 50 mm/s in the upward direction, and relative acceleration is 100 mm/s^2 in the upward direction, determine the total acceleration of a point P on link 3; link 3 slides on link 2.

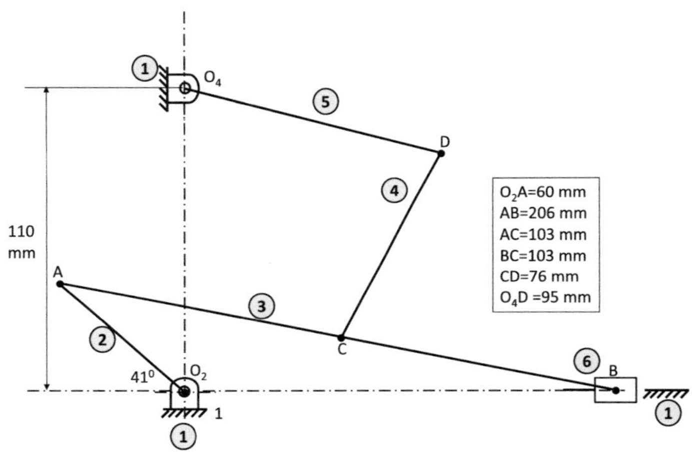

Fig. P2.32 Mechanism of exercise problem 2.32

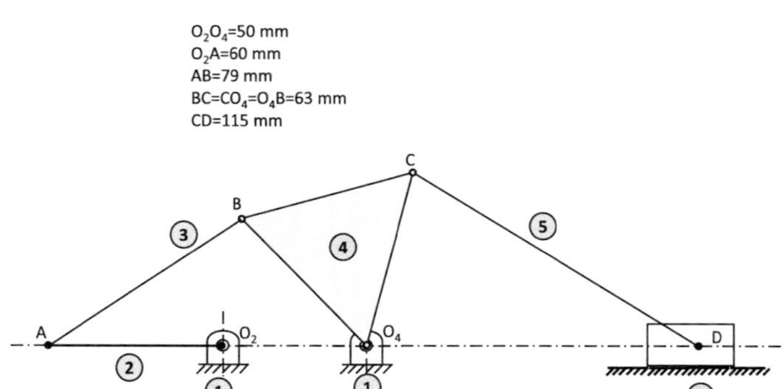

Fig. P2.33 Mechanism of exercise problem 2.33

Fig. P2.34 A slider moving up a rotating rigid link

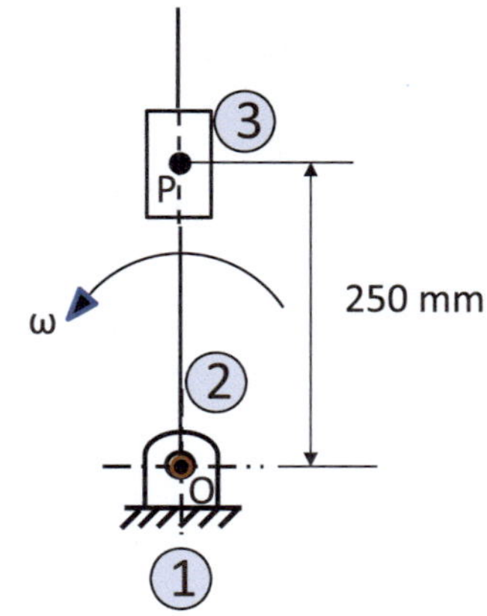

2.35 Consider the mechanism shown in Fig. P2.35 that was earlier discussed in Exercise Problem 2.20 in which link $\mathcal{D}$ rotating at an angular velocity of 2 rad/s in the clockwise direction and angular acceleration of 10 rad/s^2 in the clockwise direction, is pushing link $\mathcal{B}$ by sliding action. Determine the angular acceleration of link $\mathcal{B}$ and the relative sliding acceleration between the contact points.

2.36 In the mechanism shown in Fig. P2.36 that was earlier discussed in Exercise Problem 2.21, point A moves towards left at a velocity of 75 mm/s and an acceleration of 5 mm/s^2. Determine the angular acceleration of link 2 and the relative acceleration of point A with respect to link 2.

2.37 Consider the mechanism shown in Fig. P2.37 that was earlier discussed in Exercise Problem 2.22. If the slotted links are driven independently with the following angular velocities and accelerations, determine the absolute acceleration of point P on link 4 and its relative acceleration with the respective slotted links: $\omega_2 = 20$ rad/s (CCW), $\omega_3 = 25$ rad/s (CCW); $\alpha_2 = 5$ rad/s^2 (CCW), $\alpha_3 = 25$ rad/s^2 (CCW).

Fig. P2.35 Mechanism of exercise problem 2.35

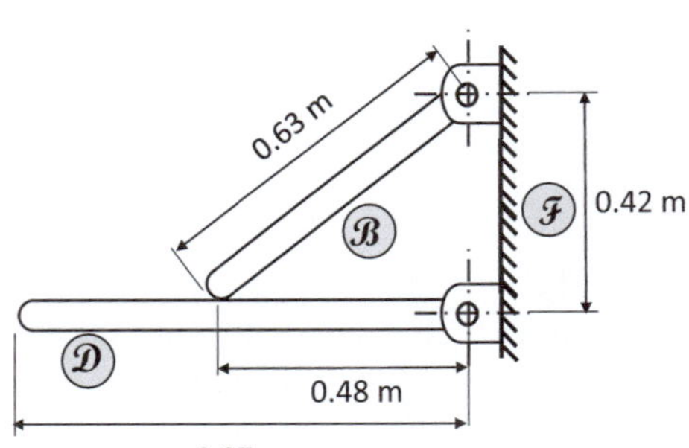

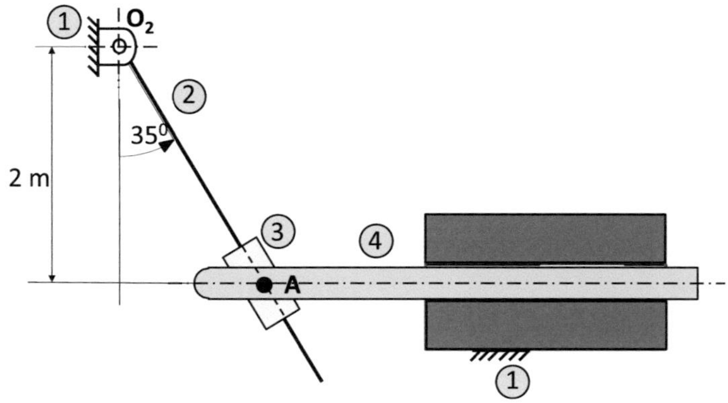

Fig. P2.36 Mechanism of exercise problem 2.36

Fig. P2.37 Mechanism of exercise problem 2.37

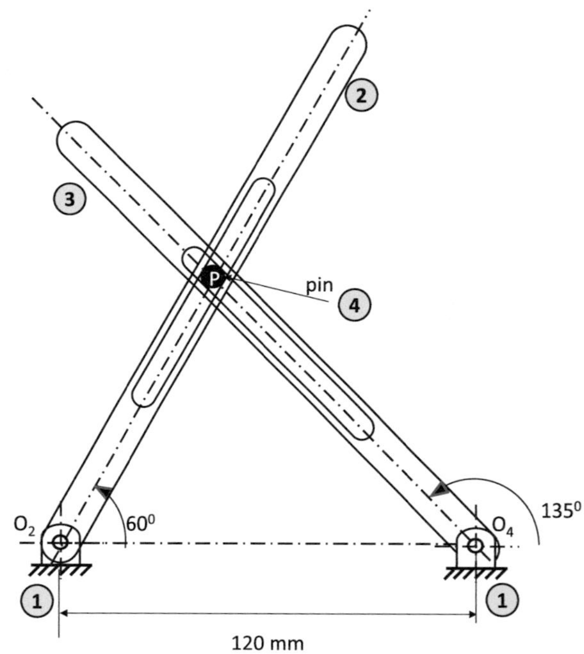

2.38 The mechanism shown in Fig. P2.38 consists of two slender rods, A and B, respectively, hinged at O_2 and O_4. Two collars, C_1 and C_2, hinged at C, respectively, slide along A and B. This mechanism was earlier discussed in Exercise Problem 2.23. If the rod A is rotating in the counterclockwise direction with an angular velocity of 0.5 rad/s, angular acceleration of 5 rad/s^2 in the clockwise direction, and B is rotating in the clockwise direction with an angular velocity of 0.7 rad/s, angular acceleration of 5 rad/s^2 in the counterclockwise direction,

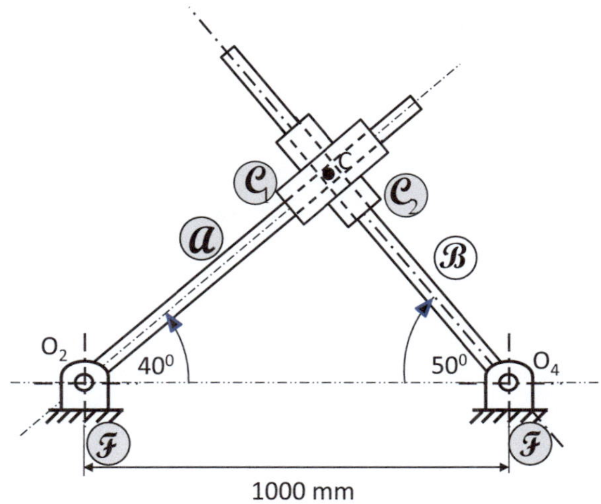

Fig. P2.38 Mechanism of exercise problem 2.38

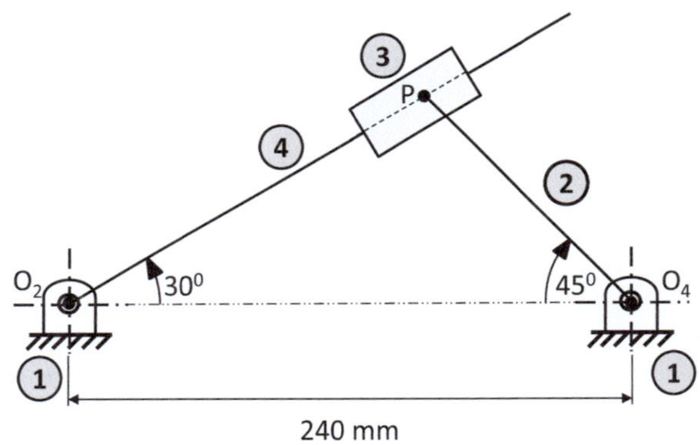

Fig. P2.39 Mechanism of exercise problem 2.39

determine the relative accelerations of the collars with respect to the two rods and the absolute acceleration of the point C.

2.39 Consider the mechanism shown in Fig. P2.39 that was earlier discussed in Exercise Problem 2.24. Link 3 (piston) slides on link 4 and pinned to link 2 (crank) at P. If the crank is rotating at an angular velocity of 15 rad/s in the clockwise direction, determine the angular acceleration of link 4 and the relative acceleration of the piston (link 3) with respect to link 4.

2.40 Consider the quick return mechanism shown in Fig. P2.40 that was earlier discussed in Exercise Problem 2.20. If link 2 (crank) is rotating at angular velocity $\omega_2 = 15$ rad/s (CCW), determine the angular accelerations of links 3 and 5, and the linear acceleration of link 6. In addition, determine the relative acceleration of link 4 with respect to link 3.

2.41 Solve Example 2.10 using graphical method.

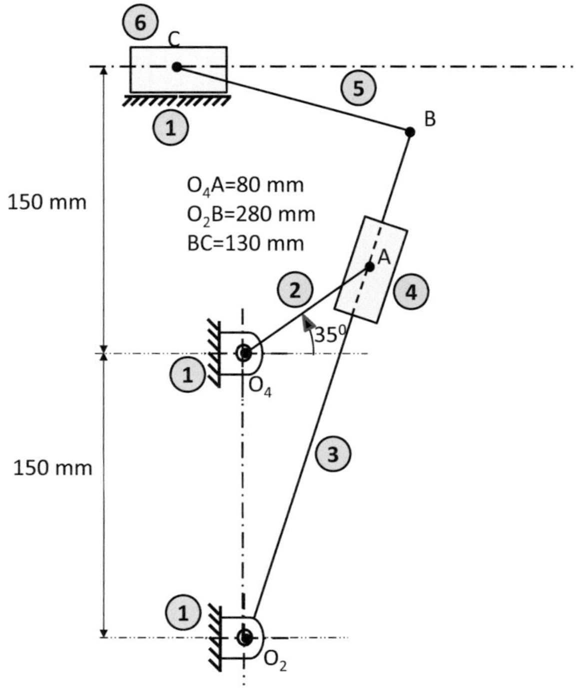

Fig. P2.40 Mechanism of exercise problem 2.40

Chapter 3
Spur Gears

3.1 Introduction

Gears are machine elements that are used for changing the rotational speed, torque, and direction from a machine to the external load. Following are the various types of gears that are used in various applications: spur, helical, bevel, worm, epicyclic, etc. Depending on the application, these various types of gears are appropriately selected, based on the unique characteristic features of each of them. However, we shall first discuss spur gears, which are the simplest of all the above gears.

Spur gears, as shown in Fig. 3.1, are used to transmit rotary motion between two parallel shafts, transmitting torques at different speeds. Hence, a pair of gears is required; the smaller one is the pinion and the large one is called the gear. Whether a pinion should be on the driving shaft and a gear on the driven shaft or vice versa will depend on the objective of increasing or decreasing speed/torque. The extent of change of torque will be inversely proportional to the extent of change of speed, which is the main objective in many applications; in addition, the input and output shafts will rotate in opposite directions. There is always a limit to the extent torque or speeds can be changed with a pair of gears. Hence, many times, there will be several stages consisting of several pairs of gears that are arranged to get the desired speed and torque at the output. They are called gear trains that are discussed in Chap. 5.

Gears are standardized so their manufacture, selection, and installation become easy and they are perhaps the most standard machine elements that can be interchangeably used in many applications. In addition, since gear pairs will have to satisfy the fundamental law of gearing, a standard tooth profile will have to be used; involute profiles are generally used and cycloidal profiles are used in some applications.

First, the fundamental law of gearing that will provide the basis for evolving an involute profile will be discussed. Then, the process of generating an involute profile and the generation of conjugate mating surfaces by unwinding a taut string between two cylinders will be discussed that will help evolve the basic concept of mating

D. N. Manik, *Fundamentals of Mechanisms and Machines*,
https://doi.org/10.1007/978-981-97-1810-8_3

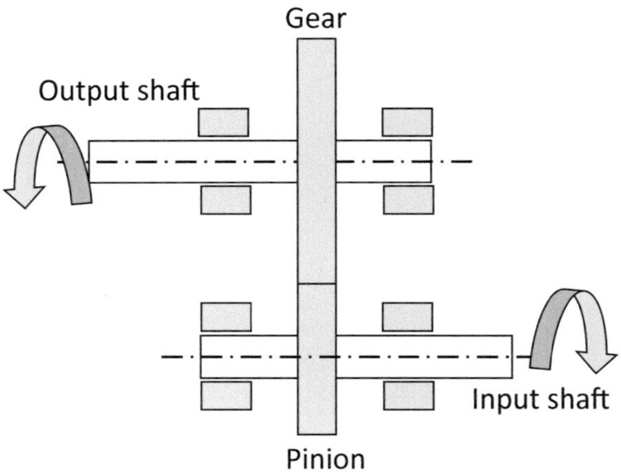

Fig. 3.1 A typical spur gear system

gears with conjugate mating surfaces. Further, path of contact and contact ratio will be discussed that will be the basis for further improving spur gears. Since gears are widely used in many applications, there is a greater standardization in their design and manufacture. AGMA (American Gear Manufacturers Association) standards are the most popular gear standards.

3.2 Historic Development

Gears evolved as positive drives and were generally used in clocks in earlier days. They were very simple devices with straight teeth and were thus not useful for high-speed machines. The involute profile development in 1868 revolutionized the use of gears in modern machines. Therefore, we shall discuss the basis for using involute profile in gears based on the basic law of gearing.

3.3 Fundamental Law of Toothed Gearing

In Sect. 2.7, in the context of discussing the instantaneous center of a four-bar linkage, we proved that the ratio of the angular velocities of the crank and follower have a definite ratio based on the positions of the instantaneous centers of the four-bar linkage. On similar lines, we can prove as follows that the ratio of angular velocities of the mating gears is inversely proportional to the ratio of their pitch diameters.

The motion transmitted by mating gears that produce rotary motion is similar to a cam and follower mechanism. In both cases, the tooth profiles or cam-follower profiles are shaped such that there is a constant angular velocity between the driver and driven shafts, then such profile surfaces are said to be conjugate. It is possible to specify a specific tooth profile for the pinion and find a conjugate tooth profile for the gear.

Consider the two conjugate mating surfaces of moving bodies 2 and 3 as shown in Fig. 3.2. Body 2 is hinged at O_2 and body 3 is hinged at O_3 and they are respectively rotating at angular velocities ω_2 and ω_3. The two bodies in the current position are contacting at the point B. A normal drawn from the point B to both the surfaces meets at C. If point C remains stationary for all points of contact as the bodies move with respect to each other, then the two mating surfaces are said to be conjugate and the ratio of angular velocities is given by

$$\frac{\omega_2}{\omega_3} = \frac{\rho_3}{\rho_2} \tag{3.1}$$

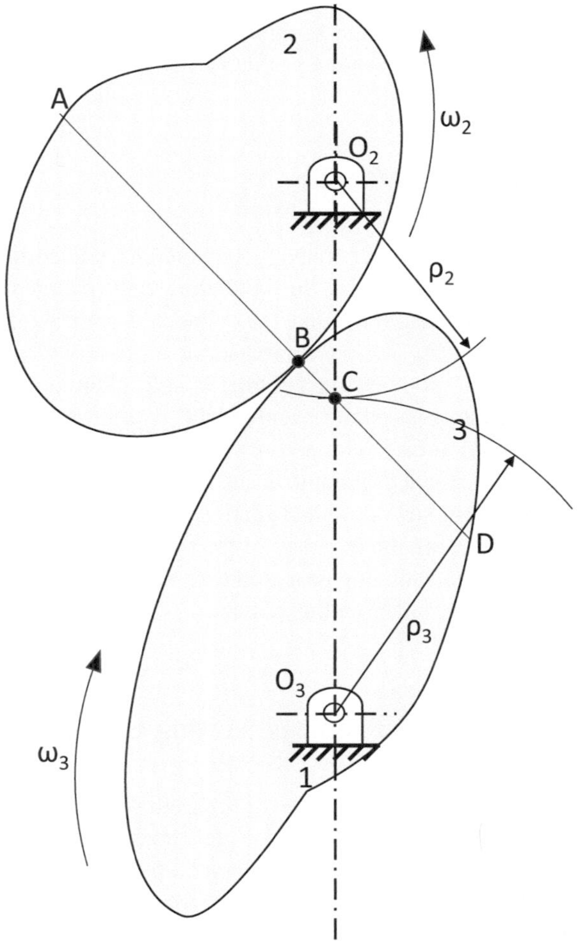

Fig. 3.2 Conjugate mating surfaces

Extending Eq. (3.1) to a mating gear set that has conjugate mating surfaces, the point P of Fig. 3.2 becomes the pitch point and the line AD is the line of action. Hence, Eq. (3.1), if the above bodies are replaced by a mating gear set, in terms of pitch diameters becomes

$$\frac{\omega_2}{\omega_3} = \frac{d_3}{d_2} \tag{3.2}$$

Hence, as per the fundamental law of gearing, as gears rotate through the meshing action, the pitch point must remain stationary on the line joining the centers for their speed ratio to be constant. Therefore, the appropriate profile that should be chosen for the mating gears should satisfy the above conditions of the fundamental law of gearing.

Just finding a conjugate profile will not make it a satisfactory choice that can be used for making gears of that profile. The profile should be able to be easily reproduced and should be robust for small changes in the center distance of the gear shafts, which an involute profile has been successful.

3.4 Involute Profile

Involute profile is generated by unwinding a taut cord wound on a cylinder as shown in Fig. 3.3. Assume that the unwinding in the counterclockwise process began when the cord was about to leave the cylinder at point A_o. As it unwinds, always keeping it taut, point A_o traces a curve and moves to point B_o, whereas the point on the cord that is about to leave the cylinder is at B. Continuing the unwinding process, point B_o now moves to point C_o and point C is about to leave the cylinder if the unwinding process continues. The path traced by cord $A_oB_oC_o$ is an involute curve and this curve, if present as conjugate surfaces on the teeth of mating gears, will satisfy the fundamental law of toothed gearing and can smoothly transmit the rotary motion as per Eq. 3.2 and the diameter of the cylinder that is used for generating the involute profile on each mating gear is known as the base cylinder and further discussion can be taken forward from this point.

3.5 Involute Profile of Mating Gears

Since the conjugate surfaces of the teeth should be on the mating gears, the concept of involute profile of Sect. 3.4 can be extended to two cylinders whose axes are, respectively, at O_1 and O_2 connected by a taut string shown in Fig. 3.4. The taut string can unwind from either of the two cylinders, generating two conjugate involute profiles. When it unwinds in the counterclockwise direction from cylinder 2, the involute profile EQF is generated. Similarly, when it unwinds in the counterclockwise

Fig. 3.3 Generating an involute profile

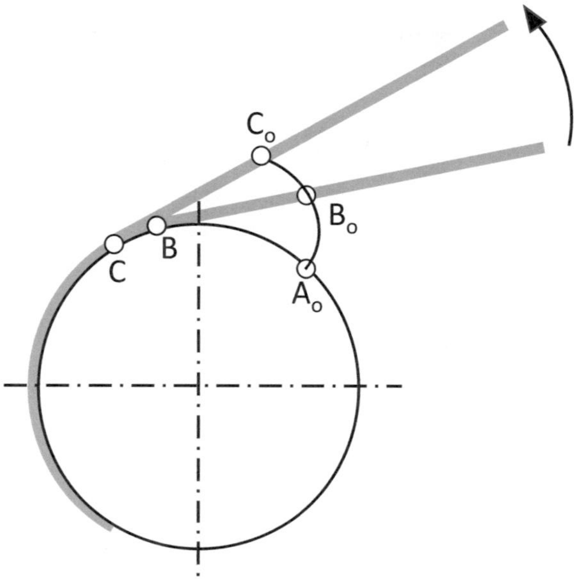

direction from cylinder 1, the involute profile CQD is generated. The two conjugate profiles are tangential at point Q. Line AB is the path of contact that is normal to the conjugate profiles and the midpoint of AB always passes through the point that later turns out to be the pitch point on mating gears. The involute pressure angle, ϕ, is the angle between O_1B and the mid point of AB, which is one of the important parameters of spur gears. The angular velocity ratio remains constant in the ratio defined by the point known as the pitch point P that is a straight line between O_1 and O_2. The circle drawn through the pitch point is known as the pitch circle. Hence, the rotation of mating gears is construed as rolling motion without slipping between two cylinders having diameters same as the pitch diameters. It is clear from Fig. 3.4 that the mating gears rotate in opposite directions. In addition, involute profile can allow small changes in the center distance of mating gears with an accompanying change in the pressure angle.

3.6 Single Tooth of a Spur Gear

The conjugate profiles discussed in the previous section need further improvement so a pair of gears can smoothly engage with each other and transfer power. Therefore, the involute profiles are symmetrically placed around an axis and a flat space is created at the top for easy meshing as shown in Fig. 3.5. Now we can have similar teeth placed around but spaced properly so the teeth of the mating gear can engage smoothly. Once the teeth are positioned, the pitch circles of both the mating gears

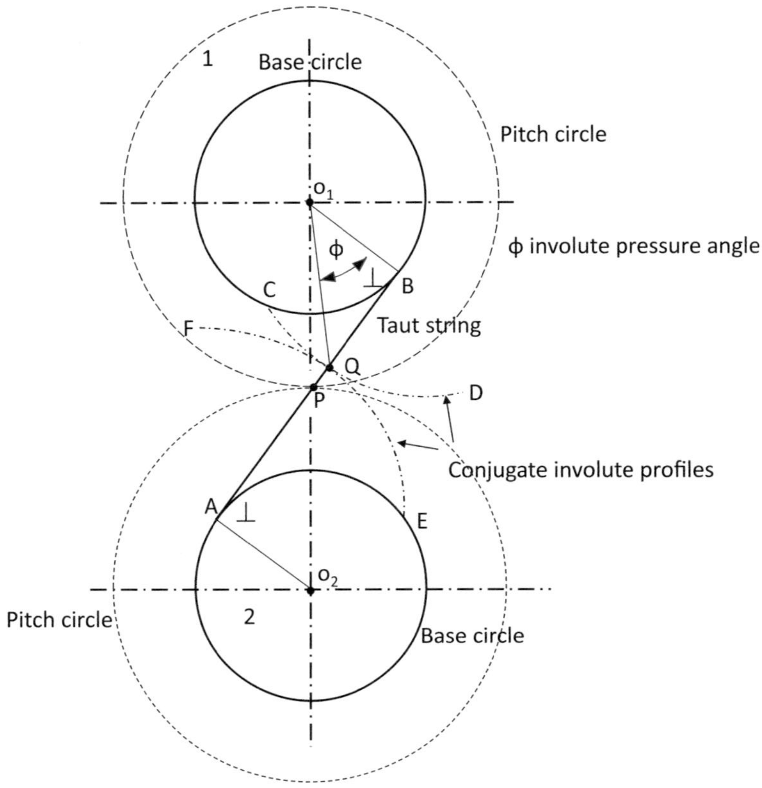

Fig. 3.4 Conjugate involute profiles between two cylinders

become the reference positions from which the proportion of the tooth and above and below will be defined.

3.7 Gear Parameters

Figure 3.6 shows additional circles of a gear that would help define the gear parameters. As discussed earlier, the base circle is the one used for the involute profile. The pitch circle of a gear is a theoretical circle on which all the calculations for its design are based upon. The pitch circles of a pair of gears (pinion and gear) are tangent to each other and the mating gears can be imagined as rolling disks without slipping. Hence, the ratio of their diameters defines the angular velocity ratio as per the law of gearing. The addendum circle defines the outermost boundary of the gears, restricting the involute profiles on either side, and dedendum circle defines the lower limit of the involute profile. Basically, all the teeth are within the domain defined by the addendum and dedendum circles. Circular pitch is the distance between the

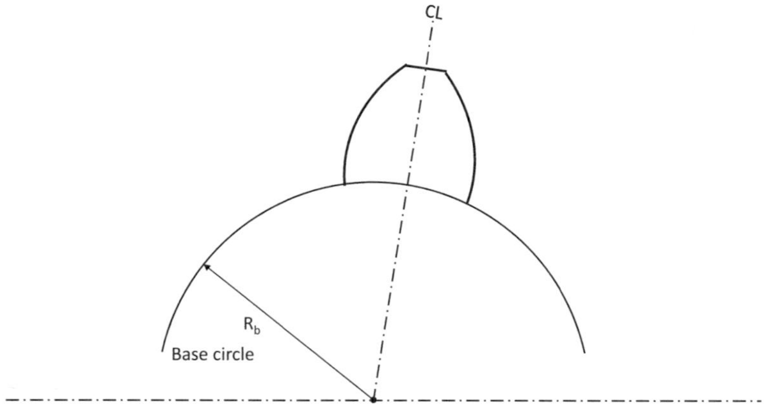

Fig. 3.5 Single tooth of a spur gear

corresponding points between two adjacent teeth on the pitch circle. Similarly, base pitch is the distance between the corresponding points of two adjacent teeth along the basic circle.

Let us denote *d* as the pitch diameter of a pinion or gear and N as the number of teeth.

Diametral pitch P is the number of teeth on the gear or pinion per inch of the pitch diameter and it is given by

$$P = \frac{N}{d} \text{ teeth/in.} \tag{3.3}$$

The diametral pitch is inversely proportional to the pitch diameter. Hence, smaller gears or pinions will have large diametral pitch and vice versa. Although SI units are

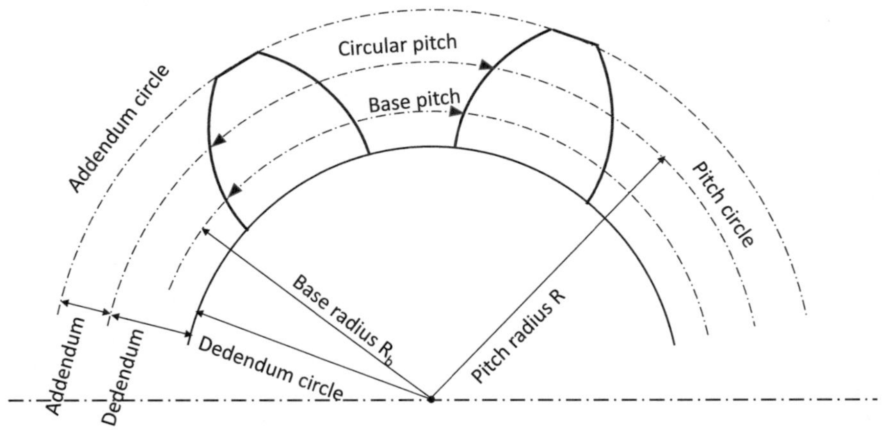

Fig. 3.6 Additional circles of a gear

Table 3.1 Module (preferred)

1, 1.25, 1.5, 2, 2.5, 3, 4, 5, 6, 8, 10, 12, 16, 20, 25, 32, 40, 50

used throughout the book, it is necessary to be aware of the diametral pitch as it is an important gear standard that is widely used in many countries.

Module m is the ratio of pitch diameter in mm to the number of teeth and it is given by

$$m = \frac{d}{N} \text{ mm/tooth} \tag{3.4}$$

And the preferred modules are shown in Table 3.1.

Circular pitch p is the distance from one tooth to the adjacent tooth measured along the pitch circle and is given by

$$p = \frac{\pi d}{N} \text{ mm} \tag{3.5}$$

From Eqs. (3.3) to (3.5), the module, circular pitch, and diametral pitch (in appropriate units) are related by

$$p = \frac{\pi}{P} = \pi m \tag{3.6}$$

Based on the discussion presented in the previous sections, the terminology of a spur gear is shown in Fig. 3.7. The various components of the terminology are noted as follows:

The addendum a is the radial distance between the pitch circle and the top land of each tooth.

The dedendum d is the radial distance between the pitch circle and the bottom land of each tooth. There is always a clearance and fillet radius at the root of each tooth to prevent interference, which is a major consideration in gear design. The tooth thickness t is measured at the pitch circle. The tooth space is measured at the pitch circle. Clearance is the distance between the dedendum of the gear and the addendum of the mating gear. The backlash is the space left by the mating in the tooth space by the mating gear. Many of the above are based on empirical formula from the standards.

Example 3.1

A 40-teeth gear has a circular pitch of 25 mm/tooth. Find the module and its pitch diameter.

From Eq. (3.6), the module is given by

$$m = \frac{p}{\pi} = \frac{25}{\pi} = 7.9577 \approx 8 \tag{1}$$

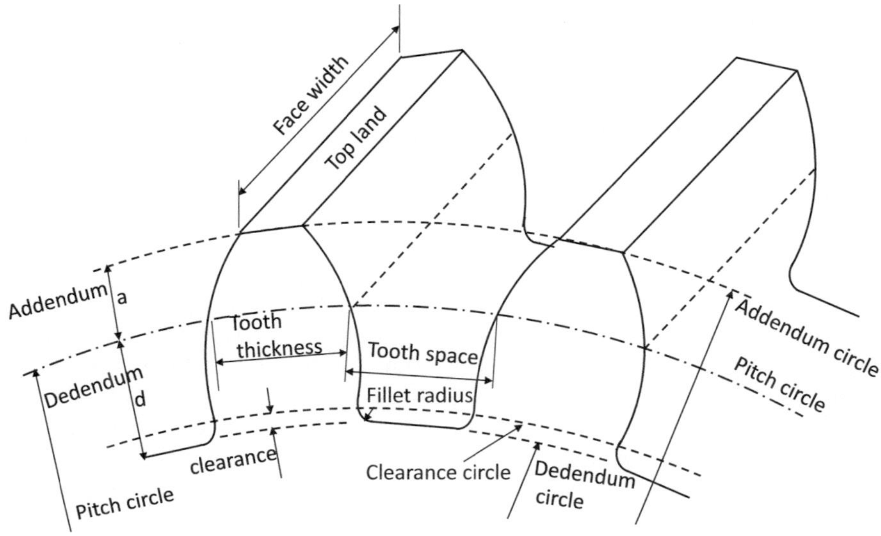

Fig. 3.7 Spur gear terminology

From Eq. (3.4), the pitch diameter is given by

$$d = mN = 8 \times 40 = 320\,\text{mm} \tag{2}$$

Example 3.2
A gear pair consisting of pinion and gear has a reduction ratio of 5, whose parallel shafts are at a distance of 150 mm. If the module of the gears is 3 mm/module, determine the pitch diameter and number of teeth on each of the gears.

Let d_1 be the pinion diameter and d_2 the gear diameter. The center distance, C, is given by

$$C = \frac{d_1 + d_2}{2} = 150\,\text{mm} \tag{1}$$

The pitch diameters of the gear and pinion are related by

$$d_2 = velocity\ ratio \times d_1 = 5d_1 \tag{2}$$

From Eqs. (1) and (2), $d_1 = 50$ mm and $d_2 = 250$ mm.
The number of teeth on the pinion is given by

$$N_1 = \frac{d_1}{m} = \frac{50}{3} \simeq 17 \tag{3}$$

The number of teeth on the gear is given by

$$N_2 = \frac{d_2}{m} = \frac{250}{3} \approx 83 \tag{4}$$

The velocity ratio with the rounded-off numbers of teeth will be 4.88 that is slightly less than 5, which should be acceptable.

3.8 Path of Contact

Now that we have defined various parameters of a gear, we can consider two mating gears G_1 and G_2 in contact with each other as shown in Fig. 3.8a. The following circles are drawn for both of them: addendum, pitch, base, and dedendum. The line of action will be along BA that is tangent to both the base circles and the pressure angle is also defined with respect to this line. The actual contact begins at point D that is an intersection of AB and the addendum circle of G_2 that is when the contact between the two gears begins. The contact ends at point C that is an intersection of AB with the addendum circle of G_1 when the contact ends. Hence, CD is the path of contact that is a straight line. The points of contact on the mating teeth at the beginning and end of contact are shown in Figs. 3.8b and 3.9.

Part of Fig. 3.8a is redrawn to derive the equation for the length of contact.

From triangles O_1CB and O_1PB,

$$CP = \sqrt{(R_1 + a)^2 - (R_1 \cos\phi)^2} - R_1 \sin\phi \tag{3.7}$$

From triangle O_2AD and O_2AP,

$$PD = \sqrt{(R_2 + a)^2 - (R_2 \cos\phi)^2} - R_2 \sin\phi \tag{3.8}$$

Contact ratio, α, is given by

$$\begin{aligned} \alpha &= \frac{CP + PD}{p_b} \\ &= \frac{\sqrt{(R_1 + a)^2 - (R_1 \cos\phi)^2} - R_1 \sin\phi + \sqrt{(R_2 + a)^2 - (R_2 \cos\phi)^2} - R_2 \sin\phi}{p_b} \end{aligned} \tag{3.9}$$

Example 3.3

A gear pair consisting of pinion and gear has a module of 3 mm/tooth, a pressure angle of 20°, and a center distance of 320 mm; there are 35 teeth on the pinion. Calculate the contact ratio.

The addendum of both the pinion and gears as per the empirical standards is given by

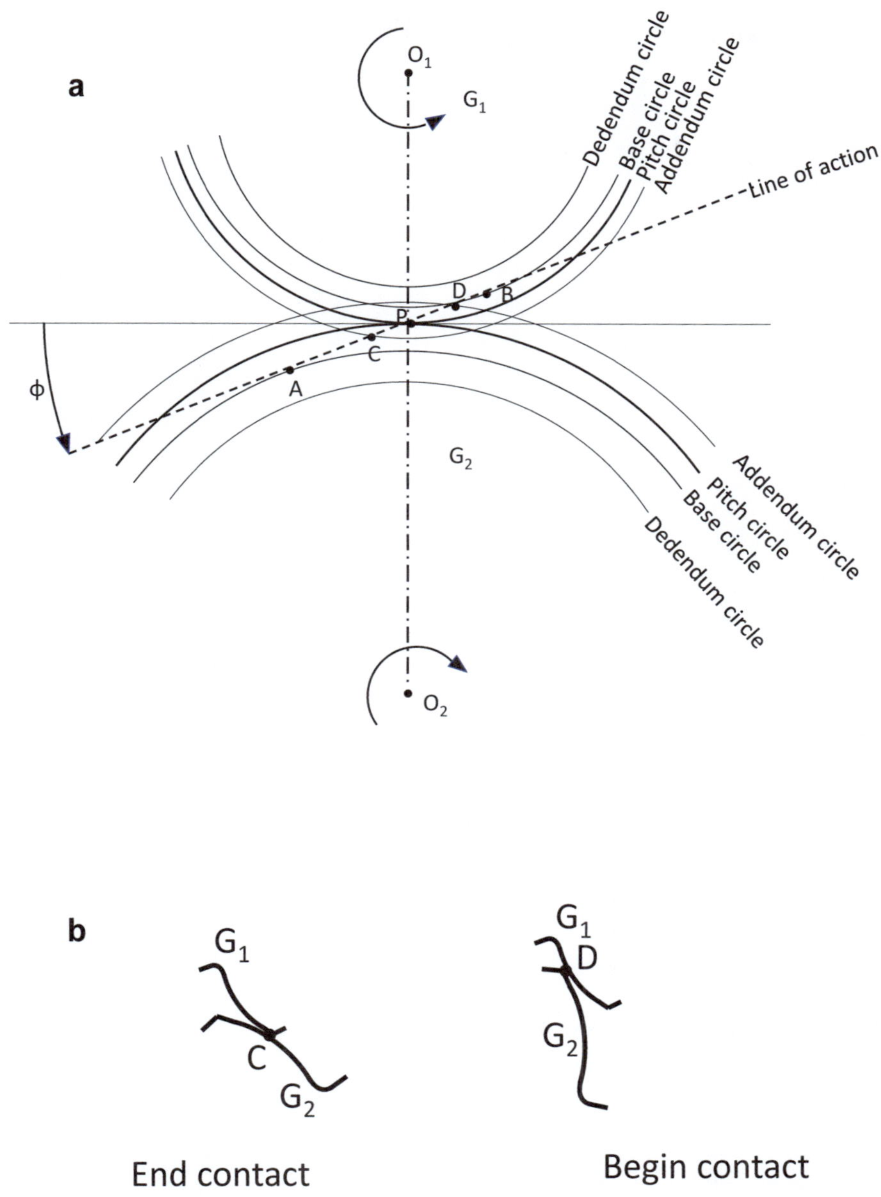

Fig. 3.8 **a** Path of contact of a spur gear. **b** Extreme points of teeth contact

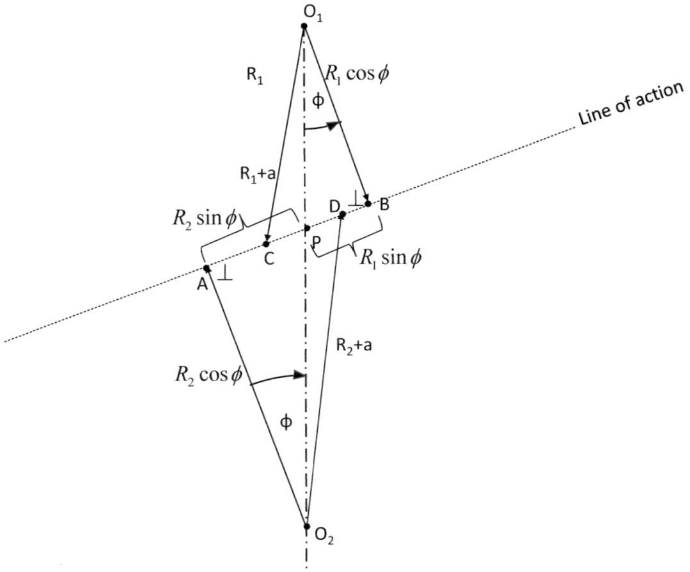

Fig. 3.9 Relevant portions of Fig. 3.8 to determine contact ratio

$$a_1 = a_2 = 0.8\,\mathrm{m} = 0.8 \times 3 = 2.4\,\mathrm{mm} \tag{1}$$

The circular pitch, p, is given by

$$p = \pi m = 3 \times \pi = 9.42\,\mathrm{mm} \tag{2}$$

The base circular pitch is given by

$$p_b = p \cos \phi = 9.4248 \times \cos 20^\circ = 8.8564\,\mathrm{mm} \tag{3}$$

The pitch radius of the pinion is given by

$$r_1 = \frac{mN_1}{2} = \frac{3 \times 35}{2} = 52.5\,\mathrm{mm} \tag{4}$$

The pitch radius of the gear is given by

$$r_2 = C - r_1 = 320 - 52.5 = 267.5\,\mathrm{mm} \tag{5}$$

From Eq. (3.9), the contact ratio is given by

$$\alpha = \frac{\sqrt{(R_1 + a)^2 - (R_1 \cos \phi)^2} - R_1 \sin \phi + \sqrt{(R_2 + a)^2 - (R_2 \cos \phi)^2} - R_2 \sin \phi}{p_b}$$

$$= \frac{\sqrt{(52.5 + 2.4)^2 - (52.5 \cos 20)^2} - 52.5 \sin 20 + \sqrt{(267.5 + 2.4)^2 - (267.5 \cos 20)^2} - 267.5 \sin 20}{8.8564}$$

$$= 1.46 \tag{3.6}$$

3.9 Conclusions

Spur gears have been discussed in this chapter. The basic objective was not to provide an exhaustive discussion on spur gears. There are many aspects of spur gears related to their design and manufacture that can be found in the references. Only relevant portions that are required for analyzing mechanisms that have gear elements were discussed here. The discussions in this chapter can be easily extended to study helical, bevel gears in Chap. 4, and in addition, gear trains in Chap. 5.

3.10 Exercise Problems

3.1 A gear pair has a pinion of 30 teeth and a gear of 80 teeth and if the center distance is 100 mm, what is the module of the gear set?

3.2 A gear has a pitch diameter of 132 mm with a module of 4 mm/tooth. What is the number of teeth and the circular pitch?

3.3 A 35-teeth gear has a circular pitch of 30 mm/tooth. Find the module and its pitch diameter.

3.4 A gear pair has a pinion of 80 mm diameter and a gear of 200 mm diameter. If the module is 1.25 mm/tooth, determine the number of teeth on the pinion and the gear.

3.5 A gear pair has a pinion having 32 teeth and has a mating gear with a center distance of 200 mm and the gear is running at 500 rpm. Determine the pitch diameters of both the gears that have a module of 4 mm/tooth and the speed of the pinion. In addition, determine the number of teeth on the gear.

3.6 A gear pair of 5 mm/tooth module has 28 teeth on the pinion and 40 teeth on the gear and has an involute profile of 20° pressure angle. Determine the addendum, dedendum, clearance, circular pitch, base pitch, base circle radii, tooth thickness, lengths of approach and recess paths, and contact ratio.

3.7 A gear pair consisting of pinion and gear has a reduction ratio of 4, whose parallel shafts are at a distance of 125 mm. If the module of the gears is 2.5 mm/module, determine the pitch diameter and number of teeth on each of the gears.

3.8 A 6 mm/tooth module pinion has 20 teeth. Find the thickness of the tooth at the base circle and at the addendum circle.

3.9 A gear pair consisting of pinion and gear has a reduction ratio of 5, whose parallel shafts are at a distance of 96 mm. If the module of the gears is 2 mm/tooth, determine the pitch diameter and number of teeth on each of the gears. In addition, determine the pitch line velocity.

3.10 A gear pair consisting of pinion and gear has a circular pitch of 12.5664 mm. If the gear has 70 teeth more than the pinion and the center distance between the pinion and the gear is 340 mm, determine the number of teeth on each of them. If the pitch line velocity is limited to 5000 mm/s, determine the rotational speeds per minute of the pinion and gear.

3.11 A gear pair consisting of pinion and gear has a module of 4 mm/tooth, angular velocity ratio of 4, and 40 teeth on the pinion. What are the tooth thicknesses at the respective base circles?

3.12 A gear pair consisting of pinion and gear has a module of 4 mm/tooth, a pressure angle of 20°, and a center distance of 300 mm; there are 40 teeth on the pinion. Calculate the contact ratio.

3.13 A gear pair consisting of pinion and gear has a module of 4 mm/tooth, a pressure angle of 20°, and a center distance of 300 mm; there are 40 teeth on the pinion. If the pinion is running at 1500 rpm, determine the extreme sliding velocities.

3.14 A gear pair consisting of pinion and gear has a module of 4 mm/tooth, pressure angle 20° (standard full-depth gear) has 50 teeth on the gear. What should be the minimum number of teeth on the pinion?

3.15 A geared system is shown in Fig. P3.15. It consists of external gears 1 and 2 and an internal gear 3, and gear 2 is an idler. The axes of gears 1 and 3 coincide. All the gears have a module of 4 mm/tooth. The gears transmit a power of 30 kW and gear 1 has a rotational speed of 1500 rpm. Determine (a) the center distances between the gears, (b) rotational speeds of gears 2 and 3, (c) torque transmitted by each gear, and (d) forces acting on each gear.

3.16 A geared system is shown in Fig. P3.16. All the gears have a module of 4 mm/tooth. The gears transmit a power of 30 kW and gear 1 has a rotational speed of 1500 rpm. Determine (a) the center distances between the gears, (b) rotational speeds of gears 2, 3, and 4, (c) torque transmitted by each gear, and (d) forces acting on each gear.

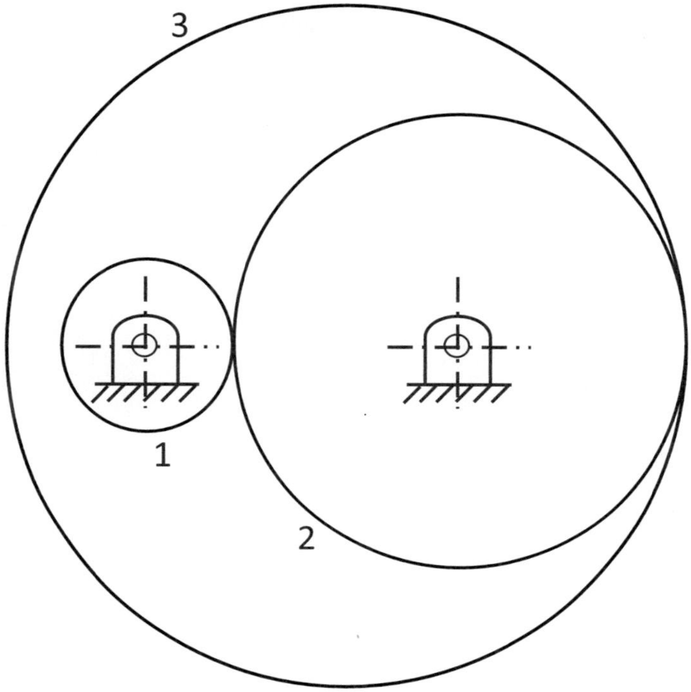

Fig. P3.15 Gear train of exercise problem 3.15

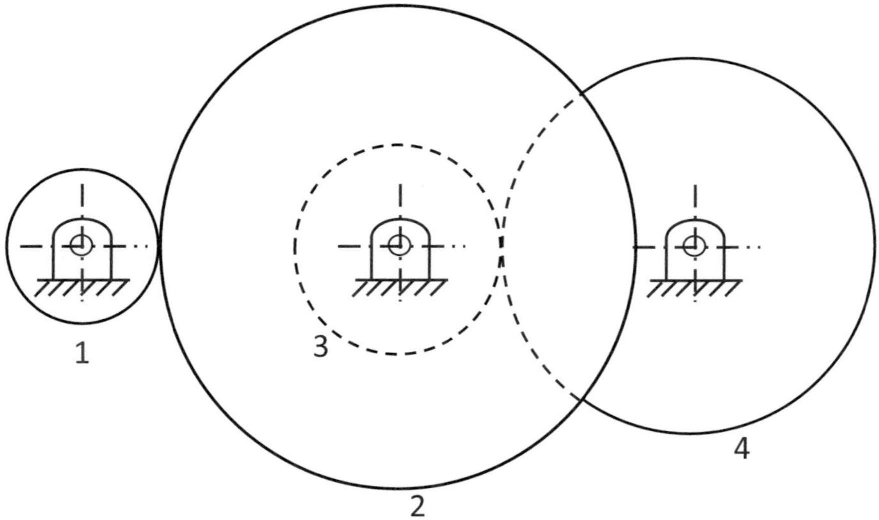

Fig. P3.16 Gear train of exercise problem 3.16

Chapter 4
Helical, Bevel, and Worm Gears

4.1 Introduction

In Chap. 3, spur gears were discussed that represent the class of mechanisms that are used as positive drives. However, spur gears, as their teeth are parallel to their rotation axis, can only be used in speed reduction between parallel shafts. In addition, spur gears have high noise and vibration. A slight modification of spur gears in which the teeth are inclined to the gear rotation axis results in helical gears. Helical gears not only reduce noise and vibration, they provide some flexibility in connecting nonintersecting shafts that are at an angle. However, for both spur and helical gears, the teeth are cut on a cylindrical surface; whereas, if the teeth are cut on a conical surface, a new class of gears result that are known as bevel gears. Bevel gears are extensively used for speed reduction between intersecting shafts. Furthermore, when very high-speed reduction ratios are required, a screw-like component known as worm meshing with a toothed wheel is used. They are known as worm gears that find applications in lifts and hoists that need self-locking capability. Except helical gears, bevel and worm gears are not that efficient but provide higher flexibility in various applications. The above class of gears is discussed in the following sections.

4.2 Helical Gears

Helical gears are a result of the improvements on spur gears to increase efficiency and to reduce vibration and noise at high speeds. The term helix refers to the three-dimensional curve on a cylinder; the helical gear teeth inclined to the gear axis resemble a portion of that helix. Hence, the basic principle of cutting helical teeth from a cylindrical blank remains the same similar to spur gears, but the teeth are cut at an angle inclined to the gear rotation axis This not only increases the average number of teeth that are in contact with the mating gears but also changes the manner in which

D. N. Manik, *Fundamentals of Mechanisms and Machines*,
https://doi.org/10.1007/978-981-97-1810-8_4

contact between mating gears occurs; whereas the spur gear teeth always have a line contact of the same length throughout the face width between the mating set of teeth, for helical gears the contact begins as a point and then becomes a line of increasing length that will help in gradual engagement, leading to improved load distribution. This action also helps in averaging out the surface imperfections in helical gears that are inherent in any manufacturing process, thus leading to the decreased vibration and associated noise. Furthermore, helical gears have the advantage of using them for either parallel or crossed-axis shafts, which provides flexibility in their application. The only unintended consequence of using helical gears is the resulting axial thrust that will require bearings that can withstand axial forces, in addition to the radial forces. However, helical gears with a double helix, also known as herringbone gears, can be used if the axial thrust becomes large for higher helix angles.

If the teeth are cut such that their helix angle is counterclockwise with respect to the rotation axis, they are called as left-handed gears; if they make a clockwise angle, they are called as right-handed gears. The mating helical gears should be of opposite hand in case of parallel helical gears, which are discussed first; such helical gears have non-intersecting rotation axes and their faces are in the same plane. On the other hand, cross-helical gears, although they have non-intersecting rotation axes, the rotation axes can make an angle with respect to each other. They are known as crossed helical gears that can have the same or opposite hands of their helix angles.

4.2.1 Profile of Helical Gears

First, we shall discuss parallel axis helical gears in which the rotation axes of mating gears are parallel to each other. As already discussed in Chap. 3, the involute profile of a spur gear can be generated by unwrapping a tightly wound paper around a cylinder by a line *ab* parallel to the axis of the cylinder, as shown in Fig. 4.1. Similarly, the profile of a helical gear can be generated by unwinding a tightly wound paper around a cylinder by a line cd, inclined at an angle ψ to the axis of the cylinder. The surface obtained when every point on the line cd generates an involute surface called involute helicoid. The wrapping and unwrapping from a base of one cylinder to the base of another cylinder generate two involute helicoids that mesh as conjugate tooth shapes, similar to a spur gear.

The above involute helicoid tooth profile of helical gears ensures that the contact between the mating teeth of a pair helical gears starts as a point and then becomes a line. This results in gradual engagement, smooth transfer of load and averaging the inaccuracy of the surface during tooth manufacture, thus resulting in lower vibration and noise.

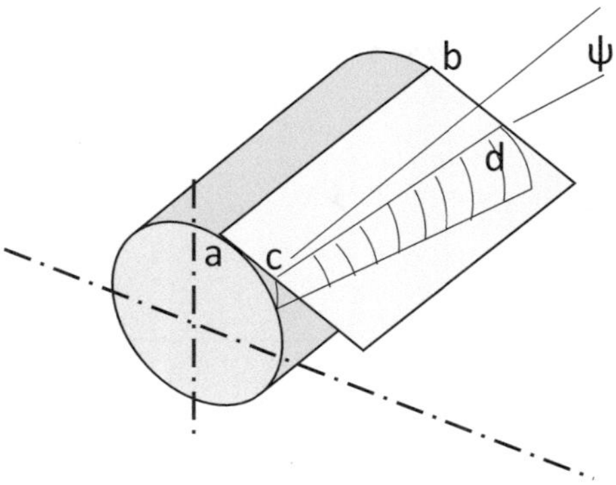

Fig. 4.1 Generation of involute profile for helical teeth

4.2.2 Helical Gear Tooth Parameters

The helical gear teeth are inclined at an angle ψ, known as helix angle, with respect to a line perpendicular to its axis of rotation as shown in Fig. 4.2a. The sections of this gear along a transverse direction AA and another along a direction BB perpendicular to the gear teeth are also shown in this figure. The transverse pitch, p_t, measured between corresponding tooth positions along the side face of the gear is as shown. In section AA, the transverse pressure angle, φ_t, is shown. In section AA, the normal pressure angle, φ_n, is shown. The axial pitch, p_x, and its relation to the helix angle is shown in Fig. 4.2b.

The normal and transverse circular pitch is related by

$$p_n = p_t \cos \psi \tag{4.1}$$

The points C and D marked on Fig. 4.2 represent the transverse pitch. If the point C is extended vertically up and D along the helix, as shown in Fig. 4.2b, the relationship between the axial and transverse circular pitch can be obtained as

$$p_x = \frac{p_t}{\tan \psi} \tag{4.2}$$

The normal diametral pitch, P_n, and the normal circular pitch, p_n, are related by

$$p_n P_n = \pi \tag{4.3}$$

The transverse diametral pitch, P_t and the transverse circular pitch, p_t, are related by

$$p_t P_t = \pi \tag{4.4}$$

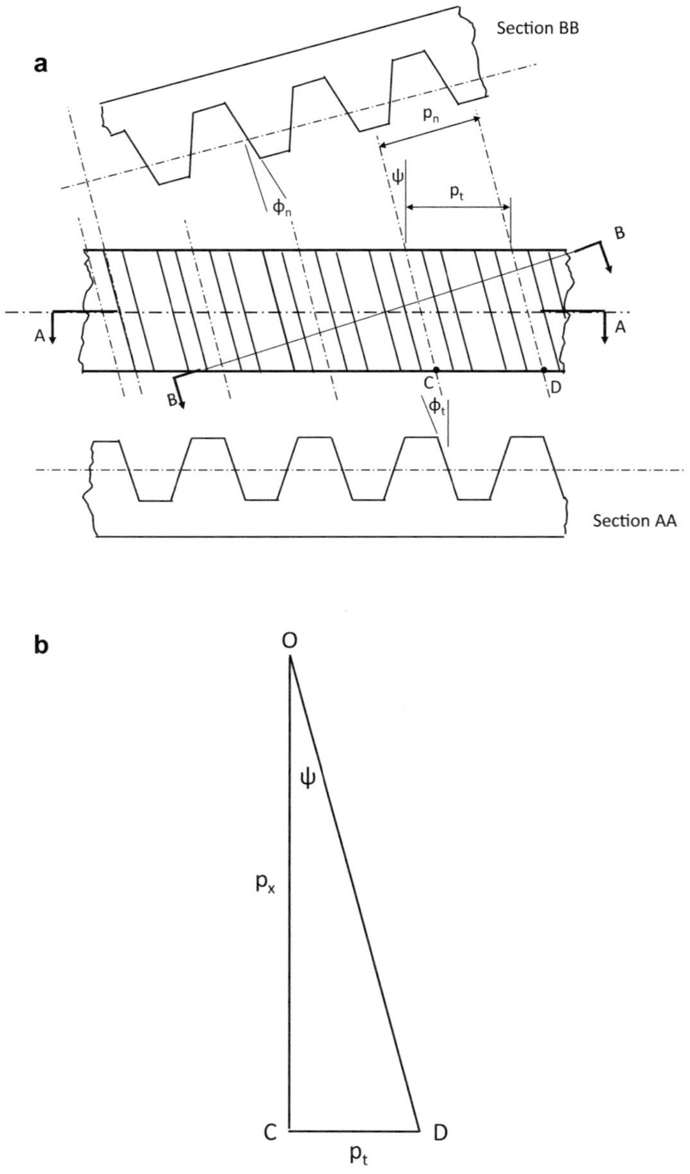

Fig. 4.2 Helical gear **a** cross-section **b** transverse and circular pitch

From Eqs. (4.3) and (4.4), the normal diametral pitch and the transverse diametral pitch are related by

$$P_n = \frac{P_t}{\cos \psi} \tag{4.5}$$

4.2.3 Equivalent Pitch Radius

Consider a helical gear of helix angle ψ and pitch diameter D as shown in Fig. 4.3. Draw a line passing through intersection of the axis and the center line of the face width inclined at the helix angle ψ. Extend the helical gear at the pitch diameter to meet at A and B. The cylinder through A and B represents the base cylinder that has a larger diameter cylinder than the actual gear and the plane of AB represents an ellipse of semimajor axis $OQ = \frac{D}{2\cos\psi}$ and semi-minor axis OP. Let R_e be the radius of the arc that tangentially passes through the point P that represents the pitch point of an equivalent spur gear.

Equation for the ellipse of Fig. 4.3 is given by

$$\left(\frac{x}{OQ}\right)^2 + \left(\frac{y}{OP}\right)^2 = 1 \tag{4.6}$$

where the x and y axes are aligned along the axes of the ellipse as shown.

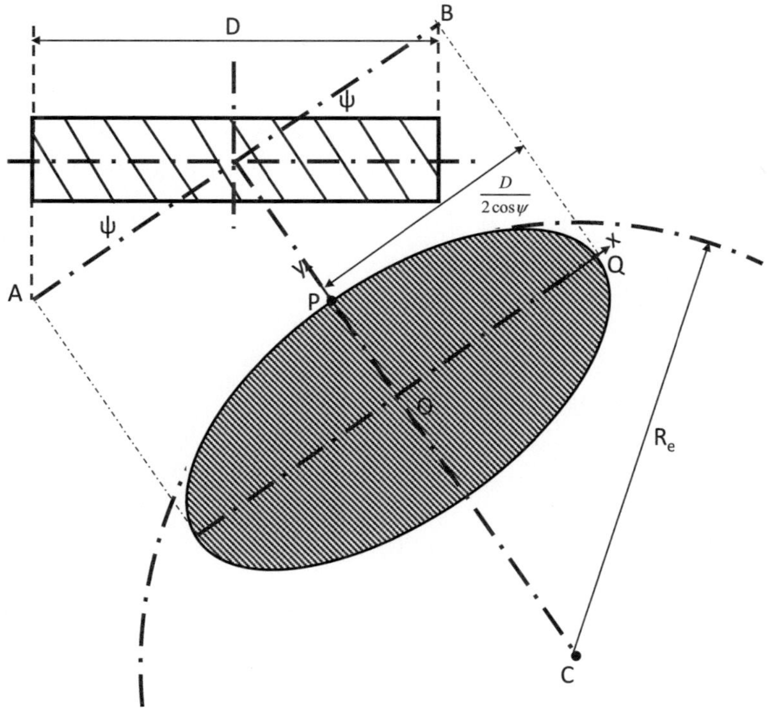

Fig. 4.3 Equivalent pitch radius of a spur gear

From Eq. 4.6,

$$y = OP\sqrt{1 - \left(\frac{x}{OQ}\right)^2} \tag{4.7}$$

The radius of curvature R_e is given by

$$R_e = \frac{\left[1 + \left(\frac{dy}{dx}\right)^2\right]^{1.5}}{\frac{d^2y}{dx^2}} \tag{4.8}$$

Applying Eqs. (4.7) and (4.8) at the intersection of the ellipse and the equivalent radius arc at P, the equivalent pitch radius of a spur gear

$$R_e = \frac{OQ^2}{OP} \tag{4.9}$$

Using $OQ = \frac{D}{2\cos\psi}$ and OP = D/2, the equivalent pitch radius becomes

$$R_e = \frac{OQ^2}{OP} = \frac{\left(\frac{D}{2\cos\psi}\right)^2}{\frac{D}{2}} = \frac{D}{2(\cos\psi)^2} \tag{4.10}$$

From Eqs. (4.5) and (4.10), the equivalent number of spur gear teeth is given by

$$N_e = 2R_eP_n = 2\left[\frac{D}{2(\cos\psi)^2}\right]\frac{P_t}{\cos\psi} = \frac{N}{\cos^3\psi} \tag{4.11}$$

4.2.4 Angle Relationships

Although analyzing the helical forces is not the main objective, it is relevant to present discussion on the angle relationships related to the forces.

Figure 4.4 shows a helical gear of helix angle ψ and the various forces acting at a point on one of its teeth. F_t is the tangential force acting on the gear that is responsible for transmitting power, F_a is the axial force resulting due to the inclination of the teeth, F_s is the separation force that tends to move away the mating gears and F_n is the normal force.

The transverse pressure angle, ϕ_t, is given by

$$\tan\phi_t = \frac{F_s}{F_t} \tag{4.12}$$

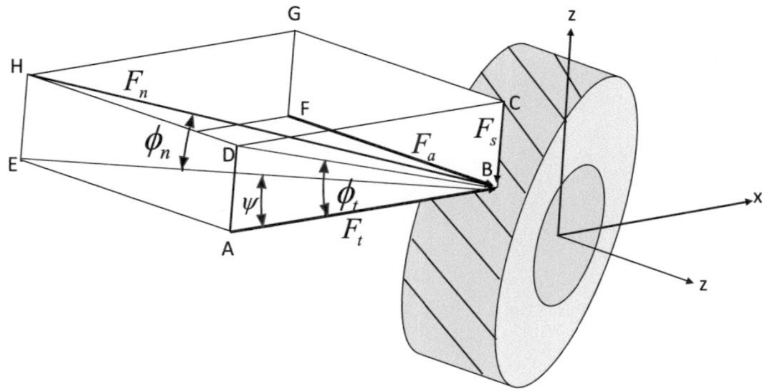

Fig. 4.4 Helical gear forces and angles

The normal pressure angle, ϕ_n, is given by

$$\tan \phi_n = \frac{F_s}{OE} \tag{4.13}$$

From Fig. 4.4, OE is given by

$$OE = \frac{F_t}{\cos \psi} \tag{4.14}$$

From Eqs. (4.12) to (4.14)

$$\tan \phi_n = \tan \phi_t \cos \psi \tag{4.15}$$

4.2.5 Crossed Helical Gears

The helical gears we discussed thus far are called parallel axis helical gears in which the axes of rotation of the two gears are parallel and their faces are in the same plane. Sometimes, the faces of the gears are in different planes although their axes are non-intersecting, when their shafts at an angle. They are known as cross helical gears as shown in Fig. 4.5. The only condition is that they must have the same normal pitches and modules, their helix angles can be different and as mentioned earlier, they can be either left-handed or right-handed.

Consider a pair of crossed helical gears shown in Fig. 4.5 and they have different helix angles, ψ_1 and ψ_2. The axes of their shafts are inclined at an angle of Σ. Their angular velocities, number of teeth, pitch diameters, and helix angles are related by

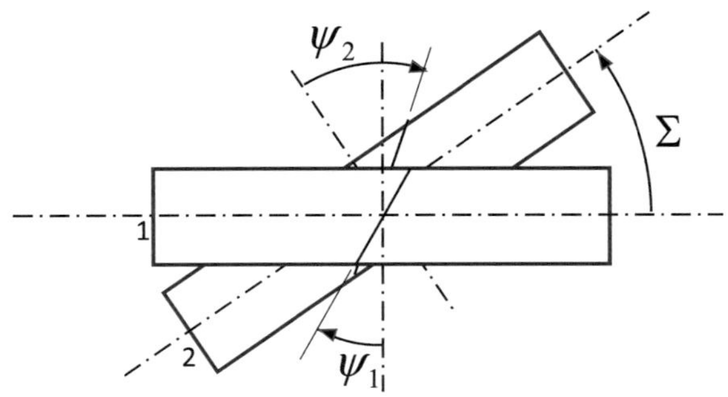

Fig. 4.5 Crossed helical gears

$$\frac{\omega_1}{\omega_2} = \frac{N_2}{N_1} = \frac{D_2 \cos \psi_2}{D_1 \cos \psi_1} \tag{4.16}$$

Σ is the angle between the shafts of the crossed helical gears, it is related to their helix angles as follows:

$$\Sigma = \psi_1 \pm \psi_2 \tag{4.17}$$

The positive sign is used when both the crossed helical gears have the same hand (left or right), negative for opposite hand. Sometimes, 2 and 3 are used as subscripts in Eq. 4.17. In any case, they should be understood as helix angles of the mating gears.

It should be noted that although crossed helical gears have the advantage of maintaining an angle between non-intersecting shafts, they cannot take very high loads since they have a point contact.

Example 4.1

A pair of crossed helical gears of the same hand are mounted on shafts that make an angle of 75°. There are 17 teeth on the pinion that have a helix angle of 40°. If both the pinion and gear have a normal module of 5 mm/tooth, and the angular velocity ratio is 5, determine (a) the number of teeth on the gear (b) pitch diameters of the pinion and gear (c) shaft center distance.

The number of teeth on the helical gear is given by

$$N_3 = N_2 \times \text{velocity ratio} = 17 \times 5 = 85 \tag{1}$$

The helix angle of the pinion, $\psi_2 = 40°$ and the angle between the shafts, $\Sigma = 75°$. Hence, from Eq. 4.17, the helix angle of the gear is given by

$$\psi_3 = \Sigma - \psi_2 = 70° - 40° = 30° \tag{2}$$

Pitch diameter of the pinion is given by

$$D_2 = \frac{m_n N_2}{\cos\psi_2} = \frac{5 \times 17}{\cos 40} = 110.96\,\text{mm} \tag{3}$$

Pitch diameter of the gear is given by

$$D_3 = \frac{m_n N_3}{\cos\psi_3} = \frac{5 \times 85}{\cos 30} = 518.83\,\text{mm} \tag{4}$$

The center distance between the gears is given by

$$C = \frac{D_1 + D_2}{2} = \frac{110.96 + 518.83}{2} = 314.9\,\text{mm} \tag{5}$$

4.3 Bevel Gears

Contrary to spur and helical gears, the teeth of a bevel gear are cut on a conical surface. This helps in using them in applications for speed reduction between intersecting shafts. The angle between the gear axis and teeth is known as pitch angle or cone angle that is an important parameter of bevel gears. There are two types of bevel gears: straight bevel gears that have straight teeth and taper toward the apex of the cone and spiral bevel gears that have teeth in a spiral manner around the cone. Straight bevel gears are used for intersecting shafts at a single point whereas spiral bevel gears are used when the axes of rotation are offset. Bevel gears operate smoothly but they are much less efficient than spur and helical gears, and are more complex to manufacture. Bevel gears are generally used in differentials of automobiles. All the discussions in this section are relevant only for straight bevel gears.

4.3.1 Bevel Gear Parameters

Figure 4.6 shows two mating bevel gears with intersecting shafts at 90°. The shaft angle $\sum$ is the angle between intersecting shafts, which is equal to 90° in the present case. The pitch surface of the mating gears is a cone and the cone angle of each gear is known as the pitch angle. However, although the cone angles are different, both the cones have a common apex. The cones roll together without slipping and have spherical motion. In addition, each point on these gears remains at a constant distance from the common apex.

Because the pitch cones roll together without slipping, the angular velocity ratio is inversely proportional to the diameter of the base of the cones. Hence, the base diameter of these cones become the pitch diameters. Very similar to spur gears, the

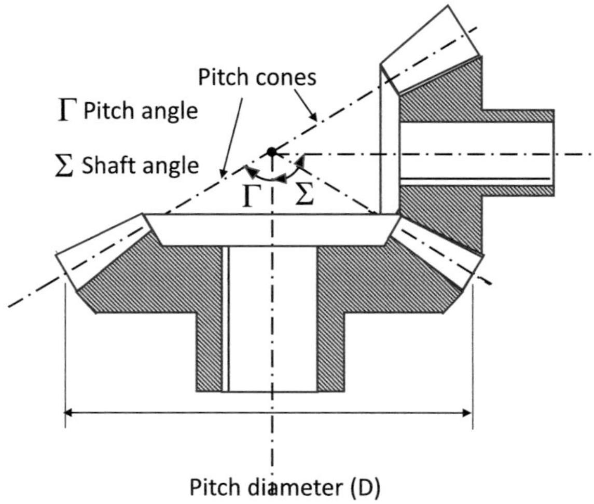

Fig. 4.6 Bevel gears at 90° shaft angle

module of each gear of the bevel set is a ratio of the pitch diameter and the number of teeth.

Hence, the module of both bevel gears is the same and it is given by

$$m = \frac{D}{N} \tag{4.18}$$

As discussed before, the pitch cones of a pair of bevel gears have spherical motion. Due to difficulties in accurately accounting for this spherical motion, an approximation can reduce this problem to that of spur gears, known as Tredgold's approximation. In this approximation, a back cone is formed that is tangential to the spherical surface formed by both the gears. However, only one of the gears is shown in Fig. 4.7 that should be sufficient to explain the concept of an equivalent pitch diameter and number of teeth of bevel gear with respect to a spur gear.

From Fig. 4.7, the equivalent pitch diameters of the pinion and gear of a bevel gear set can be expressed as

$$D_{e_2} = \frac{D_2}{\cos\Gamma_2}; D_{e_3} = \frac{D_3}{\cos\Gamma_3} \tag{4.19}$$

Based on the above equivalent pitch diameters, the equivalent number of teeth on pinion and gear of a bevel gear set can be expressed as

$$N_{e_2} = \frac{\pi D_2}{p_c}; N_{e_3} = \frac{\pi D_3}{p_c} \tag{4.20}$$

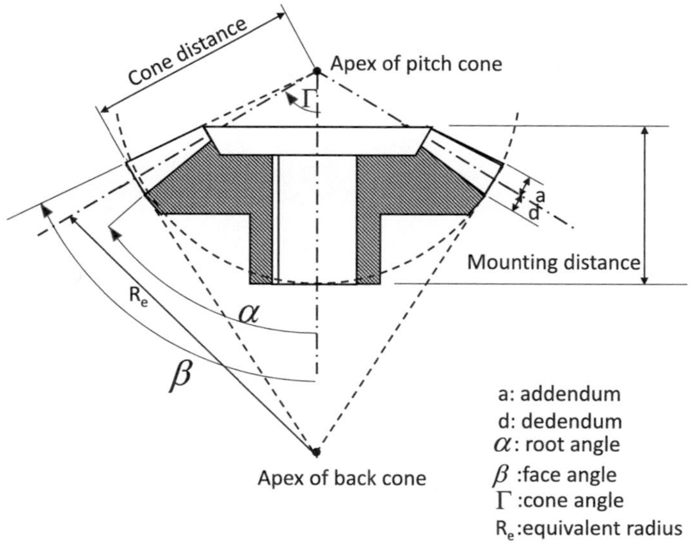

Fig. 4.7 Back cone of a bevel gear

4.3.2 Shaft and Pitch Cone Angles

Consider the pitch cones of two mating bevel gears of shafts shown in Fig. 4.8 with an intersecting angle other than 90°. $\sum$ is the angle between the intersecting shafts and Γ_2, Γ_3 are respectively the pitch cone angles of the pinion and gear; D_2 and D_3 are, respectively, the pitch diameters of the pinion and gears; N_2 and N_3 are respectively the number of teeth on the pinion and gear.

The absolute ratio of the angular velocities of gear and pinion is given by

$$\left|\frac{\omega_3}{\omega_2}\right| = \frac{D_2}{D_3} = \frac{N_2}{N_3} \tag{4.21}$$

Q is the point at which the base of two pitch cones of the pinion and gears meet, and O is the point is the common apex of the two pitch cones. The distance OQ is related to the ratio of pitch diameter and pitch angles of the mating pinion and gear as follows:

$$OQ = \frac{D_2}{2\sin\Gamma_2} = \frac{D_3}{2\sin\Gamma_3} \tag{4.22}$$

Since OQ is common to both the pinion and gear

$$\sin\Gamma_2 = \frac{D_2}{D_3}\sin\Gamma_3 = \frac{D_2}{D_3}\sin(\Sigma - \Gamma_2) \tag{4.23}$$

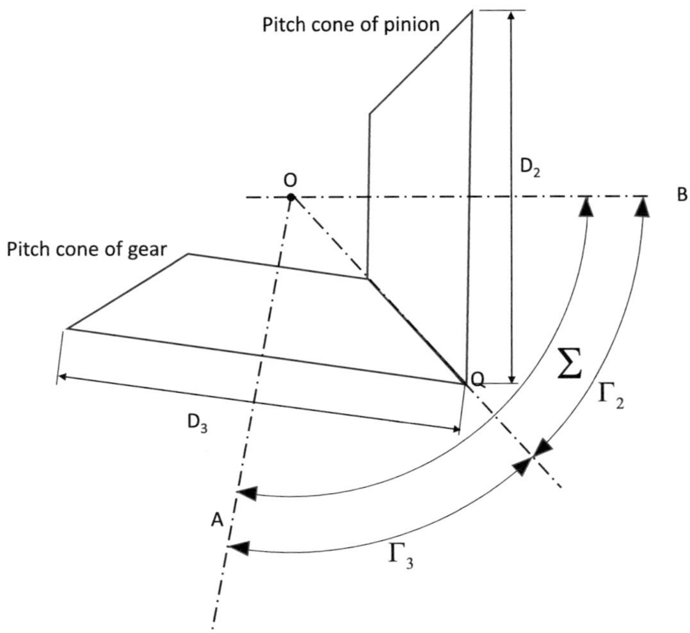

Fig. 4.8 Pitch cones of mating bevel gears for a general shaft intersecting shaft angle

Expanding the third term of Eq. (4.23)

$$\sin\Gamma_2 = \frac{D_2}{D_3}(\sin\Sigma\cos\Gamma_2 - \cos\Sigma\sin\Gamma_2) \tag{4.24}$$

Dividing both sides of Eq. (4.24) by $\cos\Gamma_2$

$$\frac{\sin\Gamma_2}{\cos\Gamma_2} = \tan\Gamma_2 = \frac{D_2}{D_3}(\sin\Sigma - \cos\Sigma\tan\Gamma_2) \tag{4.25}$$

Collecting terms of $\tan\Gamma_2$, Eq. (4.25) becomes

$$\tan\Gamma_2\left(1 + \frac{D_2}{D_3}\cos\Sigma\right) = \frac{D_2}{D_3}\sin\Sigma \tag{4.26}$$

Hence, Eq. (4.26) becomes

$$\tan\Gamma_2 == \frac{\sin\Sigma}{\frac{N_3}{N_2} + \cos\Sigma} \tag{4.27}$$

Similarly,

$$\tan \Gamma_3 = \frac{\sin \Sigma}{\frac{N_2}{N_3} + \cos \Sigma} \tag{4.28}$$

For a shaft angle $\sum = 090°$, Eq. (4.27) becomes

$$\tan \Gamma_2 = \frac{N_2}{N_3} \tag{4.29}$$

Similarly

$$\tan \Gamma_3 = \frac{N_3}{N_2} \tag{4.30}$$

Example 4.2

Two straight-tooth bevel gears that have a shaft angle of 120° are to be designed that will have 20 teeth on the pinion and 35 teeth on the gear and a normal module of 5 mm/tooth, determine the pitch angles, cone distance and equivalent pitch diameter of spur gears.

From Eq. (4.27), pitch cone angle of the pinion is given by

$$\Gamma_2 = \tan^{-1}\left(\frac{\sin \Sigma}{\frac{N_3}{N_2} + \cos \Sigma}\right) = \tan^{-1}\left(\frac{\sin 120}{\frac{35}{20} + \cos 120}\right) = 34.72° \tag{1}$$

The pitch cone angle of the gear is given by $\Gamma_3 = \Sigma - \Gamma_2 = 120° - 34.72° = 85.28°$.

As a check, from Eq. (4.28), the pitch cone angle of the gear is given by

$$\tan \Gamma_3 = \tan^{-1}\left(\frac{\sin \Sigma}{\frac{N_2}{N_3} + \cos \Sigma}\right) = \tan^{-1}\left(\frac{\sin 120°}{\frac{20}{35} + \cos 120°}\right) = 85.28° \tag{2}$$

The pitch diameter of the pinion is given by

$$D_2 = mN_2 = 5 \times 20 = 100\,\text{mm} \tag{3}$$

The pitch diameter of the gear is given by

$$D_3 = mN_3 = 5 \times 35 = 175\,\text{mm} \tag{4}$$

From Eq. (4.22), the cone distance is given by

$$OQ = \frac{D_2}{2 \sin \Gamma_2} = \frac{100}{2 \sin 34.72°} = 87.79\,\text{mm} \tag{5}$$

From Eq. (4.19), the equivalent pitch diameter of a spur gear pinion is given by

$$D_{e_2} = \frac{D_2}{\cos\Gamma_2} = \frac{100}{\cos 34.72^\circ} = 121.66\,\text{mm} \quad (6)$$

and, the equivalent pitch diameter of a spur gear is given by

$$D_{e_3} = \frac{D_3}{\cos\Gamma_3} = \frac{175}{\cos 85.28^\circ} = 212.7\,\text{mm} \quad (7)$$

4.4 Worm Gears

Worm gears, as they are popularly called, have two components. The driving component, which is screw like, called the worm that mates with a driven component that is a toothed gear, which could be either a spur or helical gear. It has a very high reduction ratio in a compact space and the planes of rotation of the worm and the worm gear are perpendicular to each other. Although they are much less efficient than spur gears, its compactness and self-locking capability is a great advantage in high torque low-speed applications such as lifts and hoists.

Consider a worm shown in Fig. 4.9a of axial pitch p_x lead angle λ, helix angle ψ and pitch diameter D.

The lead of the worm, L, the axial pitch, p_x and the number of threads, N, wrapped on the pitch cylinder of the worm are related by

$$L = p_x N \quad (4.31)$$

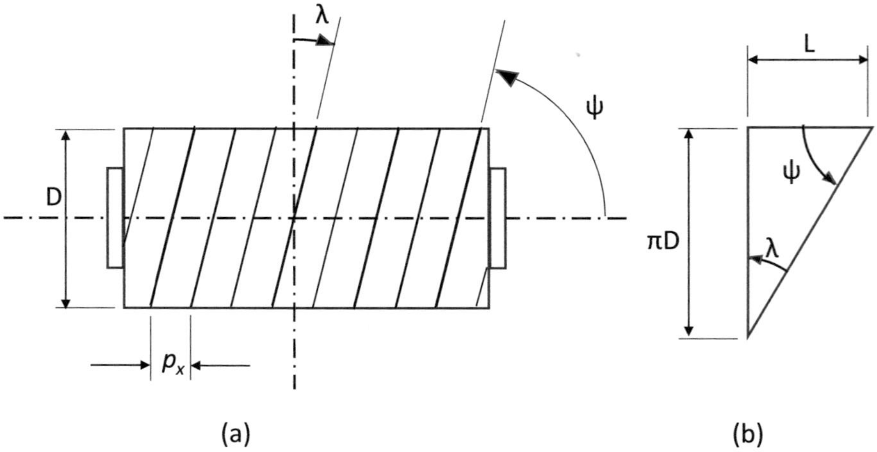

Fig. 4.9 **a** Worm **b** Lead and pitch

A worm can have up to 10 threads and it is similar to having a number of teeth on pinion. The threads will mesh with the corresponding teeth on the worm gear, similar to a spur gear set. Lower number of threads will increase the velocity ratio but cannot take large loads, whereas larger number of threads decreases the velocity ratio but can take large loads.

If a complete revolution of a worm and worm gear is unwrapped, a triangle as shown in Fig. 4.9b results. The pitch diameter, lead, and lead angle are thus related by

$$\tan\lambda = \frac{L}{\pi D} \tag{4.32}$$

The worm gear is distinctively different from other spur and helical gears in that its face has a concave surface as shown in Fig. 4.10 for a single enveloping worm and worm gear set.

Example 4.3

A worm has two threads and has a lead angle of 30° and an axial pitch 10 mm/tooth and a velocity ratio of 25. Determine the pitch diameters of the worm and worm gear, number of teeth on the worm gear, its pitch diameter and the center distance between the worm and the worm gear.

The lead of the worm is given by

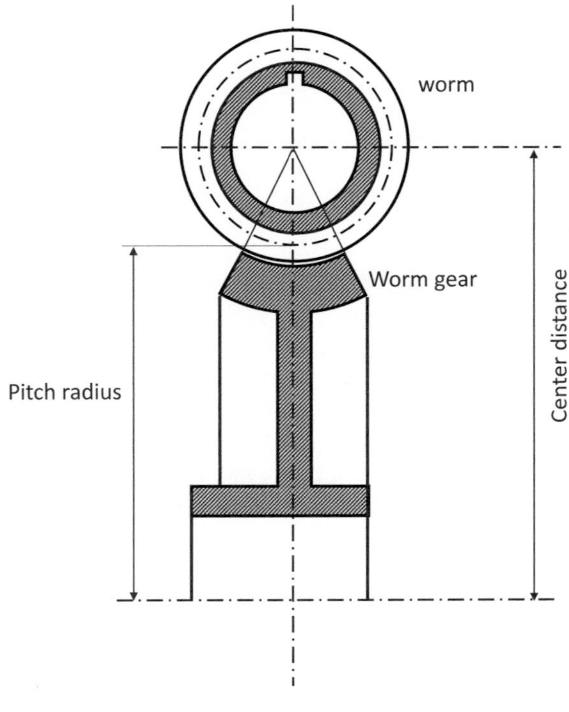

Fig. 4.10 Single enveloping worm and worm gear set

$$L = N_2 p_x = 2 \times 10 = 20\,\text{mm} \tag{1}$$

Pitch diameter of the worm is given by

$$D_2 = \frac{L}{\pi \tan \lambda} = \frac{20}{\pi \tan 30} = 11.027\,\text{mm} \tag{2}$$

Pitch diameter of the gear is given by

$$D_3 = D_2 \times \text{velocity ratio} = 11.027 \times 25 = 275.68\,\text{mm} \tag{3}$$

The center distance between the worm and worm gear is given by

$$C = \frac{D_2 + D_3}{2} = \frac{11.027 + 275.68}{2} = 143.35\,\text{mm} \tag{4}$$

4.5 Conclusions

Helical, bevel, and worm gears have been discussed in this chapter. It was not an exhaustive discussion and only relevant portions as much as required for integrating these elements into mechanisms were discussed. Further information can be obtained from references.

4.6 Exercise Problems

Helical Gears

4.1 Consider a pair of parallel axis helical gears of normal pressure angle of 14.5° and a helix angle of 50° and a module of 5 mm/tooth. There are 17 teeth on the pinion and 28 teeth on the gear. Calculate the following: (a) transverse pitch, (b) normal pitch, (c) normal module, (d) pitch diameters of the pinion and gear, (e) equivalent tooth numbers.

4.2 Consider a pair of parallel axis helical gears of normal pressure angle of 14.5° and a helix angle of 35° and a module of 2 mm/tooth. There are 20 teeth on the pinion and 42 teeth on the gear. Calculate the following: (a) transverse pitch, (b) normal pitch, (c) axial pitch, (d) pitch diameters of the pinion and gear, (e) pitch diameters of the equivalent spur gears.

4.3 Consider a pair of parallel axis helical gears of normal pressure angle of 14.5° and a helix angle of 40° and a module of 4 mm/tooth. There are 20 teeth on the pinion and 42 teeth on the gear and the face width of both the pinion and gear is 25 mm. Calculate the following: (a) transverse pitch, (b) normal pitch, (c) axial pitch, (d) pitch diameters of the pinion and gear, (e) pitch diameters of the equivalent spur gears (f) base helix angle.

4.4 Consider a pair of parallel axis helical gears of transverse pressure angle of 14.5° and a helix angle of 35° and a module of 5 mm/tooth and the center distance between the gears is 100 mm; the angular velocity ratio is 2. There are 20 teeth on the pinion and 42 teeth on the gear. Calculate the following: (a) transverse pitch, (b) normal pitch, (c) axial pitch, (d) pitch diameters of the equivalent spur gears, (e) base helix angle, (f) percentage change in the center distance due to rounding off the number of teeth.

4.5 Consider a pair of parallel axis helical gears of normal pressure angle of 20° and a helix angle of 30°, the center distance between the gears is 300 mm; the pinion has 18 teeth and runs at 2000 RPM whereas the gear runs at 500 RPM. Calculate the following: (a) transverse pitch, (b) normal pitch, (c) axial pitch, (d) gear tooth number, (e) pitch diameters of the equivalent spur gears, (f) percentage change in the center distance due to rounding off the number of teeth on the gear.

4.6 Consider a pair of parallel axis helical gears of normal pressure angle of 20° and a helix angle of 30°. The number of teeth on the pinion and gear are respectively 20 and 45 and the normal module is 5 mm/tooth. Calculate the following: (a) transverse pitch, (b) normal pitch, (c) axial pitch, (d) pitch diameters of the pinion and gear, (e) pitch diameters of the equivalent spur gears, (f) base helix angle.

4.7 A pair of crossed helical gears of the same hand are mounted on shafts that make an angle of 85°. There are 15 teeth on the pinion that has a helix angle of 50°. If both the pinion and gear have a normal module of 4 mm/tooth, and the angular velocity ratio is 4, determine (a) the number of teeth on the gear, (b) pitch diameters of the pinion and gear, (c) pitch diameters of the equivalent spur gears, (d) shaft center distance.

4.8 A pair of crossed helical gears of opposite hand are mounted on shafts that make an angle of 90°. There are 10 teeth on the pinion that has a helix angle of 65°. If both the pinion and gear have a normal module of 3 mm/tooth, and the angular velocity ratio is 8, determine (a) the number of teeth on the gear, (b) pitch diameters of the pinion and gear, (c) pitch diameters of the equivalent spur gears, (d) shaft center distance, (e) helix angle of the gear.

4.9 A pair of crossed helical gears of the same hand are mounted on shafts with a center distance of 300 mm that make an angle of 60°. If both of the pinion and gear have a normal module of 3 mm/tooth and the same helix angle, and the angular velocity ratio is 4, determine (a) the number of teeth on the gear and pinion, (b) pitch diameters of the pinion and gear, (c) pitch diameters of the equivalent spur gears, (d) percentage change in the center distance due to rounding off the number of teeth on the gear.

Bevel Gears

4.10 Two straight-tooth bevel gears that have a shaft angle of 90° are to be designed for an angular velocity ratio of 5. If there are 20 teeth on the pinion, determine the pitch angles and the number of teeth on the gear.

4.11 Two straight-tooth bevel gears that have a shaft angle of 75° are to be designed for an angular velocity ratio of 5. If there are 20 teeth on the pinion, determine the pitch angles and the number of teeth on the gear.

4.12 Two straight-tooth bevel gears that have a shaft angle of 110° are to be designed that will have 18 teeth on the pinion and 38 teeth on the gear, determine the pitch angles.

4.13 Two straight-tooth bevel gears that have a shaft angle of 105° are to be designed that will have 20 teeth on the pinion and 38 teeth on the gear and a normal module of 10 mm/tooth, determine the pitch angles, cone distance and equivalent pitch diameter of spur gears.

4.14 Two straight-tooth bevel gears that have a shaft angle of 115° are to be designed that will have 21 teeth on the pinion and 38 teeth on the gear and a normal module of 4 mm/tooth, determine the pitch angles, cone distance and equivalent pitch diameter of spur gears.

Worm Gears

4.15 A worm has four threads and a lead of 35 mm. If the angular velocity ratio is 10 and the center distance is 60 mm, determine the pitch diameters of the worm and worm gear.

4.16 A worm has three threads and an axial pitch of 14 mm/tooth and a center distance of 150 mm. If the angular velocity ratio is 50, determine the pitch diameters of the worm and worm gear.

4.17 A worm has three threads and an axial pitch 35 mm/tooth and the diameter of the worm gear is 50 mm and has 45 teeth. determine the pitch diameters of the worm and worm gear, lead, and lead angle of the worm.

4.18 A worm has three threads and has a lead angle of 25° and an axial pitch 12 mm/tooth and a velocity ratio of 20. Determine the pitch diameters of the worm and worm gear, number of teeth on the worm gear, its pitch diameter and the center distance between the worm and the worm gear.

4.19 A worm and worm gear have their shafts at 90°, a center distance of 125 mm, angular velocity ratio of 20 and lead angle of the worm of 15°. Considering that worms from 1 to 10 threads are available, specify the acceptable designs.

4.20 A worm and worm gear have their shafts at 90° that have an angular velocity ratio of 10, with an axial pitch of 20 mm and lead angle of the worm of 30°. Calculate the number of teeth on the worm gear, pitch diameters and center distance.

Chapter 5
Gear Trains

Gear trains are a group of gears that can provide rotary motion of the desired direction, speed, torque, and angle of shaft, within a compact space that is desirable in most applications. Hence, they are an integral part of machines in many practical applications. Some well-known applications are automobiles (gearbox and differential), manufacturing machines, airplanes, helicopters, etc. Although all types of gears are used in gear trains (spur, helical, bevel, worm, etc.), we shall, however, focus on spur gears only in this chapter.

First, we shall discuss the following gear types (1) fixed, parallel axis gear trains in which gear elements are rotating about a fixed axis, (2) planetary gear trains in which gears rotate about their axes and also revolve around other gears.

5.1 Fixed, Parallel Axis Gear Trains

In this type of gear train, meshing gears are mounted on parallel shafts that rotate on fixed axes. A series of meshing gears are mounted on the shafts to obtain an appropriate speed and torque at the output shaft when the input shaft is connected to a power source. When different values of reductions are to be obtained as in the case of an automobile, suitable arrangements will have to be made to move the gears so they mesh with the corresponding gears to obtain the desired reduction.

A very important consideration in the discussion of any type of gear train is the concept of *kinematic coefficient.* It can be defined as the ratio of the angular velocities of the last and the first gear of a gear train, given by

$$\alpha = \frac{\omega_L}{\omega_F} \tag{5.1}$$

D. N. Manik, *Fundamentals of Mechanisms and Machines*,
https://doi.org/10.1007/978-981-97-1810-8_5

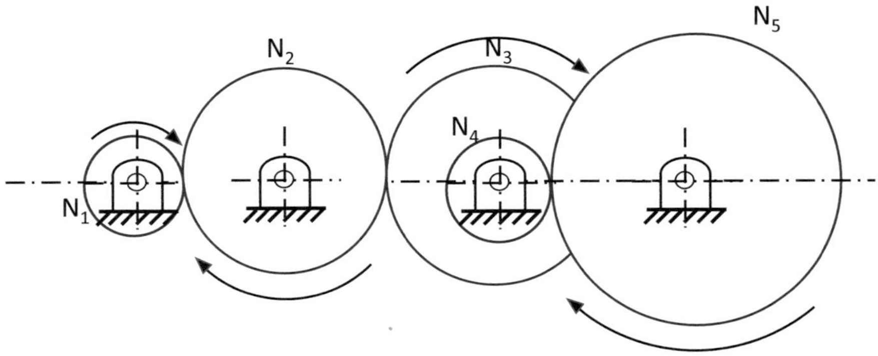

Fig. 5.1 Spur gear train

where, ω_L is the angular velocity of the last gear (output gear), and ω_F is the angular velocity of the first gear (input or driving gear).

To demonstrate the usefulness of the concept of kinematic coefficient, consider a gear train shown in Fig. 5.1. If ω_1 is the angular velocity of the input gear, and ω_5 is the angular velocity of the output gear, the kinematic coefficient of the gear train is given by

$$\alpha = \frac{\omega_L}{\omega_F} = \frac{\omega_5}{\omega_1} = \frac{N_1 N_2 N_4}{N_2 N_3 N_5} = \frac{\text{Driving gears}}{\text{Driven gears}} \tag{5.2}$$

Equation (5.2) gives a convenient form of calculating the kinematic coefficient for any type of gear train.

A typical example of a gear train of the fixed, parallel axis gear train is discussed in Example 5.1. It consists of two pairs of gears mounted on parallel shafts. The input and output gears are mounted on two independent shafts but have a common axis; the intermediate gears are mounted on the same shaft. For a given reduction ratio expected from the gear train, the reduction is generally uniformly distributed between the two stages and the module and number of teeth are chosen such that the center distance of both stages is exactly the same, so the gear train can smoothly operate. Some minor changes in the distribution of gear reduction ratios can be made to ensure exactly the same center distance for all the stages.

Example 5.1

A 350 kW motor running at 3000 RPM is connected to the input of a two-stage speed reduction gearbox using spur gears that is expected to deliver 20,070 N m torque at its output as shown in Fig. 5.2. The module of the teeth is 8 mm for the first stage and 5 mm for the second. The pressure angle is 20° for both gear sets and the minimum number of teeth on both pinions is 18. Determine (a) the number of teeth, pitch circle diameters and center distance (b) output speed and pitch line velocity for both stages.

Let N_{p1}, number of teeth on the first pinion = 18.

Module of the first pinion, m_1 = 8 mm.

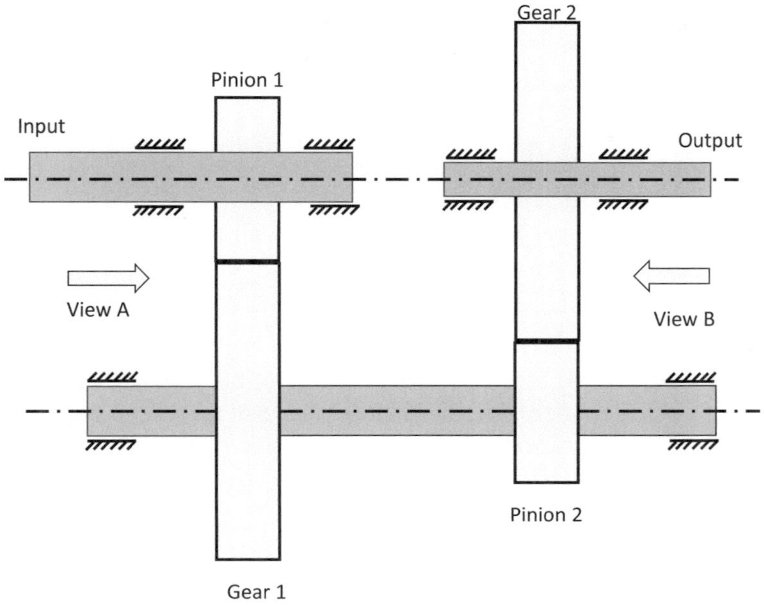

Fig. 5.2 Two-stage gear box

Module of the second pinion, $m_2 = 5$ mm.

The input speed to the gearbox, $N_1 = 3000$ RPM.

The input angular speed to the gearbox, $\omega_1 = \frac{2\pi N_1}{60} = \frac{2\pi 3000}{60} = 314\,\text{rad/s}$.

Torque input to the gearbox is given by, $T_{in} = \frac{P}{\omega_1} = \frac{350{,}000}{314} = 1115\,\text{N m}$.

Torque at the output shaft, $T_{out} = 20{,}070$ N m.

Total reduction ratio $= \frac{T_{out}}{T_{in}} = \frac{20{,}070}{1115} = 18$.

The above reduction ratio can be equally divided between the two stages as $n_1 = n_2 = \sqrt{18} = 4.2426$.

n_1: speed reduction ratio of the first stage.

n_2: speed reduction ratio of the second stage.

Based on the number of teeth on the first pinion and the modules of both pinions, the number of teeth on the second pinion is given by

$$N_{p2} = \frac{N_{p1} m_1}{m_2} = \frac{18 \times 8}{5} = 29 \tag{1}$$

The number of teeth on the first gear is given by

$$N_{g1} = n_1 N_{p1} = 4.2426 \times 18 = 76 \tag{2}$$

The number of teeth on the second gear is given by

$$N_{g2} = n_2 N_{p2} = 4.2426 \times 29 = 123 \tag{3}$$

The center distance between the gears of the first stage is given by

$$c_1 = \frac{N_{p1} m_1 (1 + n_1)}{2} = \frac{18 \times 8(1 + 4.2426)}{2} = 377.47\,\text{mm} \tag{4}$$

The center distance between the gears of the second stage is given by

$$c_2 = \frac{N_{p2} m_2 (1 + n_2)}{2} = \frac{29 \times 5(1 + 4.2426)}{2} = 380.09\,\text{mm} \tag{5}$$

From Eqs. (4) and (5), since the center distance of the two stages deviates from each other, which is not desirable, the number of teeth of the second gear is adjusted so the center distances of both stages will be the same. This may result in a slight change of the reduction ratio of the second stage, which should be acceptable within reasonable limits.

$$c_2 = c_1 = 377.47 = \frac{N_{p2} m_2 (1 + n_2')}{2} = \frac{29 \times 5(1 + n_2')}{2}\,\text{mm} \tag{6}$$

where n_2' is the revised reduction ratio of the second stage that can be obtained by solving Eq. (6) as 4.2065. Hence, Eq. (3) has to be revised to obtain the number of teeth of the second stage gear as, $N_{g2} = n_{22}' N_{p2} = 4.2065 \times 29 = 122$.

The center distance of the second stage gears can be verified as follows:

$$c_2 = \frac{N_{p2} m_2 (1 + n_2')}{2} = \frac{29 \times 5(1 + 4.2065)}{2} = 377.47\,\text{mm} \tag{7}$$

Equations (4) and (7) confirm that both stages of the gearbox have the same center distance (Table 5.1).

Diameter of the first stage pinion is given by

$$d_{p1} = m_1 N_{p1} = 8 \times 18 = 144\,\text{mm} \tag{8}$$

Diameter of the first stage gear is given by

$$d_{g1} = m_1 N_{g1} = 8 \times 76 = 608\,\text{mm} \tag{9}$$

Table 5.1 Configuration of gears and pinions of Example 5.1

	Stage I		Stage II	
	Teeth	Module	Teeth	Module
Pinion	18	8	29	5
Gear	76	8	122	5

Diameter of the second stage pinion is given by

$$d_{p2} = m_2 N_{p2} = 5 \times 29 = 145\,\text{mm} \tag{10}$$

Diameter of the second stage gear is given by

$$d_{g2} = m_2 N_{g2} = 5 \times 122 = 610\,\text{mm} \tag{11}$$

RPM of the output shaft is given by

$$N_2 = \frac{N_1}{n_1 n_2'} = \frac{3000}{4.2426 \times 4.2065} = 168.10\,\text{RPM} \tag{12}$$

Angular velocity of the output shaft in rad/s is given by

$$\omega_1 = \frac{2\pi N_1}{60} = \frac{2\pi\, 168.10}{60} = 17.6\,\text{rad/s} \tag{13}$$

The torque output is given by

$$T_{out} = T_{in} n_1 n_2' = 1115 \times 4.2426 \times 4.2065 = 19{,}899\,\text{N m} \tag{14}$$

Pitch line velocity of the first stage is given by

$$v_{p1} = \frac{d_{p1}\omega_1}{2} = \frac{144 \times 314}{2} = 22{,}608\,\text{mm/s} \tag{15}$$

Pitch line velocity of the second stage is given by

$$v_{p2} = \frac{d_{p2}\omega_2}{2} = \frac{145 \times 17.6}{2} = 1276\,\text{mm/s} \tag{16}$$

Many variations of a speed reduction gearbox can be obtained in a similar manner.

5.2 Epicyclic Gear Trains

Epicyclic gear trains are useful over traditional gearboxes when the speed reduction ratios are very high. In addition, they have the advantages of compact and lightweight design. Furthermore, epicyclic gear trains have the advantage of high efficiency, resulting in minimum loss of power, mainly due to the fact the teeth contacts occur over a large number of teeth, thus averaging the manufacturing errors of gear teeth and their defects that are developed during service.

An elementary epicyclic gear train is shown in Fig. 5.3a. It consists of a central spur gear 2, also known as sun gear, an epicyclic spur gear 4, also known as planet

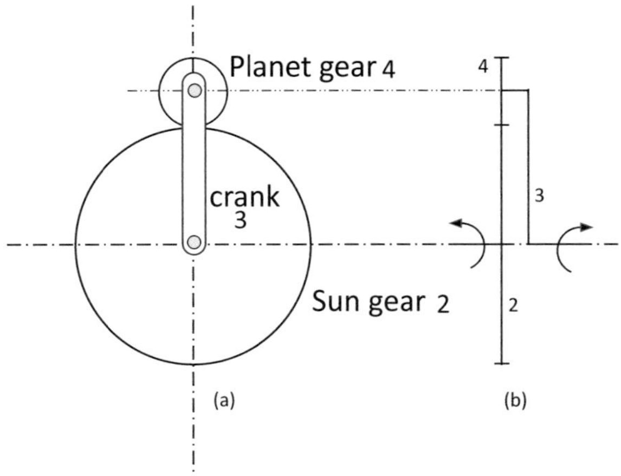

Fig. 5.3 **a** Elementary epicyclic gear train. **b** Schematic line diagram

gear that performs epicyclic motion by rolling around the central gear. This motion of the planet gear is supported by a crank 3 that ensures meshing action between the planet and sun gears. Although the crank and the sun gear have a common axis, they could be independent of each other.

Figure 5.3b shows the schematic line diagram of the epicyclic gear train shown in Fig. 5.3a. The schematic diagram has the advantage of providing clarity regarding the bearing support of various elements. The crank 3 and the sun gear 2 have independent bearing supports, which can be used for giving input or taking the output; the planet gear 4, on the other hand, does not have any bearing support as it performs epicyclic motion around the sun gear 2. The extension of the axes between the figures gives a clear relationship between them. The meshing external spur gears 2 and 4 are represented by a small horizontal line.

The modified version of the elemental epicyclic gear train is a simple epicyclic gear train that has practical applications, as shown in Fig. 5.4. The modification with respect to the elemental epicyclic gear train has been made with the addition of an internal gear 5 that has a common axis with the sun gear 2 and the crank 3, although they independently rotate with respect to each other. Many such modifications can be made to obtain epicyclic gear trains for various applications that would help in distributing the forces and to obtain outputs at an appropriate speed and torque.

Analysis of planetary gear trains

Consider the simple epicyclic gear train shown in Fig. 5.4a. The relative angular velocity between the sun gear 2 and the crank 3 is given by

$$\omega_{2/3} = \omega_2 - \omega_3 \tag{5.3}$$

Similarly, the angular velocity of the internal gear 5 with respect to the crank 3 is given by

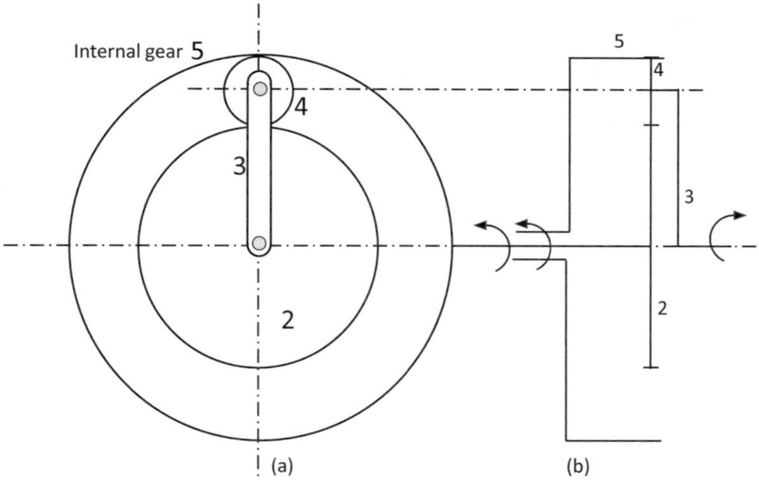

Fig. 5.4 **a** Simple epicyclic gear train. **b** Schematic line diagram

$$\omega_{5/3} = \omega_5 - \omega_3 \tag{5.4}$$

From Eqs. (5.3) and (5.4)

$$\frac{\omega_{5/3}}{\omega_{2/3}} = \frac{\omega_5 - \omega_3}{\omega_2 - \omega_3} \tag{5.5}$$

Equation (5.3) expresses the relative angular velocities of gears 5 and 2, irrespective of the angular velocity of the crank 3. This ratio is proportional to the teeth numbers and is the same if the crank is rotating or not. Hence, it is the first-order kinematic coefficient of the gear train, which from Eq. (5.2) can be written as

$$\alpha_{5/3} = \frac{\omega_{5/3}}{\omega_{2/3}} = \frac{\omega_5 - \omega_3}{\omega_2 - \omega_3} \tag{5.6}$$

The above equation can be generalized as

$$\alpha_{L/F} = \frac{\omega_L - \omega_A}{\omega_F - \omega_A} \tag{5.7}$$

where, ω_F is the angular velocity of the first gear in the train, ω_L is the angular velocity of the last gear in the train, ω_A is the angular velocity of the arm.

Example 5.2
Consider an epicyclic gear train shown in Fig. 5.5a and its schematic representation in Fig. 5.5b. It consists of an arm 3, that is present on both sides and mounted on independent bearings. Sun gear 2 is receiving an external input at a speed of 200 RPM

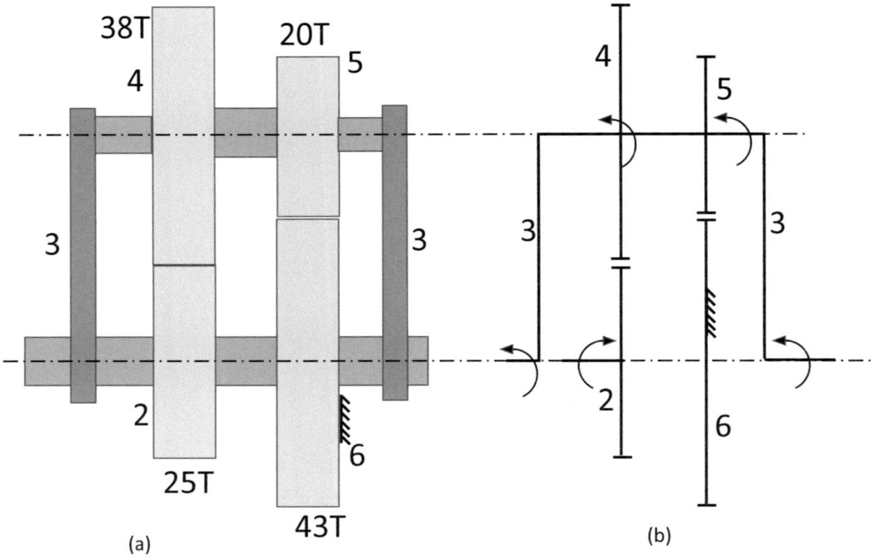

Fig. 5.5 **a** Planetary gearbox of Example 5.2. **b** Schematic line diagram

in the clockwise direction. Planetary gears 4 and 5 are mounted on the same shaft and the shaft center is hinged to the arm 3 so the planetary gears perform epicyclic motion respectively around gears 2 and 6; however, the sun gear 6 is stationary. Determine the rotational speeds of the remaining gears and the arm.

Given $\omega_2 = -200$ RPM (clockwise).

From Eq. (5.2), the kinematic coefficient of the epicyclic gear system is given by

$$\alpha_{6/2} = \frac{N_2 N_5}{N_4 N_6} = \frac{25 \times 20}{38 \times 43} = 0.3060 \tag{1}$$

It may be noted that gear 6, although fixed for the given problem, is accounted for in the kinematic coefficient of the epicyclic gear system of the above equation.

From Eq. (5.7), the kinematic coefficient of the epicyclic gear system is also given by

$$\alpha_{6/2} = 0.3060 = \frac{\omega_L - \omega_3}{\omega_F - \omega_3} \frac{0 - \omega_3}{-200 - \omega_3} \tag{2}$$

Solving Eq. (2), $\omega_3 = 88.18$ RPM (counterclockwise).

The relative angular velocity between the sun gear 2 and the arm 3 is given by

$$\omega_{2/3} = \omega_2 - \omega_3 = -200 - 88.18 = -288.18\,\text{RPM} \tag{3}$$

The angular speeds of the gears 4 and 5 are given by

$$\omega_4 = \omega_5 = \omega_3 - \omega_{2/3}\frac{N_2}{N_4}$$
$$= 88.18 - (-288.18)\frac{25}{38} = 277.77\,\text{RPM} \tag{4}$$

An alternate method of determining the angular speeds of various elements of a planetary gear system is the use of superposition principle in which the motion of the system, when all the gears are locked to the arm is superimposed on the motion of the system when only the arm is locked, in addition to any other restrictions placed on other elements of the system. It is carried out as follows:

1. Lock all the gears to the arm, including any of locked gears and allow the arm to rotate with an angular velocity ω_A. Therefore, all the elements rotate with the same angular velocity ω_A.
2. Now free all the above constraints, enforce the locking condition of the given system and allow some reference gear to rotate with an angular velocity relative to the arm and compute the apparent angular velocities of the remaining gears in terms of the relative angular velocity of the reference gear relative to the arm.
3. Add the angular velocities of the above steps and by using the given angular velocities of any of the gears or the arm, determine the unknown angular velocities.

The above method is known as the tabular method.

Example 5.3
Use the tabular method to determine the angular velocities of the epicyclic gear train system of Example 5.2.

Since the speeds of gear 2 and gear 6 are given, they can be used to determine the other unknown speeds. Based on the angular velocity of gear 2, the angular velocity of the arm, ω_3, can be expressed in terms of the relative speed $\omega_{2/3}$ as follows (see Table 5.2):

$$\omega_2 = -200 = \omega_3 + \omega_{2/3}$$
$$\Rightarrow \omega_3 = -\omega_{2/3} - 200 \tag{1}$$

Substituting the angular velocity of the arm, ω_3, into the equation of the angular velocity of gear 6 from Table 5.2

$$-\omega_{2/3} - 200 + \omega_{2/3}\frac{25}{38}\frac{20}{43} = 0 \tag{2}$$

Solving Eq. (2), $\omega_{2/3} = -288.18$ RPM (clockwise). Hence, from Eq. (1), $\omega_3 = -(-288.18) - 200 = 88.18$ RPM (counterclockwise).

From Table 5.1, the angular velocities of gears 4 and 5 are given by

$$\omega_4 = \omega_5 = \omega_3 - \omega_{2/3}\frac{N_2}{N_4} = -88.18 - (-288.18)\frac{25}{38} = 277.77\,\text{RPM} \tag{3}$$

Example 5.4

An epicyclic gear train shown in Fig. 5.6 consists of an arm (3), external gears (2, 4, 5, 6) and an internal gear (7). Internal gear (7), external gear (2), and the arm (A) are mounted on independent bearings.

The number of teeth of various gears is as follows: $N_2 = 30$, $N_4 = 25$, $N_5 = 45$, $N_6 = 50$, $N_7 = 200$. If the speed of gear 7 is 30 rpm (CCW) and that of the arm 3 is −25 rpm (clockwise), determine the speed of all the gears using the tabular method.

From Table 5.3, since the angular velocity ω_7 is known

$$\omega_7 = \omega_3 + \omega_{2/3}\frac{N_2}{N_5}\frac{N_4}{N_6}\frac{N_6}{N_7}$$

Table 5.2 Tabular analysis of Example 5.2

	Gear 2	Arm 3	Gear 4	Gear 5	Gear 6
Gear fixed to arm	ω_3	ω_3	ω_3	ω_3	ω_3
Arm fixed and the gears free to rotate	$\omega_{2/3}$	0	$-\omega_{2/3}\frac{N_2}{N_4}$	$-\omega_{2/3}\frac{N_2}{N_4}$	$\omega_{2/3}\frac{N_2}{N_4}\frac{N_5}{N_6}$
Total	$\omega_3 + \omega_{2/3}$	ω_3	$\omega_3 - \omega_{2/3}\frac{N_2}{N_4}$	$\omega_3 - \omega_{2/3}\frac{N_2}{N_4}$	$\omega_3 + \omega_{2/3}\frac{N_2}{N_4}\frac{N_5}{N_6}$
Given speeds	−200				0

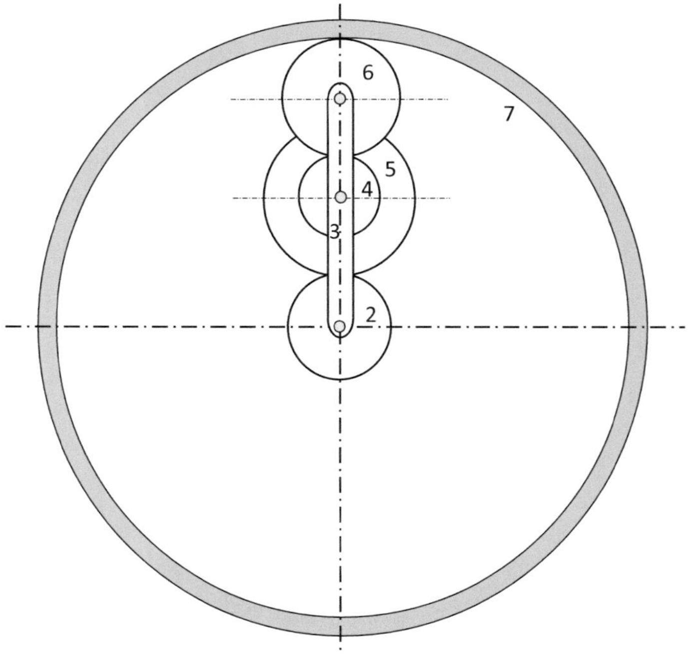

Fig. 5.6 Planetary gearbox of Example 5.4

$$\Rightarrow 30 = -25 + \omega_{2/3} \frac{30}{45}\frac{25}{50}\frac{50}{200} \quad (1)$$

Solving Eq. (1), $\omega_{2/3} = 660$ RPM.
The angular velocities of gears 4 and 5 are equal and given by

$$\begin{aligned}\omega_4 = \omega_5 &= \omega_3 - \omega_{2/3}\frac{N_2}{N_5} \\ &= -25 - 660\frac{30}{45} = -817\,\text{RPM}\end{aligned} \quad (2)$$

The angular velocity of gear 6 is given by

$$\begin{aligned}\omega_6 &= \omega_3 + \omega_{2/3}\frac{N_2}{N_5}\frac{N_4}{N_6} \\ &= -25 + 660\frac{30}{45}\frac{25}{50} = 195\,\text{RPM}\end{aligned} \quad (3)$$

Since all other speeds were derived based on the speeds of the arm and gear, they can be computed for any other speeds of the elements governed by independent equations. For example, if the speeds of gears 4 and 5 are known, all other speeds can be computed using the same set of equations.

From Eq. (5.2), kinematic coefficient is given by

$$\begin{aligned}\alpha_{L/F} &= \frac{\text{Product of driving teeth numbers}}{\text{Product of driven teeth numbers}} = \frac{N2 \times N4 \times N6}{N5 \times N6 \times N7} \\ &= \frac{30 \times 25 \times 50}{45 \times 50 \times 200} = 0.083\end{aligned} \quad (4)$$

From Eq. (5.7), kinematic coefficient is also given by

Table 5.3 Tabular analysis of Example 5.4

Steps	Operation	Speed of elements in rpm					
		2	3	4	5	6	7
1	System is locked, revolving with ω_3 rpm	ω_3	ω_3	ω_3	ω_3	ω_3	ω_3
2	Arm is fixed, gear 2 is given $\omega_{2/3}$ revolution	$\omega_{2/3}$	0	$-\omega_{2/3}\frac{N_2}{N_5}$	$-\omega_{2/3}\frac{N_2}{N_5}$	$\omega_{2/3}\frac{N_2}{N_5}\frac{N_4}{N_6}$	$\omega_{2/3}\frac{N_2}{N_5}\frac{N_4}{N_6}\frac{N_6}{N_7}$
3	Add the speeds from steps 1 and 2	$\omega_3 + \omega_{2/3}$	$\omega_3 =$ -25	$\omega_3 - \omega_{2/3}\frac{N_2}{N_5}$	$\omega_3 - \omega_{2/3}\frac{N_2}{N_5}$	$\omega_3 + \omega_{2/3}\frac{N_2}{N_5}\frac{N_4}{N_6}$	$\omega_3 + \omega_{2/3}\frac{N_2}{N_5}\frac{N_4}{N_6}\frac{N_6}{N_7} =$ 30

$$\alpha_{L/F} = \frac{\omega_L - \omega_A}{\omega_F - \omega_A} = \frac{\omega_7 - \omega_3}{\omega_2 - \omega_3} = \frac{30 - (-25)}{635 - (-25)} = 0.083 \quad (5)$$

Example 5.5

Consider the epicyclical gearbox shown in Fig. 5.7. Gear 2 is fastened to its shaft and is driven at 300 rpm (CW). Gears 4 and 5 are planet gears that rotated together but are free to turn on the shaft carried by the arm. In addition, gear 5 turns the internal gear 7. If gear 6 Is stationary, determine the speed of all the gears using tabular method (Table 5.4).

From the angular speed of gear 6

$$\omega_3 + \omega_{2/3}\frac{N_2}{N_4}\frac{N_5}{N_6} = 0$$
$$\Rightarrow \omega_3 + \omega_{2/3}\frac{25}{38}\frac{43}{20} = 0 \quad (1)$$

From the angular speed of gear 2

$$\omega_2 = \omega_3 + \omega_{2/3} = -300 \quad (2)$$

From Eqs. (1) and (2), $\omega_3 = -1023.8\,\text{RPM}$ $\omega_{2/3} = 723.81\,\text{RPM}$.

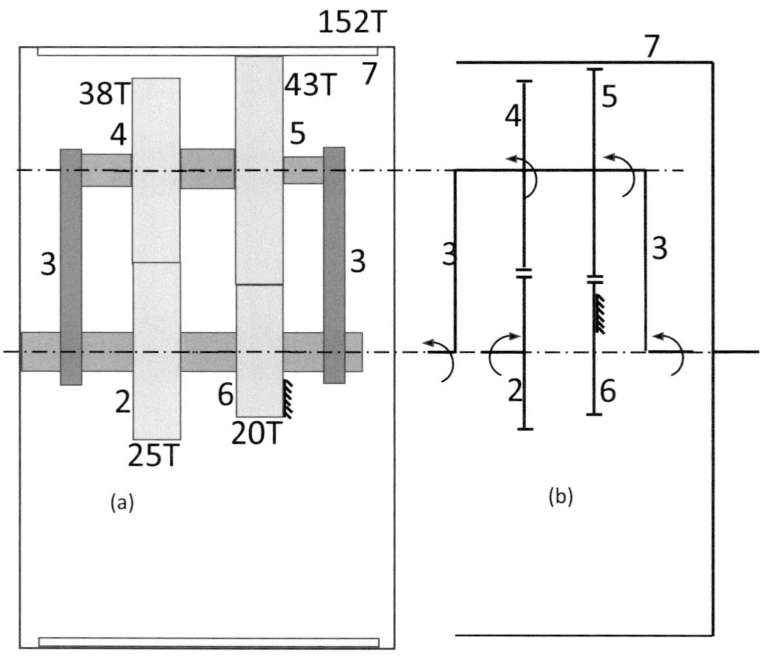

Fig. 5.7 **a** Planetary gearbox of Example 5.5. **b** Schematic line diagram

Table 5.4 Tabular analysis of Example 5.5

	Gear 2	Arm 3	Gear 4	Gear 5	Gear 6	Gear 7
Gear fixed to arm	ω_3	ω_3	ω_3	ω_3	ω_3	ω_3
Arm fixed	$\omega_{2/3}$	0	$-\omega_{2/3}\frac{N_2}{N_4}$	$-\omega_{2/3}\frac{N_2}{N_4}$	$\omega_{2/3}\frac{N_2}{N_4}\frac{N_5}{N_6}$	$-\omega_{2/3}\frac{N_2}{N_4}\frac{N_5}{N_7}$
Total	$\omega_3 + \omega_{2/3}$	ω_3	$\omega_3 - \omega_{2/3}\frac{N_2}{N_4}$	$\omega_3 - \omega_{2/3}\frac{N_2}{N_4}$	$\omega_3 + \omega_{2/3}\frac{N_2}{N_4}\frac{N_5}{N_6}$	$\omega_3 - \omega_{2/3}\frac{N_2}{N_4}\frac{N_5}{N_7}$
Given speeds	−300				0	

The angular speeds of gear 4 and 5 are given by

$$\omega_4 = \omega_5 = \omega_3 - \omega_{2/3}\frac{N_2}{N_4} = -1023.8 - (723.81)\frac{25}{38} = -1500\,\text{RPM} \tag{3}$$

The angular speed of gear 7 is given by

$$\omega_7 = \omega_3 - \omega_{2/3}\frac{N_2}{N_4}\frac{N_5}{N_7} = -1023.8 - 723.81\frac{25}{38}\frac{43}{152} = -1158.5\,\text{RPM} \tag{4}$$

From Eq. (5.2), kinematic coefficient is given by

$$\alpha_{L/F} = -\frac{\text{Product of driving teeth numbers}}{\text{Product of driven teeth numbers}} = -\frac{N2 \times N5}{N4 \times N7} = -\frac{25 \times 43}{38 \times 152} = -0.1861 \tag{5}$$

From Eq. (5.7), kinematic coefficient is also given by

$$\alpha_{L/F} = \frac{\omega_L - \omega_A}{\omega_F - \omega_A} = \frac{\omega_7 - \omega_3}{\omega_2 - \omega_3} = \frac{-1158.5 + 1023.8}{-300 + 1023.8} = -0.1861 \tag{6}$$

5.3 Conclusions

A brief discussion on traditional gearboxes and epicyclic gearboxes is presented in this chapter. The information presented in this chapter can be used to design any practical gear train using any type of gear. Hence, there are unlimited options to ingeniously design compact, lightweight gear trains for any application. It should be noted that gearbox design is a detailed task and the examples provided in this chapter can only be useful in obtain the basic design to start with.

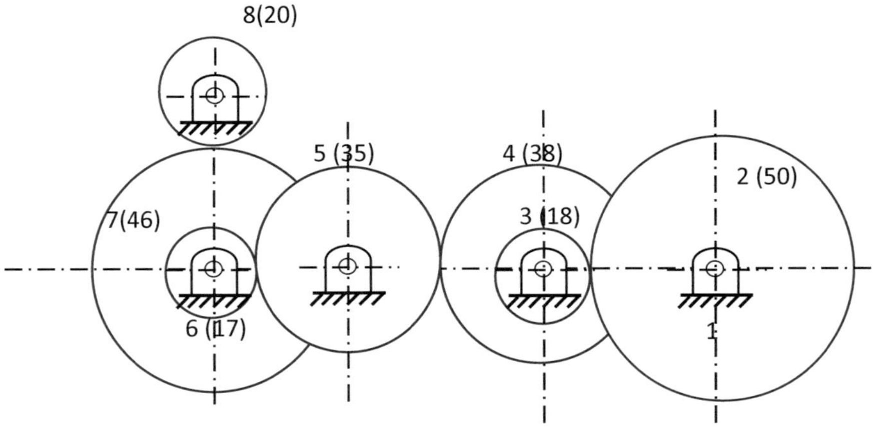

Fig. P5.1 Gear train of exercise problem 5.1

5.4 Exercise Problems

5.1 The input gear (8) of the above spur gear train shown in Fig. P5.1 is driven at 2500 RPM with a 10 kW motor. Determine the torque and speeds of all the gears of this system.

5.2 A four-speed gear box has to be designed to obtain gear reduction ratios of 1, 1.6, 2.56 and 4.096. All the gears must have a module of 4 mm and the smallest pinion must have a minimum of 18 teeth. Find the number of teeth of all gears of the four-speed gear box.

5.3 The four-speed gear box of the exercise problem 5.2 is used in an automobile that has an engine of 60 kW power running at a constant 4000 RPM. If the tyre diameter is 0.6 m, determine the speed of the vehicle at the given RPM of the engine for the third and fourth gears.

5.4 A spur gear train has to be designed to obtain a speed reduction of 200 in four steps. No pinion should have less than 18 teeth. The modules of the pinion gear combination in the ascending order of the steps are as follows: 3, 4, 5, and 6 mm. Choose the number of teeth in such a way that the center distances are accurate to four decimal places and are not recurring.

5.5 The input shaft of the gearbox is shown in Fig. P5.9 rotates at 3000 RPM. If P1 has 18 teeth and G2 has 60 teeth, what are the maximum and minimum speeds that can be obtained from the above gearbox by choosing an appropriate number of teeth for G1 and P2, such that the number of teeth can be only between 18 and 60.

5.6 Lathes are conventional manufacturing machines that are capable of carrying out a variety of jobs. One such job is the cutting of threads by making use of the gear trains that are present in the lathe in conjuction with the lead screw that helps in the movement of the tool carriage. The lead screw of one such lathe has a right-hand single thread of 5 mm pitch, the smallest wheel of the

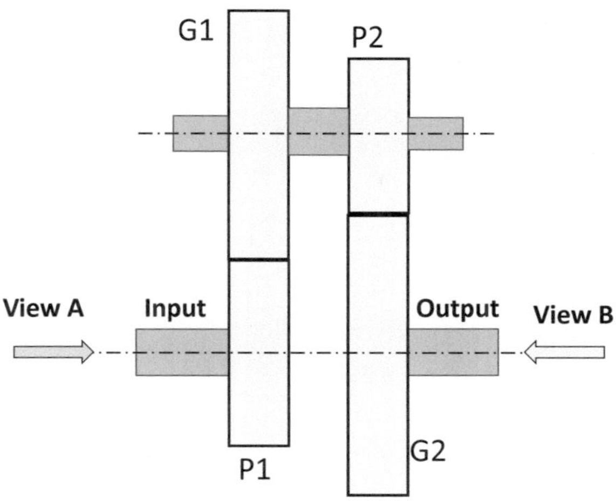

Fig. P5.5 Gearbox of exercise problem 5.5

gear train has 18 teeth, the largest 108 teeth and the numbers of teeth on intermediate gears can be assigned as per convenience. Find a gear train of one such combination for connecting the spindle and the lead screw, when (a) a right-hand screw with 2 thread/mm has to be cut and (b) a left-hand screw with 1 threads/mm has to be cut.

5.7 Figure P5.7 shows an aircraft propeller speed reduction system using an epicyclic gearbox. Determine the propeller speed if the engine runs at 3000 RPM.

5.8 In the planetary gear system shown in Fig. P5.8, the engine speed is 650 RPM. If the propeller rotates at 400 RPM in the same direction of the engine, determine the speed of the arm 5.

5.9 The planetary system shown in Fig. P5.9 is a twin output system as shown. Determine the input speed and its direction.

5.10 In the planetary gear system shown in Fig. P5.10, link 2 is the driving member and the sun gear 3 is the driven member and the internal gear 4 is stationary and there are three planetary gears (2). The pitch diameter of the internal gear is 300 mm. Determine the number of teeth on all the spur gears that are of 3 mm module, 20° pressure angle, if the sun gear is expected to rotate three times the speed of link 2.

5.11 In the planetary system shown in Fig. P5.14, the sun gear 3 is fixed, link 2 is the driven member and link 4 is the driving member. The sun gear (3) has as a diameter of 50 mm and the planet gears' (1) diameter is 25 mm; both are 20° pressure angle spur gears. If an input torque of 15 kN m at 250 RPM is applied to the inner gear (4), determine the speeds of all the elements and the torque transmitted by link 2.

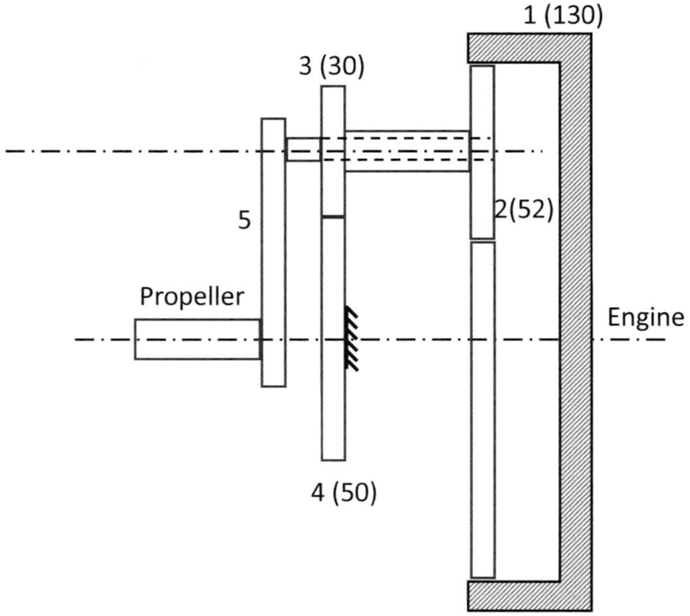

Fig. P5.7 Epicyclic gearbox of exercise problem 5.7

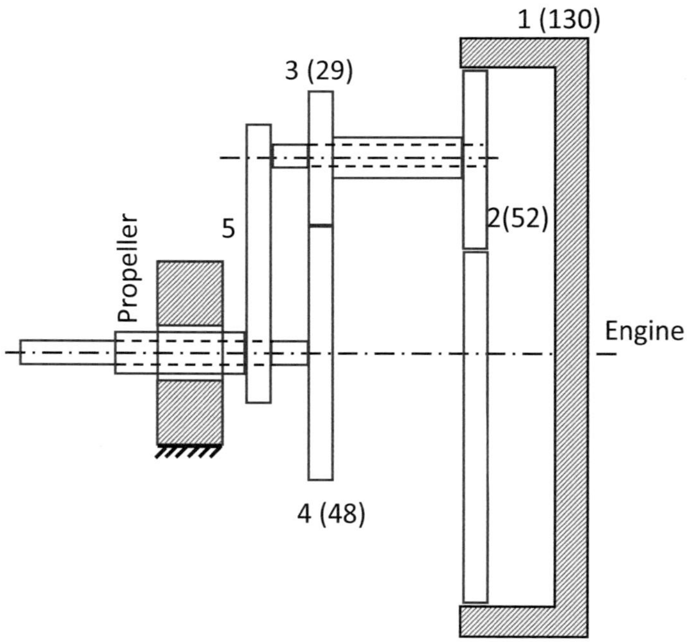

Fig. P5.8 Epicyclic gearbox of exercise problem 5.8

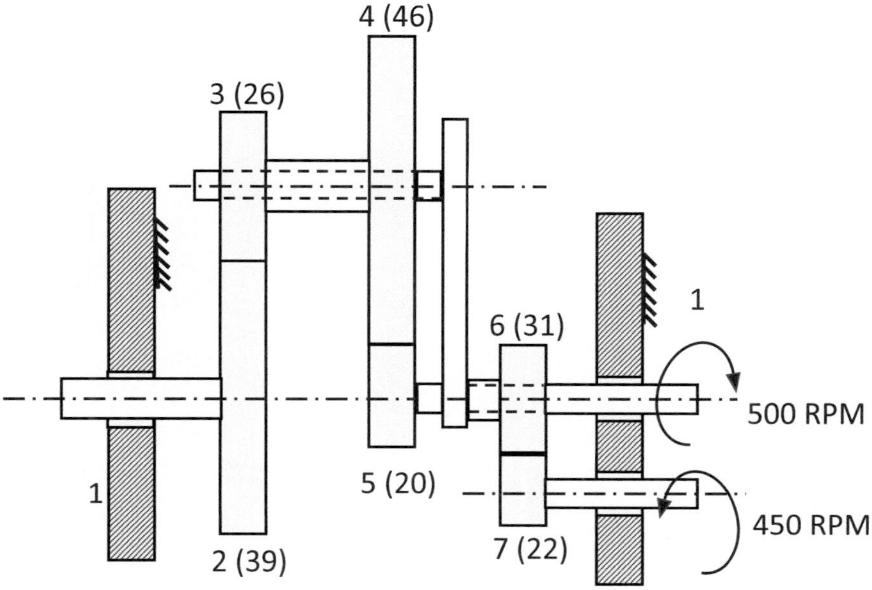

Fig. P5.9 Twin output epicyclic gearbox of exercise problem 5.9

Fig. P5.10 Planetary gear system of exercise problem 5.10

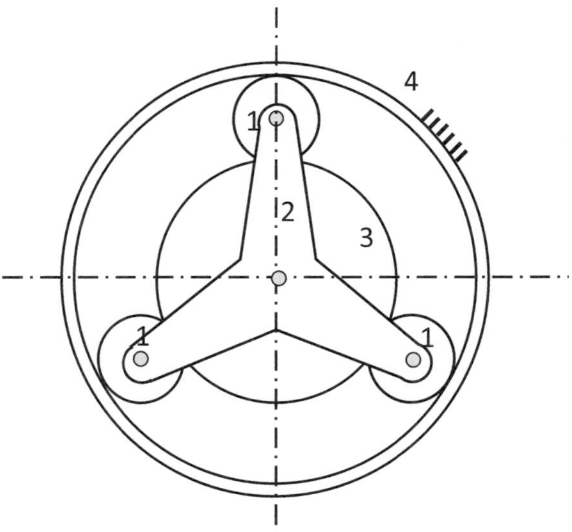

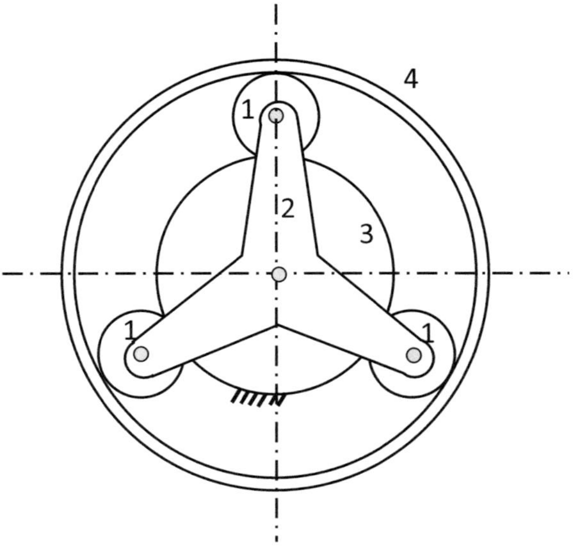

Fig. P5.11 Planetary gear system of exercise problem 5.11

Chapter 6
Cam-Follower Mechanism

A cam-follower mechanism is one in which there is at least one higher pair; the links of this pair are connected either along a line or a point. This higher pair consists of a device that inputs the motion, known as cam and the other is the follower that is the output link. Unlike the approximate function generators constructed from four-bar mechanisms, the cam-follower mechanism is an exact function generator. Hence, it is extensively used as a timing device mechanism in many applications that require precise time control. One such application is the valve operating mechanism of IC engines that requires opening and closing of input and output of engine valves for a specific instant or duration as required. The other important higher pair mechanism, gears were discussed in Chaps. 3, 4, and 5. Similar to the planar mechanisms with lower pairs, the cam-follower mechanism can also be studied as a planar mechanism.

6.1 Classification of Cams

Cam-follower mechanisms can be classified based on various criteria. One of them is based on the type of input movement that is discussed as follows.

6.1.1 Plate Cam with Roller Follower

If the input link, which is generally called as cam, rotates at uniform angular velocity, then it is known as a disc cam or plate cam. As shown in Fig. 6.1, the cam is a flat plate of uniform thickness and it imparts output movement to the roller follower based on the contour of the plate that must be specifically designed, based on the type of output motion required; that is known as cam profile design. Details of that design will be discussed in Sect. 6.5. The axis of rotation of the plate cam and the

D. N. Manik, *Fundamentals of Mechanisms and Machines*,
https://doi.org/10.1007/978-981-97-1810-8_6

follower are along a straight line. But as we shall see later, there will be an offset in practical cams between the axis of rotation of the plate cam and the axis of the follower in order to account for the forces acting on it.

The block diagram of the kinematic pairs of a plate cam with roller follower is shown in Fig. 6.1b. The plate cam, designated as 2, is hinged to the fixed link 1 and forms a revolute pair. The roller, link 3, forms a higher pair with the plate cam and the roller follower, link 4, forms a revolute pair with the roller and the follower forms a prismatic pair with the fixed link 1. Uniform rotary motion of the cam results in linear oscillation of the follower in the vertical direction. The extent of movement of the follower is totally dependent on the cam profile that is specific to a particular application. There could be an upward movement (rise), downward movement (fall), and no movement at all (dwell) and the time interval of these movements is dependent on a specific application. Hence, a cam being an exact function generator is capable of delivering the desired movement for the follower. The roller motion is redundant and irrespective of how much angular motion takes place in the roller, depending on

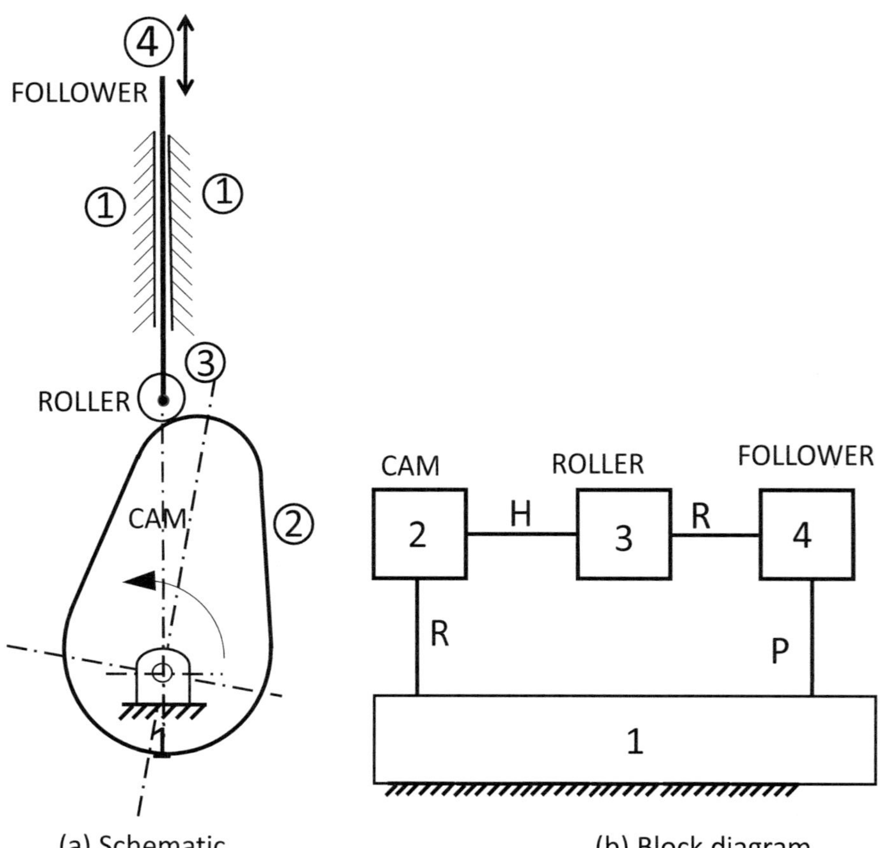

Fig. 6.1 Plate cam

the friction between the cam and the roller, the output motion of the follower remains the same.

6.1.2 Plate Cam with Flat-Faced Follower

As shown in Fig. 6.2a, the plate cam imparts output movement to the flat-faced follower, based on the contour of the plate that must be specifically designed based on the type of output motion required.

The block diagram of the kinematic pairs of a plate cam with a flat-faced follower is shown in Fig. 6.2b. The plate cam, designated as 2, is hinged to the fixed link 1 and forms a revolute pair with it. The flat-faced follower, link 3, forms a higher pair with the plate cam and forms a prismatic pair with the fixed link 1. For a uniform angular motion of the cam, the follower oscillates in the vertical direction. It is important to

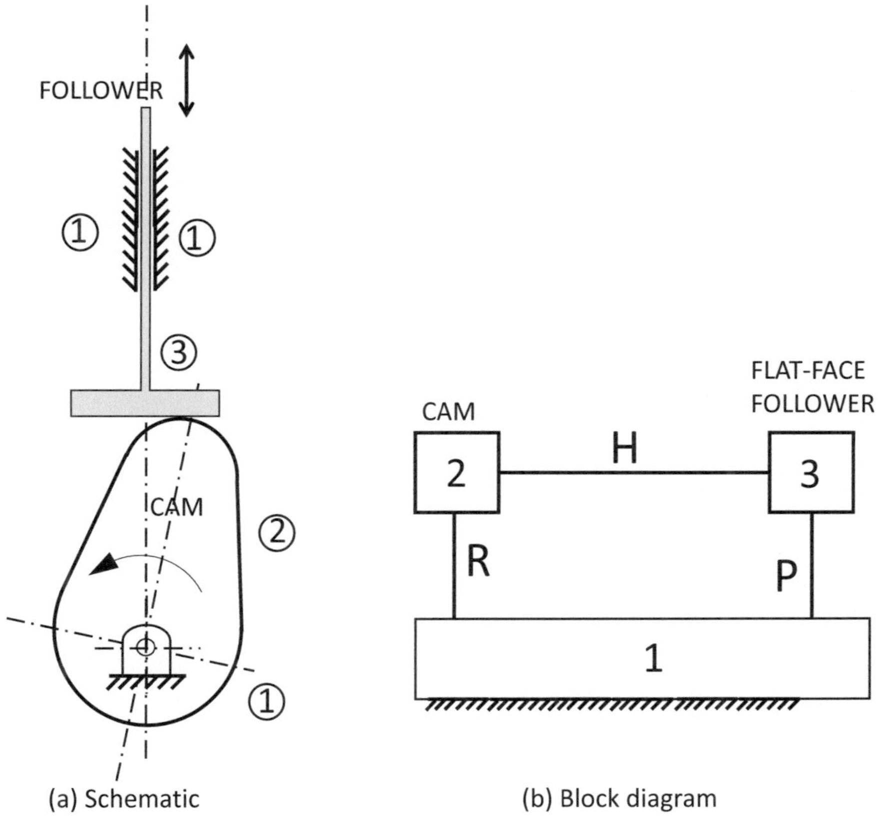

Fig. 6.2 Plate cam with flat-face follower

note that the flat face is always tangential to the cam profile and the contact can take place at any point on the flat face.

6.1.3 Wedge Cam with Roller Follower

If the input motion is a linear to and from movement as shown in Fig. 6.3a, then it is known as a wedge cam. It consists of a wedge of a particular slope that moves to and fro in the horizontal direction, against a roller follower as shown in the figure. The kinematic pairing diagram is shown in Fig. 6.3b. It consists of a wedge, link 2, that forms a prismatic pair with the fixed link 1 by sliding on it. The roller follower, link 3, forms a higher pair with the wedge and is hinged to the follower, link 4, forming a revolute pair with it; the follower, link 4, forms a prismatic pair with the fixed link 1. Corresponding to the horizontal movement of the wedge, depending on its surface, the follower moves in the vertical direction.

Cams are also classified based on the type of output motion: (1) translational and (2) oscillating.

We have already discussed the translational type of output motion in Sects. 6.1.1 and 6.1.2, and we shall discuss the output oscillating motion in Sect. 6.1.4. In the translational type of followers, we have only discussed radial followers that have their axis of translation passing the axis of rotation of the cam. But the axis of the

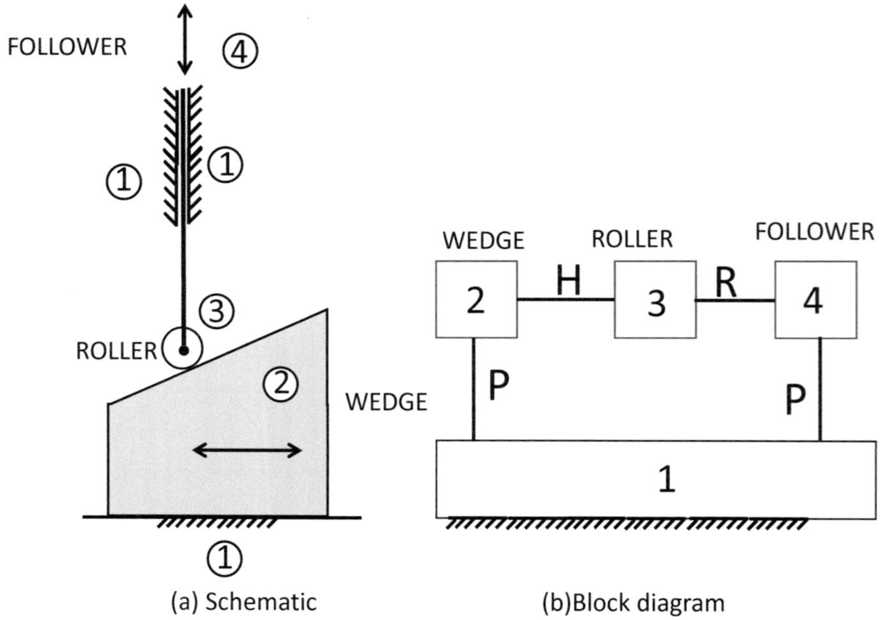

Fig. 6.3 Wedge cam

follower translation and the axis of rotation of the cam could be offset based on practical considerations that are discussed later.

6.1.4 Oscillating Follower

An oscillating follower is shown in Fig. 6.4. Figure 6.4a shows the mechanism that consists of a plate cam that rotates with a uniform angular velocity. An oscillating roller-follower moving on the plate cam performs oscillating angular motion as shown. The block diagram of the interconnected kinematic pairs of the mechanism is shown in Fig. 6.4b. The plate cam, link 2, hinged to the fixed link 1, forms a revolute pair with it. The roller, link 3, forms higher pair with the plate cam. The oscillating follower, link 4, forms a revolute pair with the roller and the fixed link 1.

The type of follower surface is also important. We have already discussed the roller follower and flat-face follower. But the follower surface could be a knife edge, which is not practical due to its high friction.

6.2 Retainer Springs for Cams

In all the cam designs, we have discussed thus far, they can drive the follower only during the rise period against gravity. However, during the fall period, the follower will be moving in the direction of gravity. Hence, although the cam losing contact is more critical during the downward movement, depending on the forces acting on the cam-follower interface, it is possible that the follower can lose contact with the cam,

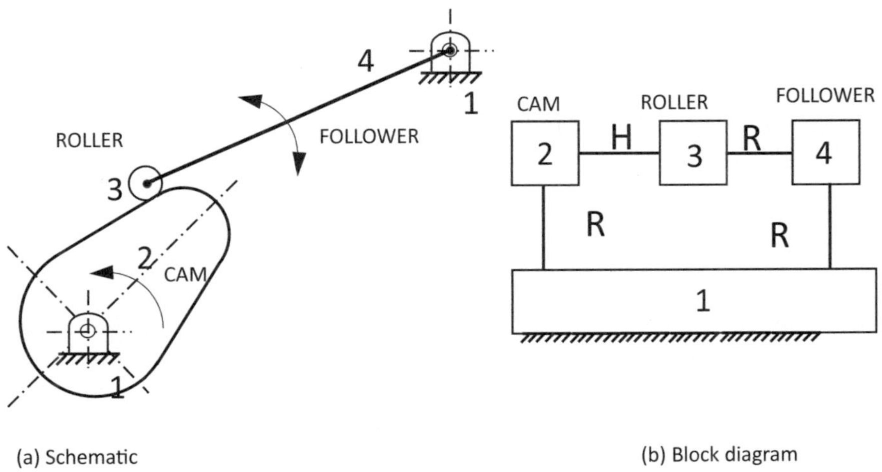

Fig. 6.4 Oscillating roller-follower

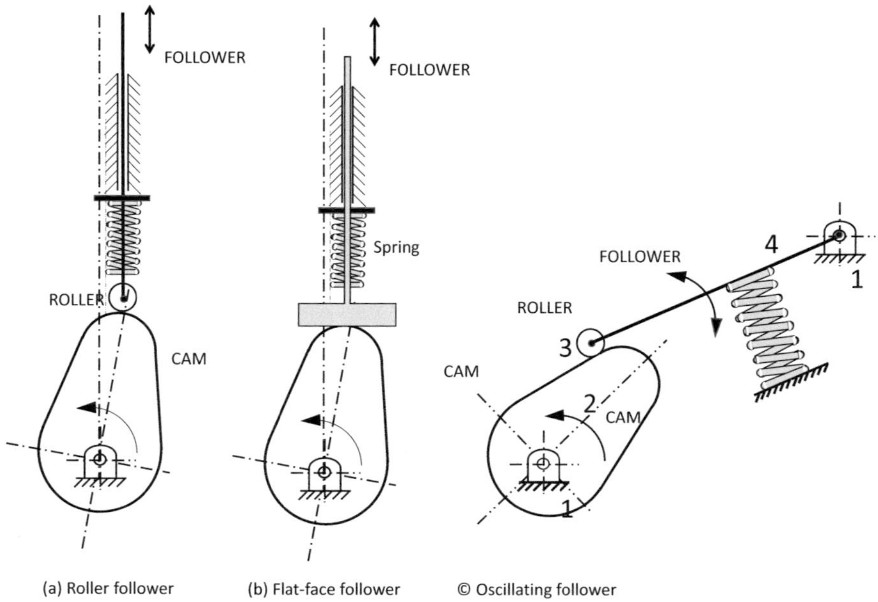

Fig. 6.5 Plate cam with retainer spring

both during the upward and downward movement and hence may not generate the desired function. Therefore, a retainer spring will be essential to any cam-follower mechanism.

Figure 6.5 shows a retainer spring in some of the plate cam designs. In addition, for Fig. 6.5a, b, there is an offset between the axis of translation and the axis of rotation of the cam for roller followers and flat-face followers. The reason for providing such an offset will be explained later. An oscillating follower also has a retainer spring as shown in Fig. 6.5c. In all the above cases, the spring is connected such that during the rise period the spring gets compressed due to which it helps the follower to maintain contact during the fall period. By keeping a retainer spring, the acceleration during the fall movement need not be restricted for the fear of losing contact.

6.3 Cam Nomenclature

Figure 6.6a shows an offset plate cam with a roller follower. The base circle is the smallest circle at the shaft axis and is tangential to the cam profile. Trace point is a point on the follower that describes the follower movement. Hence, the trace point of the roller follower is the center of the roller and hence the movement of the follower can be described in terms of the movement of the roller center. Pitch curve is the locus traced by the roller center when the cam is fixed and the follower is moved around in the clockwise direction keeping contact with the cam profile. Prime circle,

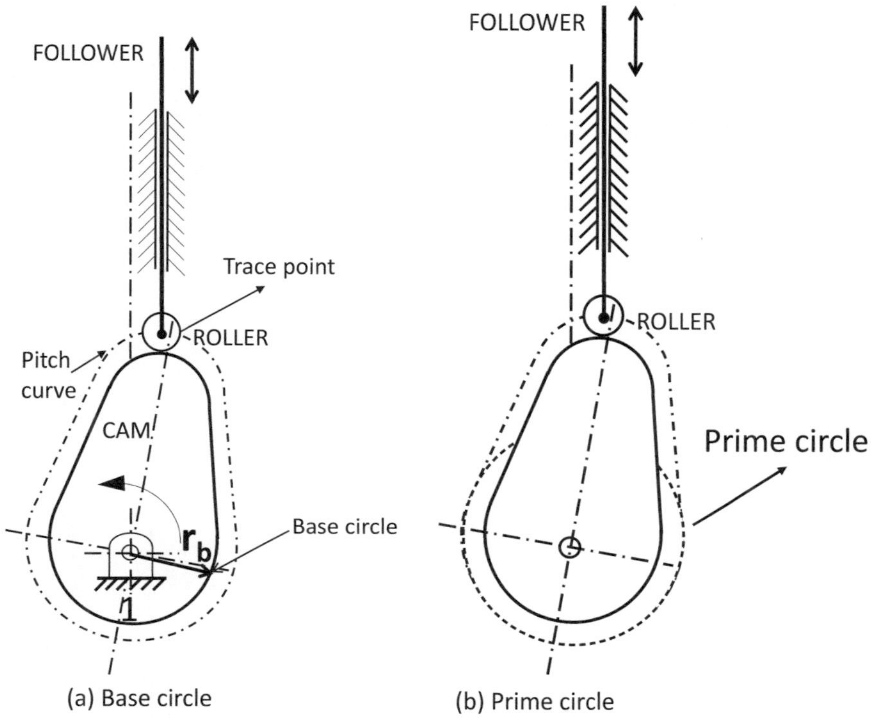

Fig. 6.6 Plate cam nomenclature

as shown in Fig. 6.6b, is the smallest circle within the pitch curve tangential to it with its center coinciding the cam shaft axis.

An oscillating plate cam-follower mechanism is shown in Fig. 6.7. Tracer point for this mechanism, similar to the radial cam described in Fig. 6.6, is the roller center point on the follower when the follower face is in contact with the cam surface.

6.4 Pressure Angle

The pressure angle of a plate cam is shown in Fig. 6.8. It is defined as the angle between the direction of the normal force acting at the cam-follower interface to the direction of movement of the follower.

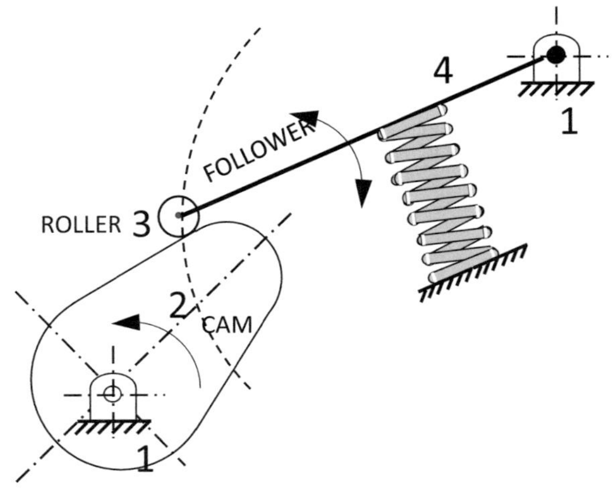

Fig. 6.7 Tracer point of an oscillating follower

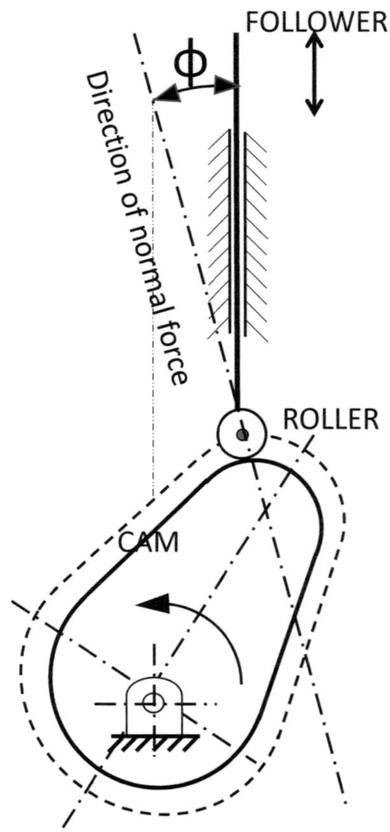

Fig. 6.8 Pressure angle-roller follower

6.4.1 Roller Follower

A plate cam with an offset roller follower and a retainer spring is shown in Fig. 6.9a, and it is assumed to move in the counterclockwise direction; the axis of translation of the follower is offset to the right by e. The normal reaction force acts at the pressure angle ϕ with respect to the vertical. The follower has a prismatic pair with the fixed link for a distance of ℓ_1 and the center of the roller is at a distance of ℓ_2 from the fixed guide at the instant shown. An exaggerated position of the follower is shown in figure (b). Since the cam is rotating counterclockwise, the follower is assumed to tilt slightly toward right as shown. Hence, there will be normal forces N_1 and N_2 at the edges of the guide. Since the follower is moving upwards, the corresponding friction forces are respectively, μN_1 and μN_2 acting downwards. All the forces acting on the roller are shown in figure (c). Since the retainer spring is compressed as the cam is trying to push roller upwards, there is a spring force F_s acting downwards. The normal reaction force F_n of the cam is acting at the pressure angle, and it is balanced against the reaction force and the spring force.

Summing the horizontal forces

$$N_1 = N_2 + F_n \sin\phi \tag{6.1}$$

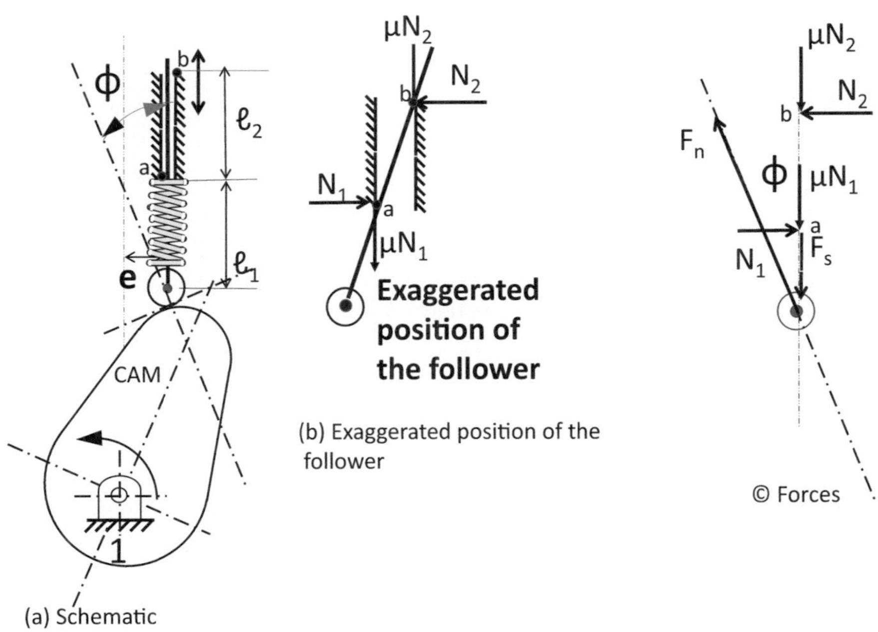

Fig. 6.9 Reaction forces on a plate cam with roller follower

Summing the vertical forces

$$F_n \cos\phi = F_s + \mu(N_1 + N_2) \tag{6.2}$$

Taking moments about the roller center

$$N_1\ell_1 = N_2(\ell_1 + \ell_2) \tag{6.3}$$

From Eq. (6.3)

$$N_1 = N_2\left(1 + \frac{\ell_2}{\ell_1}\right) \tag{6.4}$$

From Eqs. (6.1) and (6.5)

$$\begin{aligned} &N_2\left(1 + \tfrac{\ell_2}{\ell_1}\right) = N_2 + F_n \sin\phi \\ &\Rightarrow N_2 = \tfrac{\ell_1}{\ell_2} F_n \sin\phi \end{aligned} \tag{6.5}$$

From Eqs. (6.4) and (6.5)

$$N_1 = F_n \sin\phi\left(1 + \frac{\ell_1}{\ell_2}\right) \tag{6.6}$$

From Eqs. (6.4), (6.5), and (6.6)

$$F_n = \frac{F_s}{\cos\phi - \mu \sin\phi\left(1 + \frac{2\ell_1}{\ell_2}\right)} \tag{6.7}$$

From Eq. (6.7), it is clear that the offset (by increasing the pressure angle) and length of guide reduce driving effort as they increase the normal reaction.

6.4.2 *Plate Cam with Flat-Face Follower*

Figure 6.10a is an offset plate cam with a flat-face follower that has a retainer spring. The cam is assumed to be rotating at a uniform angular speed in the counterclockwise direction. The axis of translation has an offset e with respect to the axis of rotation of the plate cam. The length of the guide is ℓ_1 and the distance between the follower guide and the flat face is ℓ_2. At the instant shown, the flat face is contacting the cam at a distance ε from the axis of translation of the follower, and F_n is the normal force acting on the flat face by the cam.

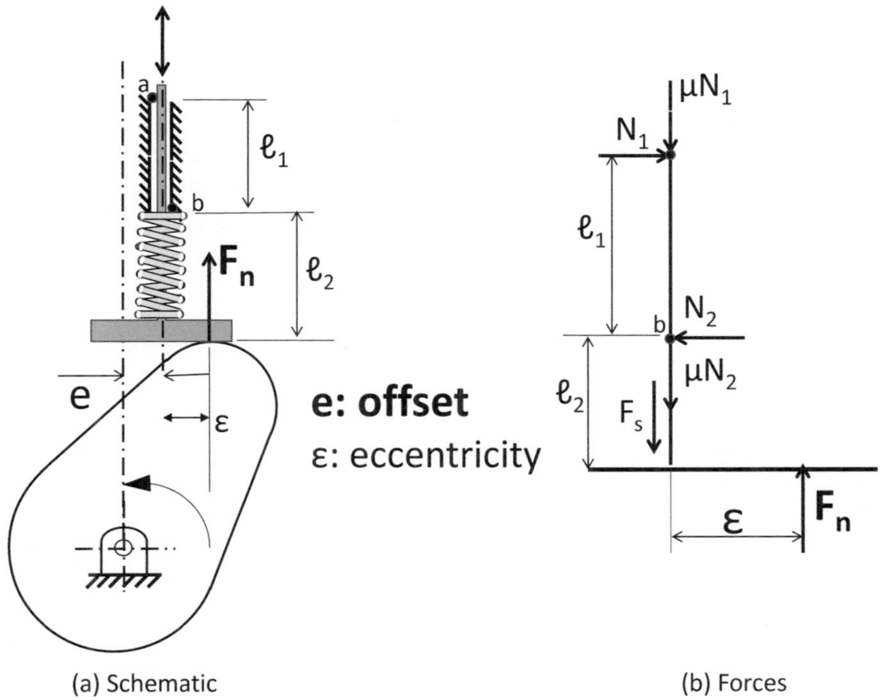

Fig. 6.10 Forces on flat-face plate cam

Figure 6.10a shows the forces acting on the follower. N_1 and N_2 are the normal forces acting at the edges of the guide and the corresponding friction forces, μN_1 and μN_2 acting downwards (since the follower is moving up at the given instant) are shown. In addition, the retainer spring force is also acting on the follower in the downward direction.

Summing horizontal forces

$$N_1 = N_2 \tag{6.8}$$

Summing vertical forces

$$F_n = F_s + \mu(N_1 + N_2) \tag{6.9}$$

Taking moments about the upper surface of the flat face

$$\begin{aligned} F_n\varepsilon &= N_1(\ell_2 + \ell_1) - N_2\ell_2 \\ &= N_1\ell_1 \end{aligned} \tag{6.10}$$

From Eqs. (6.8) and (6.10)

$$F_n = \frac{F_s}{1 - \frac{2\mu_n \varepsilon}{\ell_g}} \tag{6.11}$$

From the above Eq. (6.11), it can be inferred that if eccentricity ε increases, the driving effort F_n increases and offset decreases eccentricity. Hence, offset helps in decreasing the θ.

6.5 Cam Profile

For obtaining a cam profile, basic information presented in Fig. 6.11 is required that contains angular positions of the cam with respect to the cam output. It consists of the (i) low-dwell portion wherein there the cam output is at its minimum position, (ii) rise portion in which the cam output moves from its minimum to the maximum position, (iii) high-dwell portion in which the cam output is held at its maximum position, (iv) fall portion in which the cam output moves from its maximum position and returns to the minimum position. The maximum cam output is also known as rise or lift of the cam and it could be in mm or degrees, depending on the nature of the follower movement. Hence, the angular displacement of the cam in one revolution is divided into many parts as follows: (i) 0-θ_1, low-dwell, (ii) θ_1 to θ_2, rise, (iii) θ_2 to θ_3, high-dwell, (iv) θ_3 to θ_4, fall. The exact values of θ_1 to θ_4 depend on the functional requirements of the application and can be obtained based on the speed of rotation of the cam and time duration of each of the events shown in Fig. 6.11. The path taken during rise and fall will result in different velocity, acceleration and jerk (time derivative of acceleration) throughout the cam motion. But the nature of these values during transition is very critical in deciding the cam profile. The rise and fall of the cam can have a variety of dynamic variation of various parameters: displacement, velocity, acceleration, and jerk. Each of these variations has their own advantages and disadvantages as discussed in the following sections.

6.5.1 Constant Velocity During Rise and Fall

A constant velocity of the follower during rise and fall is one of the easier assumptions to start with. As shown in Fig. 6.12, the velocity is constant in the positive direction from θ_1 and θ_2, and constant in the negative (reverse) direction from θ_3 and θ_4. During the low dwell and high dwell, the following parameters—velocity, acceleration, and jerk—are equivalently zero. Based on the velocity profile, displacement can be obtained by integration during the rise and fall; similarly, acceleration and jerk can be obtained by successive differentiation.

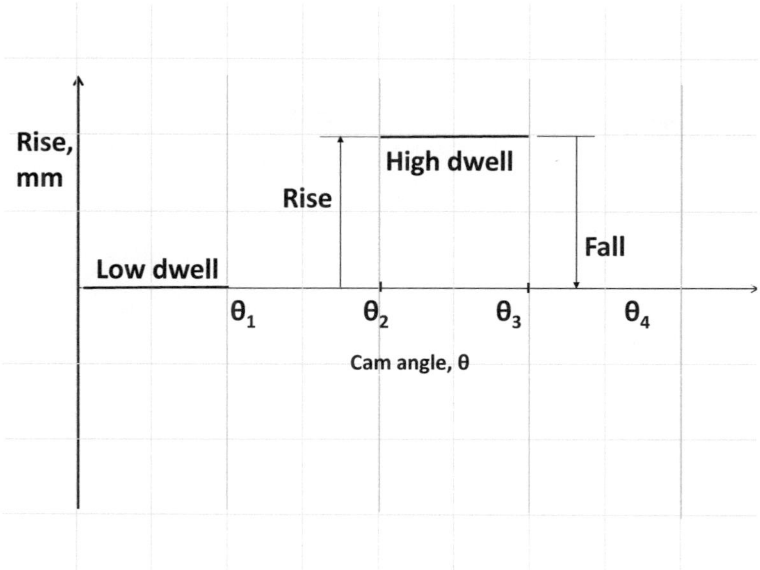

Fig. 6.11 Angular positions of a cam with respect to output

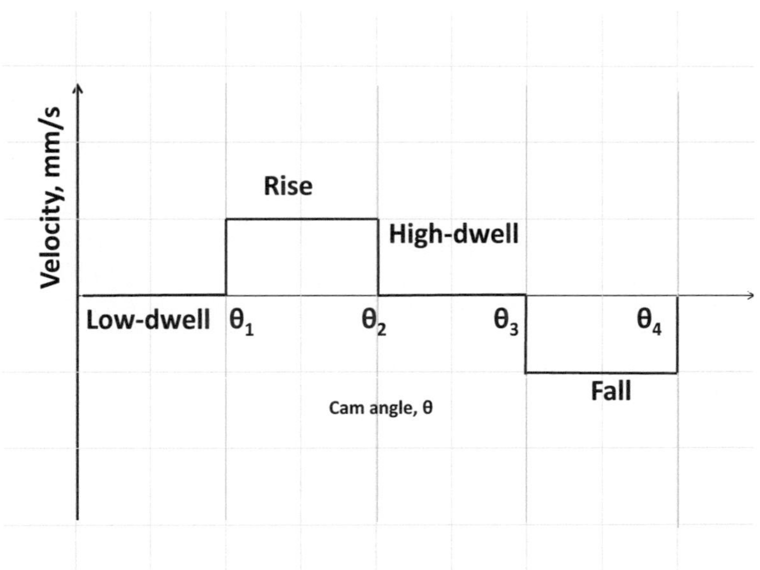

Fig. 6.12 Velocity of the follower for constant velocity rise and fall

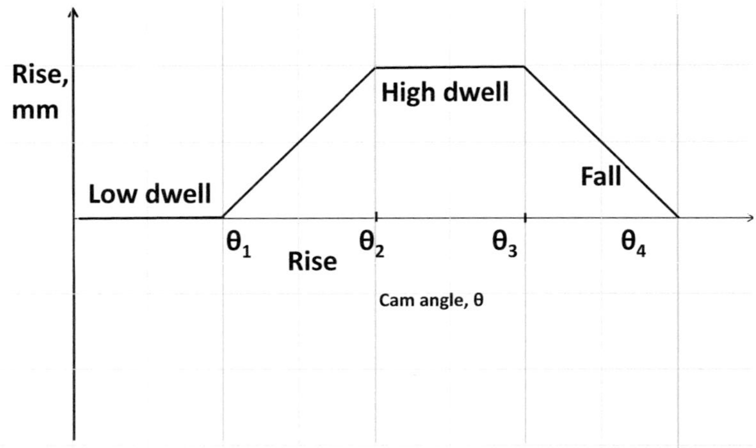

Fig. 6.13 Displacement of the follower for constant velocity

From the velocity diagram of Fig. 6.12, the displacement diagram of the follower can be obtained by integration as shown in Fig. 6.13; the integration constants can be obtained by using the actual values of rise and angular positions. Since the velocity is constant during rise in the positive direction, the displacement of the follower linearly increases with time from the low dwell position at θ_1 of the angular position of the cam and after reaching the maximum displacement the follower at the angular position θ_2 of the cam, it remains in the same position during the high dwell. Then, the follower returns to its initial position linearly in the negative direction from θ_3 to θ_4 of the angular position of the cam.

From the velocity diagram of the follower, its acceleration can be computed as shown in Fig. 6.14 by differentiation. Since the velocity of the follower is either constant or zero, the acceleration is zero throughout except at the transition points at which velocity changes as follows:

(1) From zero to a constant value that results in a very large value of acceleration at the angular position of the cam θ_1.
(2) From a constant value to zero that results in a very large value of deceleration at the angular position of the cam θ_2.
(3) From zero to a negative constant value that results in a very large value of deceleration at the angular position of the cam θ_3.
(4) From a negative constant value to zero that results in a very large value of acceleration at the angular position of the cam θ_4.

Very large values of accelerations at certain instants will result in large forces at the interface of the follower and cam and will possibly result in undesirable effects

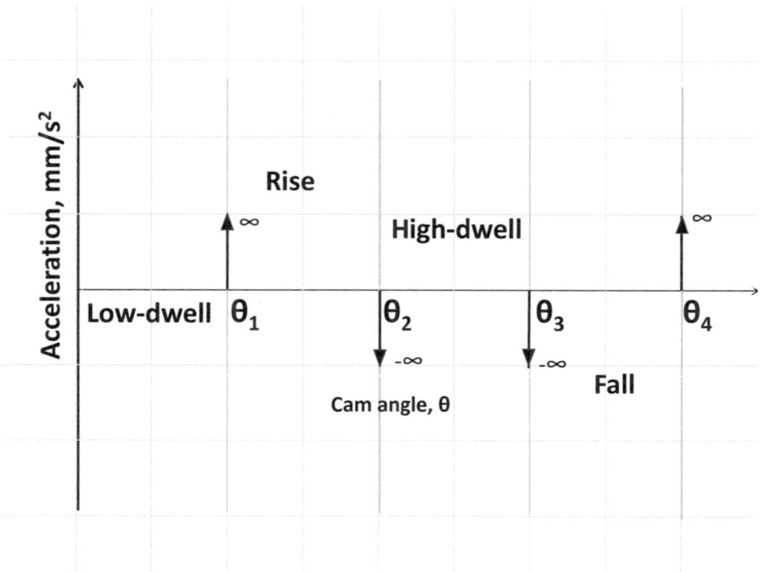

Fig. 6.14 Acceleration for constant velocity rise and fall

at the interface as well as other components of the cam and other components of the machine. Hence, the assumed constant velocity profile of the follower is not desirable for practical applications (Fig. 6.15).

Since the acceleration is infinite at certain instants of time of the cam machine, the jerk being the time derivative of acceleration goes from a high positive value to a high negative value or vice versa at the same instants of the time at which the acceleration was very large. Although the acceleration has spikes in one direction at certain instants. The jerk has spikes in both direction that would result in much worse undesirable effects on the cam mechanism and other components of the machines. Hence, a new displacement profile of the follower has to be configured to obtain a smooth operating cam mechanism.

6.5.2 Simple Harmonic Displacement

Another possible solution to the infinite spikes obtained in acceleration and jerks for a constant velocity follower profile of the cam, discussed in the previous section, would be to assume a harmonic function that has derivatives of any order and hence, even the jerk function that is the third derivative of displacement would contain finite values. Let us try a harmonic function for displacement motion of the follower and check whether it would really work.

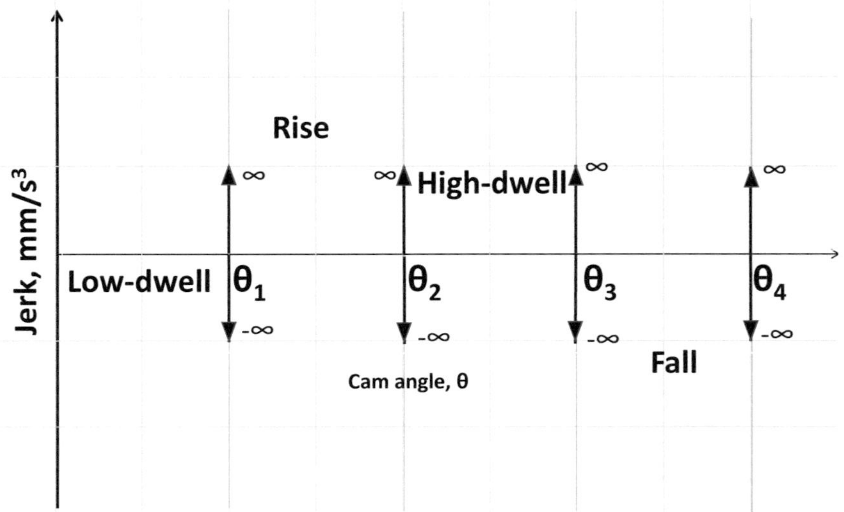

Fig. 6.15 Jerk of the follower for constant velocity rise and fall

SHM displacement, rise

A sinusoidal curve is shown in Fig. 6.16. The first quarter of the curve can be examined as a possible displacement curve for the rise portion of the follower movement that can take place due to the angular rotation of the cam from 0 to β_1. It can be expressed as

$$y = h \sin \frac{\pi\theta}{2\beta_1} \tag{6.12}$$

where h is the maximum distance moved by the follower. The above equation satisfies the condition of $y = 0$ at $\theta = 0$ and $y = h$ at $\theta = \beta_1$. However, it does not make the displacement curve to be uniformly distributed over the duration of the cam rotation during rise since $y = 0.707$ h at $\theta = \beta_1/2$. That may be a minor point but a more important consideration is that the velocity should first increase as the follower moves toward the maximum position and then gradually decrease as the follower comes to a rest, after reaching the maximum height is not satisfied by the above equation.

Hence, Eq. 6.12 is slightly modified to obtain the following displacement equation that contains a harmonic function and satisfies the requirements of the velocity curve

$$y = \frac{h}{2}\left(1 - \cos \frac{\pi\theta}{\beta_1}\right) \tag{6.13}$$

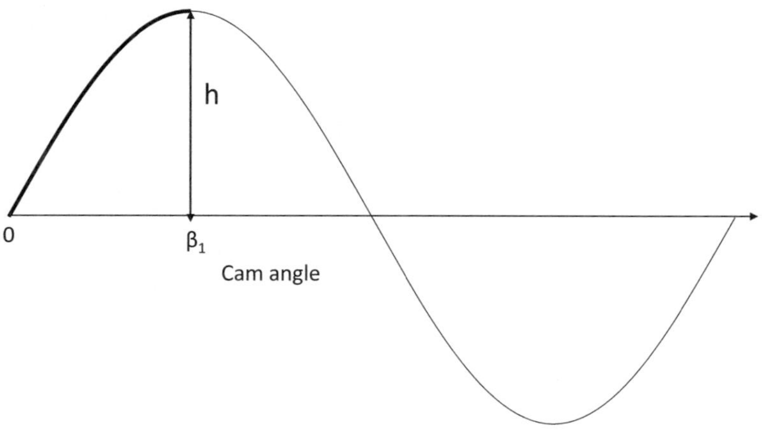

Fig. 6.16 Sinusoidal rise

The displacement y = 0 at θ = 0 and y = h at θ = β_1. In addition, y = h/2 at θ = $\beta_1/2$. Hence, the follower displacement is uniformly distributed through the rise motion. The follower displacement as per Eq. 6.13 is shown in Fig. 6.17.

Differentiating Eq. 6.13, the follower velocity during rise is given by

$$v = \frac{\pi \omega h}{2\beta_1} \sin \frac{\pi\theta}{\beta_1} \tag{6.14}$$

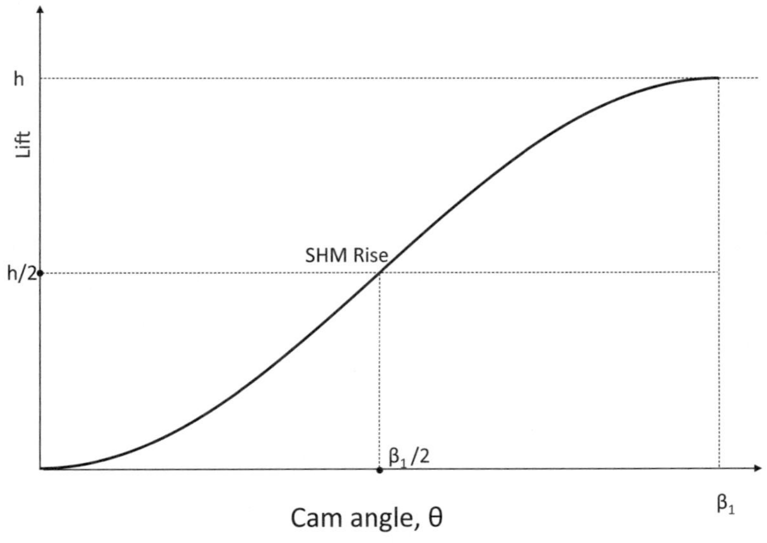

Fig. 6.17 Displacement of a follower as per Eq. 6.13 (SHM displacement, rise)

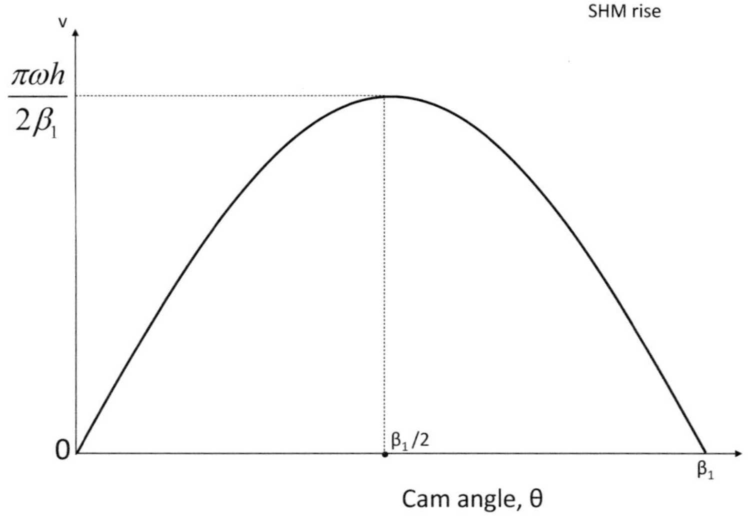

Fig. 6.18 Velocity of the follower as per Eq. 6.14 (SHM displacement, rise)

The follower velocity during rise of Eq. 6.14 is plotted in Fig. 6.18. It starts from 0 and reaches a maximum value of $\frac{\pi\omega h}{2\beta_1}$ at $\theta = \beta_1/2$ and then reduces to zero before reaching the maximum position of the follower.

By differentiating Eq. 6.14, acceleration of the follower during rise can be obtained as

$$a = \left(\frac{\pi\omega}{\beta_1}\right)^2 \frac{h}{2} \cos\frac{\pi\theta}{\beta_1} \tag{6.15}$$

And it is plotted in Fig. 6.19. The follower has an acceleration of $\left(\frac{\pi\omega}{\beta_1}\right)^2 \frac{h}{2}$ during the start of the rise phase and a deceleration of $\left(\frac{\pi\omega}{\beta_1}\right)^2 \frac{h}{2}$ after completing the rise phase. Since the rise phase is in between the low dwell and high dwell, nonzero acceleration of the follower at the beginning and end of this phase creates undesirable forces at these instants. Hence, the harmonic displacement assumed in Eq. (6.13) does not satisfy the requirements of a smooth cam mechanism.

Jerk of the follower during rise can be obtained by differentiating Eq. 6.15

$$j = -\left(\frac{\pi\omega}{\beta_1}\right)^3 \frac{h}{2} \sin\frac{\pi\theta}{\beta_1} \tag{6.16}$$

and is plotted in Fig. 6.20. Although acceleration of the follower does not meet the requirements of a smooth mechanism, the jerk is zero at the beginning and end of the rise phase. The harmonic displacement of Eq. 6.13 can still be used if the low dwell and high dwell phases are not required for the cam.

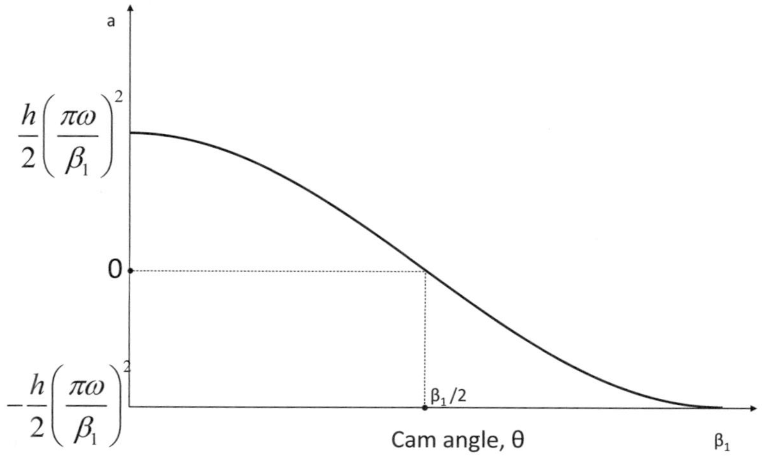

Fig. 6.19 Acceleration of the follower as per Eq. 6.15 (SHM displacement, rise)

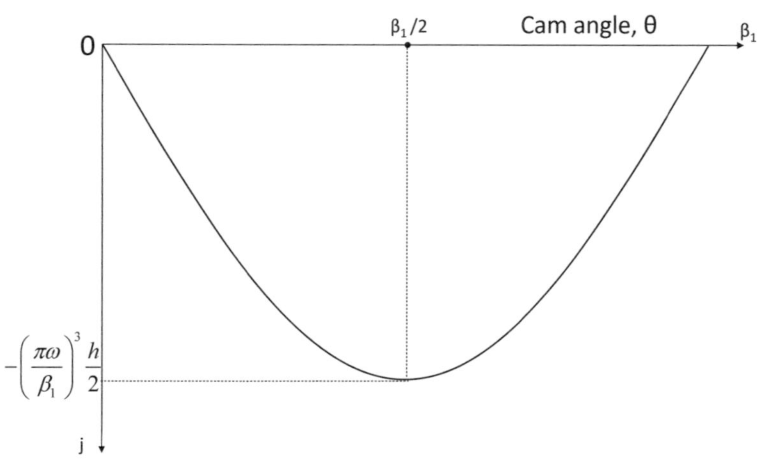

Fig. 6.20 Jerk of the follower as per Eq. 6.16 (SHM displacement, rise)

SHM Fall

Figure 6.21 shows a sinusoidal function and if β_2 is the angle traversed during the fall phase, the above portion of the sinusoidal function can be used. However, due to reasons discussed for the rise phase, the displacement function is slightly modified as follows

$$y = \frac{h}{2}\left(1 + \cos\frac{\pi\theta}{\beta_2}\right) \tag{6.17}$$

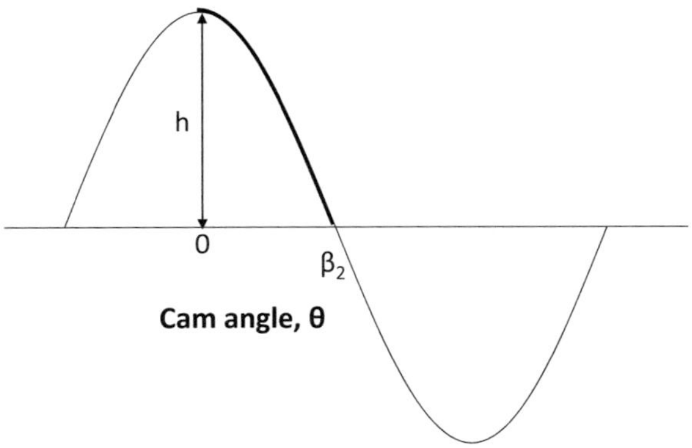

Fig. 6.21 Displacement portion during fall of a sinusoidal function

The displacement y = h at θ = 0 and y = 0 at θ = β_2. In addition, y = h/2 at θ = β_2/2. Hence, the follower displacement is uniformly distributed through the fall motion. The follower displacement as per Eq. 6.17 is shown in Fig. 6.22.

Differentiating Eq. 6.17, the follower velocity during fall is given by

$$v = -\frac{\pi \omega h}{2\beta_2}\sin\frac{\pi\theta}{\beta_2} \tag{6.18}$$

The follower velocity of Eq. 6.18 during fall is plotted in Fig. 6.23. It starts from 0 and reaches a maximum value of $\frac{\pi\omega h}{2\beta_2}$ at θ = β_2/2 in the negative direction of follower movement and then reduces to zero before reaching the minimum position of the follower.

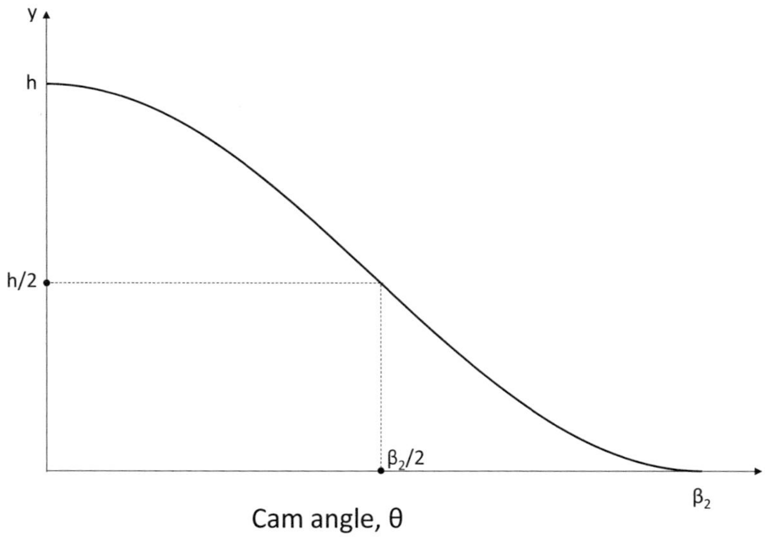

Fig. 6.22 Displacement during fall of the follower as per Eq. 6.17 (SHM displacement, fall)

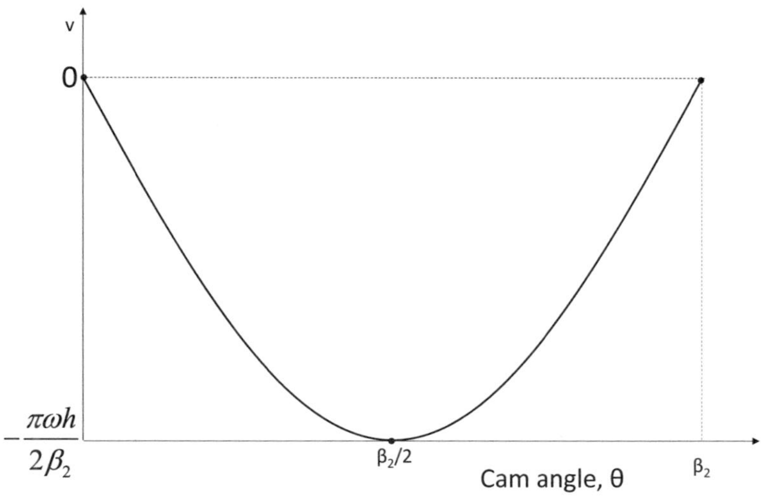

Fig. 6.23 Velocity of the follower as per Eq. 6.18 (SHM displacement, fall)

By differentiating Eq. 6.18, acceleration of the follower during fall can be obtained as

$$a = -\left(\frac{\pi\omega}{\beta_2}\right)^2 \frac{h}{2} \cos\frac{\pi\theta}{\beta_2} \tag{6.19}$$

and it is plotted in Fig. 6.24. The follower has a deceleration of $\left(\frac{\pi\omega}{\beta_2}\right)^2 \frac{h}{2}$ during the start of the fall phase and a deceleration of $\left(\frac{\pi\omega}{\beta_2}\right)^2 \frac{h}{2}$ after completing the fall phase. Since the fall phase is in between the high dwell and low dwell, nonzero acceleration of the follower at the beginning and end of this phase creates undesirable forces at these instants. Hence, the harmonic displacement does not satisfy the requirements of a smooth cam mechanism during the fall phase also.

Jerk of the follower during the fall phase can be obtained by differentiating Eq. 6.19

$$j = \left(\frac{\pi\omega}{\beta_2}\right)^3 \frac{h}{2} \sin\frac{\pi\theta}{\beta_2} \tag{6.20}$$

and is plotted in Fig. 6.25. Although acceleration of the follower does not meet the requirements of a smooth mechanism during the fall phase, the jerk is zero at the beginning and end of the fall phase. The harmonic displacement of Eq. 6.17 can still be used if the low dwell and high dwell phases are not required for the cam.

From the above discussion, it is clear that the assumption of harmonic displacement functions will not help in obtaining a smooth operating cam mechanism for both rise and fall phases. Hence, our discussion next is focused on starting with a harmonic acceleration function and then obtaining velocity and displacement functions.

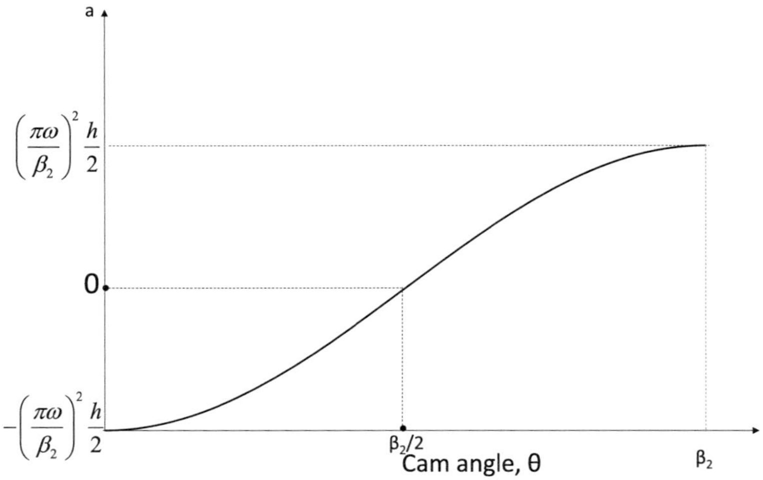

Fig. 6.24 Acceleration of the follower as per Eq. 6.19 (SHM displacement, fall)

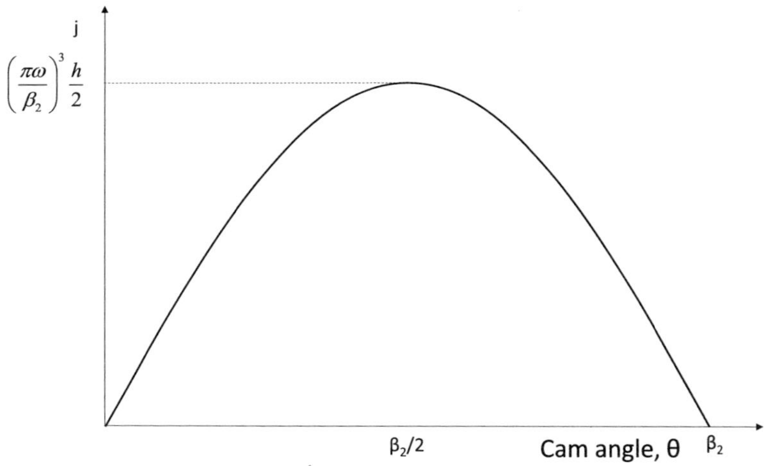

Fig. 6.25 Jerk of the follower as per Eq. 6.20 (SHM displacement, fall)

6.5.3 Cycloidal Displacement

Since the acceleration is very large at the end of rise and dwell, the above harmonic displacement functions are not satisfactory. Hence, we have to choose a displacement function that gives minimum acceleration and jerk at the ends of the rise and dwell periods. So, instead of beginning with a displacement function, we can begin with a sinusoidal acceleration function and then integrate it to get displacement and velocity and then differentiate to obtain the jerk function. Such an acceleration function will have zero acceleration at the beginning and end of the rise or dwell periods.

Sinusoidal acceleration function will result in a cycloidal displacement that will be discussed later. We shall assume in this section that the angular speed of the cam is unity so it helps to derive the equations easily , which would help in the qualitative discussion. Towards the end of this section, suggestions for necessary changes to be made for obtaining values of velocity, acceleration and jerk for actual values of angular speeds of the cams will be made.

Cycloidal displacement (rise)

Let us choose the following sinusoidal acceleration function during rise:

$$a = A \sin \frac{2\pi\theta}{\beta_1} \tag{6.21}$$

where A is the amplitude of acceleration that is yet to be decided and β_1 is the total angle traversed by the cam during the rise motion and θ is the angle at any instant.

Integrating Eq. 6.21, the velocity function during rise is given by

$$v = -\frac{A\beta_1}{2\pi} \cos \frac{2\pi\theta}{\beta_1} + C_1 \tag{6.22}$$

The constant C_1 in Eq. 6.22 can be evaluated by assuming zero velocity of the follower at the beginning of the rise. Putting $v = 0$ at $\theta = 0$, the constant C_1 is given by

$$C_1 = \frac{A\beta_1}{2\pi} \tag{6.23}$$

Hence, equation for velocity of the follower during rise now becomes

$$v = \frac{A\beta_1}{2\pi}\left(1 - \cos \frac{2\pi\theta}{\beta_1}\right) \tag{6.24}$$

Incidentally, from Eq. 6.24, the velocity is zero at $\theta = \beta_1$, which is a desirable feature.

Integrating Eq. 6.24 with respect to θ

$$y = \frac{A\beta_1}{2\pi}\left(\theta - \frac{\beta_1}{2\pi} \sin \frac{2\pi\theta}{\beta_1}\right) + C_2 \tag{6.25}$$

Since the displacement is zero at the beginning of the rise, $y = 0$ at $\theta = 0$. Hence, the constant C_2 is zero. The displacement at $\Theta = \beta_1$ should match the lift h. From Eq. 6.25

$$h = \frac{A\beta_1^2}{2\pi} \tag{6.26}$$

Equation 6.26 gives the amplitude of acceleration, given by

$$A = \frac{2\pi h}{\beta_1^2} \tag{6.27}$$

From Eq. 6.27, the equations for acceleration, velocity, and displacement can be, respectively, written as:

$$a = \frac{2\pi h}{\beta_1^2} \sin \frac{2\pi\theta}{\beta_1} \tag{6.28}$$

$$v = \frac{h}{\beta_1}\left(1 - \cos \frac{2\pi\theta}{\beta_1}\right) \tag{6.29}$$

$$y = h\left(\frac{\theta}{\beta_1} - \frac{1}{2\pi} \sin \frac{2\pi\theta}{\beta_1}\right) \tag{6.30}$$

The above displacement equation is a sum of a straight line and negative of a sinusoidal function. Since the sine function is wrapped around a linear function, it is like a cycloidal function. Hence, it was called as a cycloidal displacement function.

Differentiating Eq. 6.28, the equation for jerk is given by

$$j = \frac{(2\pi)^2 h}{\beta_1^3} \cos \frac{2\pi\theta}{\beta_1} \tag{6.31}$$

Although the cycloidal displacement function is satisfactory for zero acceleration of the follower at the beginning and end of rise, there is a finite jerk at beginning and end as shown in Fig. 6.29, which should be acceptable in most cases. However, the peak value of velocity and acceleration during the rise phase must be paid attention (Figs. 6.26, 6.27 and 6.28).

Cycloidal displacement, fall

Let us choose the following sinusoidal acceleration function for the fall phase:

$$a = -A \sin \frac{2\pi\theta}{\beta_2} \tag{6.32}$$

where A is the amplitude of acceleration that is yet to be decided and β_2 is the total angle traversed by the cam during the fall motion and θ is the angle at any instant.

Integrating Eq. 6.32, the velocity function is given by

$$v = \frac{A\beta_2}{2\pi} \cos \frac{2\pi\theta}{\beta_1} + C_1 \tag{6.33}$$

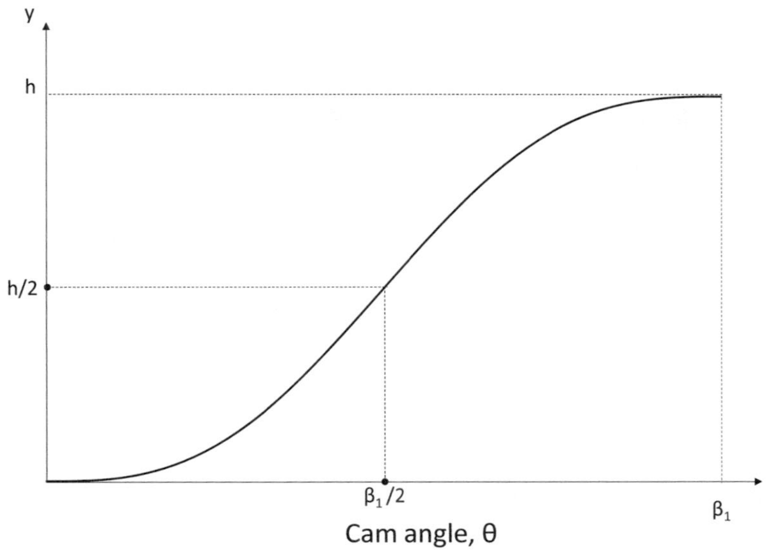

Fig. 6.26 Displacement of the follower as per Eq. 6.30 (cycloidal displacement, rise)

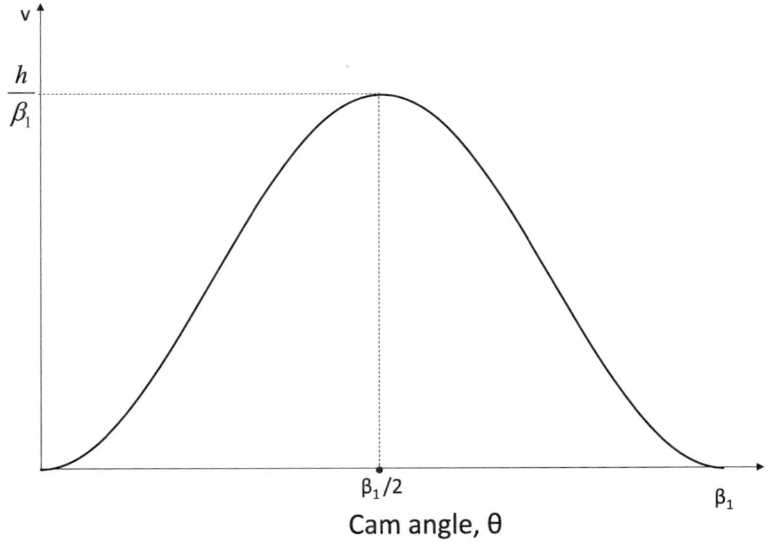

Fig. 6.27 Velocity of the follower as per Eq. 6.29 (cycloidal displacement, rise)

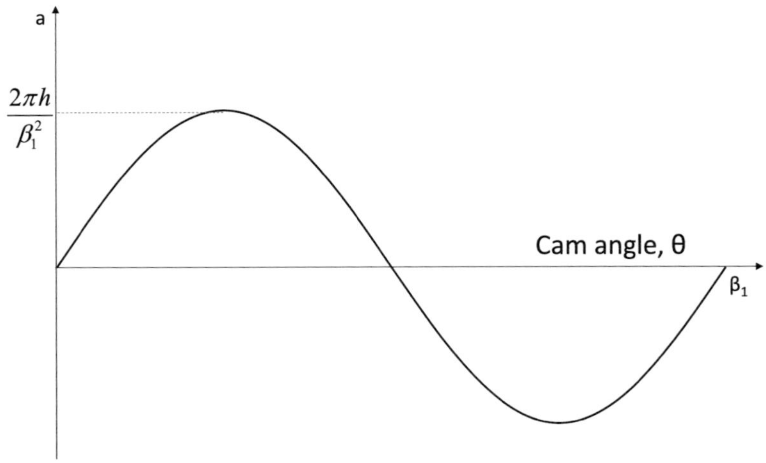

Fig. 6.28 Acceleration of the follower as per Eq. 6.28 (cycloidal displacement, rise)

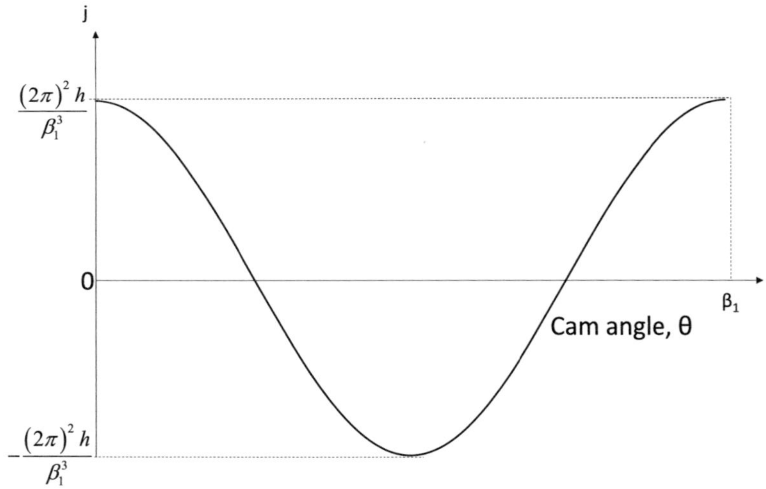

Fig. 6.29 Jerk of the follower as per Eq. 6.31 (cycloidal displacement, rise)

The constant C_1 in Eq. 6.33 can be evaluated by assuming zero velocity at the beginning of the rise. Putting $v = 0$ at $\theta = 0$, the constant C_1 is given by

$$C_1 = -\frac{A\beta_2}{2\pi} \tag{6.34}$$

Hence, equation for velocity now becomes

$$v = -\frac{A\beta_2}{2\pi}\left(1 - \cos\frac{2\pi\theta}{\beta_1}\right) \tag{6.35}$$

Incidentally, from Eq. 6.33, the velocity is zero at $\theta = \beta_2$, which is a desirable feature.

Integrating Eq. 6.35 with respect to θ

$$y = -\frac{A\beta_2}{2\pi}\left(\theta - \frac{\beta_2}{2\pi}\sin\frac{2\pi\theta}{\beta_2}\right) + C_2 \tag{6.36}$$

Since the displacement is equal to h at the beginning of the rise, $y = h$ at $\theta = 0$. Hence the constant $C_2 = h$. The displacement at $\Theta = \beta_2$ should be equal to zero. From Eq. 6.36

$$h = \frac{A\beta_2^2}{2\pi} \tag{6.37}$$

Equation 6.37 gives the amplitude of acceleration, given by

$$A = \frac{2\pi h}{\beta_2^2} \tag{6.38}$$

From Eq. 6.38, the equations for acceleration, velocity, and displacement can be respectively written as:

$$a = -\frac{2\pi h}{\beta_2^2}\sin\frac{2\pi\theta}{\beta_2} \tag{6.39}$$

$$v = -\frac{h}{\beta_2}\left(1 - \cos\frac{2\pi\theta}{\beta_2}\right) \tag{6.40}$$

$$y = -h\left(\frac{\theta}{\beta_2} - \frac{1}{2\pi}\sin\frac{2\pi\theta}{\beta_2}\right) + h \tag{6.41}$$

Differentiating Eq. 6.39, the equation for jerk is given by

$$j = \frac{(2\pi)^2 h}{\beta_2^3}\cos\frac{2\pi\theta}{\beta_2} \tag{6.42}$$

For obtaining values of velocity, acceleration and jerk for actual angular speeds of the cam, Eqs. (6.29), (6.28) and (6.31) must be respectively multiplied by ω, ω^2 and ω^3; similarly for Eqs. (6.40), (6.39) and (6.42). We have discussed many possibilities of cam profiles and how they affect different dynamic responses. Although there is some improvement with profiles generated based on cycloidal displacement, there is possibility for improvement for specific dynamic responses for an application. See exercise problem 6.8 wherein a trapezoidal acceleration profile is used for obtaining a cam profile (Figs. 6.30, 6.31, 6.32 and 6.33).

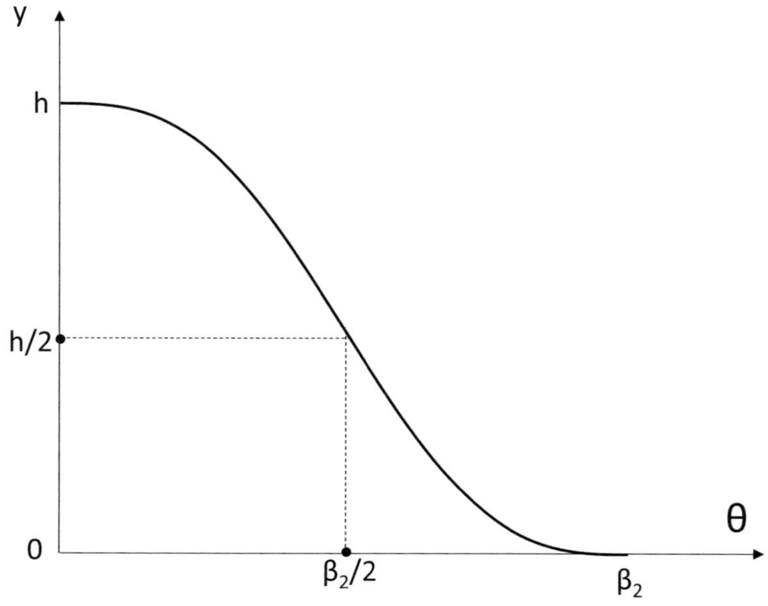

Fig. 6.30 Displacement of the follower during fall as per Eq. 6.41

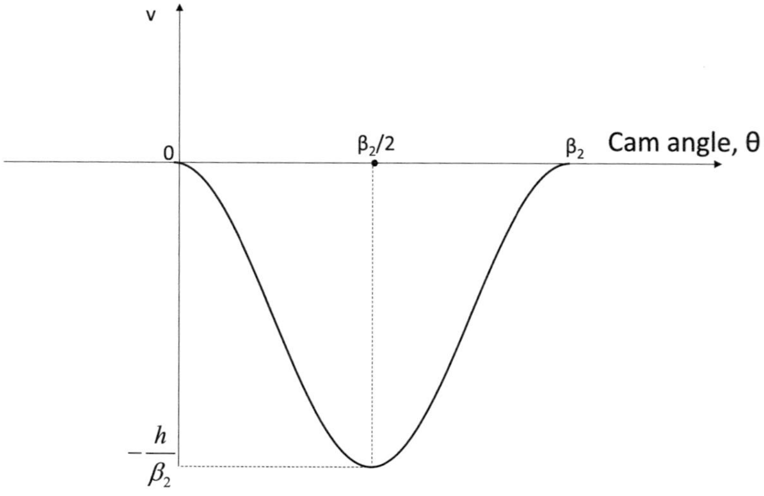

Fig. 6.31 Velocity of the follower as per Eq. 6.40

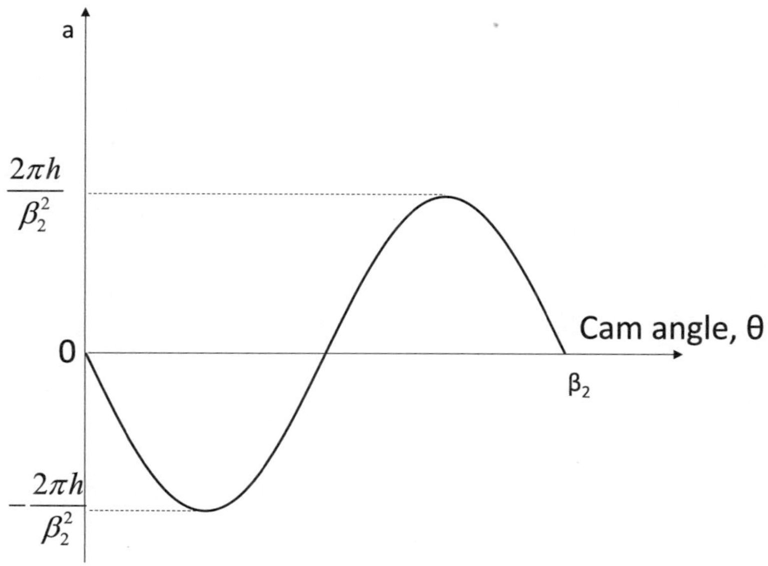

Fig. 6.32 Acceleration of the follower as per Eq. 6.39

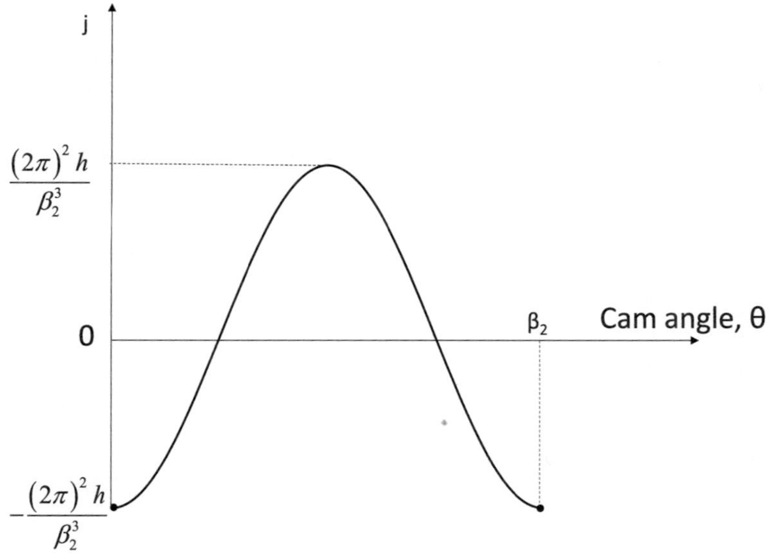

Fig. 6.33 Jerk of the follower as per Eq. 6.42

6.5.4 Procedure for Drawing Cam Profile

1. Obtain detailed information regarding the total movement of the cam follower: rise, dwell and fall durations as a function of the cam angle.
2. Decide the displacement profile during the rise and fall so the velocity and accelerations are continuous throughout and the jerk remains a finite value.
3. Obtain equations for the follower displacement for one revolution of the cam.
4. Decide on the base circle (r_b) and follower radius (r_f).
5. Draw a circle of radius r_b and divide the circle into a number of equal parts (say 10°). In addition, mark important transition points if they do not coincide with the above divisions.
6. Determine the trace point radius, $R = r_b + r_f + y$, for various cam angles and mark them on corresponding radii in the opposite direction to the direction of the cam.
7. Using r_f as the radius, draw circles at various trace point locations.
8. Join the above circles to obtain the cam profile.

6.6 Conclusions

Various types of cams that can be used as accurate function generators to help in timing control of mechanisms have been discussed in this chapter. The most challenging task of a cam design is to get an accurate profile for a specific application that will help in its smooth and accurate functioning. Interestingly, there are no example problems in this chapter as each discussion is an example itself. These discussions can be very easily applied to solve the exercise problems and as well as any practical application.

6.7 Exercise Problems

6.1 A plate cam, rotating clockwise, drives a roller follower to a maximum displacement of 50 mm and Table P6.1 gives its displacement corresponding to various angles of the cam.

Using a minimum cam radius of 40 mm and roller diameter of 25 mm, layout the cam profile and determine the maximum pressure angle.

6.2 Using the data of exercise problem 6.1, draw the cam layout for a flat-faced follower and determine the width of the flat face.

6.3 Repeat exercise problem 6.1 if the roller follower is offset by 15 mm to the left.

6.4 Repeat exercise problem 6.2 if the flat-faced follower is offset by 15 mm to the left.

6.5 A cam has the following regions of angular rotation: dwell from 0° to 90°, rise from 90° to 180°, dwell from 180° to 270°, fall from 270° to 360°. During the rise, the reciprocating follower is expected to move a total distance of

Table P6.1 Cam angle versus displacement

Cam angle (degrees)	Displacement (mm)
0	0
30	3.34
60	12.37
90	25.13
120	37.76
150	46.84
180	50
210	46.84
240	37.76
270	25.13
300	12.37
330	3.34
360	0

60 mm and the cam has a constant speed of 500 RPM. Assume a seventh-degree polynomial for determining the rise and fall profiles and determine the constants such that displacement and its derivatives up to third order are continuous functions throughout the cam motion.

6.6 A cam has the following regions of angular rotation: rise from 0° to 50°, dwell from 50° to 190°, fall from 190° to 290° and dwell from 290° to 360°. During the rise, the reciprocating follower is expected to move a total distance of 60 mm and the cam must complete one complete rotation in 10 s. Assume a suitable function, determining the rise and fall profiles, determine the constants such that the velocities are minimum throughout

6.7 Draw the displacement, velocity, acceleration, and jerk of a reciprocating plate cam follower for which the displacement versus cam angle relationship is shown in Fig. P6.7 for an angular cam speed of 1000 rpm.

6.8 A valve has a total lift of 35 mm and is operated by a reciprocating roller follower. The acceleration and deceleration corresponding to only the rise portion are shown in Fig. P6.8. Draw the velocity and displacement diagrams as a function of cam angle for the rise portion, if the cam turns counterclockwise at 200 rpm.

6.9 The displacement diagram of a reciprocating cam follower is shown in Fig. P6.9a that operates the exhaust valve of an engine to lift it by 25 mm. The corresponding velocity and acceleration diagrams are shown in Fig. P6.9b, c. If the engine runs at 500 rpm, determine the peak velocities, and accelerations of the follower during rise and fall.

6.10 Consider the displacement diagram required of a reciprocating roller follower plate cam as shown in Fig. P6.10. It consists of a SHM rise and fall equally divided over one rotation of the cam. If the roller radius is 15 mm and the prime radius is 60 mm, draw the layout of the cam profile. In addition, draw the velocity, acceleration, and jerk diagrams for a uniform clockwise angular speed of 500 rpm.

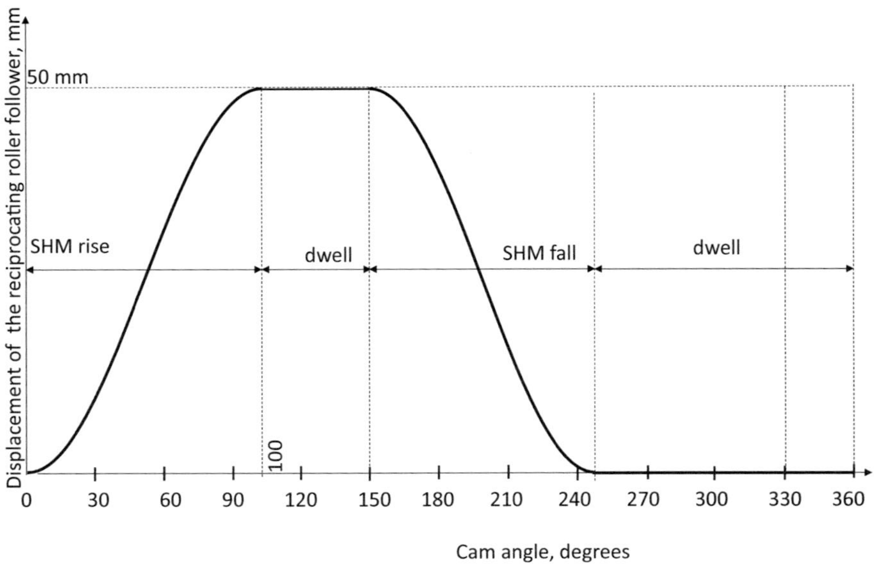

Fig. P6.7 Follower displacement diagram of exercise problem 6.7

Fig. P6.8 Follower displacement diagram of exercise problem 6.8

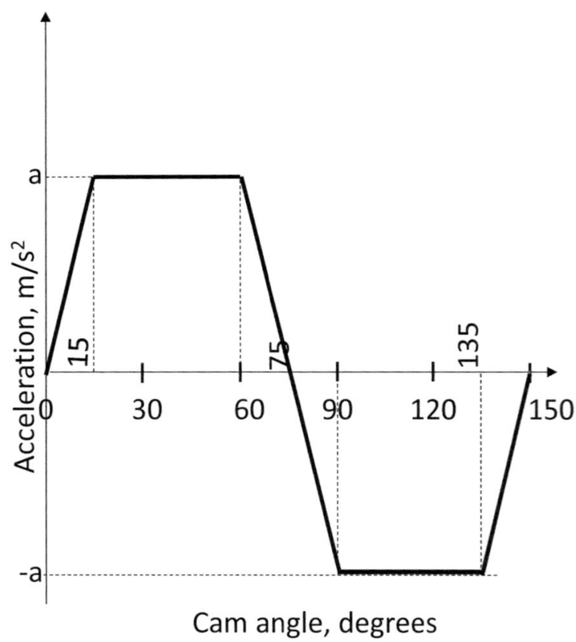

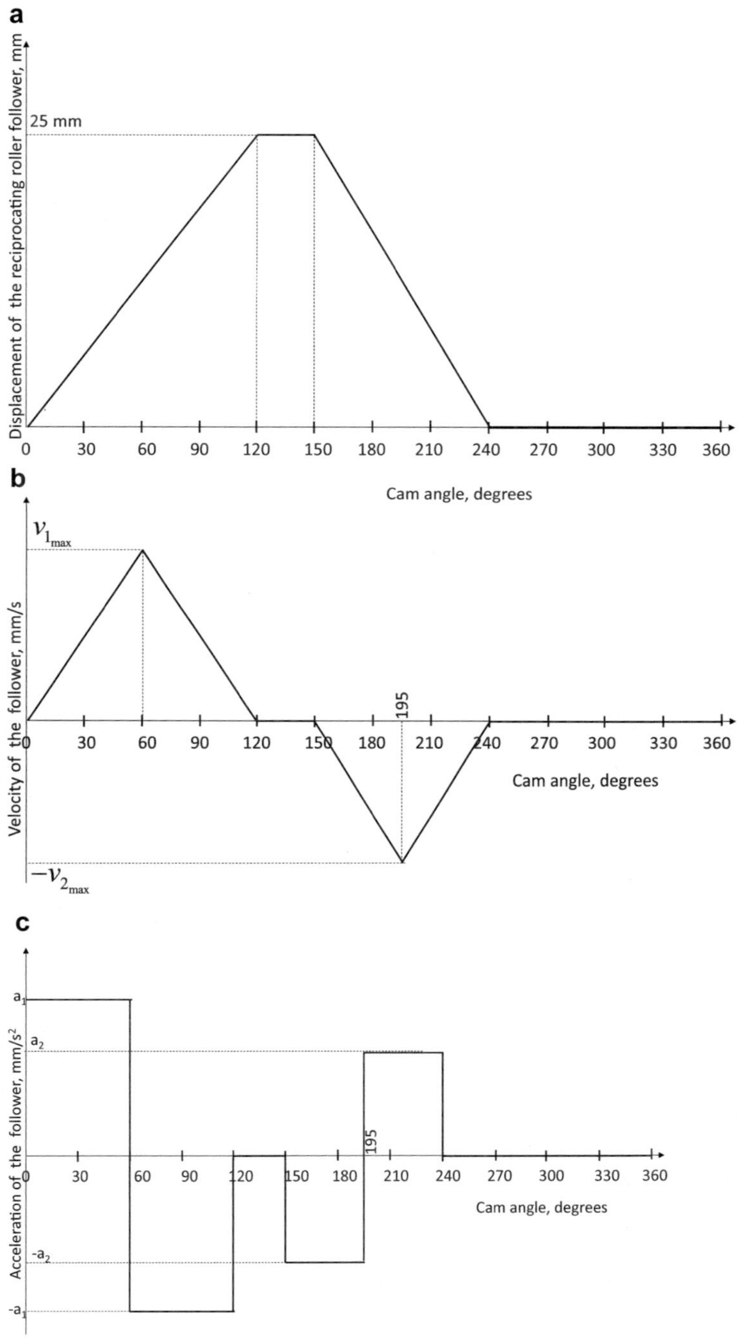

Fig. P6.9 **a** Follower displacement diagram of exercise problem 6.9. **b** Follower velocity diagram of exercise problem 6.9. **c** Follower acceleration diagram of exercise problem 6.9

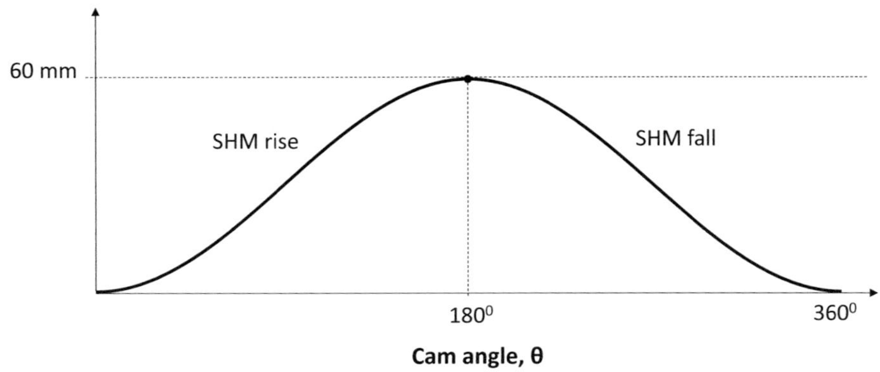

Fig. P6.10 Follower displacement diagram of exercise problem 6.10

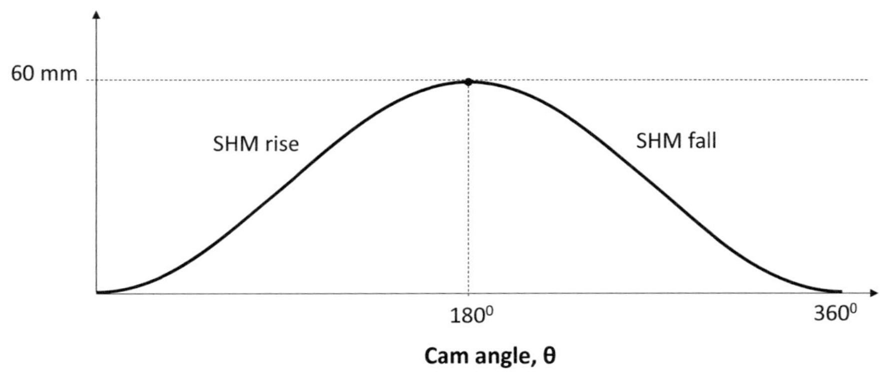

Fig. P6.11 Follower displacement diagram of exercise problem 6.11

6.11 Consider the displacement diagram required of a reciprocating flat-faced follower plate cam as shown in Fig. P6.11. It consists of a SHM rise and fall equally divided over one rotation of the cam. If the prime cam radius is 60 mm and the follower is offset 15 mm to the left, draw the layout of the cam profile. In addition, draw the velocity, acceleration, and jerk diagrams for a uniform counterclockwise angular speed of 500 rpm.

6.12 An oscillating roller follower is expected to produce the motion shown in Fig. P6.12 from a plate cam. Assume a roller radius of 15 mm, prime radius of 65 mm and the follower is pivoted 130 mm to the right of the cam axis. Draw the layout of the cam profile and in addition, draw the angular velocity, angular acceleration and jerk for a uniform counterclockwise plate cam speed of 500 rpm.

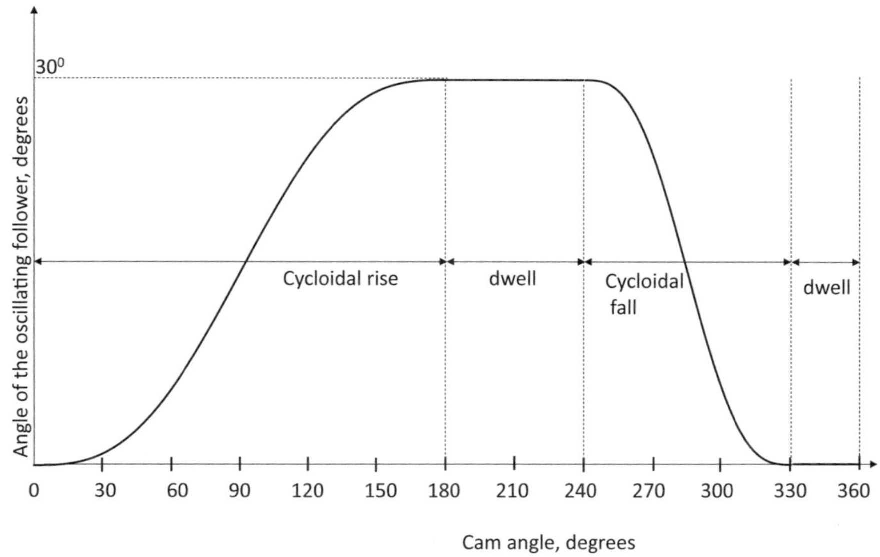

Fig. P6.12 Follower angle diagram of exercise problem 6.12

6.13 A reciprocating roller follower is expected to produce the motion shown in Fig. P6.13 from a plate cam. Assume a roller radius of 15 mm, prime radius of 65 mm and the follower is offset 15 mm to the left of the cam axis. Draw the layout of the cam profile and in addition, draw the velocity, acceleration and jerk for a uniform counterclockwise plate cam speed of 500 rpm.

6.14 An eccentric rotating at a speed of 400 rpm shown in Fig. P6.14 drives a rigid mass guided in a vertical slot. (a) Determine the angle at which the contact force becomes zero (b) Determine the speed at which the cam loses contact.

6.15 An eccentric rotating at a speed of 700 rpm shown in Fig. P6.15 drives a rigid mass guided in a vertical slot and compressing a spring. If the spring is initially compressed to 5 mm at the neutral position of the eccentric, determine the angle at which the contact force becomes zero.

6.16 An eccentric cam is operating an oscillating lever of 2 kg mass as shown in Fig. P6.16. The lever swings in the vertical plane by a maximum of 15° with respect to the equilibrium position. The radius of gyration of the lever about its mass center C is 75 mm. (a) Determine the maximum normal reaction between the lever and the cam rotating counterclockwise at 150 rpm. (b) What is the maximum speed before the cam and the lever lose contact?

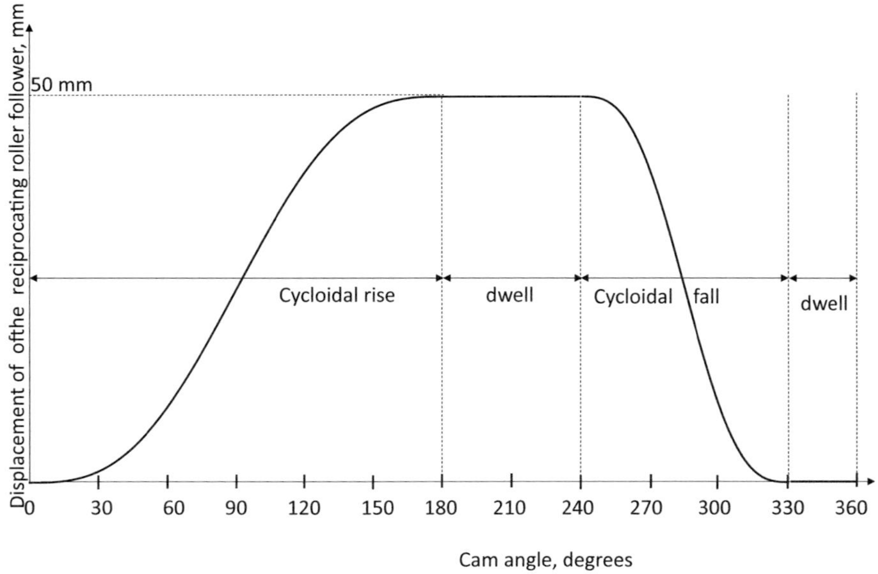

Fig. P6.13 Follower displacement of exercise problem 6.13

6.17 Consider a plate cam with roller follower as shown in Fig. P6.17a. The cam has a minimum radius of 50 mm and the (a) maximum lift is 40 mm to the follower and the roller diameter is 25 mm. For the motion shown in Fig. P6.17b,

(a) Determine the cam profile for every 30° interval (also at important transition points) and determine the pressure angle during the rise portion at the above intervals. Assume the cam is rotating counterclockwise at a speed of 350 RPM.
(b) If a cam, whose profile was obtained in part (a), is used with a spring of 100 N/m stiffness, determine the normal reaction at all the angular intervals during rise. Assume the coefficient of friction as 0.25 at the prismatic pair and that the spring is compressed by 5 mm at the beginning of the rise and mass of the follower is 0.3 kg.

6.18 Consider a plate cam with flat-faced follower as shown in Fig. P6.18. The cam has a minimum radius of 50 mm and the maximum lift is 40 mm to the follower for the motion shown in Fig. P6.18b,

Fig. P6.14 Reciprocating mass driven by an eccentric

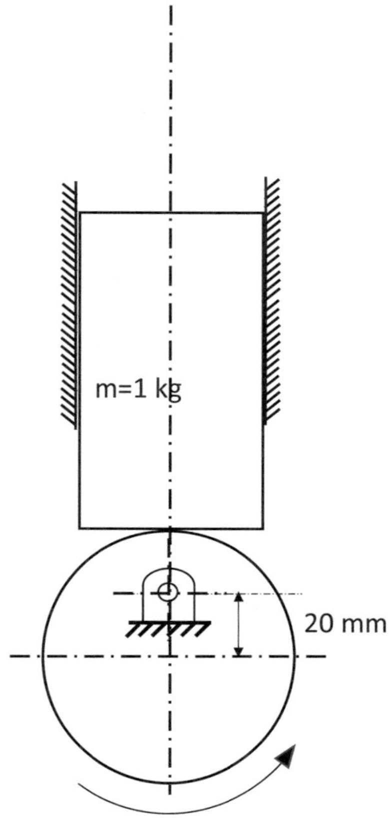

(a) Draw the cam profile for every 30° interval (also at important transition points) and determine the point of application for normal reaction during the rise portion at the above intervals. Assume the cam is rotating counterclockwise at a speed of 350 RPM.

(b) If a cam, whose profile was obtained in part (a), is used with a spring of 100 N/m stiffness, determine the normal reaction at all the angular intervals during rise. Assume the coefficient of friction as 0.2 at the prismatic pair and that the spring is compressed by 5 mm at the beginning of the rise and the mass of the follower is 0.3 kg.

6.19 The cam follower shown in Fig. P6.19 has a spring of stiffness 2 kN/m, mass of the follower is 0.4 kg, and an eccentricity of 20 mm and an angular speed of 50 rad/s. The retaining spring has a preload of 10 N. (a) Determine whether

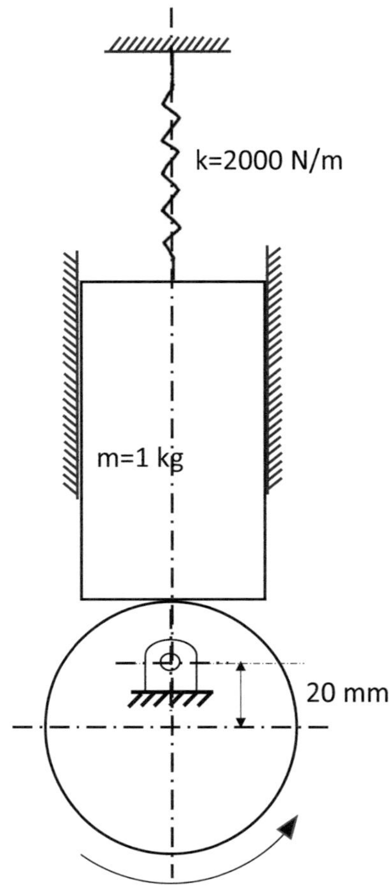

Fig. P6.15 Spring mass driven by an eccentric

there is loss of contact and if so, the angle at which it takes place. Determine the maximum and minimum values of the contact force.

6.20 The cam shown in Fig. P6.20 moves the 2 kg mass of the follower upward through a distance of 50 mm with parabolic motion of 160° of cam rotation, dwell for 40°, return to the starting position with simple harmonic motion of 120° and dwell for the remaining motion. There is no friction or damping. The spring rate is 8 kN/m and the spring preload is 20 N. Determine whether there is any loss of contact between the cam and follower and if so, at what speed? Assume the minimum cam radius of 50 mm and the roller diameter of 25 mm.

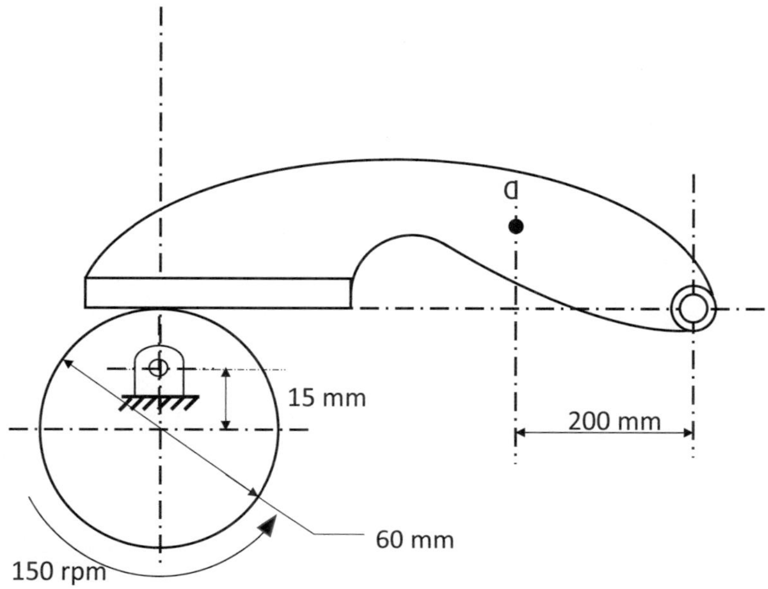

Fig. P6.16 Oscillating follower driven by an eccentric

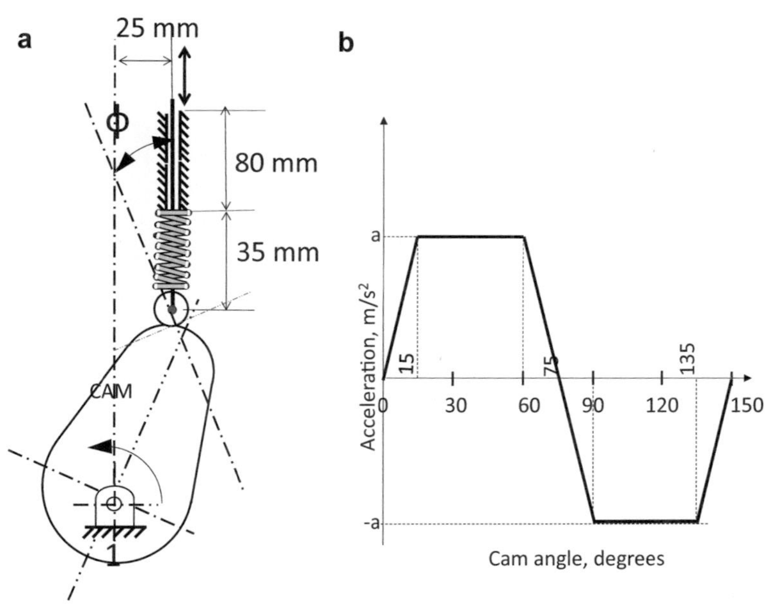

Fig. P6.17 **a** Spring loaded offset cam-roller follower. **b** Acceleration diagram of the follower

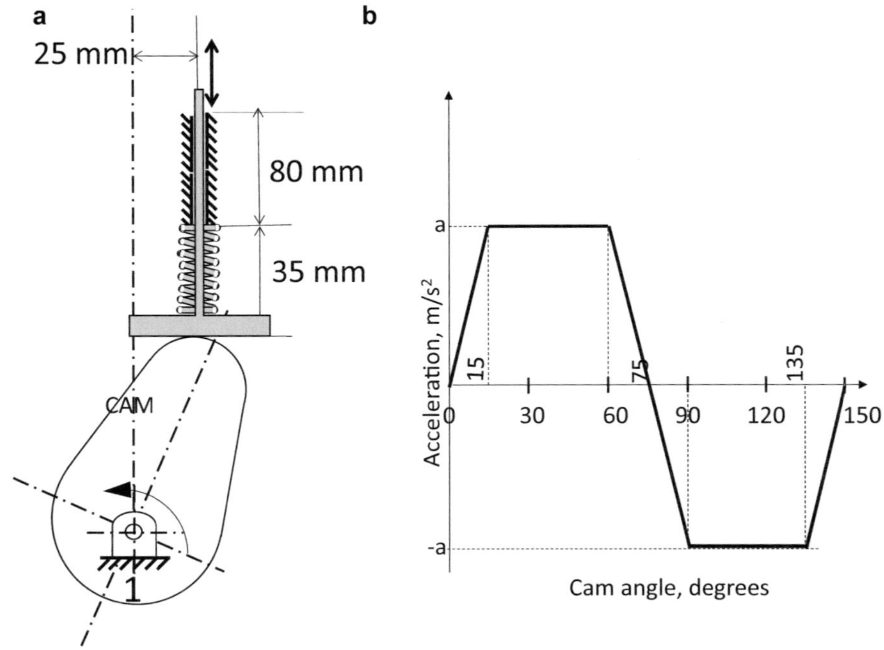

Fig. P6.18 **a** Spring loaded offset cam: flat face follower. **b** Acceleration diagram of the follower

Fig. P6.19 Eccentric driven flat-face follower

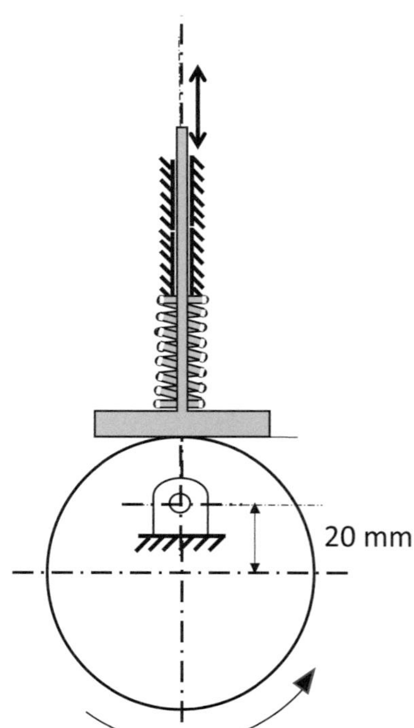

Fig. P6.20 Spring loaded offset cam-roller follower

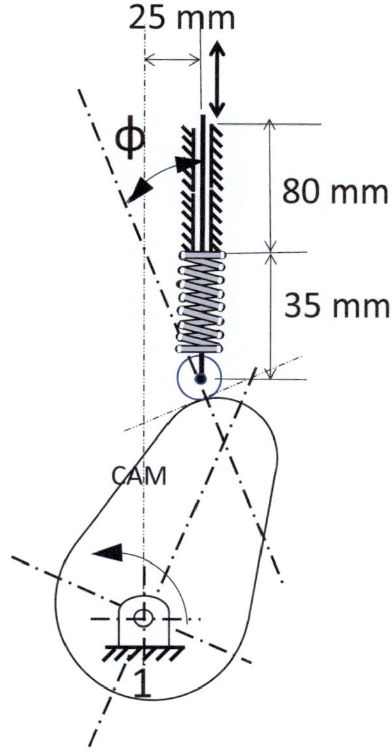

Chapter 7
Synthesis of Mechanisms

Thus far, in all the six chapters, we have discussed mechanisms whose details were well known; we only carried out their kinematic analysis and in addition, we will be analyzing the forces acting on them in Chap. 8. However, in many cases, the type of mechanism that can perform a certain task will not be known. Therefore, "synthesis" is the process by which we design and create mechanisms that can perform a certain task. For example, if we are to design the dimensions of a four-bar link that conforms to position and kinematic requirements can be called synthesis. In this chapter, we shall define some basic rules to carry out such a synthesis. Sometimes, specific function and path generation are expected from mechanisms and we shall discuss the synthesis process in such cases. In addition, we shall synthesize mechanisms subject to the constraint of kinematic parameters.

Following are various types of synthesis:

(i) Type synthesis refers to the mechanism selected: linkage, gear, cam, etc.
(ii) Number synthesis refers to the number of links and joints.
(iii) Dimensional synthesis consists of determining the dimensions of the individual links.
(iv) Function generation: output motion according to a specific function of the input.
(v) Path generation: coupler point should trace a path of prescribed shape.

7.1 Bloch's Method of Synthesis

A four-bar linkage to which the Bloch's method of synthesis is applied is shown in Fig. 7.1. The fixed link hinged at O_2 and O_4 is designated as link 1. The crank O_2A is link 2, the floating coupler link AB is link 3 and the follower O_4B is link 4. The angles made by the above links are, respectively, θ_1, θ_2, θ_3, and θ_4, measured counterclockwise with respect to horizontal. Each of the links is represented by the position vectors: $\mathbf{r}_1$, $\mathbf{r}_2$, $\mathbf{r}_3$, and $\mathbf{r}_4$. The vectors for links 1, 2, and 3 are defined such

D. N. Manik, *Fundamentals of Mechanisms and Machines*,
https://doi.org/10.1007/978-981-97-1810-8_7

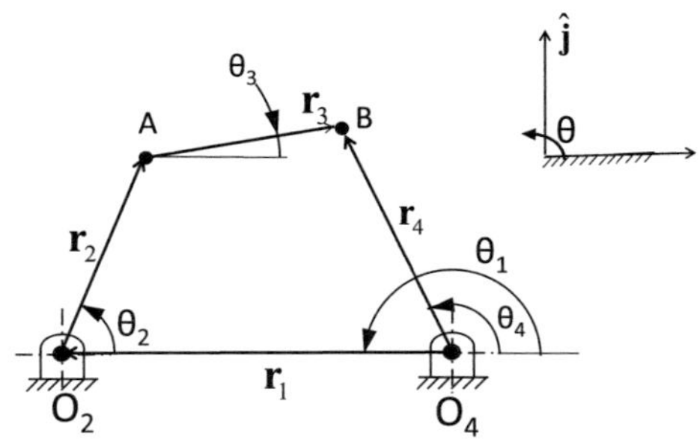

Fig. 7.1 Bloch's method of synthesis applied to a four-bar linkage

that the end of the vector of one link is the beginning of the vector for the next link. Only for the link 4, its vector is defined slightly different in the opposite direction only for the convenience of defining it and this arrangement is consistently used throughout this chapter for all four-bar mechanisms.

Based on the vector closure equation, the sum of the position vectors is equal to zero, given by

$$\mathbf{r}_1 + \mathbf{r}_2 + \mathbf{r}_3 - \mathbf{r}_4 = 0 \tag{7.1}$$

The vector $\mathbf{r}_4$ is assigned a negative sign since it is defined in the opposite direction. In complex polar coordinates, Eq. 7.1 can be written as

$$r_1 e^{j\theta_1} + r_2 e^{j\theta_2} + r_3 e^{j\theta_3} - r_4 e^{j\theta_4} = 0 \tag{7.2}$$

The first derivative of Eq. 7.2 with respect to time is given by

$$r_2\omega_2 e^{j\theta_2} + r_3\omega_3 e^{j\theta_3} - r_4\omega_4 e^{j\theta_4} = 0 \tag{7.3}$$

The second derivative of Eq. 7.2 with respect to time is given by

$$r_2\left(j\omega_2^2 + \alpha_2\right)e^{j\theta_2} + r_3\left(j\omega_3^2 + \alpha_3\right)e^{j\theta_3} - r_4\left(j\omega_4^2 + \alpha_4\right)e^{j\theta_4} = 0 \tag{7.4}$$

Note that all time derivatives of the fixed link are equal to zero.

Expressing Eqs. 7.2 through 7.4 in matrix form

$$\begin{bmatrix} 1 & 1 & -1 \\ \omega_2 & \omega_3 & -\omega_4 \\ j\omega_2^2 + \alpha_2 & \left(j\omega_3^2 + \alpha_3\right) & -\left(j\omega_4^2 + \alpha_4\right) \end{bmatrix} \begin{Bmatrix} \mathbf{r}_2 \\ \mathbf{r}_3 \\ \mathbf{r}_4 \end{Bmatrix} = \begin{Bmatrix} -\mathbf{r}_1 \\ 0 \\ 0 \end{Bmatrix} \tag{7.5}$$

The position vector of the fixed link is kept on the right-hand side so a proportionate length of other links that satisfy the given angular velocity and angular accelerations can be obtained using the following methods.

Table 7.1 Angular velocities and accelerations of Example 7.1

Angular velocity (rad/s)	Angular acceleration (rad/s^2)
$\omega_2 = 5$	$\alpha_2 = 0$
$\omega_3 = 1.2$	$\alpha_3 = 1.7$
$\omega_4 = 3.5$	$\alpha_4 = -4.4$

7.1.1 Using the Matrix Inverse

By taking the inverse of the matrix on the left-hand side of Eq. (7.5), the solution to the system of linear equations in terms of the fixed link vector can be obtained as

$$\begin{Bmatrix} \mathbf{r_2} \\ \mathbf{r_3} \\ \mathbf{r_4} \end{Bmatrix} = \begin{bmatrix} 1 & 1 & -1 \\ \omega_2 & \omega_3 & -\omega_4 \\ j\omega_2^2 + \alpha_2 & \left(j\omega_3^2 + \alpha_3\right) & -\left(j\omega_4^2 + \alpha_4\right) \end{bmatrix}^{-1} \begin{Bmatrix} -\mathbf{r_1} \\ 0 \\ 0 \end{Bmatrix} \tag{7.6}$$

Example 7.1

A four-bar linkage must be designed for the values of angular velocities and accelerations shown in Table 7.1, at a given instant of time. Assume a vector length of 200 mm in the negative x-direction for the fixed link. Use the method of matrix inversion.

From Eq. 7.6

$$\begin{Bmatrix} \mathbf{r_2} \\ \mathbf{r_3} \\ \mathbf{r_4} \end{Bmatrix} = \begin{bmatrix} 1 & 1 & -1 \\ 5 & 1.2 & -3.5 \\ j5^2 & \left(j1.2^2 + 1.7\right) & -\left(j3.5^2 - 4.4\right) \end{bmatrix}^{-1} \begin{Bmatrix} 200 \\ 0 \\ 0 \end{Bmatrix} \tag{1}$$

Solving Eq. (1), the following dimensions of the links are obtained:

$$\begin{aligned} \mathbf{r_2} &= -33.05 + j122.74 = 127.11\ \text{mm}\angle 105.1^\circ \\ \mathbf{r_3} &= 282.79 + j80.05 = 293.91\ \text{mm}\angle 15.8^\circ \\ \mathbf{r_4} &= 49.75 + j202.79 = 208.81\ \text{mm}\angle 76.71^\circ \end{aligned} \tag{2}$$

The four-bar linkage of Eq. (2) is presented in Fig. 7.2.

7.1.2 Using Cofactors

In order to solve Eq. 7.5, it would be computationally less intensive and hence easy for hand calculations, to use the method of co-factors rather than matrix inversion. But the solutions so obtained need additional steps to get the desired four-bar linkage.

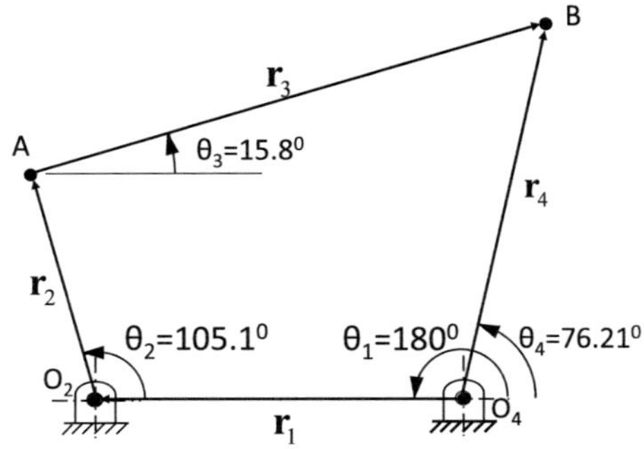

Fig. 7.2 Four-bar linkage of Example 7.1 [Eq. (2)]

Using the method of cofactors, position vector of the second linkage can be obtained as

$$\mathbf{r_2} = \frac{\begin{vmatrix} -\mathbf{r_1} & 1 & -1 \\ 0 & \omega_3 & -\omega_4 \\ 0 & \left(j\omega_3^2 + \alpha_3\right) & -\left(j\omega_4^2 + \alpha_4\right) \end{vmatrix}}{\begin{vmatrix} 1 & 1 & -1 \\ \omega_2 & \omega_3 & -\omega_4 \\ j\omega_2^2 + \alpha_2 & \left(j\omega_3^2 + \alpha_3\right) & -\left(j\omega_4^2 + \alpha_4\right) \end{vmatrix}} \tag{7.7}$$

Equation 7.7 can be simplified as

$$\mathbf{r_2} = \frac{-\mathbf{r}_1 \begin{vmatrix} \omega_3 & -\omega_4 \\ \left(j\omega_3^2 + \alpha_3\right) & -\left(j\omega_4^2 + \alpha_4\right) \end{vmatrix}}{D} \tag{7.8}$$

Where, D is the determinant of the denominator of Eq. (7.7). If $\frac{-\mathbf{r_1}}{D} = 1$, Eq. 7.8 simplifies as

$$\mathbf{r_2} = \begin{vmatrix} \omega_3 & -\omega_4 \\ \left(j\omega_3^2 + \alpha_3\right) & -\left(j\omega_4^2 + \alpha_4\right) \end{vmatrix} \tag{7.9}$$

Similarly, position vector of the third linkage can be obtained as

$$\mathbf{r}_3 = \frac{\begin{vmatrix} 1 & -\mathbf{r}_1 & 1 \\ \omega_2 & 0 & -\omega_4 \\ j\omega_2^2 + \alpha_2 & 0 & -\left(j\omega_4^2 + \alpha_4\right) \end{vmatrix}}{\begin{vmatrix} 1 & 1 & -1 \\ \omega_2 & \omega_3 & -\omega_4 \\ j\omega_2^2 + \alpha_2 & j\omega_3^2 + \alpha_3 & -\left(j\omega_4^2 + \alpha_4\right) \end{vmatrix}} = \frac{\mathbf{r}_1 \begin{vmatrix} \omega_2 & -\omega_4 \\ j\omega_2^2 + \alpha_2 & -\left(j\omega_4^2 + \alpha_4\right) \end{vmatrix}}{D} \tag{7.10}$$

Similar to Eqs. (7.9) and (7.10) simplifies as

$$\mathbf{r}_3 = -\begin{vmatrix} \omega_2 & -\omega_4 \\ j\omega_2^2 + \alpha_2 & -\left(j\omega_4^2 + \alpha_4\right) \end{vmatrix} \tag{7.11}$$

Finally, position vector of the fourth linkage is given by

$$\mathbf{r}_4 = \frac{\begin{vmatrix} 1 & 1 & -\mathbf{r}_1 \\ \omega_2 & \omega_3 & 0 \\ j\omega_2^2 + \alpha_2 & \left(j\omega_3^2 + \alpha_3\right) & 0 \end{vmatrix}}{\begin{vmatrix} 1 & 1 & -1 \\ \omega_2 & \omega_3 & -\omega_4 \\ j\omega_2^2 + \alpha_2 & \left(j\omega_3^2 + \alpha_3\right) & -\left(j\omega_4^2 + \alpha_4\right) \end{vmatrix}} = \frac{-\mathbf{r}_1 \begin{vmatrix} \omega_2 & \omega_3 \\ j\omega_2^2 + \alpha_2 & \left(j\omega_3^2 + \alpha_3\right) \end{vmatrix}}{D} \tag{7.12}$$

Similar to Eqs. (7.9) and (7.11), (7.12) simplifies as

$$\mathbf{r}_4 = \begin{vmatrix} \omega_2 & \omega_3 \\ j\omega_2^2 + \alpha_2 & \left(j\omega_3^2 + \alpha_3\right) \end{vmatrix} \tag{7.13}$$

Using the vector closure Eq. 7.1, vector of the first linkage (fixed link) is given by

$$\mathbf{r}_1 = -\mathbf{r}_2 - \mathbf{r}_3 + \mathbf{r}_4 \tag{7.14}$$

Example 7.2

Solve Example 7.1 using the method of cofactors using the angular velocity and angular acceleration of various links shown in Table 7.2.

From Eq. 7.7, position vector $\mathbf{r}_2$ of the second link is given by

$$\mathbf{r}_2 = \begin{vmatrix} \omega_3 & -\omega_4 \\ \left(j\omega_3^2 + \alpha_3\right) & -\left(j\omega_4^2 + \alpha_4\right) \end{vmatrix} = \begin{vmatrix} 1.2 & -3.5 \\ \left(j1.2^2 + 1.7\right) & -\left(j3.5^2 - 4.4\right) \end{vmatrix}$$
$$= -11.23 + j9.66 = 14.81\angle 139.3^0 \tag{1}$$

From Eq. 7.11, position vector $\mathbf{r}_3$ of the third link is given by

$$\mathbf{r}_3 = -\begin{vmatrix} \omega_2 & -\omega_4 \\ j\omega_2^2 + \alpha_2 & -\left(j\omega_4^2 + \alpha_4\right) \end{vmatrix} = -\begin{vmatrix} 5 & -3.5 \\ j5^2 & -\left(j3.5^2 - 4.4\right) \end{vmatrix}$$

Table 7.2 Angular velocities and accelerations of Example 7.2

Angular velocity (rad/s)	Angular acceleration (rad/s^2)
$\omega_2 = 5$	$\alpha_2 = 0$
$\omega_3 = 1.2$	$\alpha_3 = 1.7$
$\omega_4 = 3.5$	$\alpha_4 = -4.4$

$$= 22 + j26.25 = 34.25 \text{ mm}\angle 50.03^\circ \tag{2}$$

From Eq. 7.13, position vector $\mathbf{r}_4$ of the fourth link is given by

$$\mathbf{r}_4 = \begin{vmatrix} \omega_2 & -\omega_3 \\ j\omega_2^2 + \alpha_2 & -(j\omega_3^2 + \alpha_3) \end{vmatrix} = \begin{vmatrix} 5 & -1.2 \\ j5^2 & -(j1.2^2 + 1.7) \end{vmatrix}$$
$$= -8.5 + j22.8 = 24.33\, mm\angle 110.45^\circ \tag{3}$$

From Eq. 7.14, position vector $\mathbf{r}_1$ of the first link is given by

$$\mathbf{r}_1 = -\mathbf{r}_2 - \mathbf{r}_3 + \mathbf{r}_4 = -(-11.23 + j9.66) - (22 + j26.25) + (-8.5 + j22.8)$$
$$= -19.27 - j13.11 = 23.30 \text{ mm}\angle - 145.77^\circ \tag{4}$$

The dimensions of the links and their angular positions obtained above can be summarized as

$$\mathbf{r}_1 = 23.31\text{mm}\angle - 145.8^\circ$$
$$\mathbf{r}_2 = 14.81 \text{ mm}\angle 139.3^\circ$$
$$\mathbf{r}_3 = 34.25 \text{ mm}\angle 50.03^\circ$$
$$\mathbf{r}_4 = 24.33 \text{ mm}\angle 110.5^\circ \tag{5}$$

The above results may not be useful for practical implementation, as there was no input regarding the space available and the actual orientation of the links. So, the lengths of all the links can be proportionally increased by fixing the dimension of the fixed linkage and rotating it to align with 180°; that is by taking the fixed link length as 200 mm (as in Example 7.1) that is aligned with the 180° direction. Since the fixed length obtained in Eq. (2) is aligned with -145.8°, the correction factor that is to be added to all the angles is $180^\circ - (-145.8^\circ) = 325.77^\circ$ and the correction factor that has to be multiplied for all the vector lengths, based on the vector length of link 1 obtained in Eq. (2), is 200/23.31 = 8.58. Hence, the final dimensions of all the links and their angles are given by

$$\mathbf{r}_1 = 23.31 \times 8.58\text{mm}\angle - 145.8^\circ + 325.77^\circ = 200 \text{ mm}\angle 180^\circ$$
$$\mathbf{r}_2 = 14.81 \times 8.58 \text{ mm}\angle 139.3^\circ + 325.77^\circ = 127.11 \text{ mm}\angle 105.06^\circ$$
$$\mathbf{r}_3 = 34.25 \times 8.58 \text{ mm}\angle 50.03^\circ + 325.77^\circ = 293.91 \text{ mm}\angle 15.80^\circ$$
$$\mathbf{r}_4 = 24.33 \times 8.58 \text{ mm}\angle 110.5^\circ + 325.77^\circ = 208.81 \text{ mm}\angle 76.21^\circ \tag{6}$$

Therefore, the same results of Example 7.1 are obtained, after modifying with respect to the fixed link's length and its angular orientation. But the advantage of cofactor method is that it is much easier to compute the lengths of the links and their angular orientations is using hand calculations.

7.2 Chebychev Spacing

If the initial and final values of the angular or linear positions of the input and output links of a mechanism are given and the mechanism must be designed to generate a particular function, we will have to determine the intermediate positions that the input and output links occupy so there is an accurate generation of the function; the intermediate points are known as precision points. The difference between the predicted value and actual value of the function at the precision points of the linkages is known as structural error. An equispaced input and output precision points between the extreme positions are ruled out since there is no basis or guarantee for minimizing structural errors. Whereas, by using the unique properties of Chebychev polynomials that will optimize the location of the precision points to minimize the structural error can be worked out.

Without providing details of the proof of such a scheme, we shall present a graphical method and a direct formula that can be used to compute the precision points for minimizing structural error.

The graphical method consists of constructing a polygon of 2n sides, where n is the number of precision points, within a circle of diameter that is equal to the total range of positions between which precision points are to be spaced. The polygon is rotated such that two of its parallel sides are perpendicular to the horizontal diameter of the circle; since the polygon has even number of sides, it will have n pairs of parallel sides. Then, the vertices of the polygon are projected onto the horizontal diameter to obtain the precision points. An example of obtaining five precision points between the extreme positions of x_0 and x_n is shown in Fig. 7.3. Hence, a circle of diameter $x_n - x_o$ is first drawn. A polygon of $2 \times 5 = 10$ sides is inscribed inside the circle by noting that the exterior angle = 360/2n = 36° in this case. The polygon is inscribed in such a way that two of its parallel extreme edges are perpendicular to the horizontal diameter as shown in the figure; their intersection of the horizontal diameter corresponds to the first and final precision positions (x_1 and x_5). The other precision points are obtained by projecting the remaining vertices onto the horizontal diameter to obtain the precision points x_2, x_3, and x_4.

However, there is a general equation for the jth component of the n number of precision points between the initial point x_o and final point x_{n+1} is given by

$$x_j = \frac{1}{2}(x_{n+1} + x_o) - \frac{1}{2}(x_{n+1} - x_o)\cos\left[\frac{(2j-1)\pi}{2n}\right] \tag{7.15}$$

If the initial and final points are known, Eq. 7.15 can be used to compute the intermediate Chebyshev precision points.

Despite choosing Chebyshev precision points to synthesize a mechanism, there could still be practical problems in the successful operation of the mechanism in the desired manner. There are two possibilities of defects that can be expected: branch

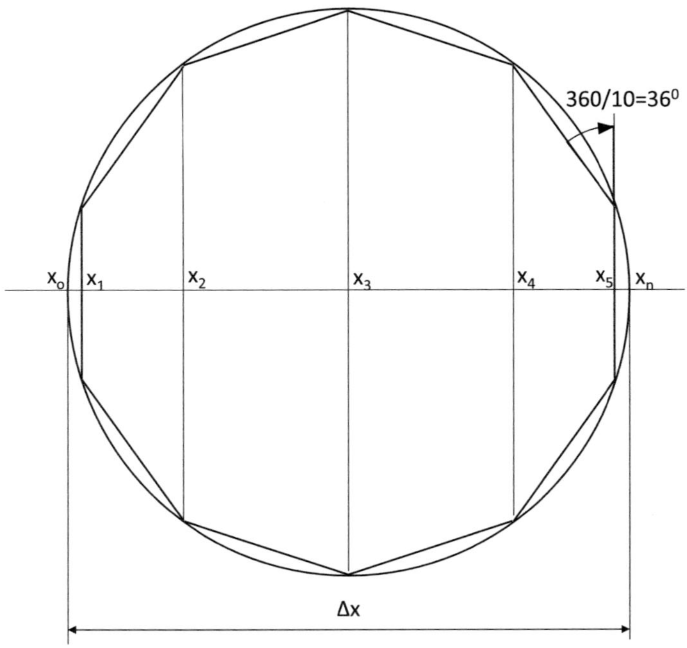

Fig. 7.3 Chebychev spacing for five points in an interval

and order defects. A branch defect is related to the inability of the mechanism to move smoothly between the precision points. The order defect is related to the improper sequence in which the precision points are reached.

Example 7.3

Determine five Chebyshev prevision points between 0 and 15.

n = 5, $x_o = 0$, $x_6 = 15$.

From Eq. 7.15, the Chebyshev points can be computed. The first point can be computed as

$$x_1 = \frac{1}{2}(15 + 0) - \frac{1}{2}(15 - 0)\cos\left[\frac{(2 \times 1 - 1)\pi}{2 \times 5}\right] = 0.37 \quad (1)$$

Similarly, other points can be computed using Eq. (7.15) and they are tabulated in Table 7.3.

Table 7.3 Chebyshev precision points of Example 7.3

Number	Variable	Chebyshev spacing	
1	x_0	0	Initial position
2	x_1	0.37	
3	x_2	3.09	
4	x_3	7.5	
5	x_4	11.91	
6	x_5	14.63	
7	x_6	15	Final position

7.3 Freudensteins's Equation

These equations are useful in obtaining the dimensions of a mechanism that can ensure that some of the linkages pass through the prescribed precision points. A four-bar linkage to which Bloch's method of synthesis was earlier applied is shown in Fig. 7.4. The fixed link is hinged at O_2 and O_4 is designated as link 1. The crank O_2A is link 2, the floating link AB is link 3 and the follower O_4B is link 4. The angles made by the above links are, respectively, θ_1, θ_2, θ_3, and θ_4, measured counterclockwise with respect to horizontal. Each of the links is represented by the position vectors: $\mathbf{r}_1$, $\mathbf{r}_2$, $\mathbf{r}_3$, and $\mathbf{r}_4$. Note that the vector of link 4 is defined slightly different as explained earlier.

The sum of the position vectors is equal to zero, given by

$$\mathbf{r}_1 + \mathbf{r}_2 + \mathbf{r}_3 - \mathbf{r}_4 = 0 \tag{7.16}$$

Equating the real parts of Eq. 7.16 to zero

$$r_1 \cos\theta_1 + r_2 \cos\theta_2 + r_3 \cos\theta_3 - r_4 \cos\theta_4 = 0 \tag{7.17}$$

Equating the imaginary parts of Eq. 7.16 to zero

$$r_1 \sin\theta_1 + r_2 \sin\theta_2 + r_3 \sin\theta_3 - r_4 \sin\theta_4 = 0 \tag{7.18}$$

Fig. 7.4 Four-bar linkage

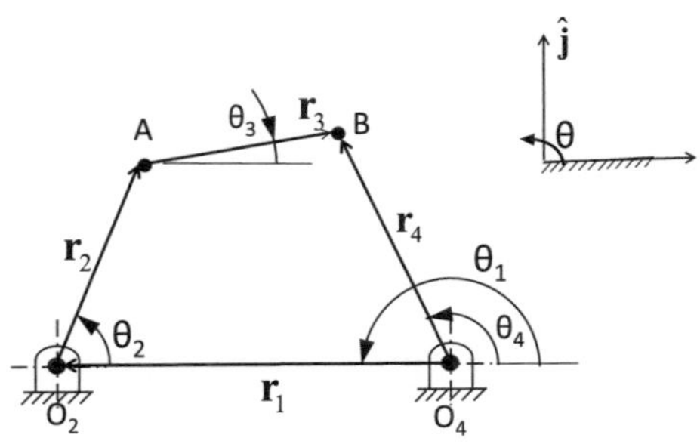

Since $\theta_1 = 180°$, Eq. (7.17) reduces to

$$-r_1 + r_2 \cos\theta_2 + r_3 \cos\theta_3 - r_4 \cos\theta_4 = 0 \tag{7.19}$$

and Eq. (7.18) reduces to

$$r_2 \sin\theta_2 + r_3 \sin\theta_3 - r_4 \sin\theta_4 = 0 \tag{7.20}$$

Now, let us eliminate θ_3, the angle of the floating link (coupler), from Eqs. 7.19 to 7.20, by writing each of them explicitly in terms of θ_3.

Rewriting Eq. 7.19

$$r_3 \cos\theta_3 = r_1 - r_2 \cos\theta_2 + r_4 \cos\theta_4 \tag{7.21}$$

Rewriting Eq. 7.20

$$r_3 \sin\theta_3 = -r_2 \sin\theta_2 + r_4 \sin\theta_4 = 0 \tag{7.22}$$

Squaring and adding Eqs. 7.21 and 7.22

$$\begin{aligned} r_3^2 = r_1^2 + r_2^2 + r_4^2 - 2r_1r_2\cos\theta_2 + 2r_1r_4\cos\theta_4 \\ - 2r_2r_4(\cos\theta_2\cos\theta_4 + \sin\theta_2\sin\theta_4) \end{aligned} \tag{7.23}$$

Since $(\cos\theta_2\cos\theta_4 + \sin\theta_2\sin\theta_4) = \cos(\theta_2 - \theta_4)$, Eq. 7.23 becomes

$$\begin{aligned} r_3^2 = r_1^2 + r_2^2 + r_4^2 - 2r_1r_2\cos\theta_2 + 2r_1r_4\cos\theta_4 \\ - 2r_2r_4\cos(\theta_2 - \theta_4) \end{aligned} \tag{7.24}$$

Dividing Eq. 7.24 throughout by $2r_2r_4$ and rearranging,

$$\frac{r_3^2 - r_1^2 - r_2^2 - r_4^2}{2r_2r_4} + \frac{r_1\cos\theta_2}{r_4} - \frac{r_1\cos\theta_4}{r_2} = -\cos(\theta_2 - \theta_4) \tag{7.25}$$

Equation 7.25 can be written in terms of constants defined with respect to the dimensions of various links as follows

$$K_1\cos\theta_2 - K_2\cos\theta_4 + K_3 = -\cos(\theta_2 - \theta_4) \tag{7.26}$$

where the constants are defined as

$$\begin{aligned} K_1 &= \frac{r_1}{r_4} \\ K_2 &= \frac{r_1}{r_2} \\ K_3 &= \frac{r_3^2 - r_1^2 - r_2^2 - r_4^2}{2r_2r_4} \end{aligned} \tag{7.27}$$

If $\theta_2 \equiv \psi(\text{crank})$, $\theta_4 \equiv \phi(\text{follower})$, Eq. 7.26 becomes

$$K_1 \cos\psi - K_2 \cos\phi + K_3 = -\cos(\psi - \phi) \tag{7.28}$$

Equation 7.28 is evaluated at three different Chebyshev precision points of the input crank and the corresponding follower angles that are determined for a definite functional relationship.

The three equations are

$$\begin{aligned} K_1 \cos\psi_1 - K_2 \cos\phi_1 + K_3 &= -\cos(\psi_1 - \phi_1) \\ K_1 \cos\psi_2 - K_2 \cos\phi_2 + K_3 &= -\cos(\psi_2 - \phi_2) \\ K_1 \cos\psi_3 - K_2 \cos\phi_3 + K_3 &= -\cos(\psi_3 - \phi_3) \end{aligned} \tag{7.29}$$

Solving the linear system of equations represented by Eq. (7.29), K_1, K_2, and K_3 are determined and dimensions of the four-bar linkage are determined from Eq. (7.27).

Example 7.4
It is required to synthesize a function generator using a four-bar linkage to generate the equation $y = \frac{1}{x^2}$ over the range of $1 \leq x \leq 2$, $1 \leq y \leq 0.25$. Assume that the range of angles of the crank as 30°–120° and the range of output angles of the follower as 60°–150°. Use three precision Chebychev points between the extreme positions of the crank and follower. Note that the input and output values of the given function are linear variables generated on the input-output links.

Using n = 3, the Chebychev precision points are generated for input and output variables. First, they are generated for x and y. Since x and y are linearly related to ψ and ϕ, the corresponding values are obtained from linear curve fitting.

First, the three Chebychev points corresponding to x and y are obtained from Eq. 7.15, for n = 3, and are presented in Table 7.4.

Since x and y are linearly related to input and output angles, the input and output angles corresponding to the precision points can be obtained by using a linear relationship based on initial and final positions.

Let the input angle ψ and x be related by

$$\psi = ax + b \tag{1}$$

Table 7.4 Initial, final, and precision points of the equation $y = \frac{1}{x^2}$ of Example 7.4

Position	x	y
Initial	1	1
1	1.067	0.88
2	1.5	0.44
3	1.93	0.27
Final	2	0.25

$\Psi = 30°$, for x = 1 and $\Psi = 120°$, for x = 2. Using these values in Eq. (7.1), a = 90 and b = −60.

Hence, Eq. (7.1) becomes

$$\psi = 90x - 60 \tag{2}$$

Let the output angle ϕ and y be related by

$$\phi = cy + d \tag{3}$$

$\phi = 60°$, for y = 1 and $\phi = 150°$, for y = 0.25. Using these values in Eq. (7.3), c = −120 and d = 180.

Hence, Eq. (7.2) becomes

$$\phi = -180y + 240 \tag{4}$$

The input and output angles corresponding to the Chebychev points are computed using Eqs. (7.2) and (7.3) and are shown in Table 7.5.

Using the values of input and output angles at the precision points in Eq. 7.29

$$\begin{aligned} K_1 \cos 36.03° - K_2 \cos 74.6° + K_3 &= -\cos(36.03° - 74.6°) \\ K_1 \cos 75° - K_2 \cos 126.67° + K_3 &= -\cos(75° - 126.67°) \\ K_1 \cos 113.97° - K_2 \cos 147.88° + K_3 &= -\cos(113.97° - 147.88°) \end{aligned} \tag{5}$$

Solving Eq. (5), $K_1 = 0.5069$; $K_2 = 0.5105$; $K_3 = -1.0563$.

From Eq. 7.27

$$r_4 = \frac{r_1}{K_1} = \frac{50}{0.5069} = 98.63\text{ mm};\ r_2 = \frac{r_1}{K_2} = \frac{50}{0.5105} = 97.95\text{ mm} \tag{6}$$

$$\begin{aligned} K_3 &= \frac{r_3^2 - r_1^2 - r_2^2 - r_4^2}{2r_2r_4} \\ -1.0563 &= \frac{r_3^2 - 50^2 - 97.95^2 - 98.62^2}{2 \times 97.95 \times 98.62} \end{aligned} \tag{7}$$

Solving Eq. (7.6), $r_3 = 37.59$ mm (Fig. 7.5). This is a rocker-rocker mechanism.

Table 7.5 Angular precision positions of the crank and follower of Example 7.4

Position	ψ (°)	ϕ (°)
Initial	30	60
1	36.03	74.6
2	75	126.67
3	113.97	147.88
Final	120	150

Fig. 7.5 Four-bar linkage synthesized for the function generator $y = 1/x^2$ of Example 7.1

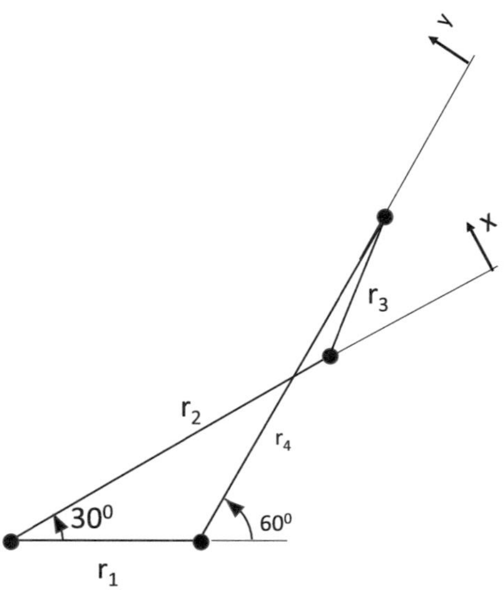

7.4 Position Synthesis

7.4.1 Two-Position Synthesis of Slider-Crank Mechanisms

Figure 7.6 shows a centered slider-crank mechanism that has already been discussed before. When it is used in most reciprocating machines, the line of action of the piston will be along the line of the crank pivot and the connecting rod length will be greater than the crank length. Consideration of forces in links and torque variations over all the crank angles will be the main objective in designing such mechanisms. The total movement of the piston will be twice the crank radius.

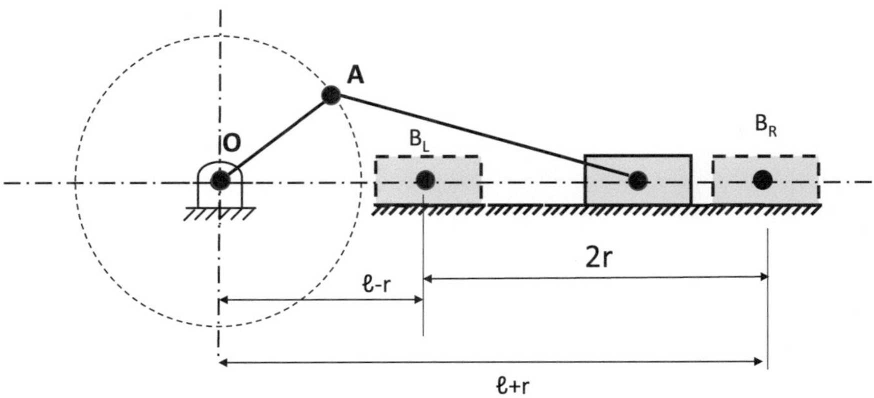

Fig. 7.6 Centered slider-crank mechanism

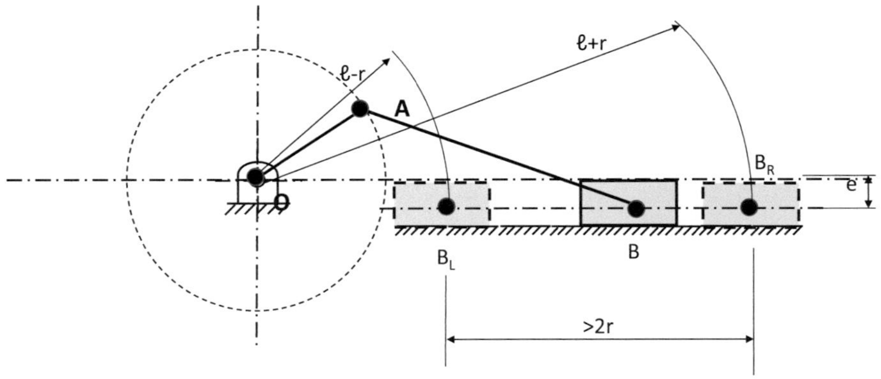

Fig. 7.7 Offset slider-crank mechanism

But if the connecting rod length and crank length are equal, it results in a special mechanism known as isosceles slider-crank mechanism. For such a mechanism, the piston movement and the crank movement will have to be in different parallel planes as the piston will cross the crank pivot during its motion. The stroke for such a mechanism will be four times the crank radius. All points on the connecting rod symmetric elliptical paths and the time taken for the forward and return motion will be the same.

An offset slider-crank mechanism is shown in Fig. 7.7. By using an offset distance between the line of action of the piston and its parallel passing through the pivot of the crank, some special effects can be obtained. For the above mechanism, the time taken for forward stroke is shorter than the return stroke. This feature can be used in a quick return mechanism required for certain machines.

7.4.2 Two-Position Synthesis of Crank and Rocker Mechanisms

Consider a four-bar crank-rocker mechanism shown in Fig. 7.8. Let r_1, r_2, r_3, r_4 be the lengths of various links; r_1 is the length of fixed link, r_2 is the length of crank, r_3 is the length of coupler, r_4 is the length of rocker. Corresponding to the extreme positions of the rocker, the crank and the coupler are along a straight line.

The rocker is initially at an angle θ_4, and it is expected to move a further angle ϕ in the counterclockwise direction. At the initial position of the rocker at θ_4, the crank and the coupler along a straight line along $O_2A_1B_1$. Corresponding to the final position of the rocker at $\theta + \phi$, as shown along a dotted line, the crank and the coupler are along the straight line $A_2O_2B_2$, also shown in a dotted line. The total angle moved by the crank is ψ during the movement of the rocker by an angle of ϕ. The total angle moved by the coupler between the extreme positions is α. During the return motion of the rocker, it moves by the same angle ϕ in the opposite direction,

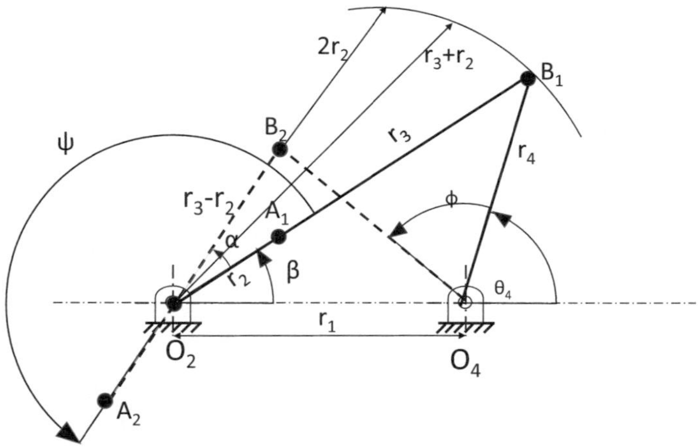

Fig. 7.8 Extreme positions of a crank-rocker mechanism

but the crank moves by an angle 360° − ψ. A crank-rocker mechanism is superior to a cam as the forces involved are much less and there is no spring retaining force. In addition, since there are only revolute pairs, the mechanism can be produced with very close tolerances.

The crank-rocker mechanism can also be used as a quick returning mechanism. If ψ is greater than 180°, the time ratio can be obtained using the following equation:

$$Q = \frac{180^\circ + \alpha}{180^\circ - \alpha} \tag{7.30}$$

7.4.3 *Two-Point Synthesis of a Four-Bar Linkage to Generate the Rocker Angle ϕ*

Consider the rocker of a crank-rocker mechanism that has to be synthesized that has a total angular movement of ϕ. Let r_4 be the length of the rocker that is hinged at the point O_4. By using the following procedure, draw the initial and final positions of the rocker and add some constructions that would help locate the hinge O_2 of the crank as shown in Fig. 7.9:

- Locate a point O_4 corresponding to the hinge of the rocker.
- Using r_4 as the length of the rocker and the angle ϕ, draw the extreme positions of the rocker and locate the corresponding positions B_1 and B_2.
- From Eq. 7.30, obtain the angle α that is required to obtain a quick return ratio. You may recall from Fig. 7.8 that α is the angle between the positions of the crank at which the crank and the coupler along a straight line.
- Through B_1 draw any line X.

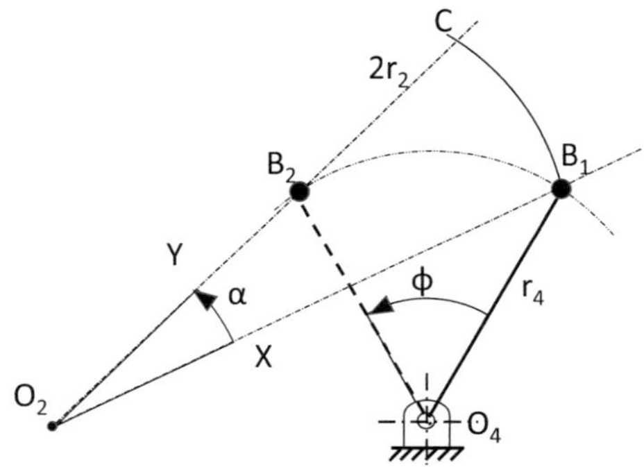

Fig. 7.9 Extreme positions of a rocker

- Through B_2 draw a line Y such that the lines X and Y make an angle α so the lines meet at the point O_2 that is the hinge point of the crank.
- With O_2 as center and O_2B_1 as radius, cut Y at C. From Fig. 7.9, $B_2C = 2r_2$ and $r_3 = O_2B_1 - r_2$. Thus, we have obtained the dimensions of the crank and coupler, as well as the hinge location of the crank, O_2.

Now that we know the dimensions of the crank and coupler, which were determined based on α that is related to the aligning positions of the crank and coupler, we can complete the crank-rocker mechanism corresponding to those positions as shown in Fig. 7.10.

Example 7.5

A crank-rocker is to be synthesized for a total rocker angle of 40° with 60 mm as the rocker length. It is required to be used as part of a quick-return mechanism that has a time ratio of 1.4. Determine dimensions of other links of the mechanism-crank, coupler and fixed links and in addition verify that these dimensions satisfy conditions of a crank-rocker mechanism.

From Eq. 7.30, the angle between the aligned positions of the crank and coupler that relate to the time ratio is given by.

$$\alpha = \frac{180^\circ(Q-1)}{Q+1} = \frac{180^\circ(1.4-1)}{1.4+1} = 30^\circ \tag{1}$$

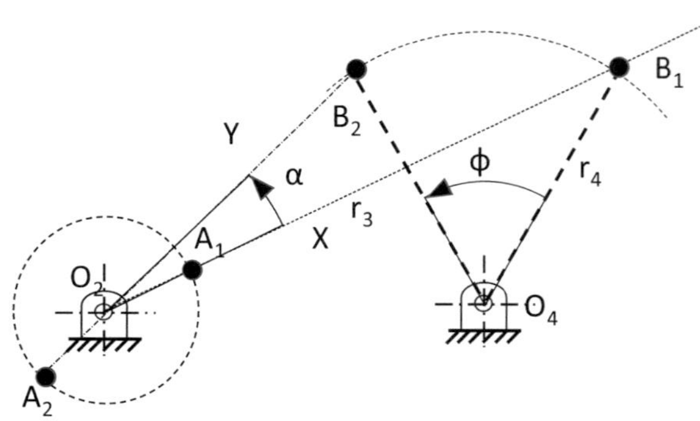

Fig. 7.10 Positions of the crank at which it along the same straight line as of the coupler

The figures shown in Fig. 7.11a–d illustrate the two-position synthesis of the crank-rocker mechanism.

As shown in Fig. 7.11a, first draw a vertical line so the 40° angular movement of the rocker is equally divided on both its sides. Then draw two lines of 60 mm inclined at 20° on either side of the vertical line that represents the initial (B_1) and final position (B_2). The two lines meet at O_4, the hinge point of the rocker.

Draw an arbitrary line X through B_1 and another line Y through B_2 but making an angle of 30° with X. The lines X and Y meet at O_2, the hinge point of the crank. This is shown in Fig. 7.11b.

With O_2B_1 as radius, draw an arc to meet the extension of O_2B_2 at C as shown in Fig. 7.11c. Measure B_2C that is twice the crank radius that is given by $r_1 = 15$ mm.

Figure 7.11d shows the initial and final configuration of all the moving linkages of the crank-rocker mechanism. Length of the coupler is given by $r_2 = O_2B_1 - r_1 = 68.25 - 15 = 53.25$ mm. Measure $O_2O_4 = r_4 = 46.8$ mm.

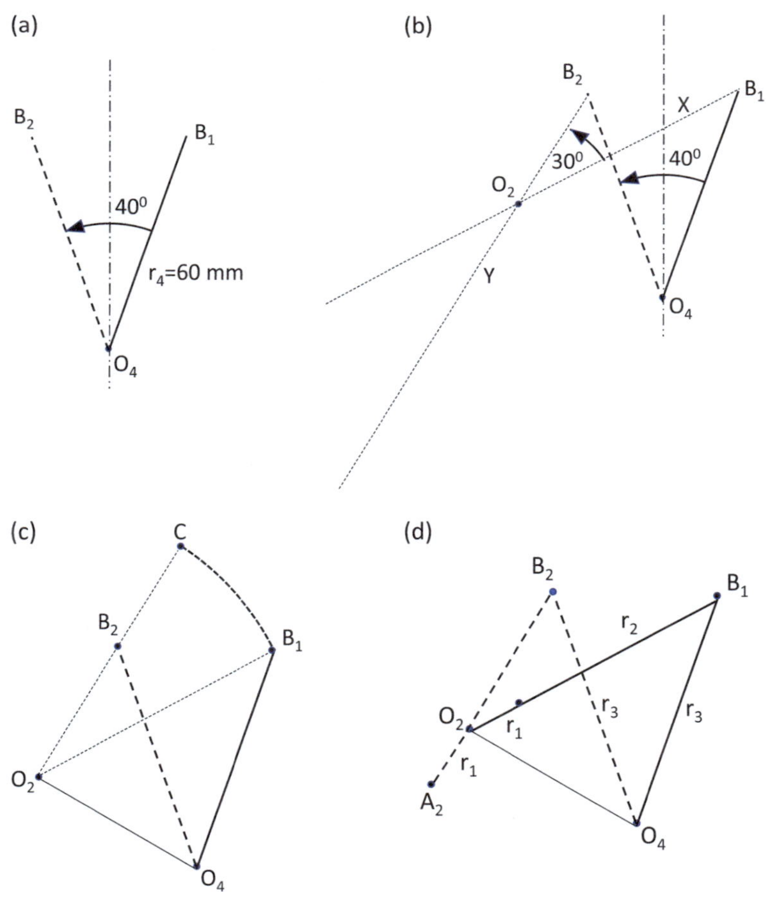

Fig. 7.11 Construction steps of Example 7.5

The above mechanism satisfies the Grashof's criterion since $r_1 + r_3 < r_2 + r_4$. The condition for a crank-rocker mechanism is satisfied since thecrank is the smallest link the adjacent the fixed link is adjacent to it.

7.5 Transmission Angles of Crank-Rocker Four-Bar Mechanisms

As shown in Fig. 7.12, the transmission angle is defined as the angle between the coupler and the rocker and it has an important bearing to the amount of torque it can transmit. The minimum transmission angle occurs when the crank is along horizontal toward right, shown with continuous lines for all the linkages; the maximum transmission angle occurs when the crank is making an angle of 180°, shown using dashed lines for all the linkages.

For unity time-ratio, Q = 1, analytical expressions can be derived from the ratio of linkage lengths if the expected value of minimum transmission angle is known.

The maximum and minimum transmission angles are related by

$$\gamma_{\min} = 180^\circ - \gamma_{\max} \tag{7.31}$$

By referring to Fig. 7.12 and using the law of cosines, the following equations can be written for the initial and final position of the rocker

$$\cos(\theta_4 + \phi) = \frac{r_1^2 + r_4^2 - (r_3 - r_2)^2}{2r_1 r_4} \tag{7.32}$$

$$\cos(\theta_4) = \frac{r_1^2 + r_4^2 - (r_3 + r_2)^2}{2r_1 r_4} \tag{7.33}$$

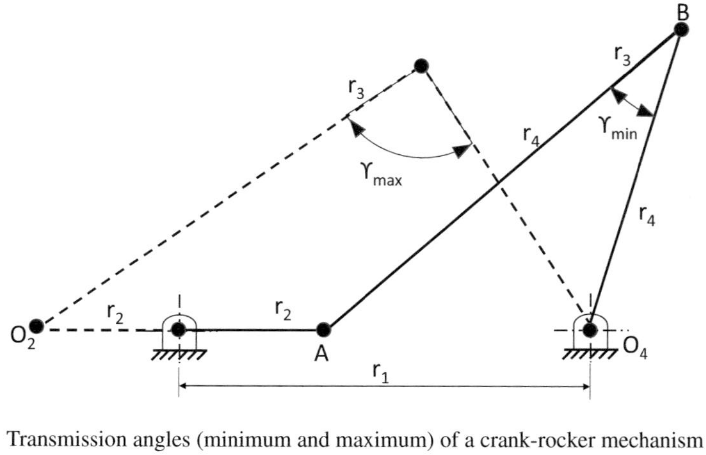

Fig. 7.12 Transmission angles (minimum and maximum) of a crank-rocker mechanism

From Fig. 7.12

$$\cos\gamma_{\min} = \frac{r_3^2 + r_4^2 - (r_1 - r_2)^2}{2r_3r_4} \tag{7.34}$$

and

$$\cos\gamma_{\max} = \frac{r_3^2 + r_4^2 - (r_1 + r_2)^2}{2r_3r_4} \tag{7.35}$$

By solving the above equations, the following ratios can be obtained

$$\frac{r_3}{r_1} = \sqrt{\frac{1 - \cos\phi}{2\cos^2\gamma_{\min}}} \tag{7.36}$$

7.5.1 When the Time-Ratio Q is Greater than 1

A larger set of the possible dimensions of the crank-rocker mechanism will have to be examined to obtain the best combination for obtaining optimum transmission angle. Therefore, if the locus of the initial and final positions of the rocker connecting point with the coupler are drawn, from which various combinations for a time-ratio-that in turn determines α-can be obtained. The transmission angle can then be examined for each combination.

Based on the range of rocker angles and time ratio, Fig. 7.13 is drawn. Draw a line O_2O_4 corresponding to the fixed link. O_4 is the hinge point for the rocker and O_2 is the hinge point of the crank. Locate the point C using ϕ and α. By drawing arcs from points O_2 and O_4 through C, locate the point C′. Hence, C and C′ are symmetrical about O_2O_4.

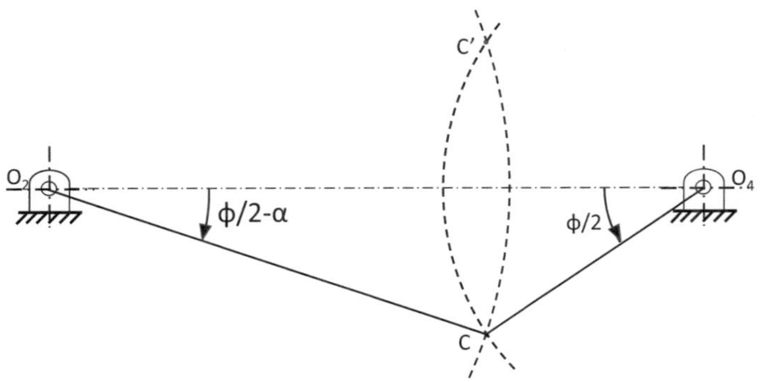

Fig. 7.13 Locating symmetric points about the fixed link of a crank-rocker mechanism for Q > 1

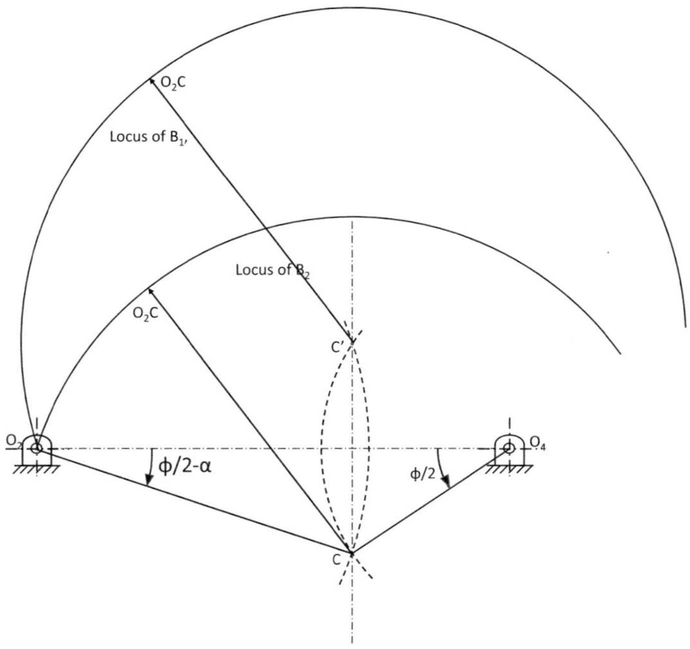

Fig. 7.14 Drawing the locus of B_1 and B_2

Figure 7.14 is extended to draw Fig. 7.15 by following these steps. With C as center and O_2C as radius, draw an arc through O_2 that defines the locus of B_2, possible final positions of the rocker; similarly, with C′ as center and with the same O_2C as the radius, draw an arc through O_2 that defines the locus of B_1, possible initial positions of the rocker.

By choosing an appropriate length, initial and final position of the rocker, one possible configuration of the crank-rocker can be obtained as shown in Fig. 7.15. The corresponding locations of the crank and coupler are also shown in the figure. As discussed earlier, the crank and the coupler align themselves corresponding to the initial and final positions of the rocker. By using the distance B_2O_4, the dimensions of the crank and coupler can be obtained, as discussed in the earlier section.

To obtain an optimum transmission angle, a variety of points like B_1 are chosen to obtain various possible configurations of the crank-rocker and for each configuration the minimum and maximum transmission angles are determined. Then, by plotting $|90° - \gamma_{min}|$ and $|90° - \gamma_{max}|$ for various values of β, the mechanism with optimum transmission angle will be the one that will be minimum in one of the above curves.

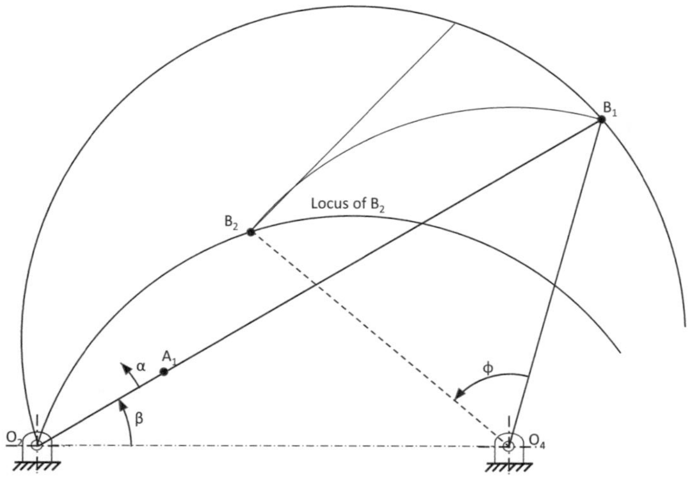

Fig. 7.15 One of the possible configurations of a crank-rocker obtained from Fig. 7.14

7.6 Three-Position Synthesis

7.6.1 Inversion Principle

Consider a rocker–rocker mechanism shown in Fig. 7.16 for which the range of angles for the input and output rockers are given, along with their input positions. A three-position synthesis can be carried out by determining the intermediate precision positions. We have already discussed the analytical method of determining the proportion of linkages. Now we shall discuss the graphical method. The basic approach of the graphical method is to use inversion as a technique for synthesis so that we can make the output rocker stationary and appropriately interpret the positions of the input rocker by using the inversion technique that will be explained as follows.

Fig. 7.16 Rocker–rocker mechanism

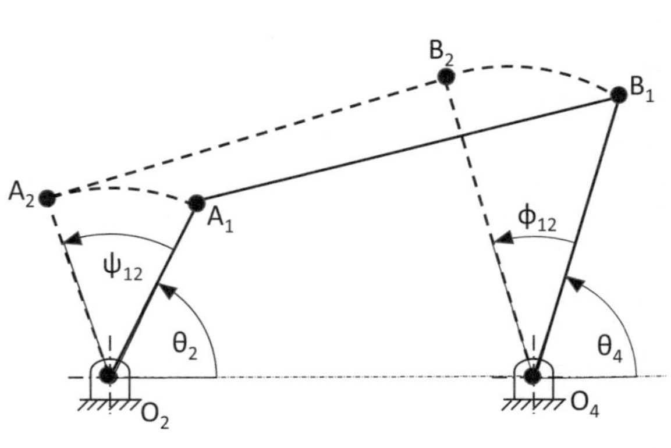

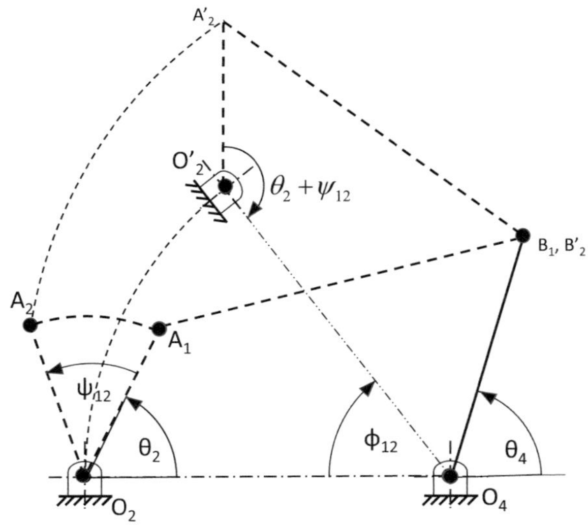

Fig. 7.17 Moving the input rocker and fixed link keeping the output rocker stationary

Figure 7.17 shows an arrangement in which the output rocker remains in the same position as the initial position whereas the input rocker and the fixed link move to achieve the same relative positions expected of them as if they moved as per Fig. 7.17. For this to happen, the fixed link has to move in the clockwise direction by ϕ_{12} so the hinge of the input rocker moves to O'_2. The input rocker moves to a position such that it makes an angle of $\theta_2 + \psi_{12}$ with respect to the new position of the fixed link O'_2O_4.

7.6.2 Application of Inversion Principle to a Rocker–Rocker Mechanism

Consider a rocker–rocker mechanism shown in Fig. 7.18 in three precision positions. The input rocker 2 is initially at the precision position making an angle θ_2, the second precision position makes an angle of ψ_{12} with respect to the first precision position and the third precision position makes an angle of ψ_{23} with respect to the second precision position and ψ_{13} with respect to the first precision position. The output rocker initial makes an angle of θ_4 with respect to the horizontal, the second precision position makes an angle of ϕ_{12} with respect to the first precision position and the third precision position makes an angle of ϕ_{13} with respect to the first precision position and ϕ_{23} with respect to the second precision position. Initially, the length of the fixed link 1, input rocker 2 are assumed; the length of the coupler and output rocker are to be determined from three-point synthesis, which is explained as follows:

As shown in Fig. 7.19, first draw the fixed link 1, O_2O_4, of a suitable length. Choose a suitable length of the input rocker and draw the first two precision positions of the hinge points A_1 and A_2 as shown. From A_2 draw a line connecting it to O_4. Draw another line from O_4 at an angle ϕ_{12} and by the inversion technique, the hinge point A_2 moves to A'_2. Join $A_1A'_2$ and draw its mid-normal as shown.

The input rocker is now moved to its third precision position as shown in Fig. 7.20 that is an extension of Fig. 7.19. Draw a line A_3O_4 and another line of the same length inclined at angle ϕ_{13} passing through O_4. By the inversion technique, the hinge point A_3 moves to A'_3. Join $A_1A'_3$ and draw its mid-normal as shown that meets other mid-normal drawn earlier at B_1. B_1 is the hinge point of the output rocker and coupler in the first precision position. Hence, it gives the dimensions of the coupler and the output rocker as shown in Fig. 7.21.

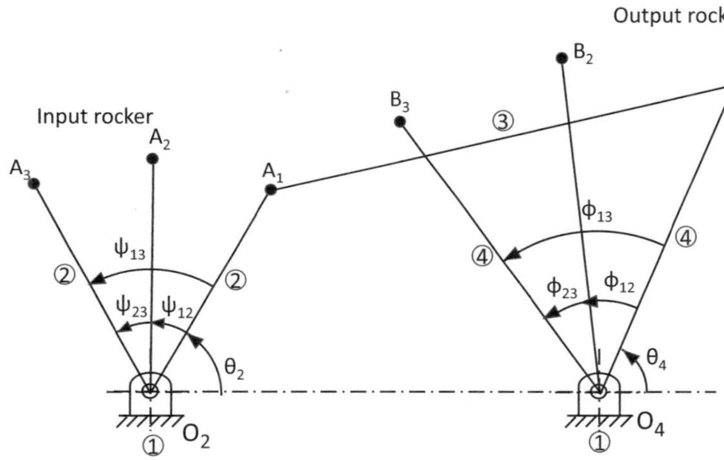

Fig. 7.18 Precision angles of a rocker–rocker mechanism

Fig. 7.19 Moving the hinge point of the second precision point

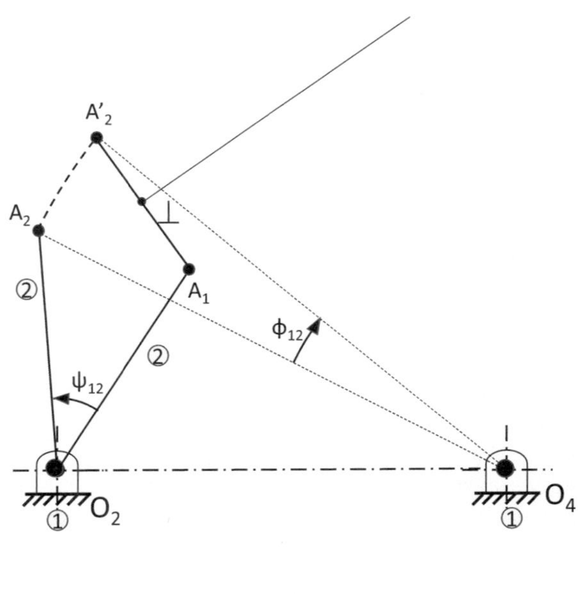

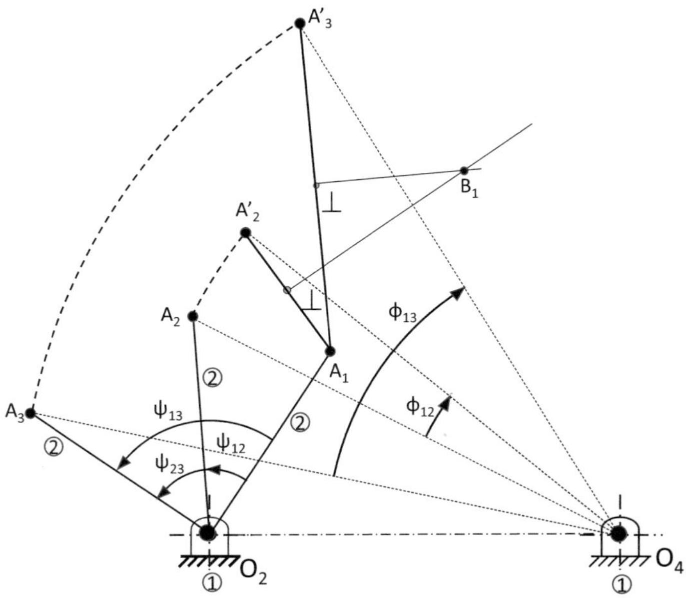

Fig. 7.20 Moving the hinge point of the third precision point

Fig. 7.21 Rocker–rocker four-bar linkage obtained from three-position synthesis

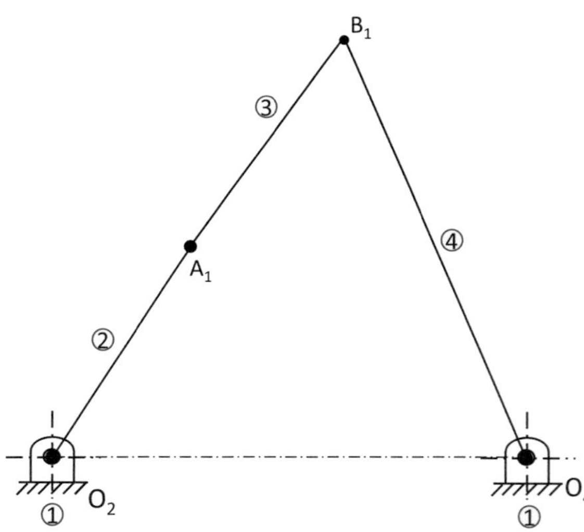

Example 7.6

Refer to Example 7.4. The input and output precision angles of a rocker–rocker mechanism are reproduced below. Using these angles and the graphical method of three-position synthesis discussed above, determine the lengths of the coupler and output rocker. Assume the solutions obtained from Example 7.5 for the lengths of the input rocker and the fixed link; $r_1 = 50$ mm and $r_2 = 98$ mm.

Table 7.6 Angular positions of the rocker–rocker mechanism of Example 7.6

Position	ψ (°)		ϕ (°)	
Initial	30		60	
1	36.03		74.6	
2	75	$\Psi_{12} = 75 - 36.03 = 38.97$	126.67	$\phi_{12} = 126.67 - 74.6 = 52°$
3	113.97	$\Psi_{23} = 113.97 - 75 = 38.97$	147.88	$\phi_{13} = 147.88 - 74.6 = 73.28°$
Final	120	$\Psi_{13} = 113.97 - 36.03 = 77.94°$	150	

Use only precision positions corresponding to positions 1, 2, and 3. The mechanism so obtained will automatically satisfy the initial and final positions as the precision points were derived based on the initial and final positions. The initial, final, and precision points of the input and output rocker are shown in Table 7.6.

Draw a line O_2O_4 representing the fixed link as shown in Fig. 7.22. Draw $O_2A_1 =$ 98 mm, first precision position of the input rocker, at a convenient angle and draw $O_2A_2 = 98$ mm, the second precision position of the input rocker O_2A_2 at angle of $\psi_{12} = 39°$.

As shown in Fig. 7.23, join A_2O_4 and with O_4A_2 as the radius and O_4 as the center draw an arc of $\phi_{12} = 72.28°$ in the clockwise direction since we are using the inversion principle, to obtain A'_2 that is the point obtained due to the inversion of A_2. Join $A_1A'_2$ and draw a mid-normal to it.

Figure 7.24 shows how the input rocker hinge of the third precision position is moved based on inversion principle. After obtaining A'_2, the construction lines for obtaining that point have been removed to improve clarity. Join A_3O_4 with O_4A_3 as radius and O_4 as the center, draw an arc corresponding to $\phi_{13} = 73.28°$ to obtain A'_3. Join $A_1A'_3$ and draw its mid-normal to meet the mid-normal drawn from $A_1A'_2$

Fig. 7.22 First and second precision positions of the input rocker

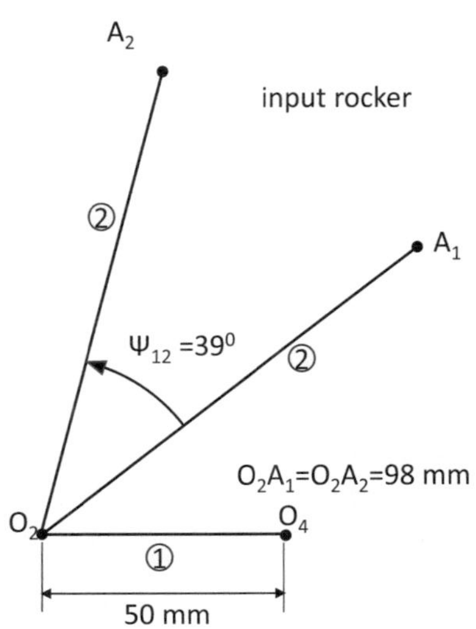

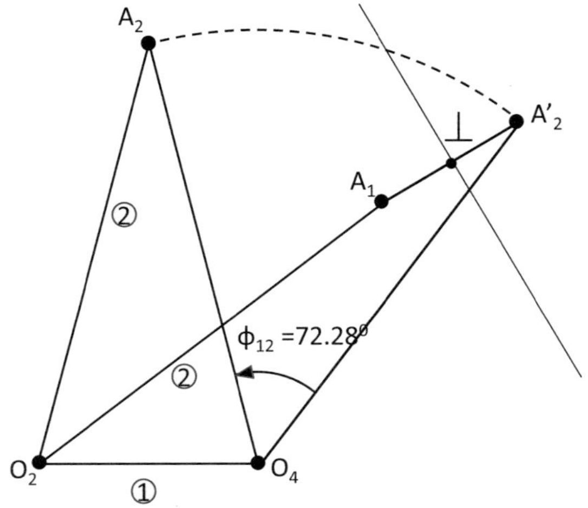

Fig. 7.23 Moving the hinge of the second precision position

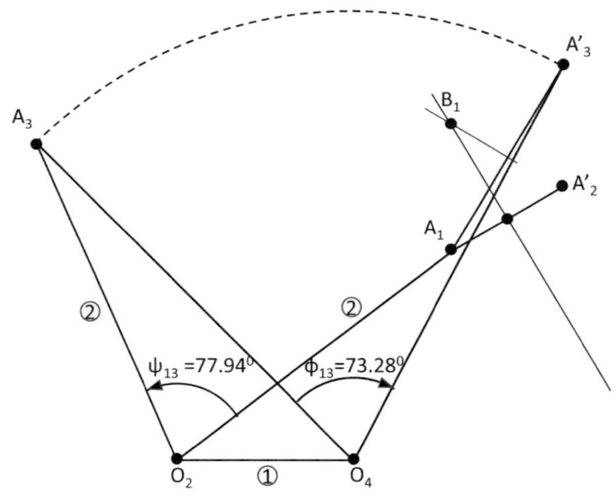

Fig. 7.24 Moving the input rocker hinge of the third precision position

at the point B_1. B_1 is the hinge point of the coupler and output rocker that remains the same for all the three precision positions of the input rocker as the inversion is carried out on the output rocker.

After obtaining the point B_1 from Fig. 7.24, the complete rocker–rocker mechanism can be obtained in the first precision position, as shown in Fig. 7.25. By direct measurement, $A_1B_1 = 38$ mm is the length of the coupler and $O_4B_1 = 98$ mm is the output rocker length that matches with the same result obtained in Example 7.5. It should be noted that all the above linkages move in parallel planes and hence do not obstruct each other's movement.

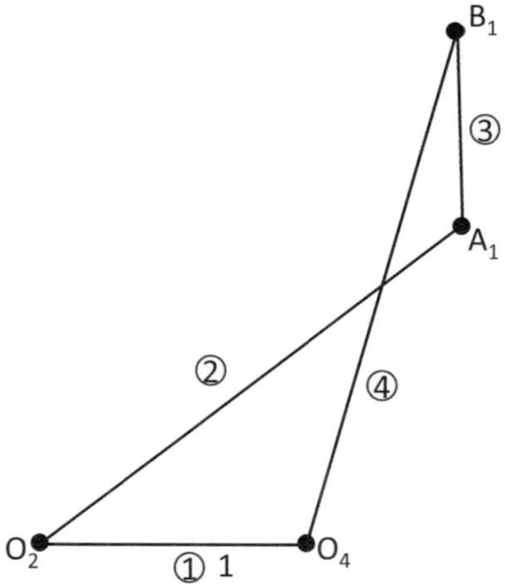

Fig. 7.25 Rocker–rocker mechanism at the first precision position

7.6.3 *Analytical Synthesis Using Complex Algebra*

By using complex algebra to the precision positions of planar linkages, it is possible to synthesize their dimensions. It is very similar to the Freudenstein's equation used in the earlier sections. Since the lengths of the linkages do not change in various positions, it is only their angles that vary. Most of these angles can be related to those obtained from the precision positions of the linkages and some missing data may have to assumed. The vector closure equations can be written in terms of the lengths of various linkages and their relative angular positions. Some additional information may be available that is exclusive to each type of mechanism. For a four-bar linkage, the fixed link has a constant magnitude and direction that can be useful in eliminating from the vector closure equations at various locations.

Four-bar rocker–rocker mechanism

To explain the analytical synthesis technique, consider a four-bar rocker–rocker mechanism shown in Fig. 7.26 that has three precision positions corresponding to the input and output rockers. Link 1 is the fixed link, link 2 is the input rocker, link 3 is the coupler and link 4 is the output rocker. The input rocker and output rockers are drawn in three precisions and the coupler is connected to the respective input and output rockers as shown.

The vector closure equation for the first precision can be written as

$$\mathbf{r}_2 + \mathbf{r}_3 + \mathbf{r}_4 = -\mathbf{r}_1 \tag{7.37}$$

The vector closure equation for the second precision can be written as

$$\mathbf{r}_2 e^{j\psi_{12}} + \mathbf{r}_3 e^{j\alpha_{12}} + \mathbf{r}_4 e^{j\phi_{12}} = -\mathbf{r}_1 \tag{7.38}$$

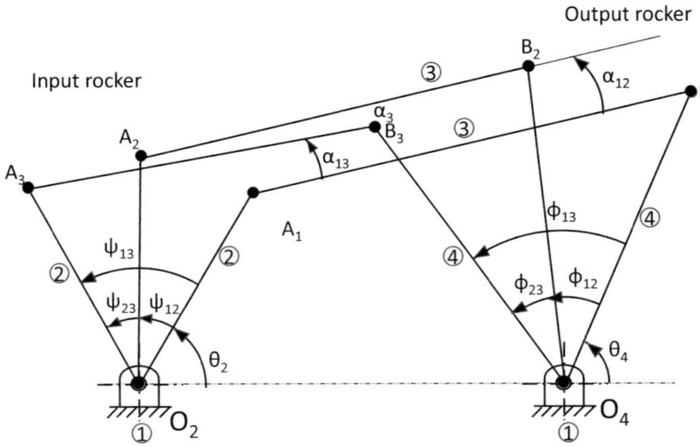

Fig. 7.26 Four-bar rocker–rocker mechanism at three precision positions

The vector closure equation for the third precision can be written as

$$\mathbf{r}_2 e^{j\psi_{13}} + \mathbf{r}_3 e^{j\alpha_{13}} + \mathbf{r}_4 e^{j\phi_{13}} = -\mathbf{r}_1 \tag{7.39}$$

Although the relative angular positions of the input and output rockers in the three precision positions are available from the precision angles, the relative angular positions of the coupler will have to be suitably assumed from experience or approximated from crude sketches of the mechanism in the precision positions.

Subtracting Eqs. (7.38) and (7.37)

$$\left(e^{j\psi_{12}} - 1\right)\mathbf{r}_2 + \left(e^{j\alpha_{12}} - 1\right)\mathbf{r}_3 + \mathbf{r}_4\left(e^{j\phi_{12}} - 1\right) = 0 \tag{7.40}$$

Subtracting Eqs. (7.39) and (7.37)

$$\left(e^{j\psi_{13}} - 1\right)\mathbf{r}_2 + \left(e^{j\alpha_{13}} - 1\right)\mathbf{r}_3 + \mathbf{r}_4\left(e^{j\phi_{13}} - 1\right) = 0 \tag{7.41}$$

Assuming that the input rocker's length and initial positions are known, along with the relative angular positions of the coupler, the above equations can be expressed as

$$\begin{bmatrix} e^{j\alpha_{12}} - 1 & e^{j\phi_{12}} - 1 \\ e^{j\alpha_{13}} - 1 & e^{j\phi_{13}} - 1 \end{bmatrix} \begin{Bmatrix} \mathbf{r}_3 \\ \mathbf{r}_4 \end{Bmatrix} = -\mathbf{r}_2 \begin{Bmatrix} e^{j\psi_{12}} - 1 \\ e^{j\psi_{13}} - 1 \end{Bmatrix} \tag{7.42}$$

After solving for the initial positions of the coupler and output rocker, the fixed link length can be obtained.

Example 7.7

Refer to Example 7.6. The input and output precision angles of a rocker–rocker mechanism are shown in Table 7.7. Using these angles and the graphical method of

Table 7.7 Angular positions of the rocker–rocker mechanism of Example 7.7

	Input rocker		Output rocker	
Position	ψ (°)		ϕ (°)	
Initial	30		60	
1	36.03		74.6	
2	75	$\Psi_{12} = 75 - 36.03 = 38.97$	126.67	$\phi_{12} = 126.67 - 74.6 = 52°$
3	113.97	$\Psi_{23} = 113.97 - 75 = 38.97$	147.88	$\phi_{13} = 147.88 - 74.6 = 73.28°$
Final	120	$\Psi_{13} = 113.97 - 36.03 = 77.94°$	150	

three-position synthesis discussed above, determine the lengths of the coupler and output rocker. Assume the solutions obtained from Example 7.5 for the lengths of the input rocker and the fixed link; Assume $r_2 = 98$ mm, $\alpha_1 = 91°$, $\alpha_{12} = 115°$, $\alpha_{13} = 188°$.

The vector of the input rocker corresponding to the initial precision position is given by

$$\mathbf{r}_2 = r_2 e^{j\psi_1} = 50 e^{j\frac{36.03\times\pi}{180}} \tag{1}$$

The vector of the output rocker corresponding to the initial precision position is given by

$$\mathbf{r}_4 = r_4 e^{j\phi_1} = r_4 e^{j\frac{74.6\times\pi}{180}} \tag{2}$$

The vector of the coupler corresponding to the initial precision position is given by

$$\mathbf{r}_3 = r_3 e^{j\alpha_1} = r_3 e^{j\frac{91\times\pi}{180}} \tag{3}$$

From Eq. 7.42, following equations are obtained

$$\begin{bmatrix} e^{j\alpha_{12}} - 1 & e^{j\phi_{12}} - 1 \\ e^{j\alpha_{13}} - 1 & e^{j\phi_{13}} - 1 \end{bmatrix} = \begin{bmatrix} e^{j\frac{115\times\pi}{180}} - 1 & e^{j52\times\frac{\pi}{180}} - 1 \\ e^{j188\times\frac{\pi}{180}} - 1 & e^{j73.28\times\frac{\pi}{180}} - 1 \end{bmatrix} \tag{4}$$

$$-\mathbf{r}_2 \begin{Bmatrix} e^{j\psi_{12}} - 1 \\ e^{j\psi_{13}} - 1 \end{Bmatrix} = -50 e^{j\frac{36.03\times\pi}{180}} \begin{Bmatrix} e^{j38.97\times\frac{\pi}{180}} - 1 \\ e^{j77.94\times\frac{\pi}{180}} - 1 \end{Bmatrix} \tag{5}$$

Using Eqs. (4) and (5) in Eq. (7.42)

$$\begin{Bmatrix} \mathbf{r}_3 \\ \mathbf{r}_4 \end{Bmatrix} = -50 e^{j\frac{36.03\times\pi}{180}} \begin{bmatrix} e^{j\frac{115\times\pi}{180}} - 1 & e^{j52\times\frac{\pi}{180}} - 1 \\ e^{j188\times\frac{\pi}{180}} - 1 & e^{j73.28\times\frac{\pi}{180}} - 1 \end{bmatrix}^{-1} \begin{Bmatrix} e^{j38.97\times\frac{\pi}{180}} - 1 \\ e^{j77.94\times\frac{\pi}{180}} - 1 \end{Bmatrix}$$

$$= \begin{Bmatrix} 1.2448 + j37.7042 \\ -29.0319 - j99.8382 \end{Bmatrix} \tag{6}$$

From Eq. (6), $r_3 = 37.72$ mm, $r_4 = 103.97$ mm
From Eq. 7.37, the fixed link is given by

$$\begin{aligned}\mathbf{r}_1 &= -(\mathbf{r}_2 + \mathbf{r}_3 + \mathbf{r}_4)\\ &= -\left[(79.2535 + j57.6445) + (1.2448 + j37.7042) + (-29.0319 - j99.8)\right]\\ &= -51.4664 + j4.4896 \text{ mm}\end{aligned} \tag{7}$$

From Eq. (7), $r_1 = 51.66$ mm.

The values obtained above compare well with those obtained in Example 7.6, except for the output rocker. Since we have obtained information about the angles of the coupler from the previous example, the values obtained are nearly accurate. However, angular positions of the coupler can be assumed as unknown, they can still be computed using the above method by converting it into an optimization problem with unknown angles of the coupler at various precision positions using the vector closure equations. That is left as an exercise. The optimization procedure is presented in the Appendix of Chapter 1.

7.7 Conclusions

In this chapter, we have discussed the synthesis of mechanisms that can generate specific functions based on position synthesis at precision points or those that satisfy the conditions of angular velocities and accelerations of various links of the mechanisms. The analysis presented in this chapter, therefore, will be useful in determining the dimensions of the links of mechanisms.

7.8 Exercise Problems

7.1. Consider the mechanism shown in Fig. P7.1. The velocities and accelerations of various hinge points, angular velocities, and angular accelerations are as follows.

$$\begin{aligned}\mathbf{v}_C &= 18\hat{\mathbf{i}} \text{ mm/s} & \mathbf{a}_C &= 100\hat{\mathbf{i}} \text{ mm/s}^2\\ \mathbf{v}_B &= -24\hat{\mathbf{j}} \text{ mm/s} & \mathbf{a}_B &= -150\hat{\mathbf{j}} \text{ mm/s}^2\\ \mathbf{v}_A &= 104\hat{\mathbf{i}} \text{ mm/s} & \mathbf{a}_A &= 200\hat{\mathbf{i}} \text{ mm/s}^2\end{aligned}$$

$$\begin{aligned}\omega_{B_4} &= 0.30 \text{ rad/s}\\ \omega_{B_5} &= 1.75 \text{ rad/s}\\ \alpha_{B_4} &= 1.8 \text{ rad/s}^2\\ \alpha_{B_5} &= 0.62 \text{ rad/s}^2\end{aligned}$$

Determine the missing dimensions of the figure and dimensions of the links so the above velocity and accelerations can be obtained from this mechanism.

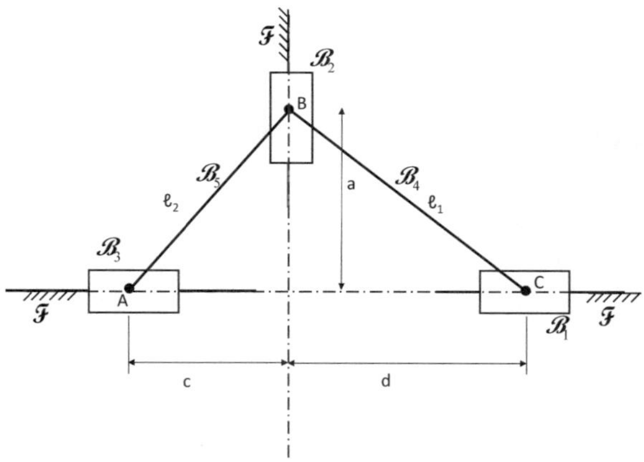

Fig. P7.1 Mechanism of exercise problem 7.1

7.2 Consider the mechanism shown in Fig. P7.2. The velocities and accelerations of various hinge points, angular velocities, and angular accelerations are as follows.

$\mathbf{v}_B = -653\hat{\mathbf{j}}$ mm/s $\quad$ $\mathbf{a}_B = 9903\hat{\mathbf{i}}$ mm/s^2

$\mathbf{v}_A = -935\hat{\mathbf{i}}$ mm/s $\quad$ $\mathbf{a}_A = -3168\hat{\mathbf{i}}$ mm/s^2

$$\omega_3 = 2.67 \text{ rad/s}$$
$$\omega_4 = 15 \text{ rad/s}$$
$$\omega_5 = -2.67 \text{ rad/s}$$
$$\alpha_3 = 135 \text{ rad/s}^2$$
$$\alpha_4 = 0$$
$$\alpha_5 = -135 \text{ rad/s}^2$$

Determine the dimensions of the links so the above velocity and accelerations can be obtained from this mechanism.

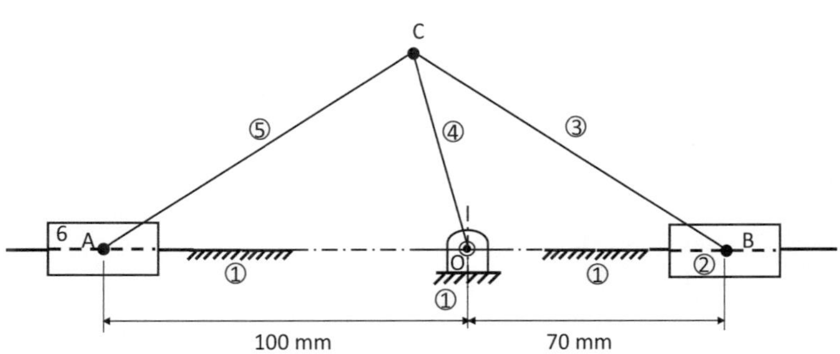

Fig. P7.2 Mechanism of exercise problem 7.2

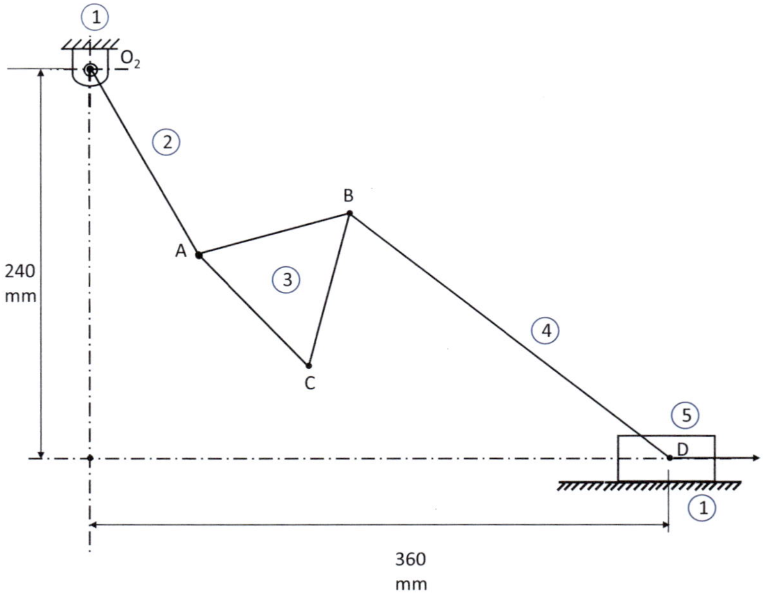

Fig. P7.3 Mechanism of exercise problem 7.3

7.3 Consider the mechanism shown in Fig. P7.3. The velocities and accelerations of various hinge points, angular velocities, and angular accelerations are as follows.

$\mathbf{v}_D = 300\,\hat{\mathbf{i}}$ mm/s $\qquad \mathbf{a}_D = 500\hat{\mathbf{i}}$ mm/s^2

$\omega_2 = 5$ rad/s

$\omega_3 = 0.15$ rad/s

$\omega_4 = -1.72$ rad/s

$\alpha_2 = 0$

$\alpha_3 = -53$ rad/s^2

$\alpha_4 = -9.59$ rad/s^2

Determine the dimensions of the links so the above velocity and accelerations can be obtained from this mechanism.

7.4 A four-bar linkage is to be designed for the following instantaneous values. If the fixed link is 100 mm long, determine the length of linkages and their relative angular positions (Table P7.4).

7.5 A four-bar linkage is represented by the following vectors as defined in Sect. 7.1.
$\mathbf{r_2} = 150 + \mathrm{j}300$ mm; $\mathbf{r_3} = 500 + \mathrm{j}250$ mm; $\mathbf{r_4} = 300 + \mathrm{j}480$ mm; $\mathbf{r_1} = -350 - \mathrm{j}70$ mm (fixed link). If $\omega_2 = 10$ rad/s and $\alpha_2 = 0$, calculate the angular velocity and angular accelerations of other moving linkages.

Table P7.4 Angular velocities and accelerations of a four-bar mechanism of the exercise problem 7.4

ω_2	5 rad/s	α_2	0	Crank
ω_3	1 rad/s	α_3	10 rad/s^2	Coupler
ω_4	4 rad/s	α_4	15 rad/s^2	Follower

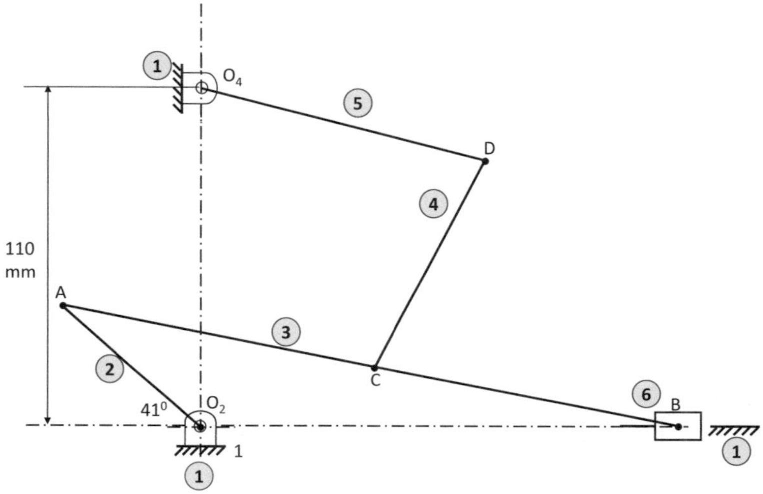

Fig. P7.6 Mechanism of exercise problem 7.6

7.6 Consider the mechanism shown in Fig. P7.6. The velocities and accelerations of various hinge points, angular velocities, and angular accelerations are as follows.

$\mathbf{v}_B = -30.54\,\hat{\mathbf{i}}$ mm/s $\quad \mathbf{a}_B = -63.47\hat{\mathbf{i}}$ mm/s^2

$\omega_2 = -1$ rad/s

$\omega_3 = 0.22$ rad/s

$\omega_4 = -0.38$ rad/s

$\omega_5 = -0.4$ rad/s

$\alpha_2 = 0$

$\alpha_3 = -0.2$ rad/s^2

$\alpha_4 = -0.6$ rad/s^2

Determine the dimensions of the links so the above velocity and accelerations can be obtained from this mechanism.

7.7 Consider the mechanism shown in Fig. P7.7. The velocities and accelerations of various hinge points, angular velocities, and angular accelerations are as follows.

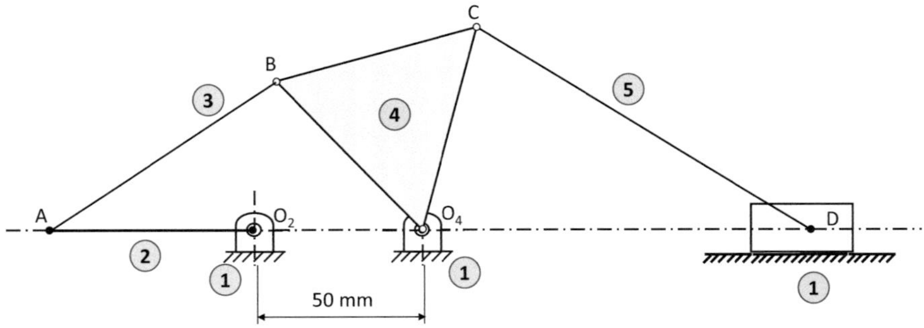

Fig. P7.7 Mechanism of exercise problem 7.7

$$\mathbf{v}_D = 580.67\,\hat{\mathbf{i}}\ \text{mm/s} \qquad \mathbf{a}_B = -63.47\hat{\mathbf{i}}\ \text{mm/s}^2$$
$$\omega_2 = -15\ \text{rad/s}$$
$$\omega_3 = -8.18\ \text{rad/s}$$
$$\omega_4 = -8.18\ \text{rad/s}$$
$$\omega_5 = 1.35\ \text{rad/s}$$
$$\alpha_2 = 0$$
$$\alpha_3 = -190\ \text{rad/s}^2$$
$$\alpha_4 = -24.28\ \text{rad/s}^2$$
$$\alpha_5 = 44.66\ \text{rad/s}^2$$

Determine the dimensions of the links so the above velocity and accelerations can be obtained from this mechanism.

7.8 Consider the mechanism shown in Fig. P7.8. The velocities and accelerations of various hinge points, angular velocities, and angular accelerations are as follows.

$$\mathbf{v}_A = -1136\,\hat{\mathbf{i}}\ \text{mm/s} \qquad \mathbf{a}_A = 19632\hat{\mathbf{i}}\ \text{mm/s}^2$$
$$\omega_2 = 15\ \text{rad/s}$$
$$\omega_3 = 3.64\ \text{rad/s}$$
$$\alpha_2 = 0$$
$$\alpha_3 = 68.5\ \text{rad/s}^2$$

Fig. P7.8 Mechanism of exercise problem 7.8

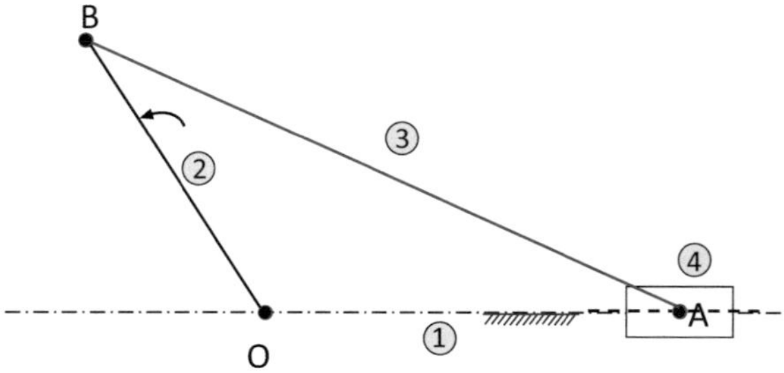

Table P7.9 Angular velocities and accelerations of a four-bar mechanism of the exercise problem 7.9

ω_2	5 rad/s	α_2	0	Crank
ω_3	1 rad/s	α_3	10 rad/s^2	Coupler
ω_4	4 rad/s	α_4	15 rad/s^2	Follower

Determine the length of the connecting rod, if the crank length is restricted to 125 mm, so the above velocity and acceleration data can be obtained at the given instant of time.

7.9 A four-bar linkage is to be designed for the following instantaneous values: For a fixed link that is 100 mm long, the following vectors are obtained for various links (Table P7.9).
$\mathbf{r}_2 = 52.72 - j11.35$ mm (crank); $\mathbf{r}_3 = 150.91 - j3.782$ mm (coupler); $\mathbf{r}_4 = 103.63 - j15.13$ mm (follower); $\mathbf{r}_1 = -100$ mm (fixed link).
If the crank vector, $\mathbf{r}_2$, changes to 45–j8 mm, determine the vector of $\mathbf{r}_3$ (coupler), if the follower and fixed links ($\mathbf{r}_4$ and $\mathbf{r}_1$) remain the same. If the angular velocity of the crank in the new configuration remains the same, determine the angular velocity of the follower.

7.10 Consider the mechanism shown in Fig. P7.10. The velocities and accelerations of various hinge points, angular velocities, and angular accelerations are as follows.

$$\mathbf{v}_D = 580.67\,\hat{\mathbf{i}}\text{ mm/s} \qquad \mathbf{a}_D = 2940\hat{\mathbf{i}}\text{ mm/s}^2$$
$$\omega_2 = 20\text{ rad/s}$$
$$\omega_3 = 25\text{ rad/s}$$
$$\alpha_2 = 5\text{ rad/s}^2$$
$$\alpha_3 = 25\text{ rad/s}^2$$
$$|a_{P_2}| = 5 \times 10^4\text{m/s}^2$$

Determine the dimensions of the links so the above velocity and accelerations can be obtained from this mechanism.

7.11 Using Freudenstein's method, determine the dimensions of a four-bar linkage (rocker–rocker mechanism) to generate a function y = tan(x), where x varies from 0 to 60°; use three precision points. The input rocker angles are: $\Psi_{min} = 80°$ and $\Psi_{max} = 170°$, whereas the output rocker angles are: $\phi_{min} = 40°$ and $\phi_{max} = 130°$. The fixed link length is 50 mm.

7.12 Using Freudenstein's method, determine the dimensions of a four-bar linkage (rocker–rocker mechanism) to generate a function y = tan^2(x), where x varies from 0 to 60°; use three precision points. The input rocker angles are: $\Psi_{min} = 80°$ and $\Psi_{max} = 170°$, whereas the output rocker angles are: $\phi_{min} = 40°$ and $\phi_{max} = 130°$. The fixed link length is 50 mm.

7.13 Using Freudenstein's method, determine the dimensions of a four-bar linkage (rocker–rocker mechanism) to generate a function y = tan^{-1}(x), where x varies from 0.5 to 2; use three precision points. The input rocker angles are: $\Psi_{min} =$

Fig. P7.10 Mechanism of exercise problem 7.10

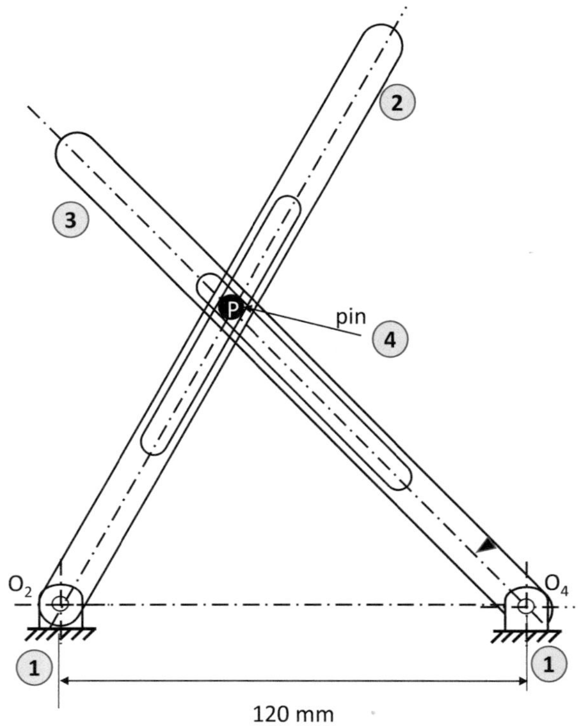

80° and $\Psi_{max} = 170°$, whereas the output rocker angles are: $\phi_{min} = 40°$ and $\phi_{max} = 130°$. The fixed link length is 50 mm.

7.14 Using Freudenstein's method, determine the dimensions of a four-bar linkage (rocker–rocker mechanism) to generate a function $y = e^x - 0.1x^2$, where x varies from 0.5 to 2; use three precision points. The input rocker angles are: $\Psi_{min} = 80°$ and $\Psi_{max} = 170°$, whereas the output rocker angles are: $\phi_{min} = 40°$ and $\phi_{max} = 130°$. The fixed link length is 50 mm.

7.15 Using Freudenstein's method, determine the dimensions of the mechanism shown in Fig. P7.15 to generate a function $y = x^2$, where x varies from 0.5 to 2; use three precision points.

7.16 Using Freudenstein's method, determine the dimensions of the mechanism shown in Fig. P7.16 to generate a function $\left(\frac{x}{a}\right)^2 + \left(\frac{y}{b}\right)^2 = 1$, where x varies from 0.5 to 2, a = 100 mm, b = 70 mm; use three precision points.

7.17 Using Freudenstein's method, determine the dimensions of the mechanism shown in Fig. P7.17 to generate a function $\theta = \tan^{-1}(x)$, where x varies from 0.5 to 2; use three precision points. The output angles are $\theta_{min} = 40°$ and $\theta_{max} = 70°$

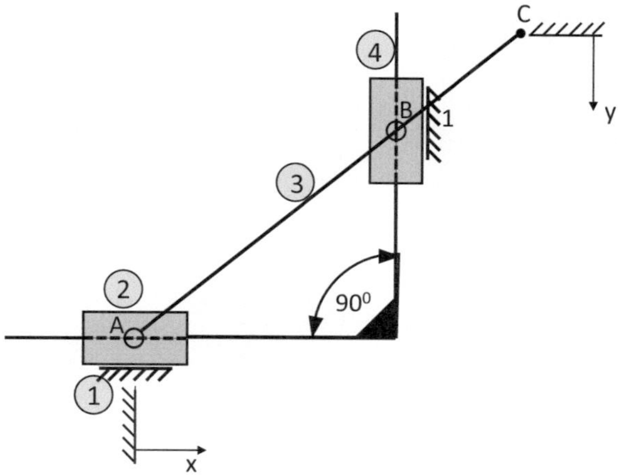

Fig. P7.15 Mechanism of exercise problem 7.15

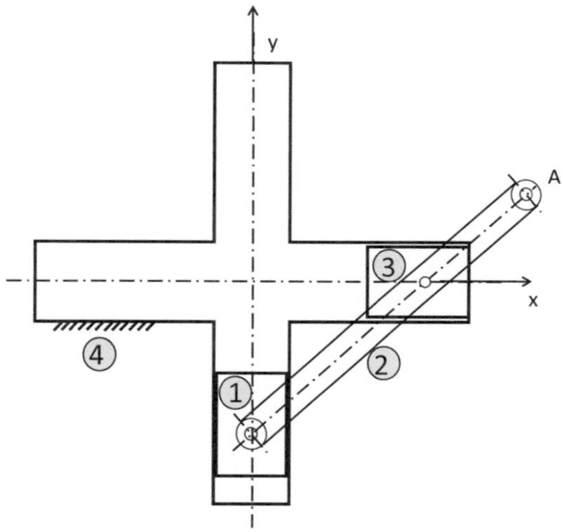

Fig. P7.16 Mechanism of exercise problem 7.16

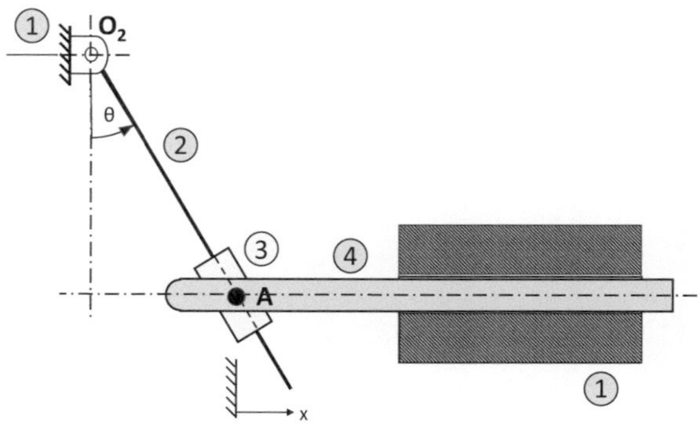

Fig. P7.17 Mechanism of exercise problem 7.17

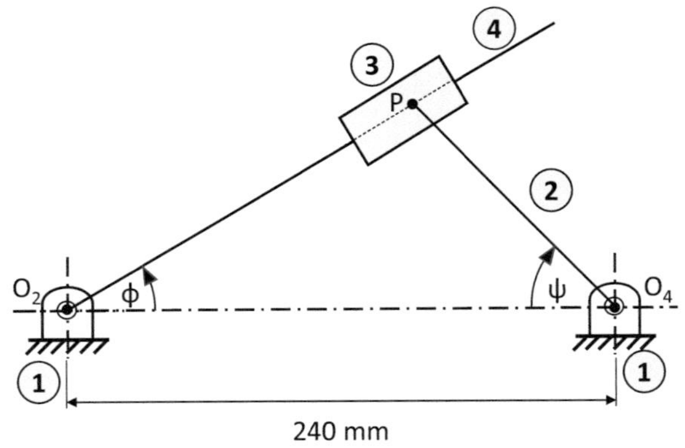

Fig. P7.18 Mechanism of exercise problem 7.18

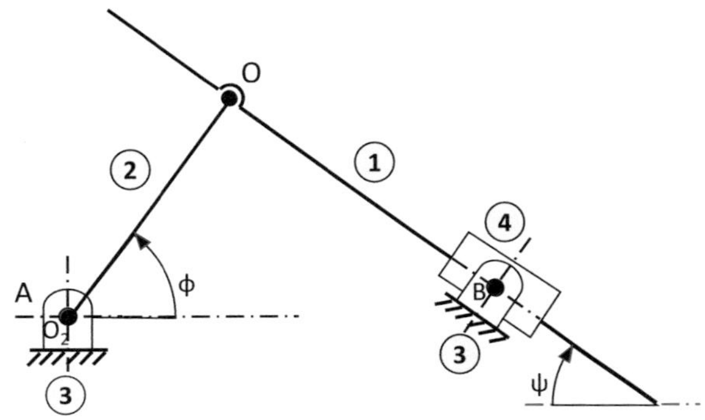

Fig. P7.19 Mechanism of exercise problem 7.19

7.18 Using Freudenstein's method, determine the dimensions of the mechanism shown in Fig. P7.18 to generate a function $y = 0.1x^2$, where x varies from 0.5 to 2; use three precision points. The range of angles are: $\Psi_{min} = 20°$ and $\Psi_{max} = 70°$, whereas the output rocker angles are: $\phi_{min} = 30°$ and $\phi_{max} = 80°$

7.19 Using Freudenstein's method, determine the dimensions of the oscillating cylinder mechanism shown in Fig. P7.19 to generate a function $y = 0.1x^2$, where x varies from 0.5 to 2; use three precision points. The range of angles are: $\Psi_{min} = 20°$ and $\Psi_{max} = 45°$, whereas the output angles are: $\phi_{min} = 30°$ and $\phi_{max} = 50°$

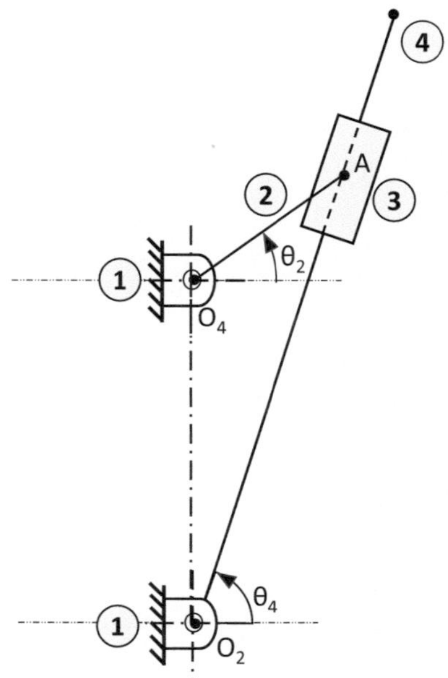

Fig. P7.20 Mechanism of exercise problem 7.20

7.20 Consider a part of crank shaper mechanism shown in Fig. P7.20. Using complex variables, obtain the functional relationship between angles θ_2, θ_4, r_2 and r_4.

Chapter 8
Force Analysis of Mechanisms

In Chap. 2, kinematic analysis of various mechanisms was discussed, without paying any attention to the forces and moments that are responsible to their motion. Kinematic analysis will help determine the dimensions of various links of the mechanism, in addition to giving an idea of velocity and acceleration of various links. However, it is important to determine the forces and moments responsible for the motion of these mechanisms under the action of load. In addition, force analysis is important in determining the power of prime movers that are required for running the mechanisms without much variations in the speed and to account for the forces transmitted to the machine foundation. In addition, the reaction forces are also important in ensuring the satisfactory functioning of the mechanism as they relate to the forces acting on individual elements. Mechanisms transmit forces to the foundation on which they are mounted that will result in vibration that can be transmitted to other parts of the machine. It is possible to control the resulting vibration by proper balancing of the mechanism that will be discussed in Chap. 9; the present chapter provides the basis for such balancing mechanisms.

8.1 Introduction

In this chapter, we shall first discuss the general relation between forces/torques to the linear/angular accelerations, mass/moments of inertia of a rigid body in plane motion. Since a mechanism is an interconnection of several rigid bodies, we shall discuss how to relate the reaction forces of various links of a mechanism. Since a slider crank mechanism is one of the most popularly used mechanisms, detailed kinetic analysis of this mechanism is presented first. In addition, a number of simplifications of a slider crank mechanism are discussed so one can get a quick estimate of reaction forces from them. In addition, a couple of mechanisms are studied from kinetic

D. N. Manik, *Fundamentals of Mechanisms and Machines*,
https://doi.org/10.1007/978-981-97-1810-8_8

analysis viewpoint so the material presented in this chapter can be used for kinetic analysis of any complex mechanism encountered in practice.

8.2 Kinetics of Rigid Body in Plane Motion

Consider a three-dimensional rigid body $\mathcal{B}$ in plane motion as shown in Fig. 8.1. Conditions for plane motion have already been discussed in Chap. 1 and, therefore, will not be repeated here. It is assumed that the rigid body is rotating about an axis $\hat{\mathbf{k}}$ with an angular velocity ω so its plane of motion is projected on the xy plane represented by unit vectors $\hat{\mathbf{i}}$ and $\hat{\mathbf{j}}$; P is the origin on the xy plane. An element mass of the rigid body dm is located at the coordinates Q′(x, y, z) and its position vector is R. The projection of this point is Q on the xy plane that has coordinates (x, y) and is represented by the position vector **r**.

From Eq. (2.58), the velocity of the point Q is given by

$$\mathbf{v}_Q = \mathbf{v}_P + \omega\hat{\mathbf{k}} \times \mathbf{r}_{PQ} \tag{8.1}$$

8.2.1 Angular Momentum

The angular momentum of an element mass about a point is the cross product of its position vector with its linear momentum. Hence, the angular momentum of the rigid body shown in Fig. 8.1 about the point P is the volume integral of the angular moment of the element mass about P and is given by

$$\mathbf{H}_P = \int_{\mathcal{B}} \mathbf{R} \times \mathbf{v}dm \tag{8.2}$$

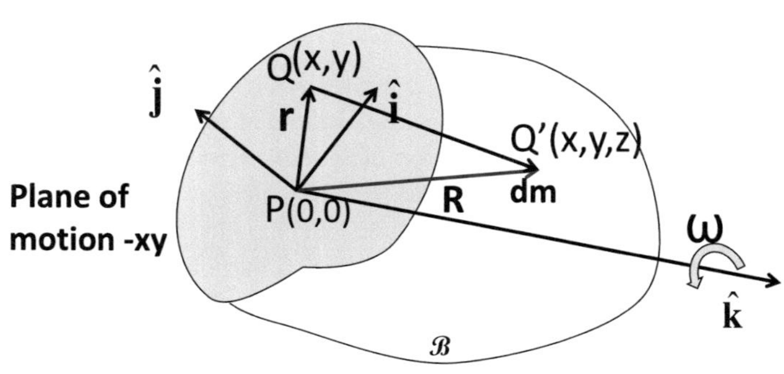

Fig. 8.1 Three-dimensional object in plane motion

Since we have assumed plane motion, $\mathbf{v} = \mathbf{v}_Q$, hence, Eq. (8.2) can be modified as

$$\mathbf{H}_P = \int_{\mathcal{B}} \mathbf{R} \times \mathbf{v}_Q dm \tag{8.3}$$

From Eqs. (8.1) and (8.3), the angular momentum of the rigid body about the point P is given by

$$\mathbf{H}_P = \int_{\mathcal{B}} \left(x\hat{\mathbf{i}} + y\hat{\mathbf{j}} + z\hat{\mathbf{k}}\right) \times \left(\mathbf{v}_P + \omega\hat{\mathbf{k}} \times \mathbf{r}_{PQ}\right) dm \tag{8.4}$$

By substituting the position vector $\mathbf{r}_{PQ}$ of the elemental mass in Eq. (8.4), the angular momentum of the body about the point P is given by

$$\mathbf{H}_P = \int_{\mathcal{B}} \left(x\hat{\mathbf{i}} + y\hat{\mathbf{j}} + z\hat{\mathbf{k}}\right) \times \left(\mathbf{v}_P + \omega\hat{\mathbf{k}} \times \left[x\hat{\mathbf{i}} + y\hat{\mathbf{j}}\right]\right) dm \tag{8.5}$$

Equation (8.5) can be written as

$$\mathbf{H}_P = \int_{\mathcal{B}} \left(x\hat{\mathbf{i}} + y\hat{\mathbf{j}} + z\hat{\mathbf{k}}\right) \times \mathbf{v}_P dm + \int_{\mathcal{B}} \left(x\hat{\mathbf{i}} + y\hat{\mathbf{j}} + z\hat{\mathbf{k}}\right) \times \left[\omega\hat{\mathbf{k}} \times \left(x\hat{\mathbf{i}} + y\hat{\mathbf{j}}\right)\right] dm \tag{8.6}$$

The first integral of Eq. (8.6) can be written as

$$\int_{\mathcal{B}} \left(x\hat{\mathbf{i}} + y\hat{\mathbf{j}} + z\hat{\mathbf{k}}\right) \times \mathbf{v}_P dm = \left[\int_{\mathcal{B}} \left(x\hat{\mathbf{i}} + y\hat{\mathbf{j}} + z\hat{\mathbf{k}}\right) dm\right] \times \mathbf{v}_P \tag{8.7}$$

If the position vector of the center of mass is $\mathbf{r}_{PC}$, then its product with the mass m can be related to the integral of the product of the position vector and elemental mass, and is given by

$$m\mathbf{r}_{PC} = \int_{\mathcal{B}} \left(x\hat{\mathbf{i}} + y\hat{\mathbf{j}} + z\hat{\mathbf{k}}\right) dm \tag{8.8}$$

From Eqs. (8.7) and (8.8)

$$\int_{\mathcal{B}} \left(x\hat{\mathbf{i}} + y\hat{\mathbf{j}} + z\hat{\mathbf{k}}\right) \times \mathbf{v}_P dm = m\mathbf{r}_{PC} \times \mathbf{v}_P \tag{8.9}$$

Following cross-product relations between the unit vectors are useful in simplifying the second integral of Eq. (8.6)

$$\begin{aligned}\hat{\mathbf{i}} \times \hat{\mathbf{i}} = 0;\ \hat{\mathbf{i}} \times \hat{\mathbf{j}} = \hat{\mathbf{k}}\\ \hat{\mathbf{j}} \times \hat{\mathbf{j}} = 0;\ \hat{\mathbf{j}} \times \hat{\mathbf{i}} = -\hat{\mathbf{k}}\\ \hat{\mathbf{k}} \times \hat{\mathbf{i}} = \hat{\mathbf{j}};\ \hat{\mathbf{k}} \times \hat{\mathbf{j}} = -\hat{\mathbf{i}}\end{aligned} \tag{8.10}$$

Using the above relations of Eq. (8.10), the second integral of Eq. (8.6) becomes

$$\int_{\mathcal{B}} \left(x\hat{\mathbf{i}} + y\hat{\mathbf{j}} + z\hat{\mathbf{k}}\right) \times \left[\omega\hat{\mathbf{k}} \times \left(x\hat{\mathbf{i}} + y\hat{\mathbf{j}}\right)\right]dm = \omega \int_{\mathcal{B}} \left[\left(x^2 + y^2\right)\hat{\mathbf{k}} - xz\hat{\mathbf{i}} - yz\hat{\mathbf{j}}\right]dm \tag{8.11}$$

Equation (8.11) contains various moments of inertia of the rigid body. So, let us define the following moments of inertia of the rigid body:

$$\begin{aligned}I_{zz}^{P} &= \int_{\mathcal{B}} \left[\left(x^2 + y^2\right)\right]dm\\ I_{YZ}^{P} &= -\int_{\mathcal{B}} yzdm\\ I_{XZ}^{P} &= -\int_{\mathcal{B}} xzdm\end{aligned} \tag{8.12}$$

From Eqs. (8.11) and (8.12)

$$\int_{\mathcal{B}} \left(x\hat{\mathbf{i}} + y\hat{\mathbf{j}} + z\hat{\mathbf{k}}\right) \times \left[\omega\hat{\mathbf{k}} \times \left(x\hat{\mathbf{i}} + y\hat{\mathbf{j}}\right)\right]dm = I_{XZ}^{P}\omega\hat{\mathbf{i}} + I_{YZ}^{P}\omega\hat{\mathbf{j}} + I_{zz}^{P}\omega\hat{\mathbf{k}} \tag{8.13}$$

From Eqs. (8.6), (8.9), and (8.13), the angular moment of the rigid body in plane motion about the point P becomes

$$\mathbf{H}_P = m\mathbf{r}_{PC} \times \mathbf{v}_P + I_{XZ}^{P}\omega\hat{\mathbf{i}} + I_{YZ}^{P}\omega\hat{\mathbf{j}} + I_{ZZ}^{P}\omega\hat{\mathbf{k}} \tag{8.14}$$

From Eq. (8.14), the angular momentum of the rigid body in plane motion about its mass center (since $\mathbf{r}_{PC} \equiv 0$, first term of Eq. 8.14 is zero) is given by

$$\mathbf{H}_C = I_{XZ}^{C}\omega\hat{\mathbf{i}} + I_{YZ}^{C}\omega\hat{\mathbf{j}} + I_{ZZ}^{C}\omega\hat{\mathbf{k}} \tag{8.15}$$

We can now use the angular momentum equations of (8.14) and (8.15) to relate the external moments acting on the rigid body that are causing its motion.

8.3 Euler's Second Law

First let us try to determine the sum of moments about the mass center; sum of moments about the mass center is equal to the time derivative of the angular momentum about the mass center. It is given by taking the time derivative of Eq. (8.15)

$$\sum \mathbf{M}_C = \frac{d\mathbf{H}_C}{dt} = \frac{d}{dt}\left(I^C_{XZ}\omega\hat{\mathbf{i}} + I^C_{YZ}\omega\hat{\mathbf{j}} + I^C_{ZZ}\omega\hat{\mathbf{k}}\right) \tag{8.16}$$

The coordinates of Eq. (8.16) are on the moving body. Hence, we may recall our earlier discussion in Chap. 2 on how we could compute the time-derivative of vectors referenced to moving coordinates.

Figure 8.2 shows a rigid body in plane motion rotating about the $\hat{\mathbf{k}}$ axis. The coordinates represented by unit vectors $\hat{\mathbf{i}}$ and $\hat{\mathbf{j}}$ are fixed on the moving body; the unit vectors and $\hat{\mathbf{J}}$ correspond to the global coordinate system. If a vector $\mathbf{A}$ is referenced to the moving coordinate system on the rigid body, the time derivative of this vector is given by

$$\dot{\mathbf{A}} = {}^{\mathcal{B}}\dot{\mathbf{A}} + \dot{\theta}\hat{\mathbf{k}} \times (\mathbf{A}) \tag{8.17}$$

By using Eq. (8.17), the derivatives of Eq. (8.16) can be obtained by noting that moments of inertia do not change with respect to time, as the coordinate system is fixed on the moving body. They are obtained as (by noting that angular acceleration of the body, $\alpha = \frac{d\omega}{dt}$)

$$\begin{aligned}\sum \mathbf{M}_C = {} & I^C_{XZ}\alpha\hat{\mathbf{i}} + I^C_{YZ}\alpha\hat{\mathbf{j}} + I^C_{ZZ}\alpha\hat{\mathbf{k}} \\ & + \omega\hat{\mathbf{k}} \times \left(I^C_{XZ}\omega\hat{\mathbf{i}} + I^C_{YZ}\omega\hat{\mathbf{j}} + I^C_{ZZ}\omega\hat{\mathbf{k}}\right)\end{aligned} \tag{8.18}$$

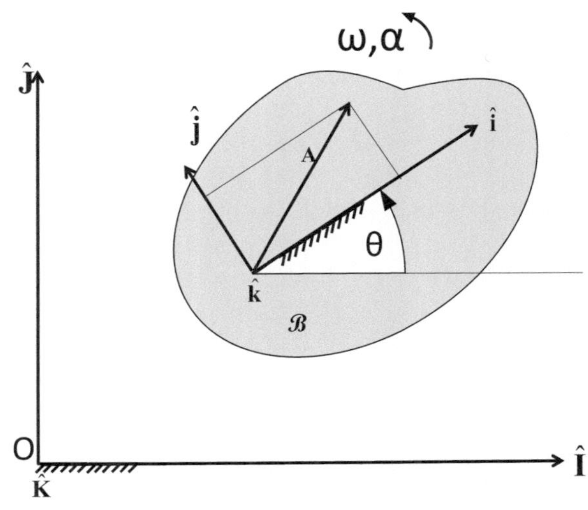

Fig. 8.2 Time derivative of vectors referenced to moving coordinates

Using Eq. (8.10), Eq. (8.18) simplifies as

$$\begin{aligned}\sum \mathbf{M}_C &= \frac{d\mathbf{H}_C}{dt} = \frac{d}{dt}\left(I_{XZ}^C\omega\hat{\mathbf{i}} + I_{YZ}^C\omega\hat{\mathbf{j}} + I_{ZZ}^C\omega\hat{\mathbf{k}}\right) \\ &= I_{XZ}^C\alpha\hat{\mathbf{i}} + I_{YZ}^C\alpha\hat{\mathbf{j}} + I_{ZZ}^C\alpha\hat{\mathbf{k}} + \omega^2 I_{XZ}^C\hat{\mathbf{j}} - \omega^2 I_{YZ}^C\hat{\mathbf{i}} \\ &= \left(I_{XZ}^C\alpha - \omega^2 I_{YZ}^C\right)\hat{\mathbf{i}} + \left(I_{YZ}^C\alpha + \omega^2 I_{XZ}^C\right)\hat{\mathbf{j}} + I_{ZZ}^C\alpha\hat{\mathbf{k}}\end{aligned} \tag{8.19}$$

Fortunately, the cross moments of inertia are zero for most of the links used in planar mechanisms. Hence, if the cross moments of inertia are equal to zero, Eq. (8.19) becomes

$$\sum \mathbf{M}_C = I_{ZZ}^C\alpha\hat{\mathbf{k}} \tag{8.20}$$

Now we can derive the equation for sum of moments about any point P that will give flexibility in deriving the dynamic equations of motion.

In addition to Eq. (8.14), the angular moment about a point P can also be written in terms of the angular momentum about the mass center as follows:

$$\mathbf{H}_P = \mathbf{H}_C + \mathbf{r}_{PC} \times m\mathbf{v}_C \tag{8.21}$$

By taking the time-derivative of Eq. (8.21), sum of moments of the rigid body about the point P is given by

$$\sum \mathbf{M}_P = \dot{\mathbf{H}}_P = \dot{\mathbf{H}}_C + \mathbf{r}_{PC} \times m\mathbf{a}_C \tag{8.22}$$

Assuming that the cross moments of inertia are zero, from Eqs. (8.20) and (8.22)

$$\sum \mathbf{M}_P = I_{ZZ}^C\alpha\hat{\mathbf{k}} + \mathbf{r}_{PC} \times m\mathbf{a}_C \tag{8.23}$$

An alternate form of Eq. (8.23) can be derived as follows:

Equation (8.14) can be rewritten as

$$\mathbf{H}_p = m\mathbf{r}_{PC} \times \mathbf{v}_P + I_{XZ}^P\omega\hat{\mathbf{i}} + I_{YZ}^P\omega\hat{\mathbf{j}} + I_{ZZ}^P\omega\hat{\mathbf{k}} \tag{8.24}$$

From Eqs. (8.21) and (8.24)

$$\mathbf{H}_C + \mathbf{r}_{PC} \times m\mathbf{v}_C = m\mathbf{r}_{PC} \times \mathbf{v}_P + I_{XZ}^P\omega\hat{\mathbf{i}} + I_{YZ}^P\omega\hat{\mathbf{j}} + I_{ZZ}^P\omega\hat{\mathbf{k}} \tag{8.25}$$

Differentiating Eq. (8.25) on both sides with respect to time

$$\begin{aligned}&\dot{\mathbf{H}}_C + \mathbf{r}_{PC} \times m\mathbf{a}_C + \dot{\mathbf{r}}_{PC} \times m\mathbf{v}_C \\ &\quad = m\dot{\mathbf{r}}_{PC} \times \mathbf{v}_P + m\mathbf{r}_{PC} \times \mathbf{a}_P + \frac{d}{dt}\left(I_{XZ}^P\omega\hat{\mathbf{i}} + I_{YZ}^P\omega\hat{\mathbf{j}} + I_{ZZ}^P\omega\hat{\mathbf{k}}\right)\end{aligned} \tag{8.26}$$

Since the first two terms of LHS of Eq. (8.26) are same as Eq. (8.22), it results in

$$\sum \mathbf{M}_P + \dot{\mathbf{r}}_{PC} \times m\mathbf{v}_C = m\dot{\mathbf{r}}_{PC} \times \mathbf{v}_P + m\mathbf{r}_{PC} \times \mathbf{a}_P + \frac{d}{dt}\left(I_{XZ}^P \omega \hat{\mathbf{i}} + I_{YZ}^P \omega \hat{\mathbf{j}} + I_{ZZ}^P \omega \hat{\mathbf{k}}\right) \tag{8.27}$$

Simplifying Eq. (8.27)

$$\sum \mathbf{M}_P = m\dot{\mathbf{r}}_{PC} \times (\mathbf{v}_P - \mathbf{v}_C) + m\mathbf{r}_{PC} \times \mathbf{a}_P + \frac{d}{dt}\left(I_{XZ}^P \omega \hat{\mathbf{i}} + I_{YZ}^P \omega \hat{\mathbf{j}} + I_{ZZ}^P \omega \hat{\mathbf{k}}\right) \tag{8.28}$$

Since $(\mathbf{v}_P - \mathbf{v}_C) = -\dot{\mathbf{r}}_{PC}$ (see Chap. 2), the first equation on the RHS vanishes and Eq. (8.28) becomes

$$\sum \mathbf{M}_P = m\mathbf{r}_{PC} \times \mathbf{a}_P + \frac{d}{dt}\left(I_{XZ}^P \omega \hat{\mathbf{i}} + I_{YZ}^P \omega \hat{\mathbf{j}} + I_{ZZ}^P \omega \hat{\mathbf{k}}\right) \tag{8.29}$$

Since the second term of the RHS of Eq. (8.27) is similar to Eq. (8.19) with the only difference being the moments of inertia computed about the point P, Eq. (8.29) becomes

$$\sum \mathbf{M}_P = m\mathbf{r}_{PC} \times \mathbf{a}_P + \left(I_{XZ}^P \alpha - \omega^2 I_{YZ}^P\right)\hat{\mathbf{i}} + \left(I_{YZ}^P \alpha + \omega^2 I_{XZ}^P\right)\hat{\mathbf{j}} + I_{ZZ}^P \alpha \hat{\mathbf{k}} \tag{8.30}$$

When the cross moments of inertia vanish, Eq. (8.30) becomes

$$\sum \mathbf{M}_P = m\mathbf{r}_{PC} \times \mathbf{a}_P + I_{ZZ}^P \alpha \hat{\mathbf{k}} \tag{8.31}$$

If the point P is a fixed point on the rigid body, which is common in many mechanisms where links are hinged to a fixed point, Eq. (8.31) simplifies to

$$\sum \mathbf{M}_P = I_{ZZ}^P \alpha \hat{\mathbf{k}} \tag{8.32}$$

In this section, we have derived various forms of equations for the sum of moments about the mass center and about an arbitrary point P. So, depending on the available data for a specific problem, any of the equations—(8.20) or (8.23) or (8.31)—can be used.

The forces and product of mass and acceleration of the center of mass are related by

$$\sum F = m\mathbf{a}_\mathbf{C} \tag{8.33}$$

From the above equations, based on kinematic analysis, reaction forces, external forces and torques related to interconnected moving bodies of planar mechanisms can be computed.

8.4 Force Analysis of a Single Rigid Body

A rigid body in plane motion subject to a number of forces and moments is shown in Fig. 8.3. But all these forces and moments can be represented by a single force **F** at its center of mass that has components in the plane of paper and a moment of magnitude M_C that has the direction of axis of rotation $\hat{\mathbf{k}}$ perpendicular to the plane of paper. Due to the action of the above moments and forces, the center of mass has an acceleration $\mathbf{a}_C$ and the rigid body has an angular acceleration of α as shown in the figure.

From Eq. (8.23), neglecting cross moments of inertia, the angular acceleration is related to the moments by

$$\alpha = \frac{M_C}{I_{ZZ}^C} \tag{8.34}$$

From Eq. (8.33), linear acceleration of the mass center is related to the resultant of all forces acting at the mass center

$$\mathbf{a}_C = \frac{\mathbf{F}}{m} \tag{8.35}$$

From Eqs. (8.34) and (8.35), it is clear that the angular and linear accelerations are a result of forces and moments acting on the rigid body. However, in many practical applications, linear and angular accelerations can be computed easily and then some of the forces (may include reaction forces) and moments (may include torque) can be determined.

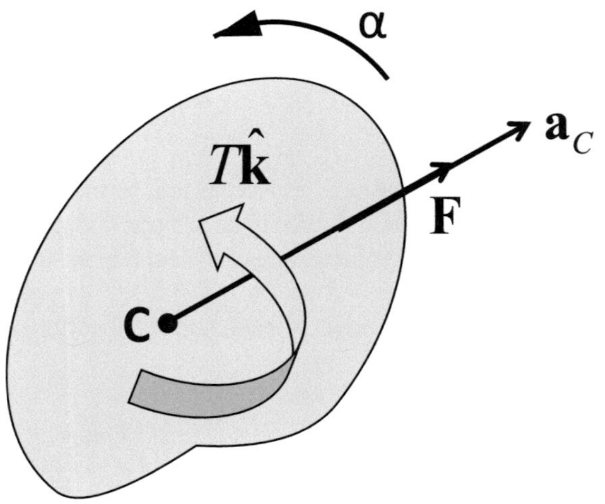

Fig. 8.3 A rigid body subjected to force and moments

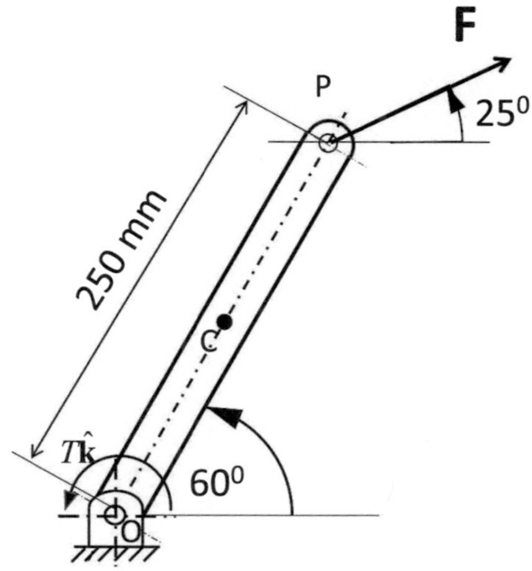

Fig. 8.4 Link of Example 8.1

Let us look at an example of a rigid body in plane motion, pivoted at a point, for which an external force and angular velocity and acceleration are known.

Example 8.1
A 250 mm long link shown in Fig. 8.4 has a mass of 2 kg and is hinged at O. Its center of gravity is located at its mid-point and it has a mass moment of inertia of 0.1 kgm^2. It is inclined at an angle of 60° at the given instant. At this instant, the angular velocity is 30 rad/s (counterclockwise) and angular acceleration 15 rad/s^2 (clockwise). A force of 200 N is acting at its free end P at an angle of 25° as shown. Determine the reaction forces at the hinge and the torque that has to be applied to satisfy the given conditions.

The position vector $\mathbf{r}_{OP}$ is given by

$$\begin{aligned}\mathbf{r}_{OP} &= 0.250\left(\cos 60°\hat{\mathbf{i}} + \sin 60°\hat{\mathbf{j}}\right)\\ &= 0.125\hat{\mathbf{i}} + 0.2165\hat{\mathbf{j}}\,\text{m}\end{aligned} \tag{1}$$

The force vector **F** is given by

$$\begin{aligned}\mathbf{F} &= 200\left(\cos 25°\hat{\mathbf{i}} + \sin 25°\hat{\mathbf{j}}\right)\\ &= 181.26\hat{\mathbf{i}} + 84.52\hat{\mathbf{j}}\,N\end{aligned} \tag{2}$$

Velocity and acceleration analysis

From Eq. (2.58), the velocity vector of the point P, $\mathbf{v}_P$ can be written as

$$\begin{aligned}\mathbf{v}_P = \mathbf{v}_O + \omega\hat{\mathbf{k}} \times \mathbf{r}_{OP} &= 30\hat{\mathbf{k}} \times \left(0.125\hat{\mathbf{i}} + 0.2165\hat{\mathbf{j}}\,\text{m}\right) \\ &= -6.48\hat{\mathbf{i}} + 3.75\hat{\mathbf{j}}\ \text{m/s}\end{aligned} \tag{3}$$

From Eq. (2.66), the acceleration vector of the point P, $\mathbf{a}_P$ can be written as

$$\begin{aligned}\mathbf{a}_P &= \mathbf{a}_O + \alpha\hat{\mathbf{k}} \times \mathbf{r}_{OP} - \omega^2\mathbf{r}_{OP} \\ &= -15\hat{\mathbf{k}} \times \left(0.125\hat{\mathbf{i}} + 0.2165\hat{\mathbf{j}}\right) \\ &\quad - 30^2\left(0.125\hat{\mathbf{i}} + 0.2165\hat{\mathbf{j}}\right) \\ &= -110.625\hat{\mathbf{i}} - 196.275\hat{\mathbf{j}}\ \ \text{m/s}^2\end{aligned} \tag{4}$$

The acceleration vector of the mass center C (since it is midway of OP), $\mathbf{a}_P$ can be written as

$$\mathbf{a}_C = \frac{1}{2}\left(-110.625\hat{\mathbf{i}} - 196.275\hat{\mathbf{j}}\right)\ \text{m/s}^2 \tag{5}$$

Refer to the free-body diagram shown in Fig. 8.5. From Eq. (8.33), equating all the forces (both applied and reaction) to the product of mass and acceleration of the mass center.

$$\begin{aligned}\sum \mathbf{F} = m\mathbf{a}_C F_x\hat{\mathbf{i}} + F_y\hat{\mathbf{j}} &+ 181.26\hat{\mathbf{i}} + 84.52\hat{\mathbf{j}} - 2 \times 9.81\hat{\mathbf{j}} \\ &= 2 \times \frac{1}{2}\left(-110.625\hat{\mathbf{i}} - 196.275\hat{\mathbf{j}}\right)\ \text{N}\end{aligned} \tag{6}$$

Grouping terms of Eq. (6) in the coordinate directions

$$\begin{aligned}&(F_x + 181.26)\hat{\mathbf{i}} + \left(F_y + 84.52 - 2 \times 9.81\right)\hat{\mathbf{j}} \\ &\quad = \left(-110.625\hat{\mathbf{i}} - 196.275\hat{\mathbf{j}}\right)\text{N}\end{aligned} \tag{7}$$

Equating components of the forces along coordinate directions

$$\begin{aligned}F_x &= -291.885\,\mathbf{N} \\ F_y &= -300.415\,\mathbf{N}\end{aligned} \tag{8}$$

From Eq. (8.23), summing moments of all forces about the pivot point O

$$\sum \mathbf{M}_O = \mathbf{r}_{OC} \times m\mathbf{a}_C + I_{ZZ}^C\alpha\hat{\mathbf{k}} \tag{9}$$

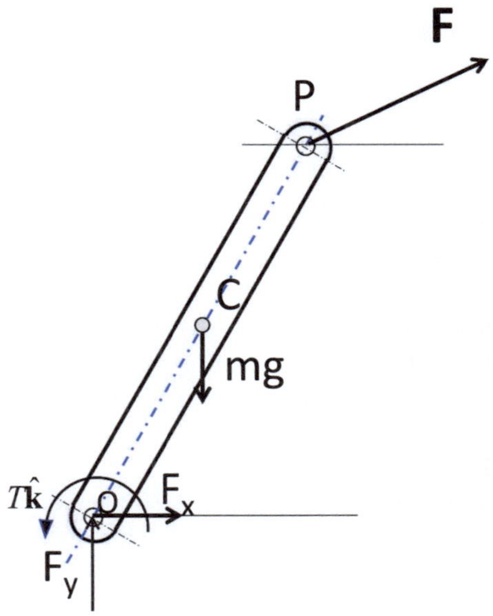

Fig. 8.5 Free-body diagram of the link shown in Fig. 8.4

From Fig. 8.5, sum of the moments about O are given by

$$\begin{aligned}\sum \mathbf{M}_O &= T\hat{\mathbf{k}} + \mathbf{r}_{OC} \times \left(-mg\hat{\mathbf{j}}\right) + \mathbf{r}_{OP} \times \mathbf{F} \\ &= T\hat{\mathbf{k}} + \frac{1}{2}\left(0.125\hat{\mathbf{i}} + 0.2165\hat{\mathbf{j}}\right) \times \left(-2 \times 9.81\hat{\mathbf{j}}\right) \\ &\quad + \left(0.125\hat{\mathbf{i}} + 0.2165\hat{\mathbf{j}}\right) \times \left(181.26\hat{\mathbf{i}} + 84.52\hat{\mathbf{j}}\right) \text{ N m}\end{aligned} \tag{10}$$

In terms of the unknown torque T, Eq. (10) simplifies to

$$\sum \mathbf{M}_O = (T - 29.9040)\hat{\mathbf{k}} \text{ N m} \tag{11}$$

The RHS of Eq. (9) is given by

$$\begin{aligned}&\mathbf{r}_{OC} \times m\mathbf{a}_C + I_{ZZ}^C \alpha \hat{\mathbf{k}} \\ &\quad = \left(0.125\hat{\mathbf{i}} + 0.2165\hat{\mathbf{j}}\right) \times 2\left(\frac{1}{2}\left(-110.625\hat{\mathbf{i}} - 196.275\hat{\mathbf{j}}\right)\right) \\ &\quad - 0.1 \times 15\hat{\mathbf{k}} = -2.0841\hat{\mathbf{k}}\end{aligned} \tag{12}$$

Equation (12) is rewritten as

$$\mathbf{r}_{OC} \times m\mathbf{a}_C + I_{ZZ}^C \alpha \hat{\mathbf{k}} = -2.0841\hat{\mathbf{k}} \tag{13}$$

Equation (11) is rewritten as

$$\sum \mathbf{M}_O = (T - 29.9040)\hat{\mathbf{k}} \text{ N m} \tag{14}$$

From Eqs. (13) and (14), the applied torque is given by

T = 27.82 N m

Many times, velocity and acceleration analysis cannot be carried out before beginning the kinetic analysis. In such cases, acceleration will depend on the forces and moments acting on the rigid body. The following example illustrates one such example in which kinetic analysis has to be carried out in order to determine angular velocity and acceleration of a link.

Example 8.2

The uniform rod of mass m is released from rest that was initially in the horizontal position indicated in Fig. 8.6. Determine the reaction at the pin as a function of the angle between the horizontal and the rod measured in the clockwise direction.

From Eq. (8.23), summing moments of all forces about the pivot point O

$$\sum M_o = I_{zz}^0 \alpha \tag{1}$$

From Eq. (1) and Fig. 8.7

$$\mathbf{r}_{OC} \times -mg\,\hat{\mathbf{j}} = \left[\frac{m\ell^2}{12} + m\left(\frac{\ell}{2}\right)^2\right]\alpha$$

$$\mathbf{r}_{OC} \times -mg\,\hat{\mathbf{j}} = \frac{m\ell^2}{3}\alpha \tag{2}$$

The position vector of the mass center C as a function of θ is given by

$$\mathbf{r}_{OC} = \frac{\ell}{2}\left(\cos\theta\,\hat{\mathbf{i}} - \sin\theta\hat{\mathbf{j}}\right) \tag{3}$$

Fig. 8.6 Link of Example 8.2

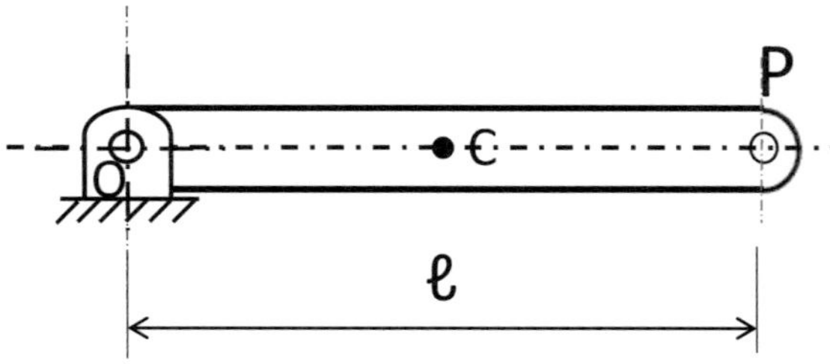

Fig. 8.7 Free-body diagram of the link shown in Fig. 8.6

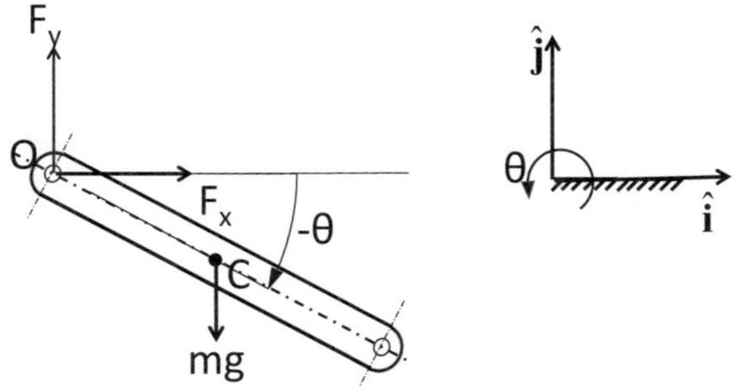

From Eqs. (2) and (3)

$$\frac{\ell}{2}(\cos\theta\,\hat{\mathbf{i}} - \sin\theta\,\hat{\mathbf{j}}) \times -mg\,\hat{\mathrm{j}} = \frac{m\ell^2}{3}\alpha \tag{4}$$

From Eqs. (3) and (4), the angular acceleration is given by

$$\alpha = \ddot{\theta} = -\frac{3g}{2\ell}\cos\theta \text{ (clockwise)} \tag{5}$$

Multiplying both sides of Eq. (5) by $\dot{\theta}$ to facilitate integration

$$\ddot{\theta}\dot{\theta} = -\frac{3g}{2\ell}\cos\theta\dot{\theta} \tag{6}$$

Integrating Eq. (6) with respect to time

$$\frac{\dot{\theta}^2}{2} = -\frac{3g}{2\ell}\sin\theta + C_1 \tag{7}$$

Based on initial conditions, $C_1 = 0$ and the angular velocity is given by

$$\dot{\theta}^2 = -\frac{3g}{\ell}\sin\theta \tag{8}$$

From Eqs. (3), (5), (8), and Eq. (2.66), acceleration of the mass center is given by

$$\begin{aligned}\mathbf{a}_c &= \mathbf{a}_o + \ddot{\theta}\,\hat{\mathbf{k}} \times \mathbf{r}_{OC} - \dot{\theta}^2\mathbf{r}_{OC}\\ &= -\frac{3g}{2\ell}\cos\theta\,\hat{\mathbf{k}} \times \frac{\ell}{2}(\cos\theta\,\hat{\mathbf{i}} - \sin\theta\,\hat{\mathbf{j}}) + \frac{3g}{\ell}\sin\theta\frac{\ell}{2}(\cos\theta\,\hat{\mathbf{i}} - \sin\theta\,\hat{\mathbf{j}})\\ &= \frac{3g}{4}[-\sin\theta\cos\theta\,\hat{\mathbf{i}} - \cos^2\theta\,\hat{\mathbf{j}}] + \frac{3g}{2}[\sin\theta\cos\theta\,\hat{\mathbf{i}} - \sin^2\theta\,\hat{\mathbf{j}}]\end{aligned} \tag{9}$$

Equation (9) simplifies to

$$\mathbf{a}_c = \frac{3g \sin\theta \cos\theta\, \hat{\mathbf{i}}}{4} - \frac{3g(1 + \sin^2\theta)\, \hat{\mathbf{j}}}{4} \tag{10}$$

Using Eq. (8.33), relating the product of mass and acceleration with forces

$$F_x\hat{\mathbf{i}} + \left(F_y - mg\right)\hat{\mathbf{j}} = \frac{3mg \sin\theta \cos\theta\, \hat{\mathbf{i}}}{4} - \frac{3mg(1 + \sin^2\theta)\, \hat{\mathbf{j}}}{4} \tag{11}$$

Equating components of the forces along the coordinate directions on both sides of Eq. (11), the reactions at the pivot are given by

$$F_x = \frac{3mg \sin\theta \cos\theta}{4}$$
$$F_y = mg - \frac{3mg(1 + \sin^2\theta)}{4} \tag{12}$$

Thus far, we have tried to relate moments and forces in a link to its accelerations. But most practical mechanisms will have more than one link. Hence, the force analysis of the entire mechanism has to be carried out. For most practical mechanisms, the kinematic analysis is carried out first and then followed by kinetic analysis. Therefore, once the kinematic information about all links of the mechanism is known, the forces and moments on each link are related to the kinematic analysis; this will help compute the reaction forces between the links as well as the applied forces and moments. Next, we shall look at such examples that will help understand the complete analysis of a mechanism. Unless stated, gravitational forces are generally neglected since they are much smaller than other forces.

In some cases, however, the kinetic analysis will be carried out first that will provide the basis for obtaining kinematic information and will be illustrated in Example 8.4.

8.5 Slider Crank Mechanism

First, we shall discuss the force analysis of a slider crank mechanism that is used in many internal combustion engines, and also in pumps, compressors, etc.; but we shall discuss its application in internal combustion engines only in this chapter. Figure 8.8 shows such a slider crank mechanism (3R-1P) that has already been discussed in Chaps. 1 and 2. This is a very important mechanism that is used in internal combustion engines for converting the reciprocating motion of the piston due to combustion pressure to rotary motion of the crank; the piston moves due to the large gas pressure arising out of combustion of the fuel. Although ignition of the fuel occurs for a small duration of time, the impulsive force due to the high pressure inside the cylinder pushes the piston to rotate the crank through the connecting rod. For two-stroke engines, the fuel ignition takes place for every cycle of the crank rotation, whereas for four stroke engines, the fuel ignition takes place for every two cycles of crank rotation. This will have implications on the variation of torque output in each cycle that will have to be smoothened out using multiple cylinders and flywheels.

Once the engine starts running, it is the inertia forces of the moving components that dominate force input to the piston. Therefore, many times the force analysis is carried without accounting for the force applied by the cylinder pressure. In addition, the flywheel attached to the engine stores excess energy at the time of combustion and releases it during the rest of the cycle. This would ensure that the crank rotates uniformly at a certain angular velocity, which makes it easier for completing the kinematic analysis of all the links before determining forces and moments that act on the mechanism and thus eventually determine the torque available at the crank.

As shown in Fig. 8.8, the slider crank mechanism consists of a crank that has purely rotational motion and a piston that has purely rectilinear motion. The connecting rod that is connected to the crank and piston at its ends has both rotation and translation and therefore is called a floating link.

The slider crank mechanism is a good representative of the simple mechanisms that are interconnected by links. Hence, a detailed study of forces and moments in a slider crank mechanism is useful in carrying out kinetic analysis of any general mechanism of interconnected links. For analyzing forces and moments of various links, each link is considered one at a time for which accelerations (both linear and angular as the case may be) are known. Reaction forces are present at points where kinematic pairing takes place; the reaction forces will be equal and opposite to each other in the paired links. Torque is either an input or an output or both, depending on the functioning of the mechanism.

Let us assume that the mass of crank, connecting rod and piston are known, in addition to the moment of inertia of the crank and connecting rod. The center of mass of the crank is assumed to be at C_2 and that of the connecting rod is assumed at C_3. Let I_2 be the moment of inertia of the crank about the axis perpendicular to the plane of motion and passing through the mass center of crank; similarly, I_3 be the moment of inertia of the connecting rod. Let P be the cylinder pressure and A, the area of cross-section of the cylinder; $F_p = A \times P$ is the force exerted by the cylinder pressure on the piston.

Fig. 8.8 Slider crank mechanism

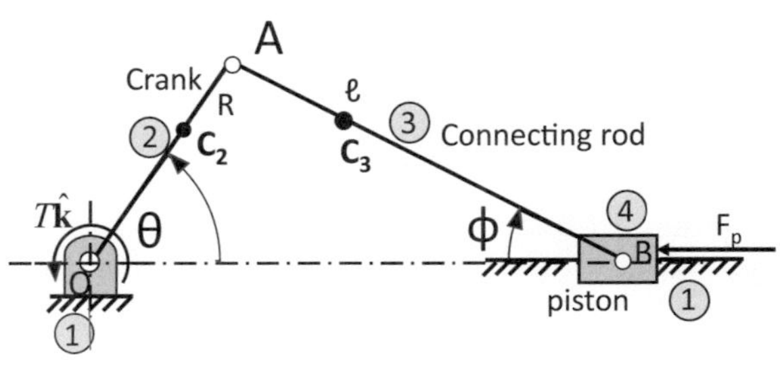

To start with, let us analyze the forces and moments of the slider crank mechanism without making any approximations. Then we can move to simpler ways of solving for forces and moments. It is assumed that the kinematic analysis of the slider crank mechanism as discussed in Chap. 2 is completed, assuming that the crank is turning uniformly at ω_2 in the counterclockwise direction.

The kinematic analysis is summarized as follows:

ω_2: angular velocity of the crank, rad/s (counterclockwise).

α_2: angular acceleration of the crank = 0.

ω_3: angular velocity of the connecting, rad/s (direction depends on the position of the crank).

α_3: angular acceleration of the connecting rod, rad/s^2 (direction depends on the position of the crank).

a_B: linear acceleration of the piston, m/s^2 (direction depends on the position of the crank).

The free-body diagram of the piston is shown in Fig. 8.9. Since the cylinder is generally well lubricated, the friction coefficient between sliding forces at the interface is extremely small and hence friction force is neglected. Therefore, the normal force F_{14} acts in the vertical direction as shown. The acceleration a_B is obtained from the kinematic analysis. There is a reaction force from the connecting rod acting on the piston. It is assumed to have components along the coordinate directions as F_{34}^x and F_{34}^y. In the vector form, the reaction force of the connecting rod on the piston can be written as

$$\mathbf{F}_{34} = F_{34}^x\,\hat{\mathbf{i}} + F_{34}^y\,\hat{\mathbf{j}} \tag{8.36}$$

From Eq. (8.33), summing all the forces on the piston in the coordinate directions

$$F_{34}^x - F_P = -m_4 a_B \tag{8.37}$$

and

$$F_{14} + F_{34}^y = 0 \tag{8.38}$$

It should be noted that the initial assignment of the sign of magnitude of reaction force vectors is arbitrary; however, the correct sign of acceleration, external forces and torques must be used in all equations. From the complete analysis and solving for the reaction forces, the correct sign of the reaction forces for the given position will be known. But we should only make sure that the reaction forces are of opposite in sign on either side of a kinematic pair.

Before we move on to the connecting rod force analysis, it should be noted that the reaction force of the piston on the connecting rod is negative of that in Eq. (8.36). It is given by

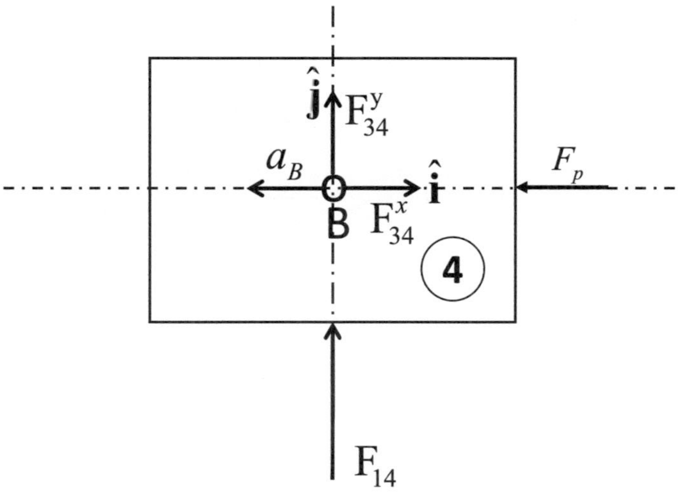

Fig. 8.9 Free-body diagram of the piston

$$\mathbf{F}_{43} = -F_{34}^{x}\hat{\mathbf{i}} - F_{34}^{y}\hat{\mathbf{j}} \tag{8.39}$$

The reaction forces on the connecting rod are shown in Fig. 8.10. From Eq. (8.33), the forces in the coordinate directions are related by

$$F_{23}^{x} - F_{34}^{x} = m_3 a_{C_3}^{x} \tag{8.40}$$

and

$$F_{23}^{y} - F_{34}^{y} = m_3 a_{C_3}^{y} \tag{8.41}$$

The reaction force at the point A can be assumed as

$$\mathbf{F}_{23} = F_{23}^{x}\,\hat{\mathbf{i}} + F_{23}^{y}\,\hat{\mathbf{j}} \tag{8.42}$$

By using Eq. (8.23), by taking moments about the point A, the following equations can be written

$$\sum \mathbf{M}_A = \mathbf{r}_{AB} \times \mathbf{F}_{43} = I_3\alpha_3\hat{\mathbf{k}} + \mathbf{r}_{AC_3} \times m_3\mathbf{a}_{C_3} \tag{8.43}$$

And by using Eq. (8.23), by taking moments about the point B, the following equations can be written

$$\sum \mathbf{M}_B = \mathbf{r}_{BA} \times \mathbf{F}_{23} = I_3\alpha_3\hat{\mathbf{k}} + \mathbf{r}_{BC_3} \times m_3\mathbf{a}_{C_3} \tag{8.44}$$

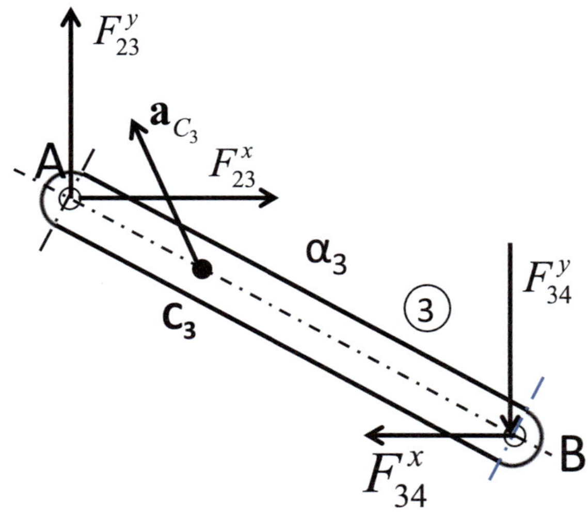

Fig. 8.10 Free-body diagram of the connecting rod

Either of the two Eqs. (8.43) or (8.44) can be used for further analysis.

Before we analyze forces on the crank, the following reaction force on it is written as follows

$$\mathbf{F}_{32} = -F_{23}^x\,\hat{\mathbf{i}} - F_{23}^y\,\hat{\mathbf{j}} \tag{8.45}$$

The reaction forces on the crank are as shown in Fig. 8.11. From Eq. (8.33), the forces in the coordinate directions are related by

$$-F_{23}^x + F_{12}^x = m_2 a_{C_2}^x \tag{8.46}$$

and

$$-F_{23}^y + F_{12}^y = m_2 a_{C_2}^y \tag{8.47}$$

The reaction force at the point O can be written as

$$\mathbf{F}_{12} = F_{12}^x\,\hat{\mathbf{i}} + F_{12}^y\,\hat{\mathbf{j}} \tag{8.48}$$

By using Eq. (8.23), the following equations can be written for moments about the point O

$$\sum \mathbf{M}_O = \mathbf{r}_{OA} \times \mathbf{F}_{23} = T\hat{\mathbf{k}} + \mathbf{r}_{OC_2} \times m_2\mathbf{a}_{C_2} \tag{8.49}$$

It should be noted that the torque available at the crank should also be accounted for in the above equation because a certain amount torque is either received by the system or given out of the system; a counterclockwise torque is assumed and it will

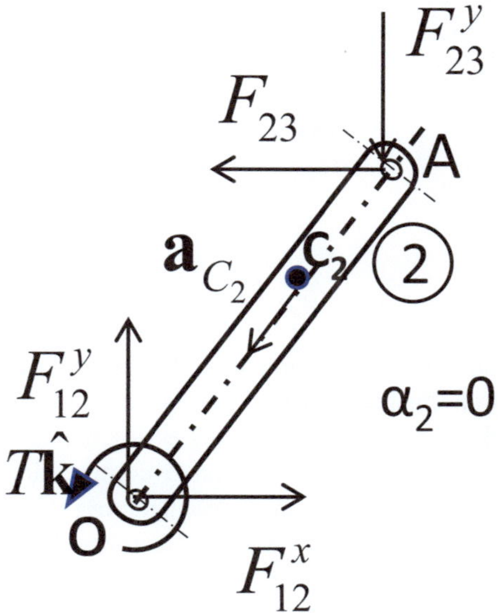

Fig. 8.11 Free-body diagram of the crank

turn out to be positive if the torque is given out by the system at the given instant. Else, the mechanism will be drawing from the external system it is driving that will give a negative value. Therefore, the torque output is the net result of inertia forces of the mechanism and the pressure acting on the system. Hence, there will be significant variation in the torque output at the crank over each cycle of the crank rotation. It can be smoothened out by using a flywheel. Using multicylinder engines that will fire in a definite sequence will also smoothen out the torque output that will be discussed in Chap. 9.

And, similarly, by taking moments about A and using Eq. (8.23), the following equation can be written

$$\sum \mathbf{M}_A = \mathbf{r}_{AO} \times \mathbf{F}_{12} = T\hat{\mathbf{k}} + \mathbf{r}_{AC_2} \times m_2 \mathbf{a}_{C_2} \tag{8.50}$$

Either of the two equations, (8.49) or (8.50) can be used for further analysis.

The above equations can be solved to get reaction forces for the entire mechanism for a given position of the crank. The above equations can be easily programmed to be solved on a computer for various crank angles to obtain the reaction forces and torque for one complete rotation of the crank. A similar approach can be used to obtain reaction forces and driving torque for any general mechanism. The above procedure can be easily applied to the exercise problem 8.11.

Although the discussion above can yield accurate values of reaction forces and torque on the crank, there are practical difficulties in setting up these equations to

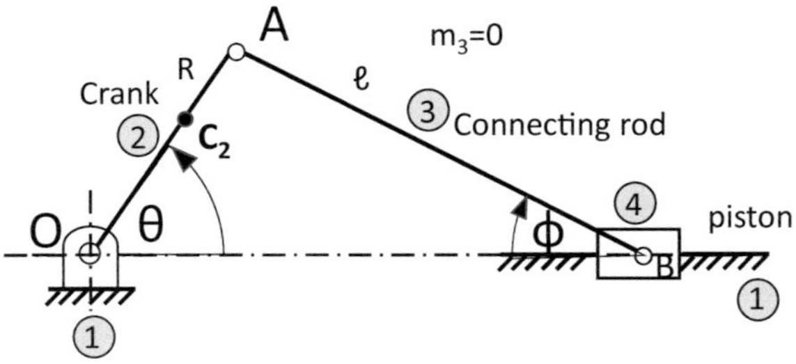

Fig. 8.12 Slider crank mechanism with a massless connecting rod

obtain a quick estimate of the reaction forces. One such simplification is the assumption that the connecting rod is massless as shown in Fig. 8.12. An important simplification that results by neglecting the connecting rod mass is that the reaction force in it can be either tension or compression. Since this will easily give the direction of reaction force in the connecting rod, the reaction force at the crank pivot can be determined easily. In addition, it would not be necessary to compute the angular acceleration and acceleration of the connecting rod mass center. Approximate acceleration of the piston can be determined by differentiating Eq. (1.22) twice.

The free-body diagram of the linkages of Fig. 8.12 is shown in Fig. 8.13. The magnitude of the reaction force of the connecting rod on the piston is given by

$$F_{34} = \frac{-m_4 a_B + F_P}{\cos\phi} = -F_{43} = F_{23} \tag{8.51a}$$

The reaction force vector of the connecting rod on the piston is given by

$$\mathbf{F}_{34} = \left(\frac{m_4 a_B + F_P}{\cos\phi}\right)\left(\cos\phi\,\hat{\mathbf{i}} - \sin\phi\,\hat{\mathbf{j}}\right) = (-m_4 a_B + F_P)\left(\hat{\mathbf{i}} - \tan\phi\,\hat{\mathbf{j}}\right) \tag{8.51b}$$

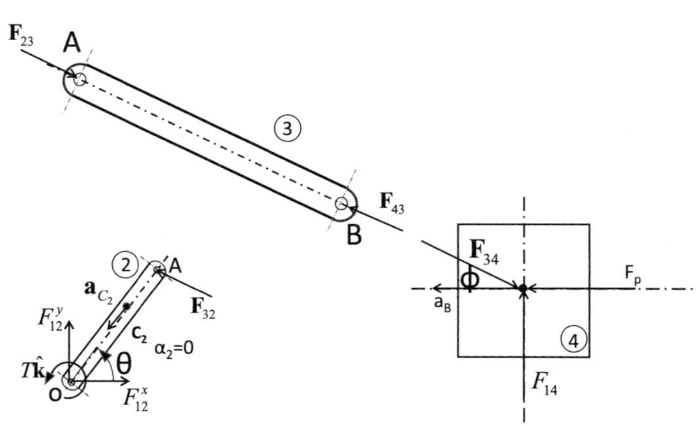

Fig. 8.13 Free-body diagram of the linkages of the slider crank mechanism shown in Fig. 8.12, neglecting the connecting rod mass

The sign of acceleration of the piston pin, a_B, must be correctly accounted for in the calculations, although it is negative for the current position.

The normal reaction force of the cylinder on the piston is given by

$$F_{14} = F_{34} \sin \phi \tag{8.52}$$

The reaction force vector of the connecting rod on the crank is given by

$$\mathbf{F}_{32} = \mathbf{F}_{43} = -\mathbf{F}_{34} = (m_4 a_B + F_P)\left(-\hat{\mathbf{i}} + \tan \phi\, \hat{\mathbf{j}}\right) \tag{8.53}$$

By using Eq. (8.23) the following equations can be written

$$\sum \mathbf{M}_O = \mathbf{r}_{OA} \times \mathbf{F}_{32} + T\hat{\mathbf{k}} = \mathbf{r}_{OC_2} \times m_2 \mathbf{a}_{C_2} \tag{8.54}$$

and

$$\sum \mathbf{M}_A = \mathbf{r}_{AO} \times \mathbf{F}_{12} + T\hat{\mathbf{k}} = \mathbf{r}_{AC_2} \times m\mathbf{a}_{C_2} \tag{8.55}$$

From any of the above two equations, the reaction force on the crank pivot and torque can be easily determined; they can be expressed in terms of algebraic equations as will be demonstrated in Example 8.3.

Point approximation of the connecting rod mass

Although neglecting the connecting rod mass makes it easier to determine the reaction forces, the error arising out of this assumption sometimes becomes unacceptable. Since the only objective of neglecting the connecting rod mass was to easily compute the reaction force that will be along its length, the same advantage can be obtained by distributing the connecting rod mass at either end of the connecting rod, by transferring one to the hinge connection between the crank and connecting rod (crank pin) and the other to the hinge connection between the connecting rod and piston (piston pin).

Consider a connecting rod as shown in Fig. 8.14, of length ℓ and mass m_3 that has a center of mass at C_3 and has a moment of inertia I_3 about an axis perpendicular to the plane of paper and passing through the mass center. If the connecting rod mass has to be distributed and transferred to points A and B, the following conditions will have to be satisfied after the transfer:

1. The center of mass of the connecting rod should remain the same ($m_A \ell_A = m_B \ell_B$)
2. The total mass of the connecting rod must remain the same ($m_A + m_B = m_2$)
3. The moment of inertia of the connecting rod, before and after transfer, must remain same ($m_A \ell_A^2 + m_B \ell_B^2 = I_2$)

All the three conditions cannot be simultaneously met. Hence, using the first two conditions mentioned above

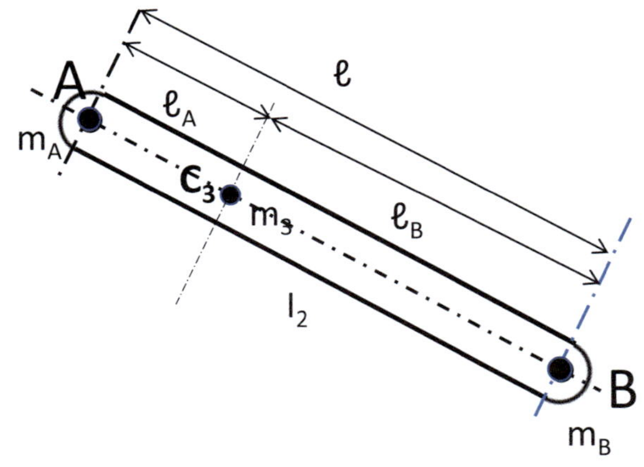

Fig. 8.14 Point mass approximation of the connecting rod

$$m_A = \left(\frac{\ell_B}{\ell_A + \ell_B}\right) m_2$$
$$m_B = \left(\frac{\ell_A}{\ell_A + \ell_B}\right) m_2 \tag{8.56}$$

Example 8.3

A slider crank mechanism is shown in Fig. 8.15. It consists of four links: (1) massless crank with an attached point mass m_D, (2) connecting rod, (3) reciprocating piston, and (4) fixed link. The crank has a total length AD = 130 mm and is pivoted midway (AO = OD) to the fixed link at "o." The connecting rod of length AB = 300 mm and total mass 1.5 kg is modeled as a two-mass system with one mass at the crank pin (point A) and the other at the piston pin (point B). The mass m_D is equal to the mass of the connecting rod transferred to the crank pin. The mass center of the connecting rod before distributing its mass is at a distance of 70 mm (AC_2) from the crank pin. The piston mass is 1 kg. When the crank angle $\theta = 45°$ and when it is rotating at 3000 RPM counterclockwise), determine the reaction forces at the crank-pivot point O (use approximate value of the piston acceleration) and the torque available at the crank shaft.

Radius of the crank r = 65 mm
Length of the connecting rod ℓ = AB = 300 mm
Mass of the connecting rod m_2 = 1.5 kg
Mass of the piston m_3 = 1 kg
AC_3 = 70 mm
Crank angle $\theta = 45°$
Speed N = 3000 RPM
The crank mass is neglected.
Angular velocity of the crank $\omega = 2\pi N/60 = 2\pi \times 3000/60 = 314.16$ rad/s.
From Eq. (2.66), acceleration of the point A is given by

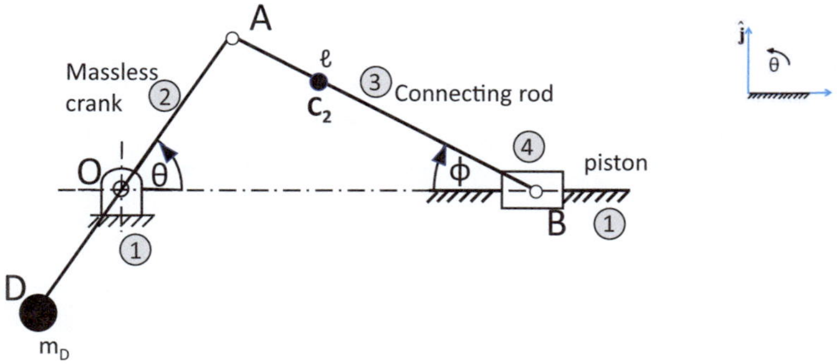

Fig. 8.15 Slider crank mechanism of Example 8.3

$$\mathbf{a}_A = \mathbf{a}_o + \alpha\hat{\mathbf{k}} \times \mathbf{r}_{OA} - \omega^2\mathbf{r}_{OA} \tag{1}$$

The position vector $\mathbf{r}_{OA}$ is given by

$$\mathbf{r}_{OA} = r(\cos\theta\,\hat{\mathbf{i}} + \sin\theta\,\hat{\mathbf{j}}) \tag{2}$$

Noting that the acceleration of the pivot point O is zero and angular acceleration of the crank is zero, Eqs. (1) and (2) result in

$$\begin{aligned}\mathbf{a}_A &= -r\omega^2(\cos\theta\,\hat{\mathbf{i}} + \sin\theta\,\hat{\mathbf{j}})\\ &= -0.065 \times 314.16^2(\cos 45^\circ\hat{\mathbf{i}} + \sin 45^\circ\,\hat{\mathbf{j}})\\ &= -4536\hat{\mathbf{i}} - 4536\,\hat{\mathbf{j}}\quad \text{m/s}^2\end{aligned} \tag{3}$$

Differentiating Eq. (1.22) twice with respect to time, approximate acceleration of the point B is given by

$$\begin{aligned}\mathbf{a}_B &\approx -r\omega^2\left(\cos\theta + \frac{r}{\ell}\cos 2\theta\right)\\ &\quad - 0.065 \times 314.16^2\left(\cos 45^\circ + \frac{65}{300}\cos 90^\circ\right)\\ &= -4536\,\hat{\mathbf{i}}\ \ \text{m/s}^2\end{aligned} \tag{4}$$

The connecting rod mass is distributed as two-point masses at the crank pin and the wrist pin, respectively, as follows

$$\mathrm{m_A} = \mathrm{m_3} \times \mathrm{BG3}/\ell = 1.5 \times 230/300 = \mathbf{1.15\,kg} \tag{5}$$

$$\mathrm{m_B} = \mathrm{m_2} \times \mathrm{AG3}/\ell = 1.5 \times 70/300 = \mathbf{0.35\,kg} \tag{6}$$

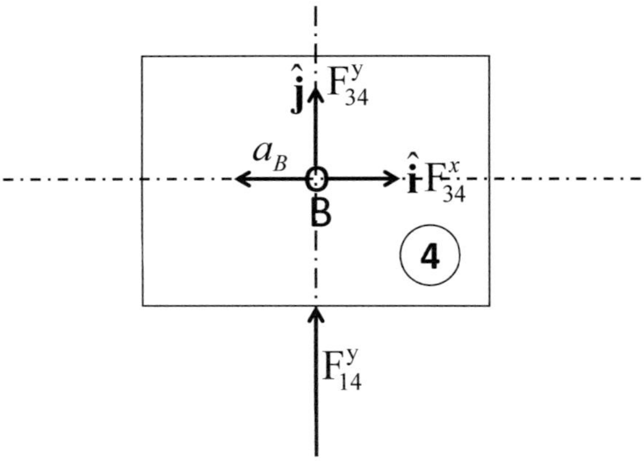

Fig. 8.16 Piston of Example 8.3

The force $\mathbf{F}_{34}$, which is the force acted by the connecting rod on the piston, is given by

$$\mathbf{F}_{34} = F^x_{34}\,\hat{\mathbf{i}} + F^y_{34}\,\hat{\mathbf{j}} \tag{7}$$

The force $\mathbf{F}_{14}$, which is the force acted by the fixed link on the piston, is given by (Fig. 8.16)

$$\mathbf{F}_{14} = F^y_{14}\,\hat{\mathbf{j}} \tag{8}$$

Equating forces and accelerations in the horizontal direction for the piston

$$\begin{aligned} F^x_{34} &= m_4 a_B \\ &= -4536\,\mathrm{N} \end{aligned} \tag{9}$$

From Eq. (8.23) equating forces and accelerations in the vertical direction for the piston

$$F^y_{34} = -F_{14} \tag{10}$$

From Eqs. (7) and (9), $\mathbf{F}_{34}$ becomes

$$\mathbf{F}_{34} = -4536\hat{\mathbf{i}} + F^y_{34}\,\hat{\mathbf{j}} \tag{11}$$

$\mathbf{F}_{43}$, which is the force acted by the piston on the connecting rod (see Fig. 8.17) is given by

$$\begin{aligned} \mathbf{F}_{43} = -\mathbf{F}_{34} &= F^x_{43}\,\hat{\mathbf{i}} + F^y_{43}\,\hat{\mathbf{j}} \\ &= 4536\hat{\mathbf{i}} + F^y_{43}\,\hat{\mathbf{j}}\ \mathbf{N} \end{aligned} \tag{12}$$

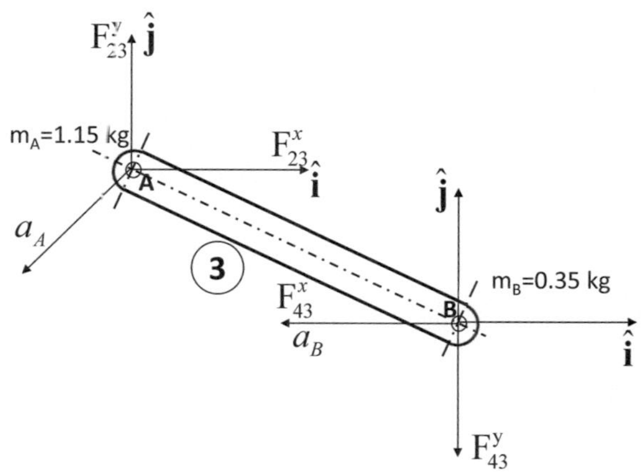

Fig. 8.17 Connecting rod of Example 8.3

$\mathbf{F}_{23}$, which is the force acted by the crank on the connecting rod is given by

$$\mathbf{F}_{23} = F^x_{23}\,\hat{\mathbf{i}} + F^y_{23}\,\hat{\mathbf{j}} \tag{13}$$

From Eq. (8.33) summing forces and accelerations of the connecting rod along the horizontal direction is given by

$$\begin{aligned} F^x_{23} + F^x_{43} &= m_B a_B + m_A a^x_A \\ &= -0.35 \times 4536 - 1.15 \times 4536 \\ &= -6804\,\mathrm{N} \end{aligned} \tag{14}$$

From Eqs. (12) and (14), F^x_{23} is given by

$$\begin{aligned} F^x_{23} &= -6804 \; - F^x_{43}\mathrm{N} \\ &= -6804 - 4536 = -\,11{,}340\,\mathrm{N} \end{aligned} \tag{15}$$

From Eq. (8.33) summing forces and accelerations of the connecting rod along the vertical direction

$$\begin{aligned} F^y_{23} + F^y_{43} &= m_A a^y_A \\ &= -1.15 \times 4536 \\ &= -5216\,\mathrm{N} \end{aligned} \tag{16}$$

The angle ϕ of the connecting rod is given by

$$\varphi = \sin^{-1}\left(\frac{r}{\ell}\sin\theta\right)$$

$$= \sin^{-1}\left(\frac{65}{300}\sin 45^\circ\right)$$
$$= 8.8^\circ \tag{17}$$

The position vector $\mathbf{r}_{AB}$ is given by

$$\mathbf{r}_{AB} = \ell\left(\cos\varphi\,\hat{\mathbf{i}} - \sin\varphi\,\hat{\mathbf{j}}\right) \tag{18}$$

For the connecting rod modeled as a two-mass system, the sum of moments about the point A is related to change of angular momentum about the point A and from Eq. (8.23)

$$\mathbf{r}_{AB} \times \mathbf{F}_{43} = \mathbf{r}_{AB} \times m_B\mathbf{a}_B$$
$$\ell\left(\cos\varphi\,\hat{\mathbf{i}} - \sin\varphi\,\hat{\mathbf{j}}\right) \times \left(F_{43}^x\,\hat{\mathbf{i}} + F_{43}^y\hat{\mathbf{j}}\right) = m_B\ell a_B\left(\cos\varphi\,\hat{\mathbf{i}} - \sin\varphi\,\hat{\mathbf{j}}\right) \times \hat{\mathbf{i}}$$
$$\Rightarrow F_{43}^y\cos\varphi + F_{43}^x\sin\varphi = m_B a_B\sin\varphi \tag{19}$$

From Eqs. (9) and (19)

$$F_{43}^y = \frac{m_B a_B\sin\varphi - F_{43}^x\sin\varphi}{\cos\varphi}$$
$$= \frac{-0.35 \times 4536 \times \sin 8.8^\circ - 4536\sin 8.8^\circ}{\cos 8.8^\circ} = -948\,\mathbf{N} \tag{20}$$

From Eqs. (16) and (20)

$$F_{23}^y = -5216 - F_{43}^y\,\mathrm{N}$$
$$= -5216 - (-948) = -4268\ \mathrm{N} \tag{21}$$

From Eqs. (12) and (18)

$$F_{32}^x = 11{,}340\,\mathrm{N}$$
$$F_{32}^y = 4268\ \mathrm{N} \tag{22}$$

The position vector $\mathbf{r}_{OD}$ of the crank (Fig. 8.18) is given by

$$\mathbf{r}_{OD} = -r\left(\cos\theta\,\hat{\mathbf{i}} + \sin\theta\,\hat{\mathbf{j}}\right) \tag{23}$$

From Eq. (3), the acceleration of the point D is equal and opposite to that of point A and it is given by

$$\mathbf{a}_D = 4536\,\hat{\mathbf{i}} + 4536\hat{\mathbf{j}}\ \ \mathrm{m/s^2} \tag{24}$$

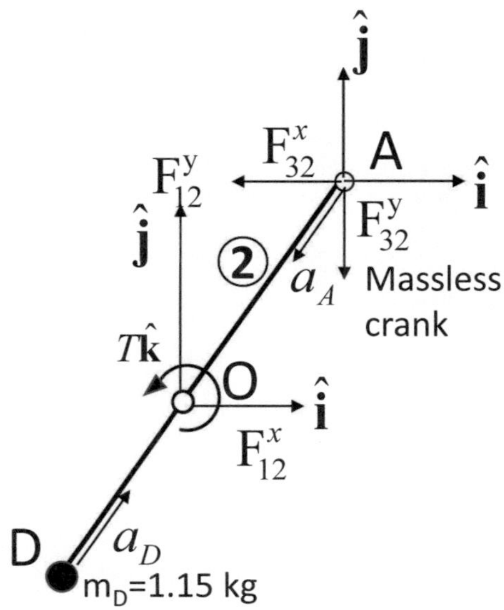

Fig. 8.18 Crank of Example 8.3

For the link 1, using Eq. (8.23) sum of the moments of force and torque are equated to the rate of change of angular momentum, given by

$$
\begin{aligned}
&\mathbf{r}_{OA} \times \mathbf{F}_{32} + T = \mathbf{r}_{OD} \times m_D \mathbf{a}_D \\
&r\left(\cos\theta\,\hat{\mathbf{i}} + \sin\theta\hat{\mathbf{j}}\right) \times \left(F^x_{32}\,\hat{\mathbf{i}} + F^y_{32}\,\hat{\mathbf{j}}\right) = -m_D r\left(\cos\theta\,\hat{\mathbf{i}} + \sin\theta\,\hat{\mathbf{j}}\right) \times \left(a^x_D\,\hat{\mathbf{i}} + a^y_D\,\hat{\mathbf{j}}\right) \\
&T = -r\cos\theta F^y_{32} + r\sin\theta F^x_{32} - m_D r\left(a^y_D\cos\theta - a^x_D\sin\theta\right) \\
&\quad = -0.065 \times \cos 45^\circ(4268) + 0.065 \times \sin 45^\circ(11340) \\
&\quad = 325\text{N m}
\end{aligned}
\tag{25}
$$

From Eq. (8.33) summing forces on link 1 and the acceleration of mass D in the horizontal direction

$$
\begin{aligned}
F^x_{12} &= -F^x_{32} + m_D a^x_D \\
&= -11340 + 1.15 \times 4536 = -6124\,\text{N}
\end{aligned}
\tag{26}
$$

Summing forces on link 1 and the acceleration of mass D in the vertical direction

$$
\begin{aligned}
F^y_{12} &= -F^y_{32} + m_D a^y_D \\
&= -4268 + 1.15 \times 4536 = 948\,\text{N}
\end{aligned}
\tag{27}
$$

An alternate way of solving this problem is to shift the connecting rod mass to its hinge points on the crank and piston and use Eqs. (8.51a, 8.51b) through (8.55). Then it becomes a slider crank mechanism with masses at the piston pin, crank pin and a balancing mass at the opposite end of the crank at the same radius. Then, mass of the piston becomes, $m_4 = m_p + m_B = 1 + 0.35 = 1.35$ kg.

From Eq. (8.51b), the reaction force vector of the connecting rod on the piston is given by

$$\begin{aligned}\mathbf{F}_{34} &= (m_4 a_B)\left(\hat{\mathbf{i}} - \tan\phi\,\hat{\mathbf{j}}\right)\\ &= -1.35 \times 4536\left(\hat{\mathbf{i}} - \tan 8.8^\circ\,\hat{\mathbf{j}}\right)\\ &= -6124\hat{\mathbf{i}} + 948\,\hat{\mathbf{j}}\,\text{N}\end{aligned} \tag{28}$$

From Eq. (28), the reaction force vector of the connecting rod acting on the crank is given by

$$\mathbf{F}_{32} = \mathbf{F}_{43} = -\mathbf{F}_{34} = 6124\,\hat{\mathbf{i}} - 948\,\hat{\mathbf{j}}\,\text{N} \tag{29}$$

The torque output of the crank is given by

$$T = -\mathbf{r}_{OA} \times \mathbf{F}_{32} = \mathbf{r}_{OA} \times \mathbf{F}_{34} \tag{30}$$

Using the value of $\mathbf{F}_{34}$ from Eq. (28) in Eq. (30), the torque output becomes

$$\begin{aligned}T &= \mathbf{r}_{OA} \times \mathbf{F}_{34} = R\left(\cos\theta\,\hat{\mathbf{i}} + \sin\theta\,\hat{\mathbf{j}}\right) \times (m_4 a_B)\left(\hat{\mathbf{i}} - \tan\phi\,\hat{\mathbf{j}}\right)\\ &= -R m_4 a_B(\cos\theta\tan\phi + \sin\theta)\\ &= -0.065 \times 1.35 \times -4536(\cos 45^\circ \tan 8.8^\circ + \sin 45^\circ) = 325\,\text{N m}\end{aligned} \tag{31}$$

The reaction force of the fixed link on the crank is given by

$$\mathbf{F}_{12} = -\mathbf{F}_{32} = \mathbf{F}_{23} = \mathbf{F}_{34} = -6124\,\hat{\mathbf{i}} + 948\,\hat{\mathbf{j}}\,\text{N} \tag{32}$$

Hence, except the interactive forces on the connecting rod, the forces acting on the machine foundation and the torque output remain the same, irrespective of the type of modeling.

Similar to the force analysis of a slider crank mechanism, force analysis of other mechanisms can be carried out. Another popular mechanism is the four-bar mechanism that is a 4R mechanism. Since only the crank can be assumed to have uniform angular velocity, angular acceleration of the coupler and follower will have to be computed before the force analysis. The next example, however, is a highly simplified version of a four-bar mechanism in which kinematic and kinetic analysis are interlinked.

Example 8.4

Linkages 2, 3, and 4 are hinged at points A and B and links 2 and 4 are hinged to the fixed frame 1 at points C and D as shown in Fig. 8.19. Link 3, the vertical bar, has a mass of 10 kg with the center of mass midway between the ends A and B. The vertical bar 3 is initially at rest when links 2 and 4 are horizontal that corresponds to $\Theta = 0$. Due to a constant torque of 2 kN m applied to the lower link at C as shown, the mechanism is set into motion. Determine the angular acceleration of the links as a function of Θ and the reaction forces at O_2, B, A and O_4 at the instant when $\Theta = 30°$. $AB = \ell_2 = 2$ m, $O_4A = O_2B = \ell_1 = 1$ m. Neglect the mass of links 2 and 4 (Fig. 8.20).

The position vector for link 4 (O_2B) is given by

$$\mathbf{r}_{O_2B} = \ell_1\left(\cos\theta\,\hat{\mathbf{i}} + \sin\theta\,\hat{\mathbf{j}}\right) \tag{1}$$

The cross product of the axis of rotation vector with the above position vector is given by

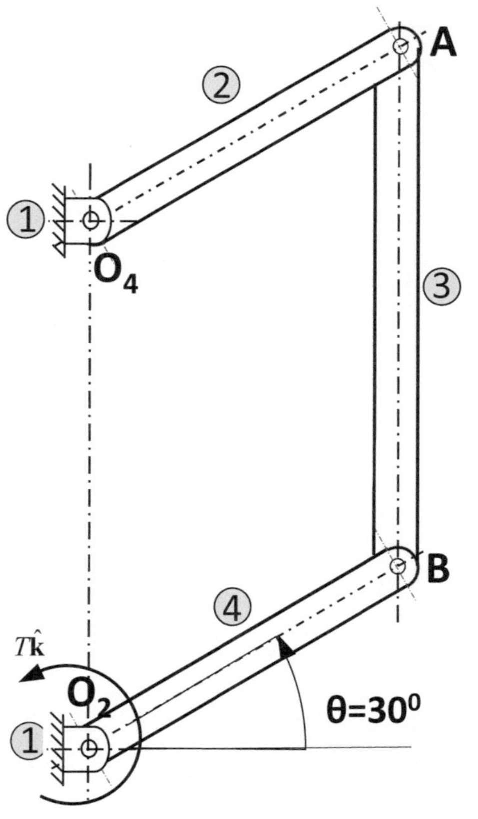

Fig. 8.19 Four-bar mechanism of Example 8.4

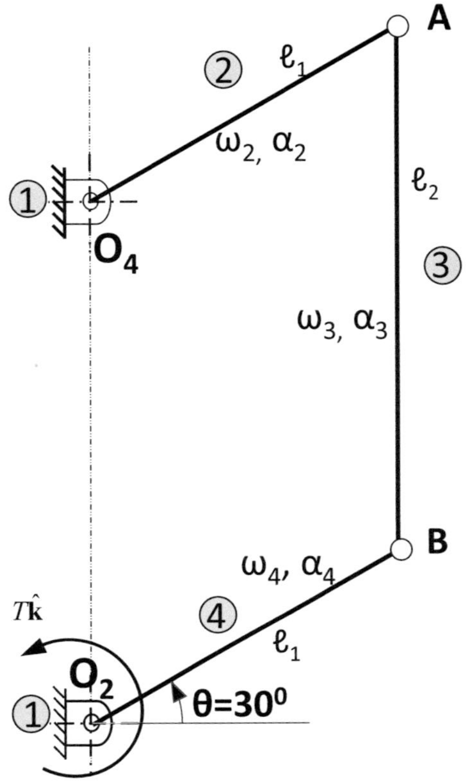

Fig. 8.20 Line diagram representation of four-bar mechanism of Example 8.4

$$\hat{\mathbf{k}} \times \mathbf{r}_{O_2B} = \ell_1\left(-\sin\theta\,\hat{\mathbf{i}} + \cos\theta\,\hat{\mathbf{j}}\right) \tag{2}$$

The position vector for link 3 (BA) is given by

$$\mathbf{r}_{BA} = \ell_2\hat{\mathbf{j}} \tag{3}$$

The cross product of the axis of rotation vector with the above position vector is given by

$$\hat{\mathbf{k}} \times \mathbf{r}_{BA} = -\ell_2\hat{\mathbf{i}} \tag{4}$$

The position vector of link 2 (DA) is given by

$$\mathbf{r}_{O_4A} = \ell_1\left(\cos\theta\,\hat{\mathbf{i}} + \sin\theta\,\hat{\mathbf{j}}\right) \tag{5}$$

The cross product of the axis of rotation vector with the above position vector is given by

$$\hat{\mathbf{k}} \times \mathbf{r}_{O_4A} = \ell_1\big(-\sin\theta\,\hat{\mathbf{i}} + \cos\theta\,\hat{\mathbf{j}}\big) \tag{6}$$

Now that we have the position vector and its cross-product for various links, they will be useful in the kinematic analysis

Velocity analysis

From Eq. (2.58) and Eq. (1), the velocity of B is given by

$$\begin{aligned}\mathbf{v}_B &= \omega_4\hat{\mathbf{k}} \times \mathbf{r}_{O_2B}\\ &= \omega_4\ell_1\big(-\sin\theta\,\hat{\mathbf{i}} + \cos\theta\,\hat{\mathbf{j}}\big)\end{aligned} \tag{7}$$

Similarly, from Eq. (2.58) and Eq. (4), the velocity of A is given by

$$\begin{aligned}\mathbf{v}_A &= \mathbf{v}_B + \omega_3\,\hat{\mathbf{k}} \times \mathbf{r}_{BA}\\ &= \omega_4\ell_1\big(-\sin\theta\,\hat{\mathbf{i}} + \cos\theta\,\hat{\mathbf{j}}\big) - \omega_3\ell_2\,\hat{\mathbf{i}}\\ &= (-\omega_4\ell_1\sin\theta - \omega_3\ell_2)\hat{\mathbf{i}} + (\omega_4\ell_1\cos\theta)\hat{\mathbf{j}}\end{aligned} \tag{8}$$

The velocity of A is also given by

$$\begin{aligned}\mathbf{v}_A &= \omega_2\hat{\mathbf{k}} \times \mathbf{r}_{O_4A}\\ &= \omega_2\ell_1\big(-\sin\theta\,\hat{\mathbf{i}} + \cos\theta\,\hat{\mathbf{j}}\big)\end{aligned} \tag{9}$$

From Eqs. (8) and (9)

$$(-\omega_4\ell_1\sin\theta - \omega_3\ell_2)\hat{\mathbf{i}} + (\omega_4\ell_1\cos\theta)\hat{\mathbf{j}} = \omega_2\ell_1\big(-\sin\theta\,\hat{\mathbf{i}} + \cos\theta\,\hat{\mathbf{j}}\big) \tag{10}$$

By equating the coefficients of the unit vectors on both sides of Eq. (10)

$$\begin{aligned}-\,\omega_4\ell_1\sin\theta - \omega_3\ell_2 &= -\omega_2\ell_1\sin\theta\\ \omega_4\ell_1\cos\theta &= \omega_2\ell_1\cos\theta\end{aligned} \tag{11}$$

From the above equations

$$\begin{aligned}&\omega_3 = 0;\;\; \alpha_3 = 0\\ &\omega_4 = \omega_2 = \omega;\;\; \alpha_4 = \alpha_2 = \alpha\end{aligned} \tag{12}$$

This means that the angular velocity of the vertical link 3 is always zero. Hence, its angular acceleration is also always zero and it is always in translation in the vertical direction.

Acceleration analysis

From Eq. (2.66), acceleration of B is given by

$$\begin{aligned}\mathbf{a}_B &= -\omega^2\mathbf{r}_{O_2B} + \alpha\hat{\mathbf{k}} \times \mathbf{r}_{O_2B} \\ &= -\omega^2\ell_1\big(\cos\theta\,\hat{\mathbf{i}} + \sin\theta\,\hat{\mathbf{j}}\big) + \alpha\ell_1\big(-\sin\theta\,\hat{\mathbf{i}} + \cos\theta\,\hat{\mathbf{j}}\big) \\ &= \big(-\omega^2\ell_1\cos\theta - \alpha\ell_1\sin\theta\big)\hat{\mathbf{i}} + \big(-\omega^2\ell_1\sin\theta + \alpha\ell_1\cos\theta\big)\hat{\mathbf{j}}\end{aligned} \tag{13}$$

Since the link AB does not rotate, the acceleration of the center of mass of AB is same for all points on it. Hence the acceleration of its mass center is given by

$$\mathbf{a}_G = \mathbf{a}_B = \big(-\omega^2\ell_1\cos\theta - \alpha\ell_1\sin\theta\big)\hat{\mathbf{i}} + \big(-\omega^2\ell_1\sin\theta + \alpha\ell_1\cos\theta\big)\hat{\mathbf{j}} \tag{14}$$

It will be useful to define a different set of coordinates that can simplify the kinetic analysis. Hence, a normal unit vector is defined parallel to the links 2 and 4 and a tangential unit vector is defined perpendicular to it as shown in Fig. 8.21. From the above figure, the new coordinates can be expressed as

$$\hat{\mathbf{e}}_n = \cos\theta\,\hat{\mathbf{i}} + \sin\theta\,\hat{\mathbf{j}} \tag{15}$$

and

$$\hat{\mathbf{e}}_t = -\sin\theta\,\hat{\mathbf{i}} + \cos\theta\,\hat{\mathbf{j}} \tag{16}$$

Equations (15) and (16) can be rewritten in terms of the fixed coordinate system as

$$\begin{aligned}\hat{\mathbf{i}} &= \cos\theta\,\hat{\mathbf{e}}_n - \sin\theta\,\hat{\mathbf{e}}_t \\ \hat{\mathbf{j}} &= \sin\theta\,\hat{\mathbf{e}}_n + \cos\theta\,\hat{\mathbf{e}}_t\end{aligned} \tag{17}$$

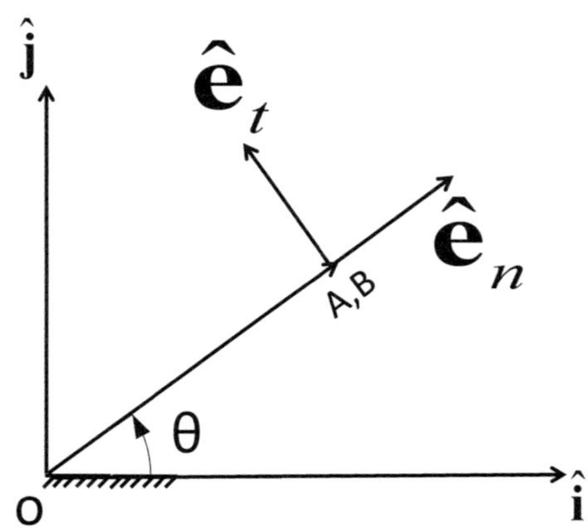

Fig. 8.21 Tangential and normal coordinates

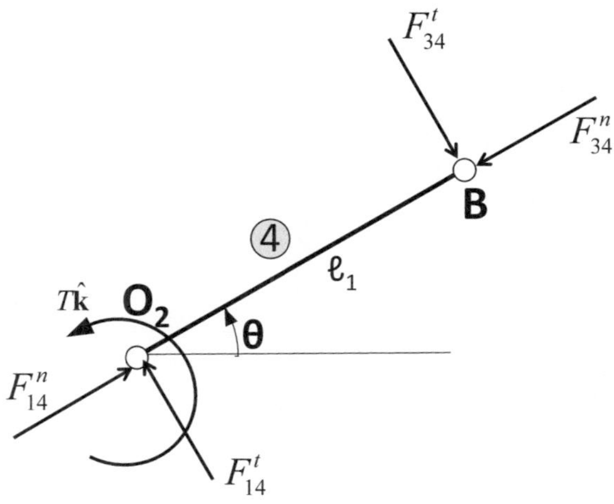

Fig. 8.22 Reaction forces, torque, and moments on link 4 of Example 8.4

The mass center acceleration of Eq. (14) can be expressed in terms of normal and tangential coordinates using Eq. (17) as

$$\begin{aligned}\mathbf{a}_G &= \left(-\omega^2\ell_1\cos\theta - \alpha\ell_1\sin\theta\right)(\cos\theta\hat{\mathbf{e}}_n - \sin\theta\hat{\mathbf{e}}_t)\\ &\quad + \left(-\omega^2\ell_1\sin\theta + \alpha\ell_1\cos\theta\right)(\sin\theta\hat{\mathbf{e}}_n + \cos\theta\hat{\mathbf{e}}_t)\\ &= \left(-\omega^2\ell_1\cos^2\theta - \alpha\ell_1\sin\theta\cos\theta - \omega^2\ell_1\sin^2\theta + \alpha\ell_1\cos\theta\sin\theta\right)\hat{\mathbf{e}}_n\\ &\quad + \left(\omega^2\ell_1\cos\theta\sin\theta + \alpha\ell_1\sin^2\theta - \omega^2\ell_1\sin\theta\cos\theta + \alpha\ell_1\cos^2\theta\right)\hat{\mathbf{e}}_t\\ &= \left(-\omega^2\ell_1\right)\mathbf{e}_n + (\alpha\ell_1)\hat{\mathbf{e}}_t\end{aligned} \tag{18}$$

Refer to Fig. 8.22. Since link 4 is a massless member, all the forces will be acting along its length. In addition, static equilibrium can be used to compute the reaction forces.

The following equations can be obtained from such considerations:

$$\begin{aligned}T &= -F^t_{34}\ell_1\\ F^n_{14} &= F^n_{34}\\ F^t_{14} &= F^t_{34}\end{aligned} \tag{19}$$

Reaction forces of all the links are shown in Fig. 8.23. It should be noted that the reaction forces are equal and opposite at the hinge point of all the hinges.

For the link 2, following equations are satisfied

$$\begin{aligned}F^n_{12} &= F^n_{32}\\ F^t_{12} &= F^t_{32} = 0 \text{ (from moment considerations)}\end{aligned} \tag{20}$$

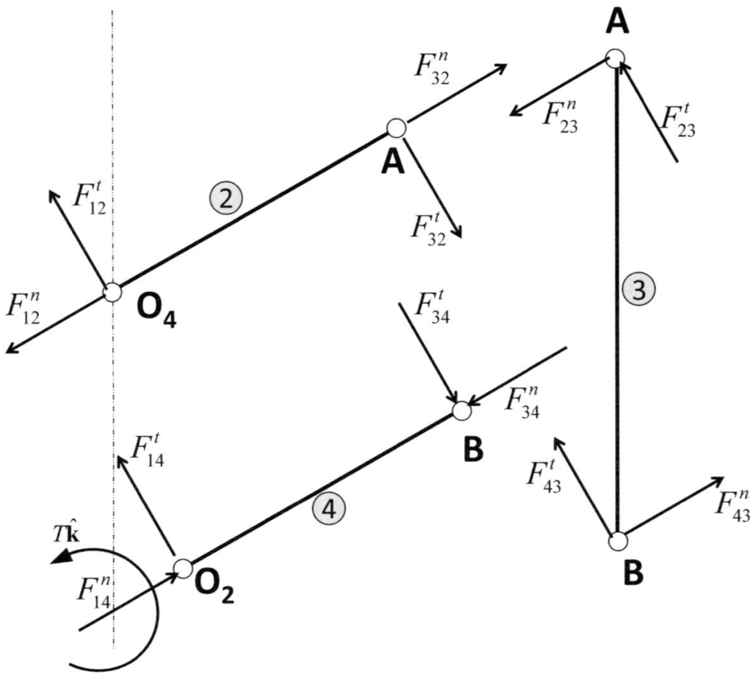

Fig. 8.23 Reaction forces of all links of Example 8.4

The weight vector of link 3 in the tangential and normal coordinate system is given by

$$-mg\,\hat{\mathbf{j}} = -mg(\sin\theta\,\hat{\mathbf{e}}_n + \cos\theta\,\hat{\mathbf{e}}_t) \tag{21}$$

For the link 3, from Eqs. (18) and (21), sum of the forces in the tangential direction are given by

$$\sum F(\hat{\mathbf{e}}_t \text{ direction}) = F_{34}^t - mg\cos\theta = m\alpha\ell_1 \tag{22}$$

From Eqs. (19) and (22)

$$\frac{T}{\ell_1} - mg\cos\theta = m\alpha\ell_1 \tag{23}$$

Rearranging Eq. (23)

$$\alpha = \frac{T}{m\ell_1^2} - \frac{g\cos\theta}{\ell_1}$$

$$= \frac{2000}{10 \times 1^2} - \frac{9.81\cos\theta}{1}$$
$$= 200 - 9.81\cos\theta \tag{24}$$

The common angular velocity and angular acceleration of links 2 and 4 are related by

$$\alpha = \frac{d\omega}{dt};\ \omega = \frac{d\theta}{dt}$$
$$\frac{\alpha}{\omega} = \frac{\frac{d\omega}{dt}}{\frac{d\theta}{dt}} \tag{25}$$

From Eq. (25)

$$\int_0^\theta \alpha d\theta = \int_0^\omega \omega d\omega \tag{26}$$

From Eqs. (24) and (26)

$$\int_0^\theta (200 - 9.81\cos\theta)d\theta = \frac{\omega^2}{2} \tag{27}$$

The angular velocity as a function θ is given by

$$\omega^2 = 400\theta - 19.62\sin\theta \tag{28}$$

The angular velocity at the given angle of θ = 30° is given by

$$\omega = \sqrt{\frac{400 \times 30\pi}{180} - 19.62\sin 30^\circ} = 14.12\,\text{rad/s} \tag{29}$$

And the angular acceleration is given by

$$\alpha = 200 - 9.81\cos 30^\circ = 191.50\,\text{rad/s}^2 \tag{30}$$

Refer to Fig. 8.24. Using Eq. (8.23), moments about the point B are equated to the cross product of inertia forces

$$\mathbf{r}_{BA} \times -F_{23}^n\,\hat{\mathbf{e}}_n = m\mathbf{r}_{BG} \times \mathbf{a}_G$$
$$\Rightarrow \ell_2(\sin\theta\,\hat{\mathbf{e}}_n + \cos\theta\,\hat{\mathbf{e}}_t) \times -F_{23}^n\,\hat{\mathbf{e}}_n$$
$$= m\frac{\ell_2}{2}(\sin\theta\,\hat{\mathbf{e}}_n + \cos\theta\,\hat{\mathbf{e}}_t) \times \left[\left(-\omega^2\ell_1\right)\mathbf{e}_n + (\alpha\ell_1)\hat{\mathbf{e}}_t\right] \tag{31}$$

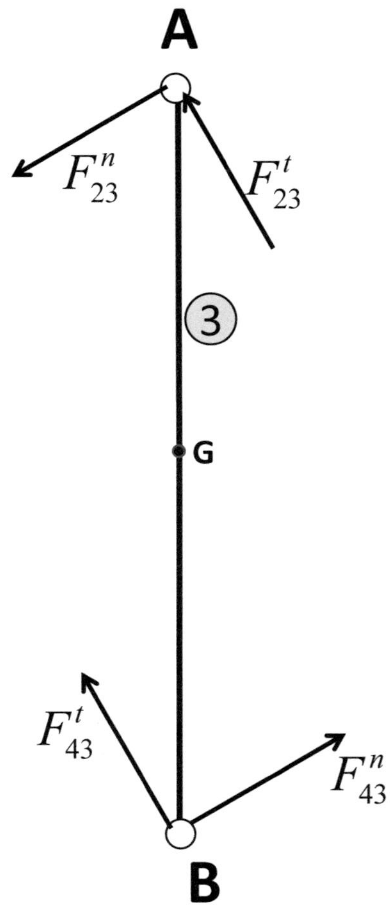

Fig. 8.24 Reaction forces of link 3 of Example 8.4

From Eq. (31), the reaction force F_{23}^n is given by

$$F_{23}^n = \frac{m\ell_1}{2\cos\theta}\left(\alpha \sin\theta + \omega^2 cos\theta\right) \tag{32}$$

At the given angle of 30°, the reaction force F_{23}^n is given by

$$\begin{aligned} F_{23}^n &= \frac{m\ell_1}{2\cos\theta}\left(\alpha \sin\theta + \omega^2 \cos\theta\right) \\ &= \frac{10}{2\cos 30^\circ}\left(191.50 \times 1 \times \sin 30^\circ + 1 \times 14.13^2 \cos 30^\circ\right) \\ &= 1551\ \mathrm{N} \end{aligned} \tag{33}$$

From Eq. (19), F_{43}^t is given by

$$F_{43}^{t} = \frac{T}{\ell_1} = \frac{2000}{1} = 2000\,\text{N} \tag{34}$$

From Eq. (8.23), moments about the point A are equated to the cross product of inertia forces

$$\begin{aligned}
&\mathbf{r}_{AB} \times F_{43}^{n}\,\hat{\mathbf{e}}_n + \mathbf{r}_{AB} \times F_{43}^{t}\,\hat{\mathbf{e}}_t = m\mathbf{r}_{AG} \times \mathbf{a}_G \\
&\Rightarrow -\ell_2(\sin\theta\,\hat{\mathbf{e}}_n + \cos\theta\,\hat{\mathbf{e}}_t) \times F_{43}^{n}\,\hat{\mathbf{e}}_n \\
&\quad - \ell_2(\sin\theta\,\hat{\mathbf{e}}_n + \cos\theta\,\hat{\mathbf{e}}_t) \times F_{43}^{t}\,\hat{\mathbf{e}}_t \\
&= -m\frac{\ell_2}{2}(\sin\theta\,\hat{\mathbf{e}}_n + \cos\theta\,\hat{\mathbf{e}}_t) \times \left[(-\omega^2\ell_1)\mathbf{e}_n + (\alpha\ell_1)\,\hat{\mathbf{e}}_t\right]
\end{aligned} \tag{35}$$

Equation (35) simplifies to

$$F_{43}^{n} = -\frac{1}{\cos\theta}\left[\frac{m\ell_1}{2}\left(\alpha\sin\theta + \omega^2\cos\theta\right) - F_{43}^{t}\sin\theta\right] \tag{36}$$

From Eq. (36), the reaction force F_{43}^{n} at $\theta = 30°$ is given by

$$\begin{aligned}
F_{43}^{n} &= -\frac{1}{\cos 30°}\left(\frac{10 \times 191.50 \times 1\sin 30°}{2} + \frac{10 \times 1 \times 14.12^2\cos 30°}{2} - 2000\sin 30°\right) \\
&= -396\,\text{N}
\end{aligned} \tag{37}$$

Example 8.5

Consider the mechanism shown in Fig. 8.25a and its block diagram representation of kinematic pairs was earlier shown in Fig. 2.52b of Example 2.13. We have already analyzed this mechanism for velocity in Example 2.12 and for acceleration in Example 2.13. For the same angular velocity of the link 4, $\omega_4 = 2$ rad/s, as in Example 2.12, and angular acceleration of link 2 $\alpha_2 = 4.31$ rad/s^2 and angular acceleration of link 4, $\alpha_4 = 2$ rad/s^2, determine the torque input to link 2 and the reaction forces. Assume BQ = 3000 mm and OR = 2500 mm as slender rods, $m_2 = 10$ kg, $m_3 = 4$ kg and $m_4 = 15$ kg. Account for gravitational forces.

From Example 2.11, $\mathbf{v}_{P/2} = -2000\,\mathbf{j}$ mm/s and $\omega_2 = 2$ rad/s.

From Example 2.13, $\mathbf{a}_P = 536\hat{\mathbf{i}} - 8928.4\hat{\mathbf{j}}$ mm/s^2, $\mathbf{a}_{P'} = -7464.1\hat{\mathbf{i}} - 6928.2\hat{\mathbf{j}}$ mm/s^2.

The moment of inertia of link 4 is given by

$$I_4 = \frac{m_4 OR^2}{12} = \frac{15 \times 3^2}{12} = 1.1.25\text{ kg m}^2 \tag{1}$$

The moment of inertia of link 2 is given by

$$I_2 = \frac{m_2 OR^2}{12} = \frac{10 \times 2.5^2}{12} = 5.2\text{ kg m}^2 \tag{2}$$

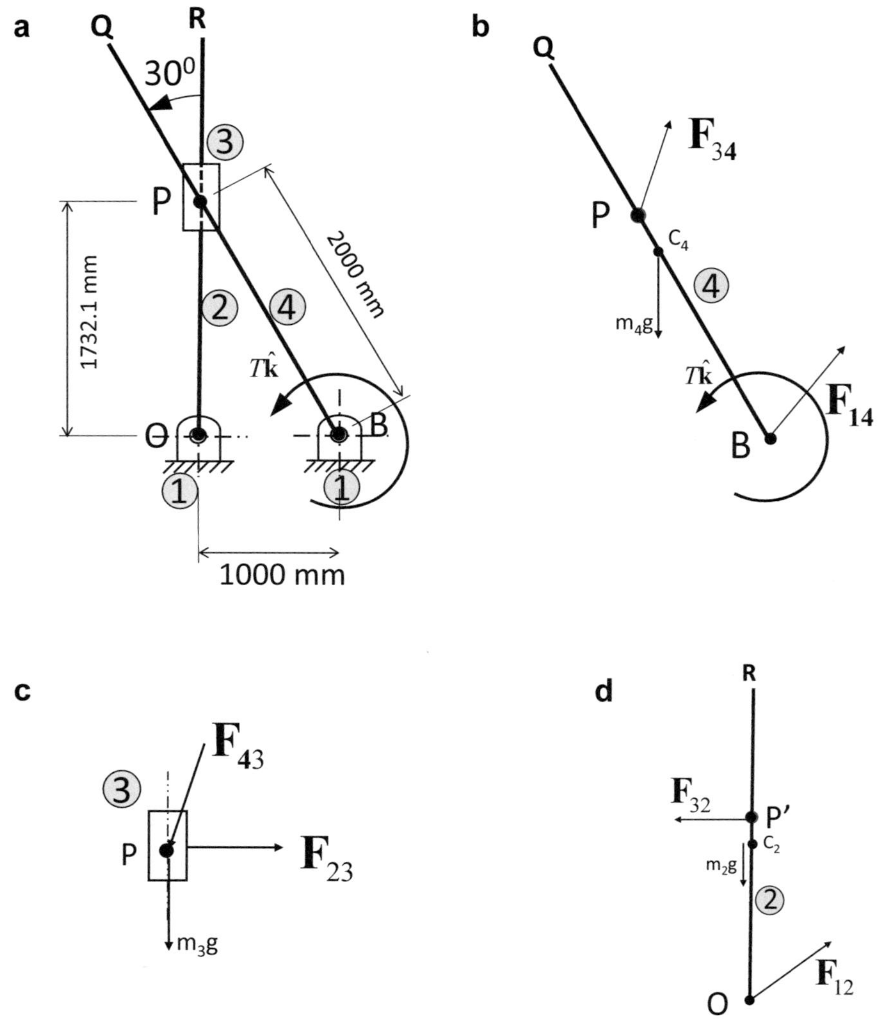

Fig. 8.25 **a** Mechanism of Example 8.5. **b** Force and torque on link 4 of Example 8.5. **c** Forces on link 3 of the mechanism of Example 8.5. **d** Forces on link 2 of the mechanism of Example 8.5

The position vector of the center of mass of link 4 is given by

$$\mathbf{r}_{BC_4} = 1.5\left(-\cos 60^\circ\,\hat{\mathbf{i}} + \sin 60^\circ\,\hat{\mathbf{j}}\right) m$$
$$\left(-0.75\,\hat{\mathbf{i}} + 1.3\,\hat{\mathbf{j}}\right) m \tag{3}$$

The position vector of the center of mass of link 2 is given by

$$\mathbf{r}_{OC_2} = 1.25\,\hat{\mathbf{j}} \tag{4}$$

The acceleration of the center of mass of link 4 is given by

$$\begin{aligned}\mathbf{a}_{C_4} &= -\omega_4^2\mathbf{r}_{BC_4} + \alpha_4\,\hat{\mathbf{k}} \times \mathbf{r}_{BC_4}\\ &= -2^2\big(-0.75\,\hat{\mathbf{i}} + 1.3\,\hat{\mathbf{j}}\big) + 2\,\hat{\mathbf{k}} \times \big(-0.75\,\hat{\mathbf{i}} + 1.3\,\hat{\mathbf{j}}\big)\\ &= (0.4)\,\hat{\mathbf{i}} + (-6.7)\,\hat{\mathbf{j}}\,\mathrm{m/s^2}\end{aligned} \tag{5}$$

The acceleration of the center of mass of link 2 is given by

$$\begin{aligned}\mathbf{a}_{C_2} &= -\omega_2^2\mathbf{r}_{OC_2} + \alpha_2\hat{\mathbf{k}} \times \mathbf{r}_{OC_2}\\ &= -2^2\big(1.25\,\hat{\mathbf{j}}\big) + 4.31\,\hat{\mathbf{k}} \times \big(1.25\,\hat{\mathbf{j}}\big)\\ &= (-5.39)\,\hat{\mathbf{i}} + (-5)\,\hat{\mathbf{j}} \quad \mathrm{m/s^2}\end{aligned} \tag{6}$$

Sum of the forces acting on link 4 is given by

$$\begin{aligned}\mathbf{F}_{14} + \mathbf{F}_{34} - m_4 g &= \big(F_{14}^x + F_{34}^x\big)\,\hat{\mathbf{i}} + \big(F_{14}^y + F_{34}^y - m_4 g\big)\,\hat{\mathbf{j}}\\ &= m_4\mathbf{a}_{C_4} = m_4\big(0.4\,\hat{\mathbf{i}} - 6.7\,\hat{\mathbf{j}}\big)\end{aligned} \tag{7}$$

Equating components of Eq. (7)

$$\begin{aligned}F_{14}^x + F_{34}^x &= 6.029\\ F_{14}^y + F_{34}^y &= 46.71\end{aligned} \tag{8}$$

Summing torque and moments about the point B

$$\begin{aligned}&\mathbf{r}_{BP} \times \mathbf{F}_{34} + \mathbf{r}_{BC_4} \times -m_4 g\,\hat{\mathbf{j}} + T\,\hat{\mathbf{k}} - m_4\mathbf{r}_{BC_4} \times \mathbf{a}_{C_4} - I_4\alpha_4\,\hat{\mathbf{k}} = 0\\ &\Rightarrow \big(-\hat{\mathbf{i}} + 1.7321\,\hat{\mathbf{j}}\big) \times \big(F_{34}^x\,\hat{\mathbf{i}} + F_{34}^y\,\hat{\mathbf{j}}\big) + \big(-0.75\,\hat{\mathbf{i}} + 1.29\,\hat{\mathbf{j}}\big) \times (-15)\big(9.81\,\hat{\mathbf{j}}\big) + T\,\hat{\mathbf{k}}\\ &\quad - 15\big(-0.75\,\hat{\mathbf{i}} + 1.29\,\hat{\mathbf{j}}\big) \times \big(0.4\,\hat{\mathbf{i}} - 6.7\,\hat{\mathbf{j}}\big) - 11.25 \times 2\,\hat{\mathbf{k}} = 0\end{aligned} \tag{9}$$

Simplifying Eq. (9)

$$-1.732F_{34}^x - F_{34}^y + T = -20.36 \tag{10}$$

Sum of the forces acting on link 3 is given by

$$\begin{aligned}&\big(-F_{34}^x + F_{23}^x\big)\hat{\mathbf{i}} + \big(-F_{34}^y - m_3 g\big)\,\hat{\mathbf{j}} = m_3\mathbf{a}_P\\ &\Rightarrow \big(-F_{34}^x + F_{23}^x\big)\hat{\mathbf{i}} + \big(-F_{34}^y - 4 \times 9.81\big)\,\hat{\mathbf{j}} = 4\big(0.54\,\hat{\mathbf{i}} - 8.93\,\hat{\mathbf{j}}\big)\end{aligned} \tag{11}$$

Equation (11) simplifies to

$$\begin{aligned}-F_{34}^x + F_{23}^x &= 2.1436\\ F_{34}^y &= -3.53\ \mathrm{N}\end{aligned} \tag{12}$$

Sum of the forces acting on link 2 is given by

$$\begin{aligned}&\left(F_{12}^{x}-F_{23}^{x}\right)\hat{\mathbf{i}}+\left(F_{12}^{y}-m_2 g\right)\hat{\mathbf{j}}=m_2\mathbf{a}_{C_2}\\&\Rightarrow\left(F_{12}^{x}-F_{23}^{x}\right)\hat{\mathbf{i}}+\left(F_{12}^{y}-10\times 9.81\right)\hat{\mathbf{j}}=10\left(-5.39\,\hat{\mathbf{i}}-5\,\hat{\mathbf{j}}\right)\end{aligned}\tag{13}$$

Equation (13) simplifies as

$$\begin{aligned}F_{12}^{x}-F_{23}^{x}&=-53.86\\F_{12}^{y}&=48.10\text{ N}\end{aligned}\tag{14}$$

Summing torque and moments about the point O

$$\begin{aligned}&\mathbf{r}_{OP}\times\mathbf{F}_{32}+\mathbf{r}_{OC_2}\times -m_2 g\,\hat{\mathbf{j}}-m_2\mathbf{r}_{OC_2}\times\mathbf{a}_{C_2}-I_2\alpha_2\,\hat{\mathbf{k}}=0\\&\Rightarrow\left(1.7321\hat{\mathbf{j}}\times -F_{23}^{x}\,\hat{\mathbf{i}}\right)-1.25\,\hat{\mathbf{j}}\times m_2 g\,\hat{\mathbf{j}}-m_2 1.25\,\hat{\mathbf{j}}\\&\quad\times\left(-5.39\,\hat{\mathbf{i}}-5\,\hat{\mathbf{j}}\right)-5.2083\times 4.31\,\hat{\mathbf{k}}=0\end{aligned}\tag{15}$$

Equation (14) simplifies to

$$1.7321F_{23}^{x}-89.78=0\tag{16}$$

From Eq. (16), $F_{23}^{x}=\frac{89.78}{1.7321}=51.83$ N.
From Eq. (14), $F_{12}^{x}=-53.86+F_{23}^{x}=-53.86+51.83=-2.03$ N.
From Eq. (12), $\begin{aligned}F_{34}^{x}&=-2.1436+F_{23}^{x}\\&=-2.1436+51.83=49.69\text{ N.}\end{aligned}$
From Eq. (10), $T=-20.36+1.732\times 49.69-3.53=62.18$ N m.

8.6 Conclusions

In this chapter, we have seen how a detailed kinetic analysis of a mechanism can be carried out. The same principle can be applied to any complex mechanism. In the exercise, a large number of mechanisms of practical interest are presented and a kinetic analysis of them will give sufficient confidence in determining reaction forces of any mechanism, however complex it might be.

8.7 Exercise Problems

8.1 A long slender rod AB shown in Fig. P8.1 of mass 20 kg and length 4 m has one end resting on the floor and the other resting on the wall, making an angle of $\theta=60°$ with the horizontal. A rope of negligible mass is tied to the end B and pulled in the horizontal direction so its acceleration is about 6.5 m/s^2 and velocity 2.5 m/s in the indicated direction. (i) Determine the value of force P that causes the above acceleration and velocity, and the normal reactions on the wall at A and B. Coefficient of friction between the rod and surfaces (wall and floor) is 0.15. Do not neglect gravitational forces. (ii) Determine the angle θ at which the slender starts slipping due to its own weight.

8.2 Consider the mechanism shown in Fig. P8.2a. Determine the reaction forces due to the acceleration diagram shown in Fig. P8.2b. In addition, determine the angular velocity and angular acceleration of link 3.

8.3 Consider the mechanism shown in Fig. P8.3. If $\mathcal{B}_1$ moves toward right with a velocity of 0.2 m/s and an acceleration of 0.5 m/s^2 determine (a) the velocities of $\mathcal{B}_2$ and $\mathcal{B}_3$ (b) the angular velocities and angular accelerations of $\mathcal{B}_4$ and $\mathcal{B}_5$. If $\mathcal{B}_2$ and $\mathcal{B}_3$ are subjected to forces as shown, determine the force to be exerted at and the reaction forces at the revolute and prismatic joints. $M_{\mathcal{B}_1} =$

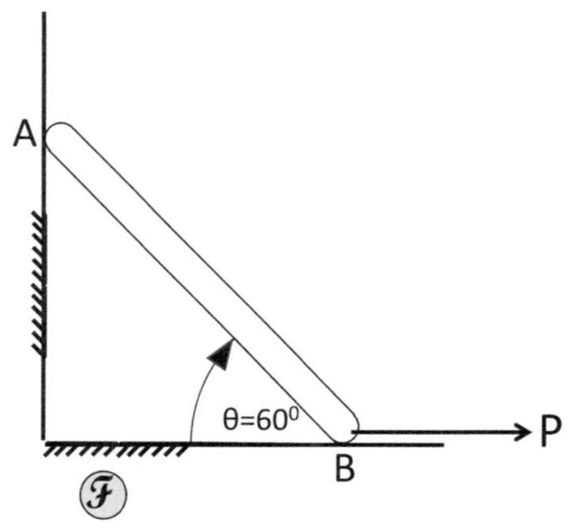

Fig. P8.1 A rigid link with its ends sliding along the vertical and horizontal surfaces

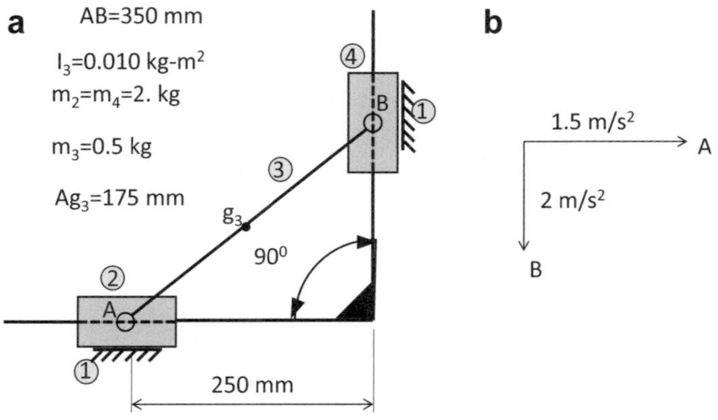

Fig. P8.2 **a** Mechanism of exercise problem 8.2. **b** Acceleration of the sliders

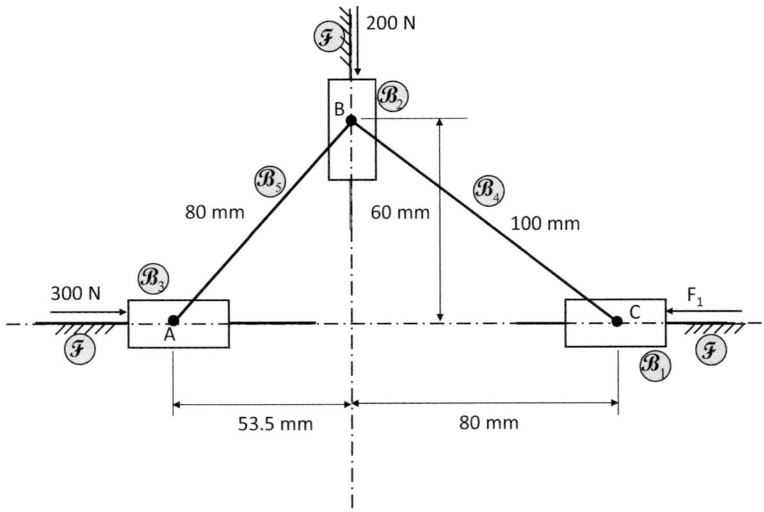

Fig. P8.3 Mechanism of exercise problem 8.3

0.2 kg, $M_{\mathcal{B}_2} = 0.2$ kg, $M_{\mathcal{B}_3} = 0.2$ kg, $M_{\mathcal{B}_4} = 0.1$ kg, $M_{\mathcal{B}_5} = 0.08$ kg. $\mathcal{B}_4$ and $\mathcal{B}_5$ can be assumed as slender rods. [Note: Refer to the exercise problem 1.1 for the block diagram and you can use relevant kinematic data computed in the exercise problem 2.27.]

8.4 For the mechanism shown in Fig. P8.4, if the link 2 is subjected to a torque of 1000 N m, determine the torques T_1 and T_2 available at the other revolute joints and all the reaction forces. Links 2 and 3 are slender rods and link 4 is a disc of radius 0.15 m. The various angular velocities of the links are: $\omega_2 =$ 3 rad/s (CCW), $\omega_3 = 2$ rad/s (CCW) and $\omega_4 = -1$ rad/s (CW). $M_2 = 0.5$ kg, $M_3 = 0.7$ kg, $M_4 = 3$ kg [Note: Refer to the exercise problem 1.3 for the block diagram and you can use relevant kinematic data computed in the exercise problem 2.29.].

8.5 If the massless crank OA of 60 mm length of the Scotch-yoke mechanism shown in Fig. P8.5 is rotating clockwise at a constant angular speed of 10 rad/s, determine the torque that has to be applied to the crank in order to resist the force applied to the yoke. In addition, determine all the reaction forces. $M_3 =$ 0.5 kg, $M_4 = 4$ kg. Neglect friction along the sliding surfaces.

8.6 Consider the mechanism shown in Fig. P8.6. If the link 4 has an angular velocity of 15 rad/s, if the crank is subjected to a torque of 500 N m, determine the resisting forces F_6 and F_2 at the pistons and reactions at all the revolute and prismatic joints. $M_2 = M_6 = 0.25$ kg; $M_3 = 0.1$ kg, $M_4 = 0.06$ kg, $M_5 =$ 0.1 kg. Assume links 3, 4 and 5 as slender rods. [Note: Refer to the exercise problem 1.2 for the block diagram and you can use relevant kinematic data computed in the exercise problem 2.28.]

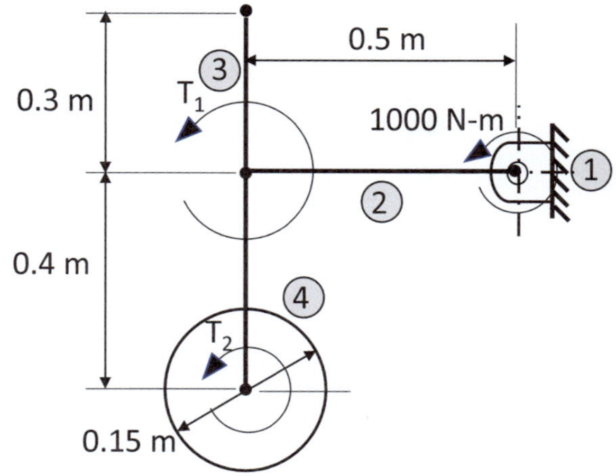

Fig. P8.4 Mechanism of exercise problem 8.4

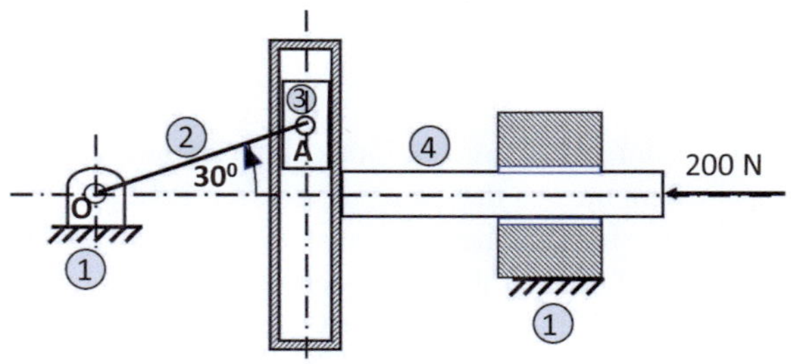

Fig. P8.5 Mechanism of exercise problem 8.5

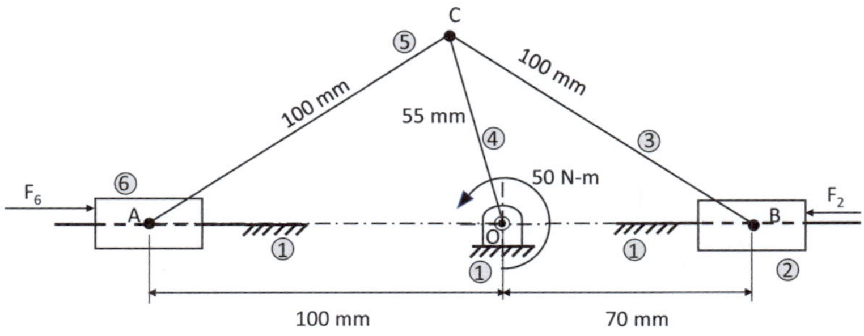

Fig. P8.6 Mechanism of exercise problem 8.6

8.7 Consider the mechanism shown in Fig. P8.7. If $\omega_2 = -1$ rad/s (clockwise), if the resistive force on link 6 is 2500 N, determine the torque applied to the link 2 and all the reaction forces. $M_2 = 0.1$ kg, $M_3 = 0.25$ kg, $M_4 = 0.150$ kg, $M_5 = 0.125$ kg, $M_6 = 0.75$ kg. Assume links 2, 3, 4 and 5 as slender rods with their centers of mass at the middle of their lengths. [Note: Refer to the exercise

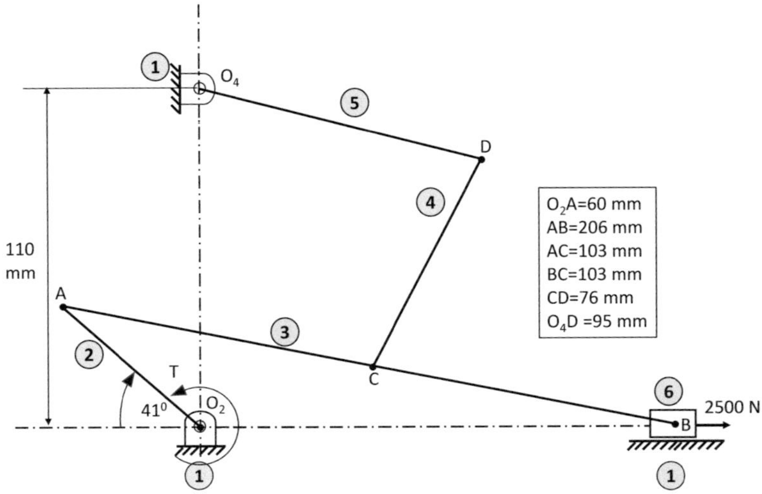

Fig. P8.7 Mechanism of exercise problem 8.7

problem 1.5 for the block diagram and you can use relevant kinematic data computed in the exercise problem 2.32.]

8.8 Consider the mechanism shown in Fig. P8.8. If the angular velocity of the link 2 is 5 rad/s in the counterclockwise direction, determine the torque applied to the link 2 against a resistive force on the piston and all the reaction forces. $M_2 = 0.15$ kg, $M_3 = 0.5$ kg, $M_4 = 0.3$ kg. Assume links 2 and 3 as slender rods with their center of mass at the middle of their lengths. [Note: Refer to the exercise problem 1.4 for the block diagram and you can use relevant kinematic data computed in the exercise problem 2.31.]

8.9 Determine the torque applied to the crank (link 2) in the given position of the drag link mechanism shown in Fig. P8.9 if the crank (link 2) is rotating clockwise at 15 rad/s to resist a force of 3000 N at the piston. In addition, determine all the reaction forces. $M_2 = 0.1$ kg, $M_3 = 0.125$ kg, $M_4 = 0.25$ kg, $M_5 = 0.15$ kg, $M_6 = 0.5$ kg. Assume links 2, 3, 5 as slender rods with their mass centers at the middle of their lengths. [Note: Refer to the exercise problem 1.6 for the block diagram and you can use relevant kinematic data computed in the exercise problem 2.33.]

8.10 Consider a cross-link mechanism shown in Fig. P8.10. It consists of a crank 2 and a follower 4 and a connecting rod 3 (AC) that are connected as shown. The crank 2 is driven at a constant angular velocity, ω_2, of 15 rad/s (counterclockwise) by a driving torque, T N m. There is a resisting force F_c at the end of connection rod in the direction shown of magnitude 200 N and the follower is hinged to a point B on the connecting rod.

The dimensions of various are as follows: $O_2A = 120$ mm, $AB = 400$ mm, $BC = 120$ mm, $O_4B = 120$ mm, $AG_3 = 250$ mm.

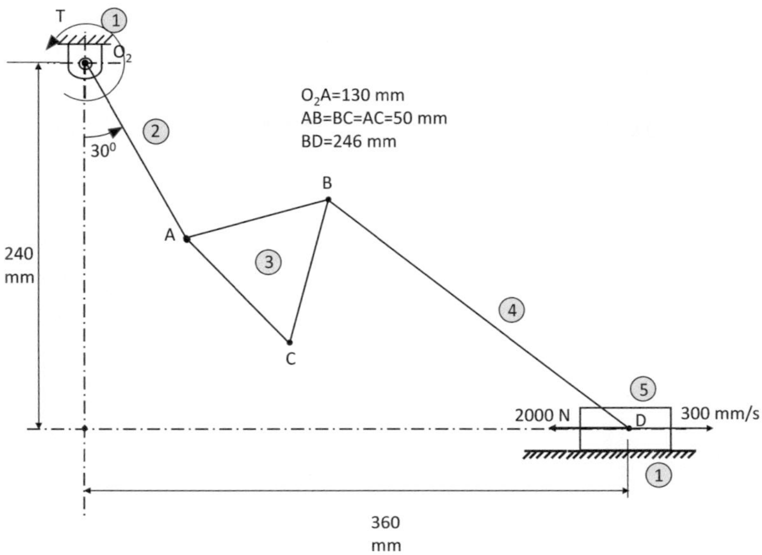

Fig. P8.8 Mechanism of exercise problem 8.8

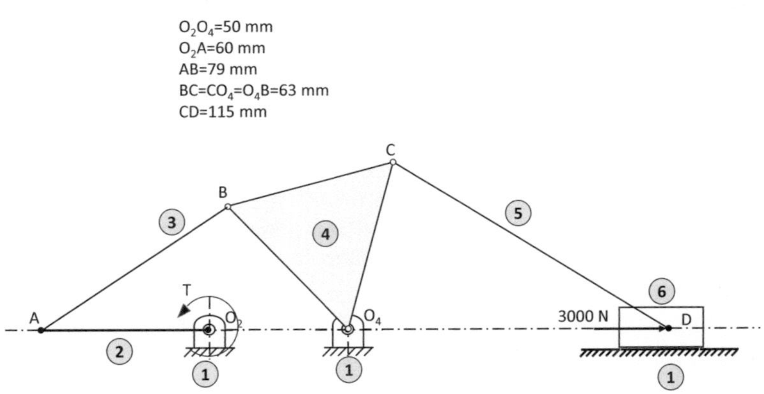

Fig. P8.9 Mechanism of exercise problem 8.9

Mass of the connecting rod, $m_3 = 2$ kg; neglect the mass of the crank and follower as they are balanced.

Moment of inertias of the crank and follower, $I_{G2} = I_{G4} = 0.01$ kg m^2.

Moment of inertia of the connecting rod, $I_{G3} = 0.06$ kg m^2. Determine the driving torque and pin reactions.

8.11 A slider crank mechanism of an internal combustion engine is shown in the Fig. P8.11. At the instant shown, the combustion pressure acting on the piston is 2 MPa and diameter of the piston is 80 mm. The crank is rotating at a uniform angular velocity of 2500 RPM (CCW). Following are the parameters of this

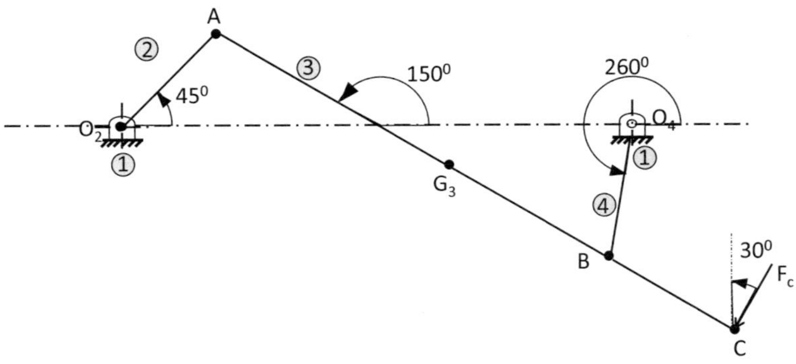

Fig. P8.10 Mechanism of exercise problem 8.10

Fig. P8.11 Mechanism of exercise problem 8.11

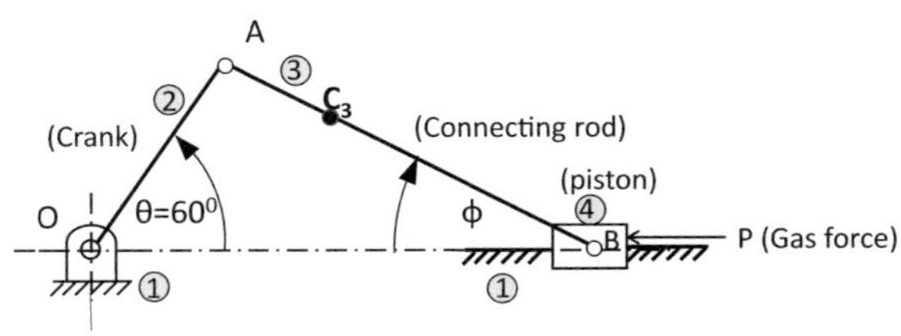

mechanism: OA = 60 mm, AB = 240 mm, AC_3 = 55 mm (location of the connecting rod mass center from the crank pin), m_4 = 1 kg (piston mass), m_3 = 1.25 kg (connecting rod mass), I_3 = 0.01 kg m^2 (moment of inertia of the connecting rod). Assuming that the crank is massless, determine all the reaction forces and the torque available at the crankshaft. Do not make any simplification of mass distribution of the connecting rod. Neglect forces due to gravity.

8.12 A slider crank mechanism shown in Fig. P8.12 has a crank of total length AD = 120 mm and is pivoted midway (AO = OD) to the fixed link at "o." The connecting rod of length AB = 200 mm and mass 1.5 kg is connected to a reciprocating piston of 1 kg mass. The mass center of the connecting rod is at a distance of 70 mm (AC_3) from the crank pin. In the position shown, the crank angle $\theta = 45°$ and it is uniformly rotating at 2500 RPM (counterclockwise). Distribute the connecting rod mass at the crank and piston pins and balance the crank at point D with a mass m_D equivalent to the connecting rod mass placed at the crank pin. Determine the reaction forces and torque at the crank pin.

8.13 A single cylinder engine is running at a constant speed of 3000 rpm (uniform angular speed CCW). It has a crank radius of 60 mm and a connecting rod of length 240 mm. Mass of the piston = 0.6 kg; mass of connecting rod = 1.5 kg; length of connecting rod = 280 mm; center of gravity of the connecting rod 80 mm from the crank end; radius of gyration of the connecting rod 65 mm.

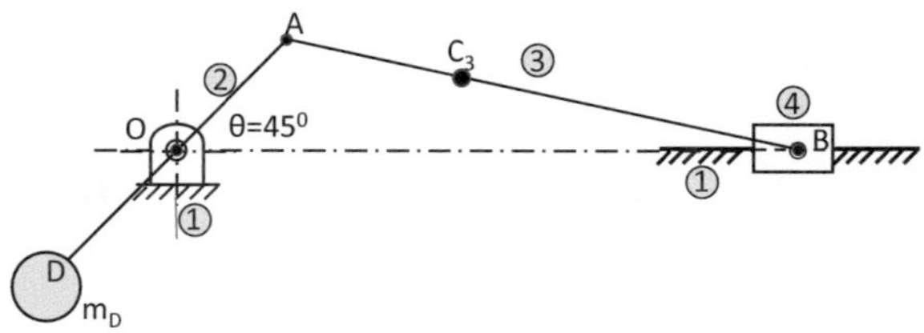

Fig. P8.12 Mechanism of exercise problem 8.12

Find all the reaction forces, torque at the crank shaft and shaking force on the frame, for one complete revolution of the crank.

8.14 A four-bar mechanism is shown in Fig. P8.14. The disk (link 2), mounted at its center $O_{4,}$ acts like a crank. The link 3 that acts as a coupler connects to the crank at A and to the follower at B in the position shown and is hinged at O_2. If the link 2 is uniformly rotating at 4000 rpm in the clockwise direction, and if the resisting force, $F_C = 2500$ N, determine to the torque T that has to be applied to the disk (link 2) and all the reaction forces. $O_2B = 88$ mm, $AB = 82$ mm, $BC = 54$ mm. [Note: you can use relevant kinematic data computed in the exercise problem 2.30.]

8.15 Consider a four-bar mechanism shown in Fig. P8.15. It consists of links 1, 2, 3, and 4. Link 1 is fixed, link 2 is the crank, link 3 the coupler and link 4 the follower. Link 2 is hinged at O_2 and an extension of link 2 is attached

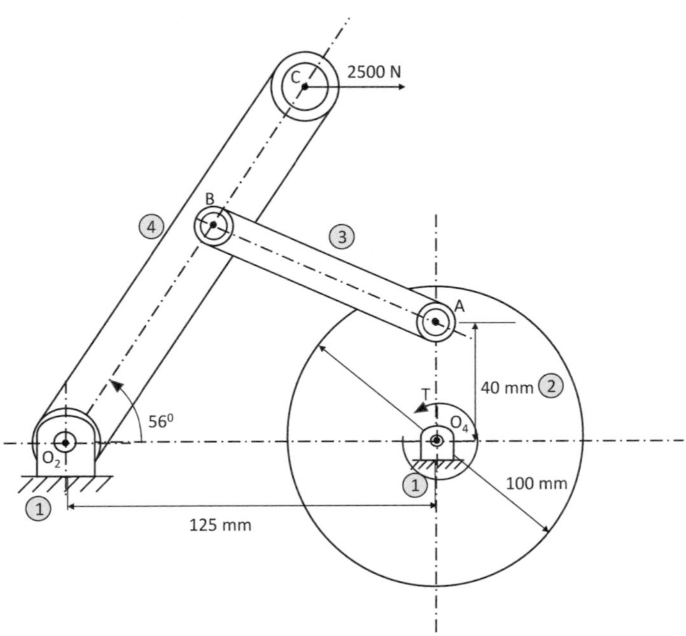

Fig. P8.14 Mechanism of exercise problem 8.14

Fig. P8.15 Mechanism of exercise problem 8.15

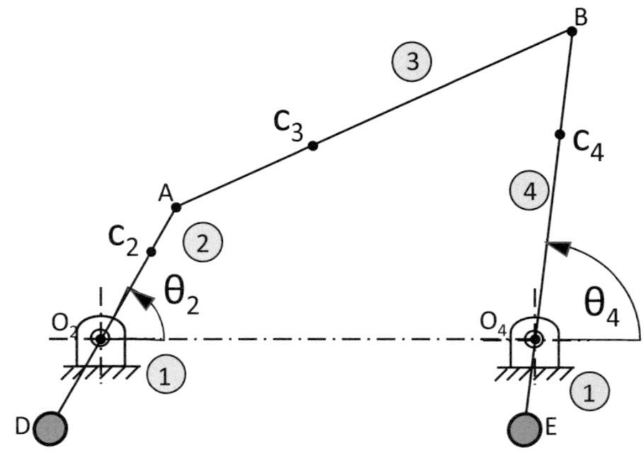

with an unknown balancing mass m_D. Similarly, link 4 is hinged at O_4 and an extension of it is attached with an unknown mass m_E. The mass center of link 2 (not accounting for the balancing mass that will be attached later) is at c_2 and similarly, the mass center of link 4 is at c_4 (not accounting for the balancing mass that will be attached later); mass center of link 3 is at c_3. The center masses of links 2, 3 and 4 are moved to the respective revolute joints at their ends. The balancing masses m_D and m_E are computed based on the balance of centrifugal forces on independent links 2 and 3, when they are not connected as part of the four-bar chain. If the crank (link 2) rotates at a constant angular velocity of 15 rad/s (counterclockwise direction), determine the reaction forces at the revolute joints on the fixed link with the balancing masses when $\theta_2 = 60°$ and the torque available at O_4.

$O_2O_4 = 200$ mm, $O_2A = 70$ mm, $O_2D = 40$ mm, $AB = 200$ mm, $O_4B = 137$ mm, $O_4E = 40$ mm $m_2 = 0.3$ kg, $m_3 = 1$ kg, $m_4 = 0.3$ kg.

8.16 Consider the part of a mechanism shown in Fig. P8.16 (its block diagram shown in Fig. P2.19b). If the angular velocity of the link 2 is 2 rad/s (CCW) and the angular acceleration is 10 rad/s^2 (CCW); the relative velocity of link 3 with respect to link 2 is 50 mm/s in the upward direction, and relative acceleration is 100 mm/s^2 in the upward direction, determine the torque applied to link 2 and all the reaction forces. $M_2 = 0.3$ kg, $M_3 = 0.8$ kg. Neglect friction at the sliding surface.

8.17 Consider the mechanism shown in Fig. P8.17 in which the link $\mathcal{D}$ rotating at an angular velocity of 2 rad/s in the clockwise direction and angular acceleration of 10 rad/s^2 in the clockwise direction is pushing link $\mathcal{B}$ by sliding action. In addition, a vertical force of 1000 N is acting downwards at the midpoint of $\mathcal{B}$. Determine the torque applied to link $\mathcal{D}$ and all the reaction forces. $\mathbf{M}_{\mathcal{D}} = 1$ kg, $\mathbf{M}_{\mathcal{B}} = 0.6$ kg. Assume both the moving links as slender rods with their mass centers at the middle of their lengths. [Note: Refer to the exercise problem 1.7 for the block diagram and you can use relevant kinematic data computed in the exercise problem 2.35.]

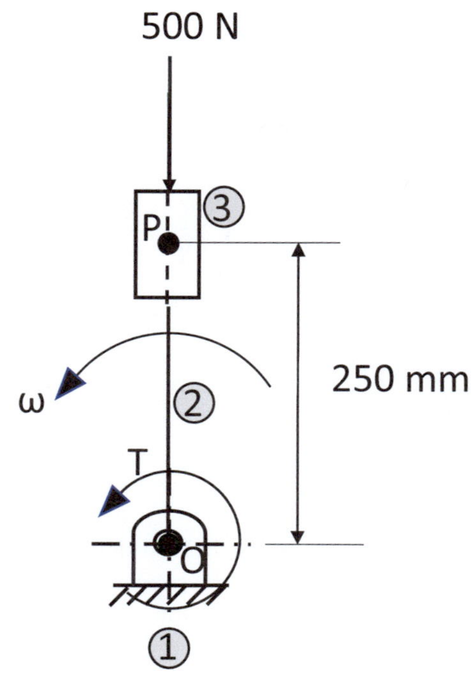

Fig. P8.16 Mechanism of exercise problem 8.16

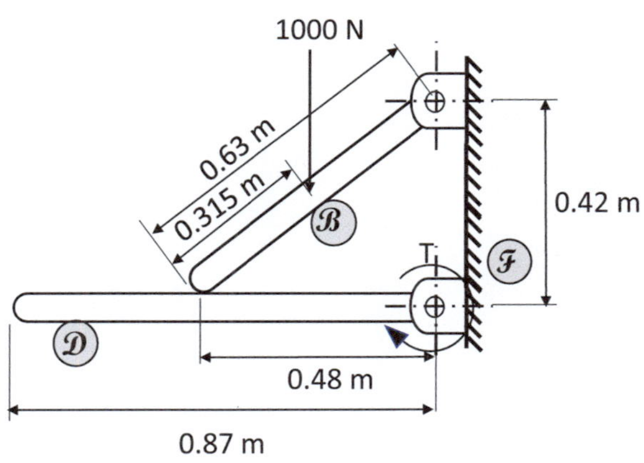

Fig. P8.17 Mechanism of exercise problem 8.17

8.18 In the mechanism shown in Fig. P8.18, the point A moves toward left at a velocity of 75 mm/s and an acceleration of 5 mm/s^2 toward left. Determine the torque applied to link 2 against the resisting force and all the reaction forces. $M_2 = 2$ kg, $M_3 = 3$ kg, $M_4 = 5$ kg. Assume links 2 and 4 as slender rods with their centers of mass at the middle of their lengths. [Note: Refer to the exercise problem 1.8 for the block diagram and you can use relevant kinematic data computed in the exercise problem 2.36.]

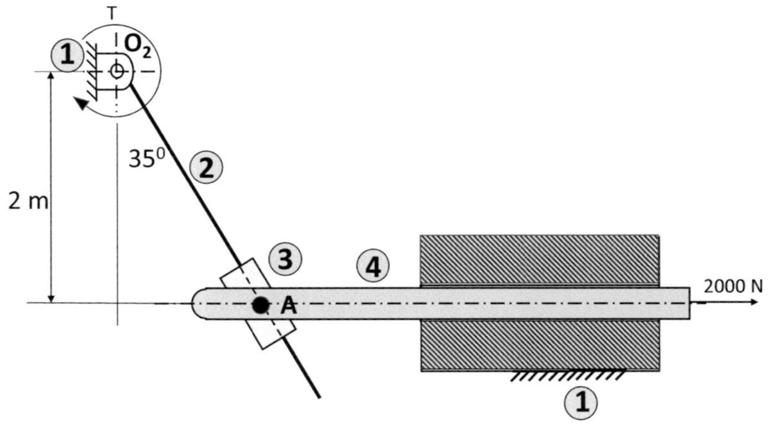

Fig. P8.18 Mechanism of exercise problem 8.18

8.19 Consider the mechanism shown in Fig. P8.19 in which the link 2 is hinged to the piston 3 and the piston is allowed to slide along the link 4. At the instant shown, link 2 is rotating counterclockwise with a uniform angular speed of 10 rad/s. Determine the reactions at various pin joints and the torque that has to be applied to link 2. Account forces due to gravity of all the links and assume that both links 2 and 4 are slender rods.

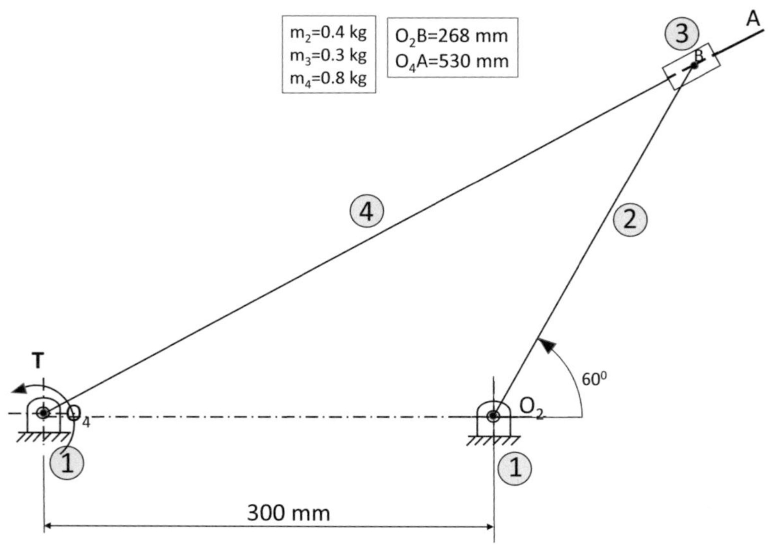

Fig. P8.19 Mechanism of exercise problem 8.19

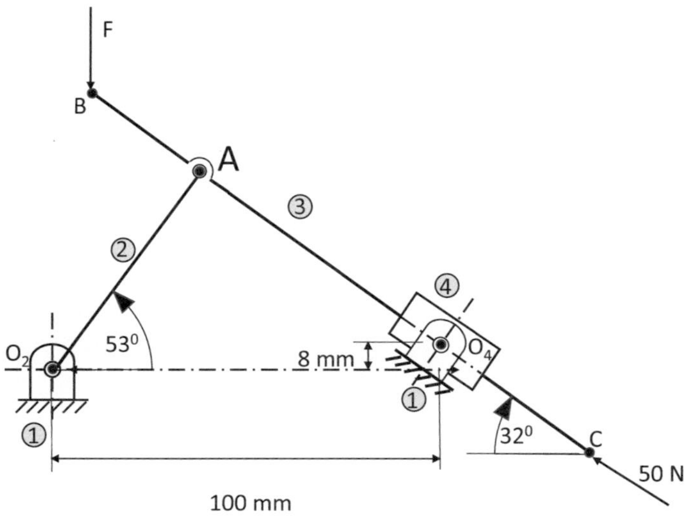

Fig. P8.20 Mechanism of exercise problem 8.20

8.20 Consider the mechanism shown in Fig. P8.20. It consists of a crank (link 2) that is pinned to link 3 at A. The piston (link 4) is pinned at the point O_4 that is located with respect O_2 at distances shown. The link 3 passes through the piston. The links 2 and 3 are in the angular positions as shown. A resisting force is acting along the link 3. If the crank is rotating clockwise at a uniform speed of 5 rad/s, determine the force F that has to be applied at the given instant. Assume links 2 and 3 to be slender rods. $O_2A = 60$ mm, BC = 130 mm, AB = 40 mm, AC = 90 mm; $m_2 = 0.1$ kg, $m_3 = 0.250$ kg, $m_4 = 0.5$ kg.

8.21 Consider the mechanism shown in Fig. P8.21 (its block diagram is shown in Fig. P1.11b). The link 3 (piston) slides on link 4 and pinned to link 2 (crank) at P. If the crank is rotating at angular velocity of 15 rad/s in the clockwise direction and if the resisting force on the piston is 1500 N as shown, determine the torque applied to the crank, and all the reaction forces. $M_1 = 0.15$ kg, $M_4 = 0.2$ kg, $M_3 = 0.35$ kg. Assume links 2 and 4 as slender rods and link 3 as a concentrated mass. [Note: you can use relevant kinematic data computed in the exercise problem 2.39.]

8.22 Consider the mechanism shown in Fig. P8.22 (its block diagram representation shown in Fig. P1.9b). If the slotted links are driven independently with the following angular velocities and accelerations, if the resisting torque for link 3 is 1500 N m in the clockwise direction, determine the torque applied to link 2 and all the reaction forces: $\omega_2 = 20$ rad/s (CCW), $\omega_3 = 25$ rad/s (CCW); $\alpha_2 = 5$ rad/s^2 (CCW), $\alpha_3 = 25$ rad/s^2 (CCW). [Note: you can use relevant kinematic data computed in the exercise problem 2.37.]

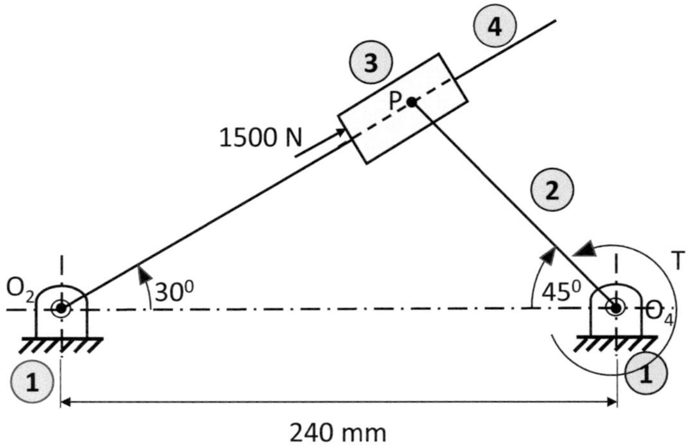

Fig. P8.21 Mechanism of exercise problem 8.21

Fig. P8.22 Mechanism of exercise problem 8.22

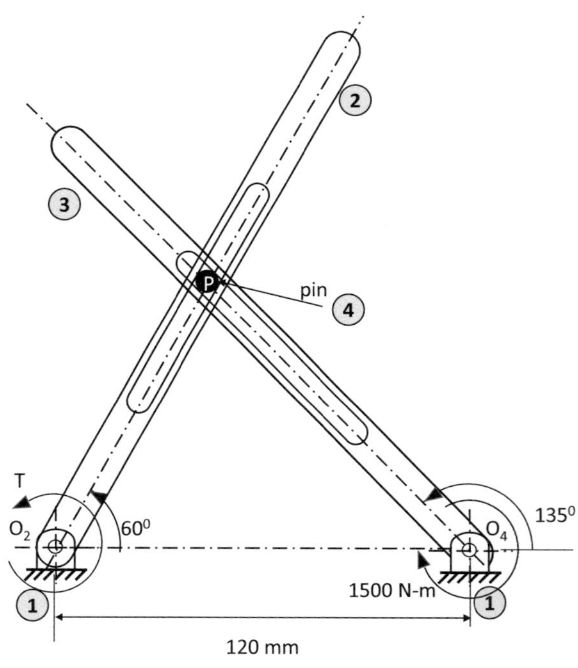

8.23 The mechanism shown in Fig. P8.23 (its block diagram shown in Fig. 1.10b) consists of two slender rods, $\mathcal{A}$ and $\mathcal{B}$, respectively hinged at O_2 and O_4. Two collars, $\mathcal{C}_1$ and $\mathcal{C}_2$, hinged at C, respectively slide along $\mathcal{A}$ and $\mathcal{B}$. If the rod A is rotating in the counterclockwise direction with an angular velocity of 0.5 rad/s, angular acceleration of 5 rad/s^2 in the clockwise direction, and

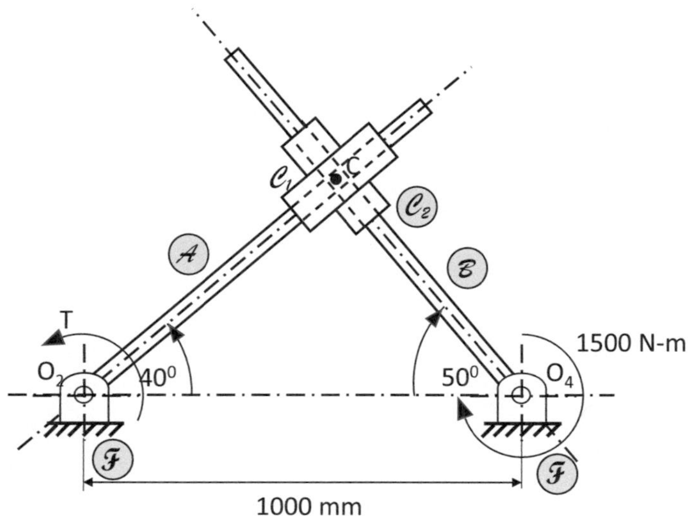

Fig. P8.23 Mechanism of exercise problem 8.23

B is rotating in the clockwise direction with an angular velocity of 0.7 rad/s, angular acceleration of 5 rad/s^2 in the counterclockwise direction, if the resisting torque for link $\mathcal{B}$ is 1500 N m in the clockwise direction, determine the torque applied to link $\mathcal{A}$ and all the reaction forces. $M_{\mathcal{A}} = M_{\mathcal{B}} = 2$ kg. $M_{\mathcal{C}_1} = M_{\mathcal{C}_2} = 0.25$ kg. Assume links $\mathcal{A}$ and $\mathcal{B}$ as slender rods and the collars as point masses. [Note: you can use relevant kinematic data computed in the exercise problem 2.38.]

8.24 Consider the quick return mechanism shown in Fig. P8.24 (its block diagram shown in Fig. P1.12b). If the link 2 (crank) is rotating at angular velocity ω_2 = 15 rad/s (CCW), determine the torque applied to the crank to resist a force of 5000 N by link 6 as shown. $M_2 = 0.1$ kg, $M_3 = 0.3$ kg, $M_5 = 0.15$ kg, $M_4 = 0.5$ kg, $M_6 = 1$ kg. Assume links 2, 3, 5 as slender rods. [Note: you can use relevant kinematic data computed in the exercise problem 2.40.]

8.25 Configure both the quick-return mechanisms, Whitworth and slotted-link, for the same time ratio, length of the stroke and angular speed of the input link and compare them in terms of kinematic and kinetic analysis?

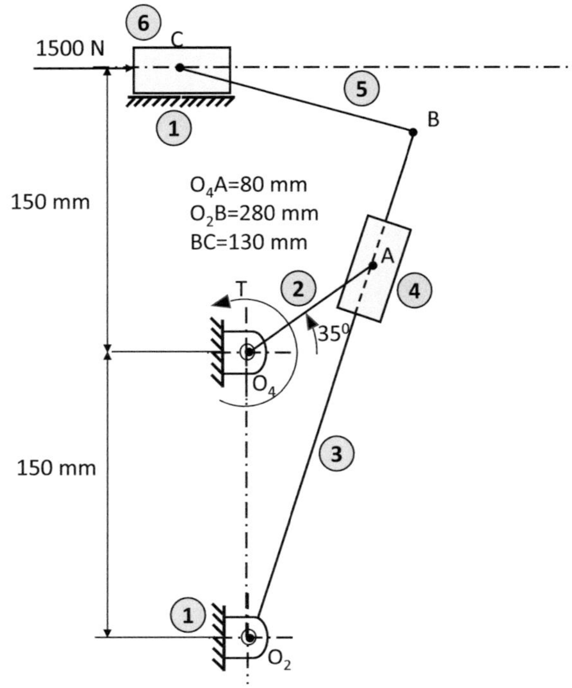

Fig. P8.24 Mechanism of exercise problem 8.24

Chapter 9
Balancing

Consider a machine shown in Fig. 9.1, which is supported on a foundation. Forces and moments generated within the machine will result in forces and moments at the foundation that are further transmitted to the supporting structures connected to it. The forces and moments generated are many times due to unbalanced forces created from concentrated masses m_1, m_2... that have accelerations a_1, a_2.... In addition, it can be presumed that the forces and moments transferred through the foundation will result in its angular acceleration and linear acceleration. In most machines, such forces and moments generated due to unbalance can be nullified by attaching balancing masses within the machine. Such a balancing is an efficient and cost-effective method of preventing the vibration of the foundation. In this chapter, we shall discuss methods of balancing a variety of machines.

Balancing of machines can be broadly classified as follows:

1. Balancing of rotating masses
2. Balancing of reciprocating masses.

9.1 Balancing of Rotating Masses

Rotating machines constitute a significant portion of the machines that practically operate in various applications and in addition, rotating machines operate in a wide range of speeds. Hence, balancing of rotating masses is important and has, in fact, achieved great success even in very high-speed machines like turbines used in jet aircraft. We shall demonstrate some simple examples that form the basis for the design of many balancing machines.

D. N. Manik, *Fundamentals of Mechanisms and Machines*,
https://doi.org/10.1007/978-981-97-1810-8_9

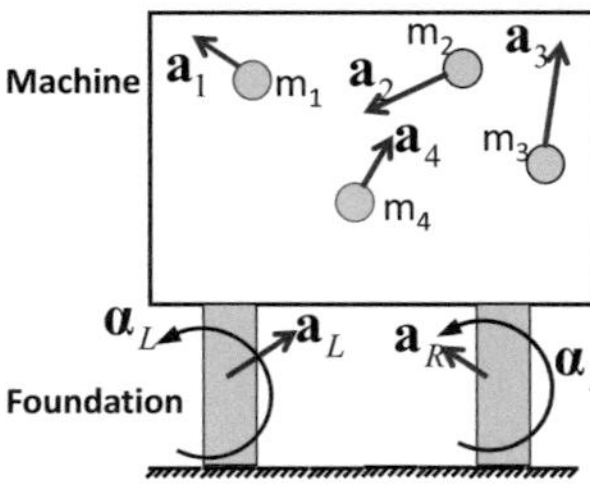

Fig. 9.1 Mass unbalances in a machine transferred to foundation

9.1.1 *Balancing in the Same Vertical Plane as the Unbalanced Mass*

Consider a uniform diameter shaft shown in Fig. 9.2, rotating at an angular velocity ω, with an unbalanced mass, represented by m at an eccentricity r, from the geometric center of the shaft. The centrifugal force due to the unbalanced mass is given by

$$F_c = mr\omega^2 \tag{9.1}$$

If no remedial action is taken against this unbalanced mass, the above centrifugal force will be equally divided as reactions on the bearings. Since the unbalance is assumed at the center of the shaft length, equal reaction forces will be transmitted to the bearings and the bearings will transmit these forces to other connected structures, thus resulting in vibration of the foundation. In addition, there will be dynamic stresses on the shaft as well.

One of the easiest remedies to the unbalance shown in Fig. 9.2 is to attach a balancing mass m_b at a distance r_b, diametrically opposite to the unbalanced mass, as shown in Fig. 9.3. The balancing mass generates a centrifugal force equal and opposite to that created by the unbalanced mass as follows.

$$mr\omega^2 = m_b r_b \omega^2 \tag{9.2}$$

Since the angular velocity is same for both masses, Eq. (9.2) reduces to

$$mr = m_b r_b \tag{9.3}$$

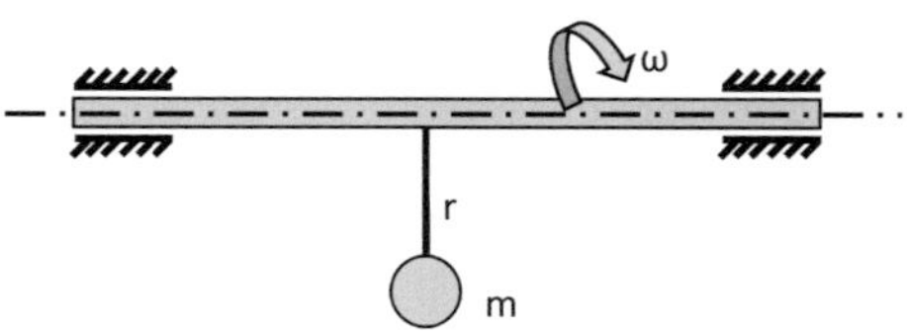

Fig. 9.2 Unbalance due to a symmetric eccentricity of a long shaft

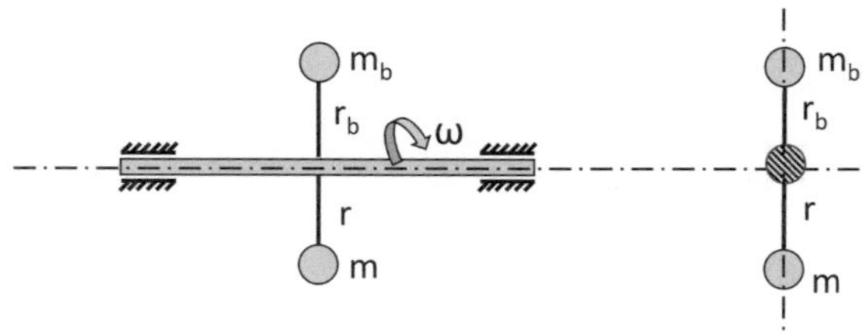

Fig. 9.3 Balancing the eccentric shaft of Fig. 9.1

Hence, if the balancing mass is placed diametrically opposite to the unbalanced mass in the same vertical plane perpendicular to the shaft and satisfying Eq. (9.3), there will not be any reaction forces at the bearings and no dynamic stresses will be developed in the shaft either.

9.1.2 Balancing Masses in Different Vertical Planes

If the balancing mass cannot be placed in the same vertical plane perpendicular to the shaft as the unbalanced mass, we may have to place the balancing masses in two different vertical planes. As shown in Fig. 9.4, the balancing masses are placed in two different planes, but diametrically opposite to the unbalanced mass.

Since the angular velocity is the same for the entire shaft, the centrifugal forces of the balancing masses and the unbalanced mass can be related as

$$mr = m_{b_1} r_{b_1} + m_{b_2} r_{b_2} \tag{9.4}$$

Similarly, the moments due to the centrifugal forces of the balancing masses and the unbalanced mass can be related as

$$mra = m_{b_1} r_{b_1} a_1 + m_{b_2} r_{b_2} a_2 \tag{9.5}$$

By choosing the balancing masses and their locations as per Eqs. (9.4) and (9.5), the reaction forces at the bearings due to the unbalanced mass can be reduced to zero. However, the dynamic stresses on the shaft will remain. It should be clear

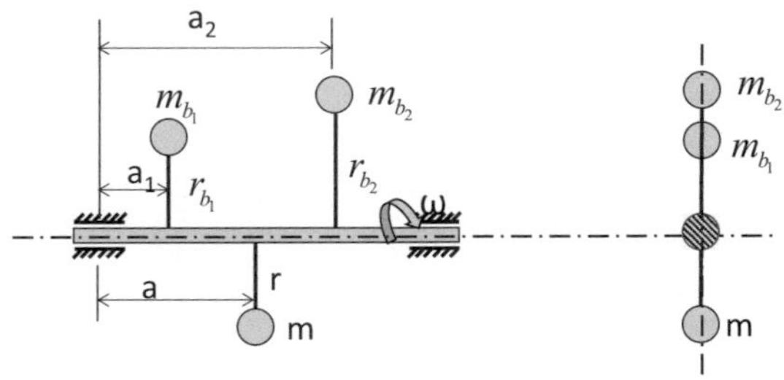

Fig. 9.4 Balancing in different planes with diametrically opposite masses

from Fig. 9.4 that the unbalanced mass as well as the balancing masses are on the same plane parallel to the shaft. But we shall see later that such a restriction can be removed so the balancing masses can be located at convenient locations, as required in specific practical situations.

9.1.3 Unbalance Masses in Different Planes Perpendicular to the Shaft

Consider the shaft system shown in Fig. 9.5 that has two equal unbalanced masses, diametrically located opposite to each other, at the same radius and in two different vertical planes perpendicular to the shaft. The system is balanced when it is stationary. However, when it starts rotating, there is a couple due to the action of centrifugal forces that will induce dynamic stresses on the shaft and will as well transmit reaction forces to the bearings. When the shaft length becomes extremely small, the mass unbalance will be restricted to a single plane and there will not be any dynamic stresses on the shaft when it is rotating and in addition, there will not be any reaction forces at the bearings during the shaft rotation.

However, if the shaft is sufficiently long and the unbalanced masses are placed sufficiently apart, the balancing masses of mass m and located at radius r will have to be added at two planes perpendicular to the shaft as shown in Fig. 9.6. Then, both forces and moments are balanced when the shaft is rotating.

The above discussions lead to the conclusion that when the unbalance is in more than one plane of a long shaft, the balancing masses also will have to be placed in two different planes that can balance both forces and moments.

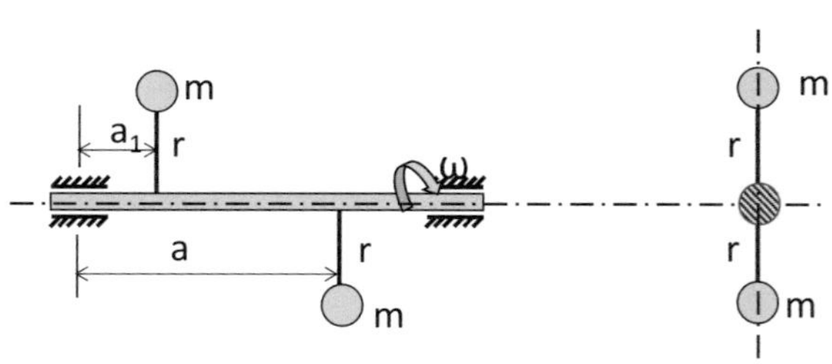

Fig. 9.5 Unbalance in different planes due to diametrically opposite eccentricity

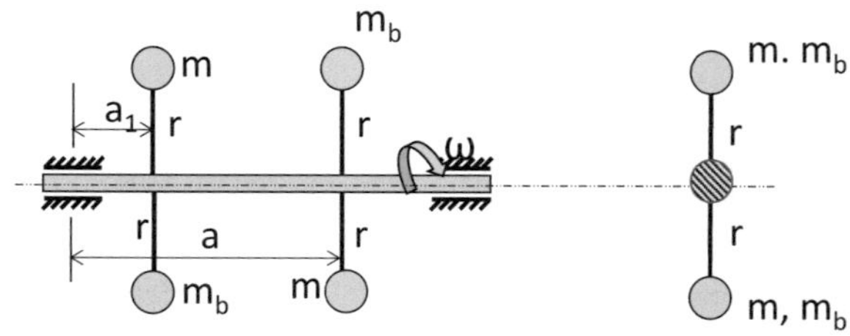

Fig. 9.6 Balancing the shaft of Fig. 9.5

9.1.4 Single Plane Balancing of Multiple Masses

Consider the general case of a very short shaft system shown in Fig. 9.7 that has several unbalanced masses m_1, m_2, and m_3, respectively located at angles θ_1, θ_2 and θ_3, and radii r_1, r_2, and r_3, on a single vertical plane perpendicular to the shaft. A single balancing mass m_b, located at a radius r_b and at an angle θ_b is assumed to balance the net unbalance created by all the masses. Figure 9.8 shows the vectors of the centrifugal forces of unbalanced masses and balancing masses. The closure of the vector diagram is an indication of balancing.

The vectors satisfy the following equation:

$$m_b\mathbf{r}_b + \sum_{i=1}^{3} m_i\mathbf{r}_i = 0 \tag{9.6}$$

Equation (9.6) can be expanded in terms of scalar equations as follows:

$$m_b r_b \cos\theta_b = -\sum_{i=1}^{3} m_i r_i \cos\theta_i \tag{9.7}$$

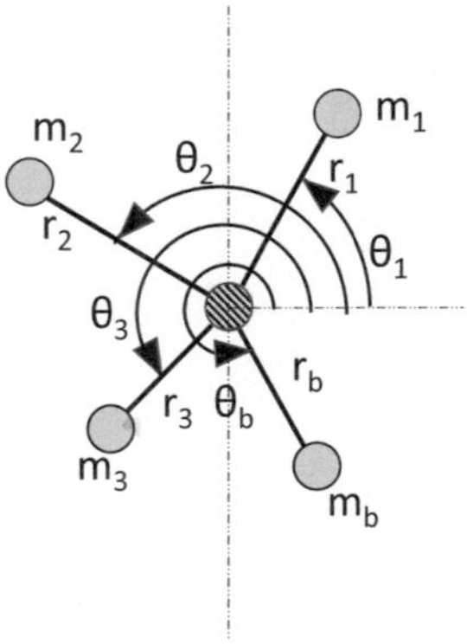

Fig. 9.7 Several unbalanced masses on a single plane

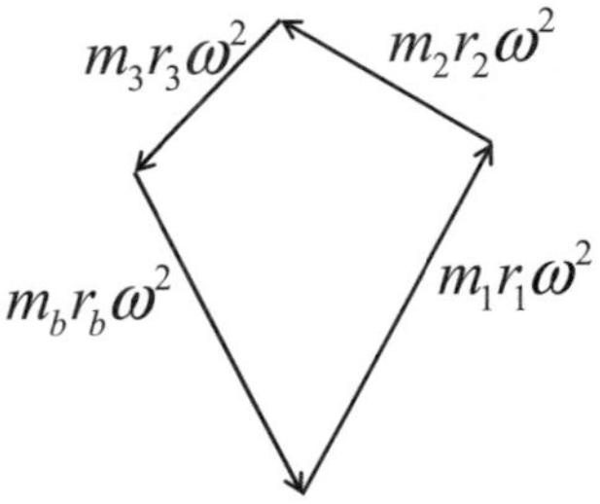

Fig. 9.8 Vectors of the centrifugal forces of unbalanced masses and balancing mass of Fig. 9.7

And

$$m_b r_b \sin\theta_b = -\sum_{i=1}^{3} m_i r_i \sin\theta_i \tag{9.8}$$

Squaring and adding Eqs. (9.7) and (9.8)

$$m_b r_b = \sqrt{\left(\sum_{i=1}^{3} m_i r_i \cos\theta_i\right)^2 + \left(\sum_{i=1}^{3} m_i r_i \sin\theta_i\right)^2} \tag{9.9}$$

Dividing Eq. (9.8) by (9.7)

$$\theta_b = \tan^{-1}\frac{-\sum_{i=1}^{3} m_i r_i \sin\theta_i}{-\sum_{i=1}^{3} m_i r_i \cos\theta_i} \tag{9.10}$$

Hence, by using a balancing mass m_b satisfying the radial and angular location in Eqs. (9.9) and (9.10), single plane balancing can be obtained. It is important to note that there is enough flexibility in choosing the location of a balancing mass. Equations (9.9) and (9.10) can be generalized to n unbalanced masses.

Example 9.1

A thin rotor shown in Fig. 9.9 has the following masses mounted on it at various radii and angular positions as shown in Table 9.1. Determine the balancing mass and its angular position at a radial position of 0.2 m.

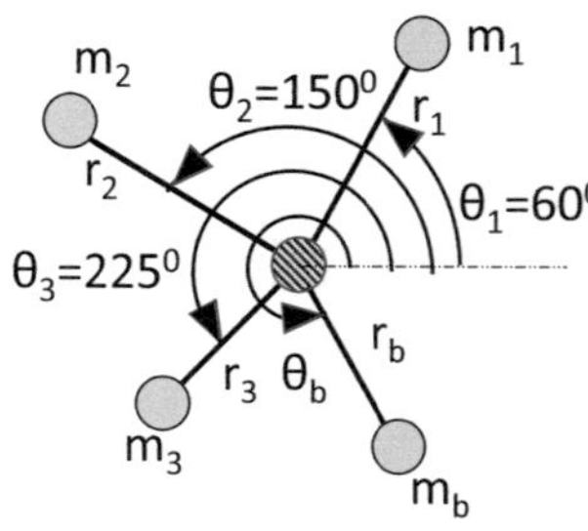

Fig. 9.9 Unbalanced masses on a single plane balanced by a mass m_b

Table 9.1 Details of unbalanced masses, and their locations

No	Mass (kg)	Radius (m)	Angle (°)
1	0.2	0.15	60
2	0.4	0.25	150
3	0.35	0.1	225

From Eq. (9.9)

$$m_{b1}r_{b1} = \frac{\sqrt{\left(\sum_{i=1}^{N} m_i r_i a_i \cos\theta_i\right)^2 + \sqrt{\left(\sum_{i=1}^{N} m_i r_i a_i \sin\theta_i\right)^2}}}{a}$$

$$= \frac{\sqrt{(104680)^2 + (-22172)^2}}{800} = 133.75 \text{ kg-mm} \qquad (1)$$

Since $r_b = 0.2$ m, the balancing mass is given by $m_b = 0.1091/0.2 = 0.5456$ kg. From Eq. (9.10)

$$\theta_b = \tan^{-1} \frac{-\sum_{i=1}^{3} m_i r_i \sin\theta_i}{-\sum_{i=1}^{3} m_i r_i \cos\theta_i}$$

$$= \tan^{-1} \frac{-(0.2 \times 0.15 \times \sin 60 + 0.4 \times 0.25 \times \sin 150 + 0.35 \times 0.1 \times \sin 225)}{-(0.2 \times 0.15 \times \cos 60 + 0.4 \times 0.25 \times \cos 150 + 0.35 \times 0.1 \times \cos 225)}$$

$$= 332^\circ \qquad (2)$$

9.1.5 Two-Plane Balancing

Consider a shaft system shown in Fig. 9.10 that has a number of unbalanced masses m_1, m_2, m_3 respectively at radial locations r_1, r_2, r_3, angular locations θ_1, θ_2, θ_3 and axial locations a_1, a_2, a_3 from a right plane (RP) chosen close to the right bearing. It is assumed that the balancing masses will be placed in the left plane (LP) close to the left bearing and in the right plane (RP), close to the right bearing. The axial distance between LP and RP, a_b, will have to be suitably chosen before calculating details of the balancing masses and their locations.

First, the moment balance of all the centrifugal forces due to balancing as well as unbalanced masses is calculated with respect to the RP. It is given by

$$\mathbf{r}_{b_1} \times m_{b_1}\mathbf{a}_b + \sum \mathbf{r}_i \times m_i\mathbf{a}_i = 0 \qquad (9.11)$$

Applying Eq. (9.11) to the given problem

$$m_{b_1} r_{b_1} a_b \cos\theta_{b_1} = -\sum m_i r_i a_i \cos\theta_i \qquad (9.12)$$

And

$$m_{b_1} r_{b_1} a_b \sin\theta_{b_1} = -\sum m_i r_i a_i \sin\theta_i \qquad (9.13)$$

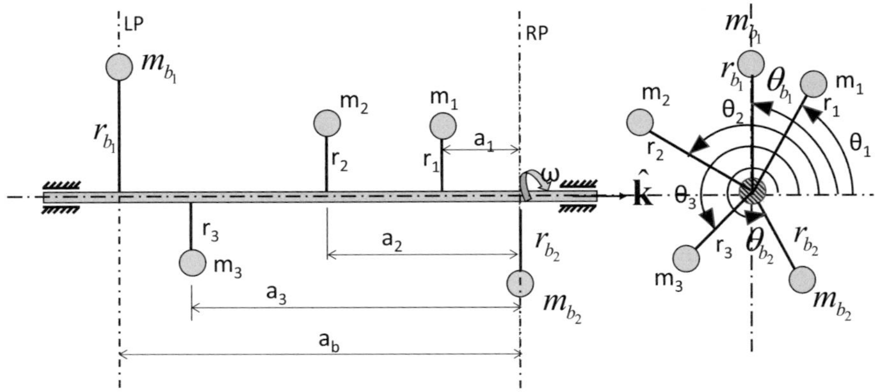

Fig. 9.10 Unbalance masses in different planes, radial, and angular locations

From Eqs. (9.12) and (9.13), the product of the balancing mass and its radial location on the LP is given by

$$m_{b_1} r_{b_1} = \frac{\sqrt{\left(\sum m_i r_i a_i \cos\theta_i\right)^2 + \left(\sum m_i r_i a_i \sin\theta_i\right)^2}}{a_b} \tag{9.14}$$

From Eqs. (9.12) and (9.13), the angular location of the balancing mass in the LP is given by

$$\theta_{b_1} = \tan^{-1} \frac{-\sum m_i r_i a_i \sin\theta_i}{-\sum m_i r_i a_i \cos\theta_i} \tag{9.15}$$

Now that we have determined the details of the balancing mass on the left plane by summing the moments, details of the balancing mass on the right plane can be obtained by summing the centrifugal forces as follows

$$m_{b_1} \mathbf{r}_{b_1} + m_{b_2} \mathbf{r}_{b_2} + \sum m_i \mathbf{r}_i = 0 \tag{9.16}$$

By writing the horizontal and vertical components of Eq. (9.16)

$$m_{b_2} r_{b_2} \cos\theta_{b_2} = -\left(m_{b_1} r_{b_1} \cos\theta_{b_1} + \sum m_i r_i \cos\theta_i\right) \tag{9.17}$$

And

$$m_{b_2} r_{b_2} \sin\theta_{b_2} = -\left(m_{b_1} r_{b_1} \sin\theta_{b_1} + \sum m_i r_i \sin\theta_i\right) \tag{9.18}$$

Squaring and adding Eqs. (9.17) and (9.18)

$$m_{b_2} r_{b_2} = \sqrt{\begin{array}{c}\left(m_{b_1} r_{b_1} \sin\theta_{b_1} + \Sigma m_i r_i \sin\theta_i\right)^2 \\ +\left(m_{b_i} r_{b_i} \cos\theta_{b_i} + \Sigma m_i r_i \cos\theta_i\right)^2\end{array}} \tag{9.19}$$

Dividing Eq. (9.18) by (9.17)

$$\theta_{b_2} = \tan^{-1}\frac{-\left(m_{b_1} r_{b_1} \sin\theta_{b_1} + \sum m_i r_i \sin\theta_i\right)}{-\left(m_{b_1} r_{b_1} \cos\theta_{b_1} + \sum m_i r_i \cos\theta_i\right)} \tag{9.20}$$

Note that it is the direction of the vector $\hat{\boldsymbol{k}}$ that decides the signs of a's.

Example 9.2

A long rotor has the following masses mounted on it at various radii, angles and axial positions. Total length of the rotor between bearings is 1000 mm and the balancing planes are 100 mm from each bearing; the left balancing plane is 100 mm to the right of left bearing and the right balancing plane is 100 mm to the left of right bearing. The axial positions are measured with respect to the right balancing plane (Fig. 9.11, Tables 9.2 and 9.3).

From Eq. (9.14), the product of balancing mass and its radial location in the left plane is given by (a_b =800 mm, distance between the balancing planes)

$$\begin{aligned} m_{b_1} r_{b_1} &= \frac{\sqrt{\left(\sum_{i=1}^{N} m_i r_i a_i \cos\theta_i\right)^2 + \left(\sum_{i=1}^{N} m_i r_i a_i \sin\theta_i\right)^2}}{a_b} \\ &= \frac{\sqrt{(104680)^2 + (-22172)^2}}{800} = 133.75 \text{ kg-mm} \end{aligned} \tag{1}$$

and from Eq. (9.15), its angular position in the left plane is given by

$$\theta_{b_1} = \tan^{-1}\frac{-\sum_{i=1}^{N} m_i r_i a_i \sin\theta_i}{-\sum_{i=1}^{N} m_i r_i a_i \cos\theta_i} = \tan^{-1}\frac{-22172}{104680} = 348^\circ \tag{2}$$

From Eq. (9.19), the product of balancing mass and its radial position in the right plane is given by

Table 9.2 Details of unbalanced masses and their locations

Mass	kg	Radius	mm	Angle	Degrees	Axial position (mm)
M1	3	R1	80	ϴ1	60	**200**
M2	2	R2	80	ϴ2	150	**500**
M3	**2**	**R3**	**60**	**ϴ3**	**225**	**700**

Table 9.3 Calculations regarding force and moment imbalances of (Fig. 9.11)

M, kg	R, mm	Deg	Axial position, mm	$m_i r_i a_i \cos\theta_i$	$m_i r_i a_i \sin\theta_i$	$m_i r_i \cos\theta_i$	$m_i r_i \sin\theta_i$
3	80	60	−200	−24,000	**−41,569**	**120**	**207.85**
2	80	150	−500	69,282	**−40,000**	**−138.56**	**80**
2	60	225	−700	59,397	**59,397**	**−84.85**	**−84.85**
			Σ	**104,680**	−22,172	−103.42	203

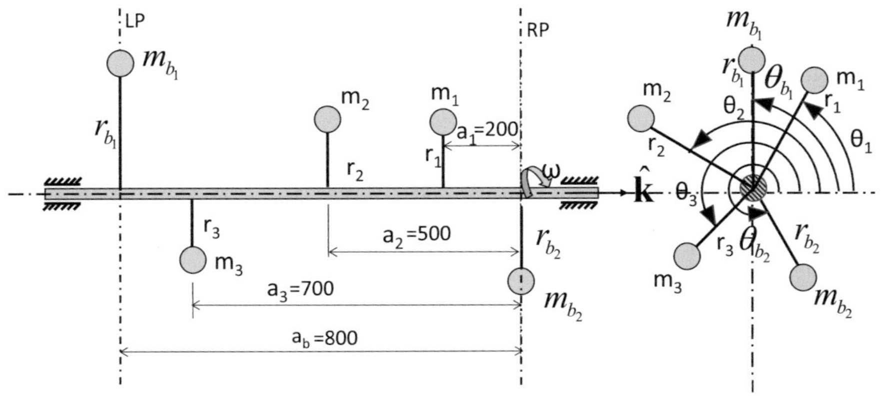

Fig. 9.11 Unbalance masses in various planes and the balancing planes

$$m_{b_2} r_{b_2} = \sqrt{\begin{array}{c}\left[\left(\sum_{i=1}^{N} m_i r_i \cos\theta_i\right) + m_{b_i} r_{b_i} \cos\theta_{b_i}\right]^2 \\ +\left[\left(\sum_{i=1}^{N} m_i r_i \sin\theta_i\right) + m_{b_i} r_{b_i} \sin\theta_{b_i}\right]^2\end{array}}$$

$$= \sqrt{\begin{array}{c}(-103.42 + 133.75\cos 348)^2 \\ +(203 + 133.75\sin 348)^2\end{array}} = 177.3\ \text{kg} - mm \tag{3}$$

And from Eq. (9.20), its angular position in the right plane is given by

$$\theta_{b_2} = \tan^{-1}\frac{-\left[\left(\sum_{i=1}^{N} m_i r_i \sin\theta_i\right) + m_{b_1} r_{b_1} \sin\theta_{b_1}\right]}{-\left[\left(\sum_{i=1}^{N} m_i r_i \cos\theta_i\right) + m_{b_1} r_{b_1} \cos\theta_{b_1}\right]}$$

$$= \tan^{-1}\frac{-(203 + 133.75\sin 348)}{-(-103.42 + 133.75\cos 348)} = 261^\circ \tag{4}$$

9.2 Balancing Machines

9.2.1 Single Plane Balancing Machine

The basic principle of a single plane balancing machine is shown in Fig. 9.12. A thin disk that has an unbalance is mounted between two sets of bearings and driven by a motor at the desired speed. Two vibration pickups are mounted on the bearings along the horizontal and vertical directions. Based on the vibration measurements obtained from these pickups, the magnitude and phase of the vibration vector can be obtained. Two sets of experiments are conducted: one without a trial mass and another with a trial mass at a certain radial and angular location. Based on the above vibration measurements, the value and location of the balancing mass that has to be mounted on the rotor can be determined. The mathematics that relates vibration measurement to the balancing mass and its location is presented as follows.

The given unbalanced rotor is operated at the rated speed and the amplitude and phase of vibration are measured by vector $\mathbf{A}_1$: A trial mass m_t is added at some location and once again the rotor is operated at the rated speed and the amplitude and phase of vibration are measured by vector $\mathbf{A}_2$. The difference $\mathbf{A}_2 - \mathbf{A}_1$ shown in Fig. 9.13a is a measure of the contribution of the trial mass. A specific balancing mass m_b placed at the same radius as the trial mass and at a specified angular location ϕ obtained from Fig. 9.13b can cancel the original unbalance.

The force due to the trial mass is given by

$$m_t \mathbf{r}_t \omega^2 \tag{9.21}$$

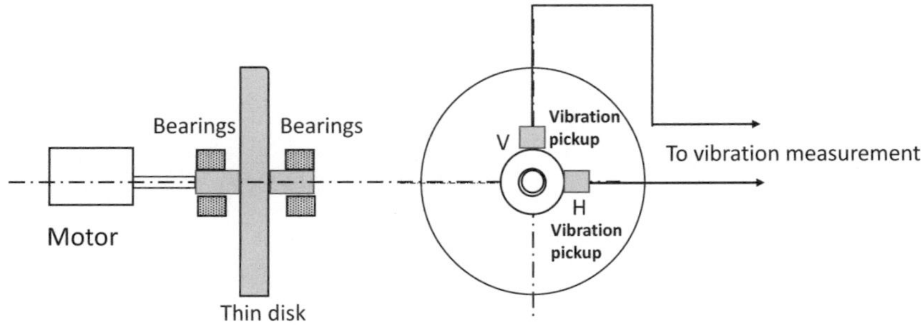

Fig. 9.12 Single plane balancing machine

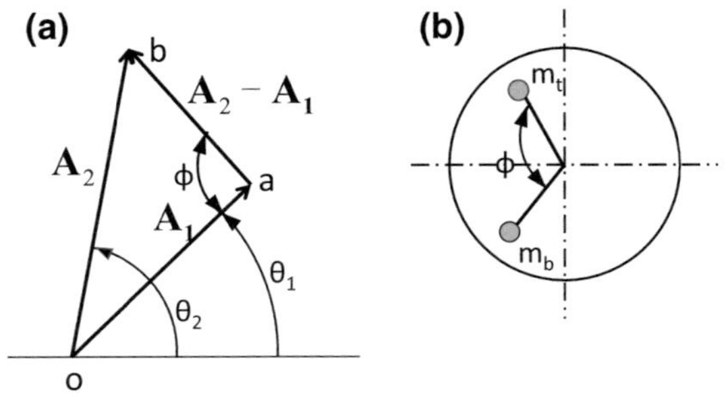

Fig. 9.13 **a** Vibration vectors due to unbalance and trial mass **b** Location of trial and balancing masses

The force due to the balancing mass is given by

$$m_b \mathbf{r}_b \omega^2 \tag{9.22}$$

The force due to the balancing mass and the trial mass can be related by a proportionality complex constant $\boldsymbol{\delta}$ as follows:

$$m_b \mathbf{r}_b \omega^2 = \boldsymbol{\delta}\left(m_t \mathbf{r}_t \omega^2\right) \tag{9.23}$$

Since the balancing mass should result in a vibration opposite to that of the vibration due to unbalance, the following equation can be written based on the vibration resulting from unbalanced vibration and the trial mass vibration

$$\boldsymbol{\delta}(\mathbf{A}_2 - \mathbf{A}_1) = -\mathbf{A}_1 \tag{9.24}$$

Therefore, the complex constant $\boldsymbol{\delta}$ is given by

$$\boldsymbol{\delta} = \frac{\mathbf{A}_1}{\mathbf{A}_1 - \mathbf{A}_2} \tag{9.25}$$

From Eqs. (9.22) through (9.25)

$$m_b \mathbf{r}_b \omega^2 = \frac{\mathbf{A}_1}{\mathbf{A}_1 - \mathbf{A}_2}\left(m_t \mathbf{r}_t \omega^2\right) \tag{9.26}$$

Equating the magnitude components of both sides of the Eq. (9.26)

$$m_b = \frac{oa}{ab} m_t \tag{9.27}$$

From Fig. 9.13a, the angle between the vectors $\mathbf{A}_1$ and $\mathbf{A}_1 - \mathbf{A}_2$ is ϕ. Therefore, the phase of the radial locations of the balancing mass and the trial mass is given by

$$\angle \frac{\mathbf{r}_b}{\mathbf{r}_t} = \varphi \tag{9.28}$$

Therefore, placing a balancing mass m_b as per Eq. (9.28) at the same radial location as the trial mass and at an angle ϕ will balance the given system.

Example 9.3

A thin disk that has unbalance is running at 300 rpm counterclockwise. It indicates an unbalanced magnitude of 3.2 mm at an angular position of 30° counterclockwise from a reference mark. When a trial mass of 70 gm is added at an angular location of 143° counterclockwise from the reference mark, it indicates an unbalance of 7 mm at an angular location of 77° counterclockwise from the reference mark. Determine the balancing mass that has to be added and its angular location with respect to the reference mark. Assume that all the tests were conducted at 300 rpm and that the trial mass and balancing mass radial locations are the same.

The unbalanced mass results in vibration given by

$$\begin{aligned}\mathbf{A}_1 &= 3.2\left(\cos 30°\hat{\boldsymbol{i}} + \sin 30°\hat{\boldsymbol{j}}\right)\\ &= 2.77\hat{\boldsymbol{i}} + 1.6\hat{\boldsymbol{j}}\ \text{mm}\end{aligned} \tag{1}$$

The unbalance and trial mass combination results in

$$\begin{aligned}\mathbf{A}_2 &= 7\left(\cos 77°\hat{\boldsymbol{i}} + \sin 77°\hat{\boldsymbol{j}}\right)\\ &= 1.57\hat{\boldsymbol{i}} + 6.82\hat{\boldsymbol{j}}\ \text{mm}\end{aligned} \tag{2}$$

The difference between the above two vectors is given by

$$\begin{aligned}\mathbf{A}_2 - \mathbf{A}_1 &= -1.2\hat{\boldsymbol{i}} + 5.22\hat{\boldsymbol{j}}\ \text{mm}\\ &= 5.365\ \text{mm}\angle 103°\end{aligned} \tag{3}$$

Figure 9.14a is drawn as follows: Vector $\mathbf{A}_1$ of Eq. (1) is drawn at an angle of 30° and vector $\mathbf{A}_2$ of Eq. (2) is drawn at an angle of 77°. The vector $\mathbf{A}_2 - \mathbf{A}_1$ is drawn from Eq. (3) that joins the vectors $\mathbf{A}_1$ and $\mathbf{A}_2\mathbf{A}_2$

Hence, from Fig. 9.14a, the phase angle between the trial mass and the balancing mass is given by

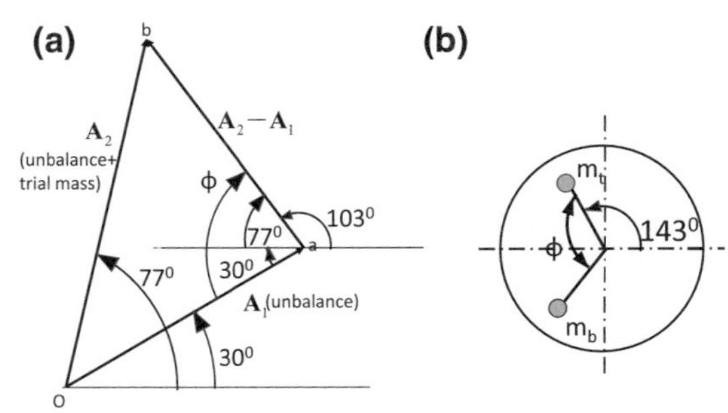

Fig. 9.14 **a** Vibration vectors due to unbalance mass and unbalance-trial mass combination **b** Locations of trial and balancing masses

$$\varphi = 77 + 30 = 107^\circ \tag{4}$$

From Eq. (9.27), the balancing mass is given by

$$m_b = \frac{oa}{ab} m_t = \frac{3.2}{5.365} \times 70 = 41.75\,\text{gm} \tag{5}$$

A balancing mass of 41.75 gm is placed at an angle of 107° (CCW) from the position of the trial mass and at the same radius, as shown in Fig. 9.14b.

9.2.2 Two-Plane Balancing Machines

When there is unbalance in long rotors, they cannot be balanced by placing balancing masses in a single plane. However, the principles that were outlined in the earlier section to balance in a single plane can be extended to determine the value of balancing masses in two reference planes.

Consider a long rotor mounted on bearings and supported by a carriage. It has two reference planes, left and the right, on which the balancing masses will be placed. Before we go for a more accurate method of two-plane balancing, let us look at the possibility of balancing the system by using a simple technique as shown in Fig. 9.15. The left reference plane is supported by a spring and the right reference plane is placed on a pivot so that the unbalance as seen from the right reference plane will not influence the deflection of the spring placed below the left reference plane. Now, balancing masses are added on the left reference plane till there is a reduction in the spring deflection. Then, the carriage is pivoted on the left reference plane and supported by springs on the right reference plane; this is shown in Fig. 9.16. Now the springs deflect only due to the unbalance referenced to the right reference plane. Balancing masses are placed on the right reference plane till the deflection almost reduces to zero. The process is carried out repeatedly till the spring deflection becomes negligible.

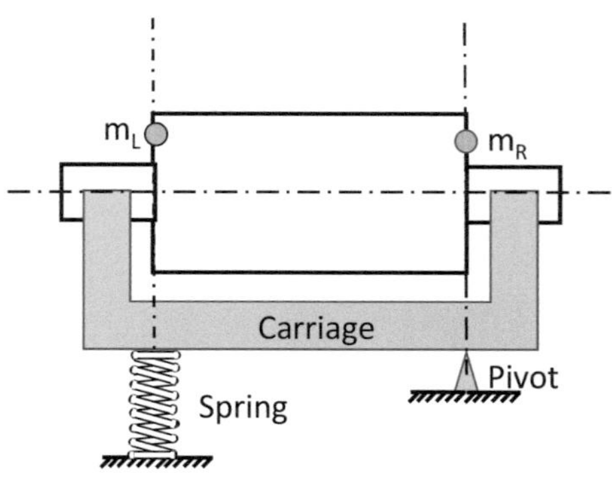

Fig. 9.15 Two-plane balancing machine: pivoted at the right reference plane

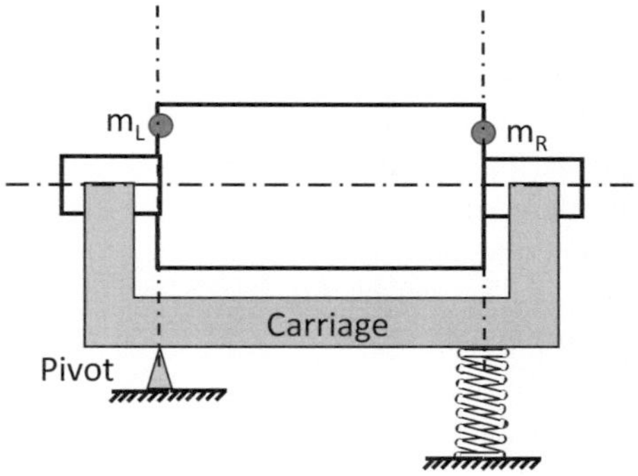

Fig. 9.16 Two-plane balancing machine: pivoted at the left reference plane

The above principles can be extended to a more precise method of balancing long rotors using vibration measurement.

The schematic of a two-plane balancing machine using vibration measurement is shown in Fig. 9.17. It consists of a long rotor mounted on bearings, left (A) and right (B). The balancing planes, left and right, are chosen as shown that will allow easy installation of balancing masses in the vertical plane parallel to the shaft. The masses, m_L and m_R are the unknown unbalancing masses located at unknown locations along the reference planes. Vibration transducers are mounted along the vertical and horizontal directions on each of the bearings as shown.

Following sets of experiments are conducted:

1. The unbalanced rotor is rotated at its rated speed and vibration measurements at both the bearings are taken from all the vibration transducers
2. Trial mass m_L at the left reference plane is placed and the corresponding vibration measurements at both the bearings are taken
3. Trial mass m_R at the right reference plane is placed and the corresponding measurements at both the bearings are taken.

The vibration measurements at the bearings are used to compute the magnitude and phase of vibration at each bearing. The vibration at the bearings will be proportional

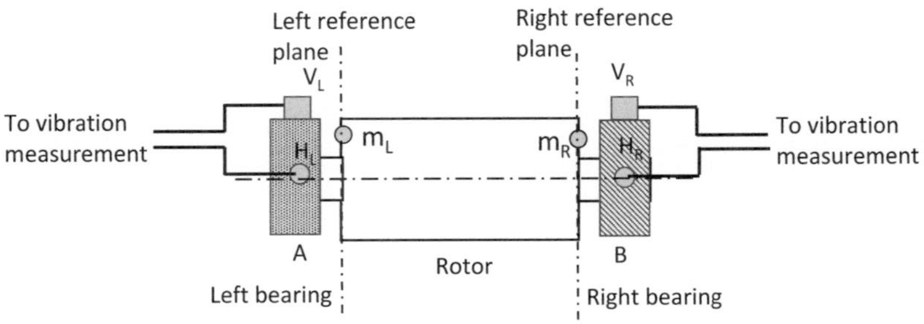

Fig. 9.17 Two-plane balancing machine through vibration measurements

to the extent of unbalance at each of the reference planes. The vibration measurements are used to determine these complex proportionality constants that will help determine the original value of unbalance and location (radial and angular) of the masses in these reference planes. Once the original unbalance is determined, suitable balancing masses are placed in these reference planes to nullify the unbalances, as outlined in the Sect. 9.1.5.

Example 9.4
The following vibration measurements were noted during the balancing of a long rotor as shown in Table 9.4. Derive the relevant equations to determine the unbalance in the reference planes shown in Fig. 9.17.

Let $\mathbf{M}_L = M_L\angle\phi_L$ and $\mathbf{M}_R = M_R\angle\phi_R$ be the original unbalances. Note that all masses, original and trial, are expressed as complex numbers that contain the value of masses and its angular positions.

The increase or decrease of vibration at the bearings due to the trial masses placed in the reference planes can be related to the respective vibration measurements with and without the trial masses.

For example, the vibration in bearing A, $\mathbf{X}_{AL}$, due to the trial mass $\mathbf{m}_L = m_L\angle\psi_L$ in the left reference plane and the vibration in bearing A without the trial mass, $\mathbf{X}_A$, can be related by

$$\mathbf{X}_{AL} = \mathbf{X}_A + \alpha_L\mathbf{m}_L \tag{1}$$

where α_L is a proportionality constant.

Similarly, the vibration in bearing B, $\mathbf{X}_{BL}$, due to the trial mass $\mathbf{m}_L = m_L\angle\psi_L$ in the left reference plane and the vibration in bearing B without the trial mass, $\mathbf{X}_B$, can be related by

$$\mathbf{X}_{BL} = \mathbf{X}_B + \beta_L\mathbf{m}_L \tag{2}$$

where β_L is a proportionality constant.

On similar lines, the following equation can be obtained for vibration in bearing A by placing a trial mass $\mathbf{m}_R = m_R\angle\psi_R$ in the right reference plane

Table 9.4 Vibration measurements from a two-plane balancing machine

Vibration at the bearings		
	Left bearing	Right bearing
	Displacement units/s	Displacement units/s
Without correction masses	$\mathbf{X}_A$	$\mathbf{X}_B$
With correction mass, m_L, left plane	$\mathbf{X}_{AL}$	$\mathbf{X}_{BL}$
With correction mass, m_R right plane	$\mathbf{X}_{AR}$	$\mathbf{X}_{BR}$

$$\mathbf{X}_{AR} = \mathbf{X}_A + \alpha_R \mathbf{m}_R \tag{3}$$

where α_R is a proportionality constant

The following equation can be obtained for vibration in bearing B by placing a trial mass $\mathbf{m}_R = m_R \angle \psi_R$ in the right reference plane B

$$\mathbf{X}_{BR} = \mathbf{X}_B + \beta_R \mathbf{m}_R \tag{4}$$

where β_R is a proportionality constant.

The vibration at the bearings can be expressed in terms of the original unbalances and the proportionality constants defined above.

They are given by

$$\mathbf{X}_A = \boldsymbol{\alpha}_L \mathbf{M}_L + \boldsymbol{\alpha}_R \mathbf{M}_R \tag{5}$$

And

$$\mathbf{X}_B = \boldsymbol{\beta}_L \mathbf{M}_L + \boldsymbol{\beta}_R \mathbf{M}_R \tag{6}$$

Equation (1) through (6) can be solved to determine the unbalances of the long rotor, mass and locations, $\mathbf{M}_L$ and $\mathbf{M}_R$. (See exercise problem 9.20.)

9.3 Balancing of Reciprocating Masses

Consider a reciprocating mechanism shown in Fig. 9.18. It consists of a piston, a connecting rod, and a crank, and the crank is rotating counterclockwise, making an angle θ at the instant shown. The main objective of this section is to derive a simple equation for the dominant shaking force resulting from the above mechanism, which contributes toward unbalanced forces that will be transferred to the foundation. The crank is assumed to be rotating at a constant speed and the approximate acceleration of the piston from Eq. (1.25) that can be obtained from the angular position of crank and its speed derived earlier will be used. In addition, the connecting rod mass is distributed to points B and C and the crank is balanced in such a way that the balancing mass that accounts for the mass of the crank and the part of the connecting rod mass relocated to B moves the mass center of the crank to the pivot point O, as per Eq. (8.56). The mass of the piston and the portion of the connecting rod mass moved to point C is designated as *m*.

The interacting and reaction forces of the reciprocating mechanism are shown in Fig. 9.19. The piston has a normal reaction F_{14}^{y} from the frame and an interacting force $\mathbf{F}_{34}$ with the connecting rod. Since the connecting rod is massless, the interacting forces on the connecting rod are along its length. Hence, F_{23} is equal and opposite to $\mathbf{F}_{43}$. The reaction force on the crank from the connecting rod, $\mathbf{F}_{32,}$ is equal and opposite to $\mathbf{F}_{23}$. The reaction force of the frame on the crank, $\mathbf{F}_{12,}$ is equal and

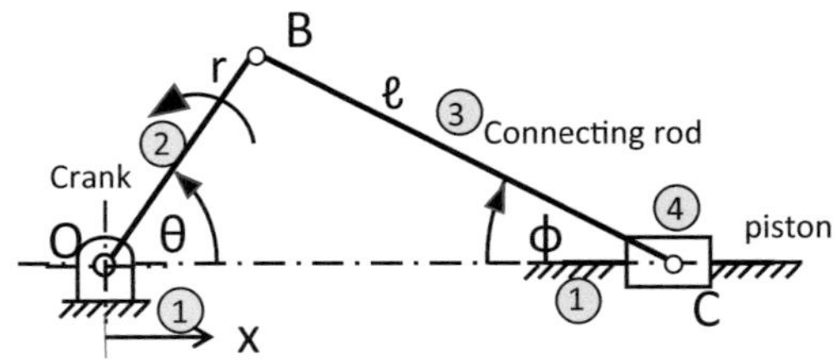

Fig. 9.18 Reciprocating mechanism of an internal combustion engine

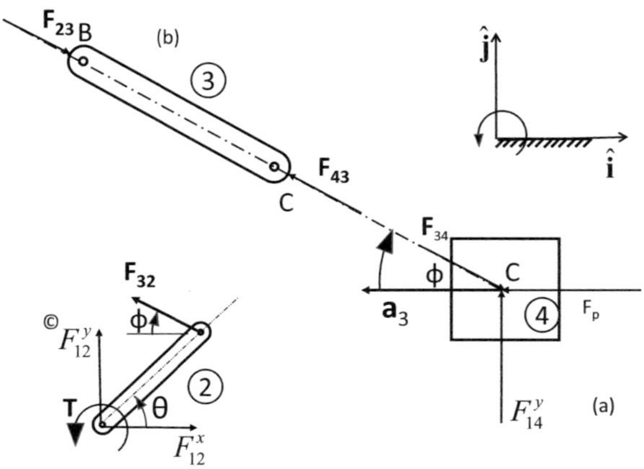

Fig. 9.19 Interacting and reaction forces of the reciprocating mechanism shown in Fig. 9.18

opposite to $\mathbf{F}_{32}$. The piston force F_p will be acting for a very short time and not significant in comparison to the contribution from the inertia of moving parts, and hence neglected in the analysis.

Equating the horizontal forces to the acceleration of the piston in the coordinate system defined in Fig. 9.19,

$$F_{34} = \frac{-ma_3}{\cos\phi} \tag{9.29}$$

Equating the vertical forces on the piston,

$$F_{14}^{y} = F_{34}^{y} = F_{34}\sin\phi = \frac{-ma_3\sin\phi}{\cos\phi} \tag{9.30}$$

The reaction force of the piston on the connecting rod is given by

$$F_{43} = -F_{34} = \frac{ma_3}{\cos\phi} \tag{9.31}$$

The reaction force of the crank on the connecting rod is given by

$$F_{23} = -F_{43} = -\frac{ma_3}{\cos\phi} \tag{9.32}$$

The reaction force of the connecting rod on the crank is given by

$$F_{32} = -F_{23} = \frac{ma_3}{\cos\phi} \tag{9.33}$$

The horizontal reaction force of the frame on the crank is given by

$$F_{12}^x = -F_{32}\cos\phi = -ma_3 \tag{9.34}$$

The vertical reaction force of the frame on the crank is given by

$$F_{12}^y = F_{32}\sin\phi = \frac{ma_3\sin\phi}{\cos\phi} \tag{9.35}$$

Since we are interested in the forces transmitted by the mechanism to the frame, from the above equations the forces acting on the frame are given by

$$F_{41}^y = \frac{ma_3\sin\phi}{\cos\phi} \tag{9.36a}$$

$$F_{21}^x = ma_3 \tag{9.36b}$$

$$F_{21}^y = -\frac{ma_3\sin\phi}{\cos\phi} \tag{9.36c}$$

From Fig. 9.18,

$$\sin\phi = \frac{r}{\ell}\sin\theta \tag{9.37a}$$

$$\cos\phi = \frac{\sqrt{(\ell^2 - r^2\sin^2\theta)}}{\ell} \tag{9.37b}$$

From Eq. (1.25), the approximate piston acceleration is given by

$$a_3 = r\omega^2\left[\cos\theta + \left(\frac{r}{\ell}\right)\cos 2\theta\right] \tag{9.38}$$

From Eqs. (9.36) through (9.38), the shaking forces on the frame at the crank pin are given by

$$F_{21}^{x} = mr\omega^2\left[\cos\theta + \left(\frac{r}{\ell}\right)\cos 2\theta\right] \tag{9.39}$$

and

$$F_{21}^{y} = \frac{-mr^2\omega^2 \sin\theta\left[\cos\theta + \left(\frac{r}{\ell}\right)\cos 2\theta\right]}{\sqrt{\left(\ell^2 - r^2\sin^2\theta\right)}} \tag{9.40}$$

The normal reaction force of the piston on the cylinder is given by

$$F_{41}^{y} = \frac{mr^2\omega^2 \sin\theta\left[\cos\theta + \left(\frac{r}{\ell}\right)\cos 2\theta\right]}{\sqrt{\left(\ell^2 - r^2\sin^2\theta\right)}} \tag{9.41}$$

The above forces are shown in Fig. 9.20. Among the reaction forces on the frame, F_{21}^{x} is the dominant shaking force that contributes toward the unbalanced force that needs to be accounted for. It acts at the crank hinge along the line of piston movement. The other forces, F_{41}^{y} and F_{21}^{y}, that are equal and opposite, form a couple; the amplitude of these forces is much less compared to F_{21}^{x}. This will be demonstrated in Example 9.5.

Although F_{41}^{y} and F_{21}^{y} are much less than F_{21}^{x}, their effect on the engine frame vibration can be reduced if all the rotational unbalance and 50% of the reciprocating unbalance are balanced using a balancing mass at a suitable distance opposite to the crank.

The torque at the crankshaft is given by

$$T = -r\cos\theta F_{12}^{y} + r\sin\theta F_{12}^{x} \tag{9.42}$$

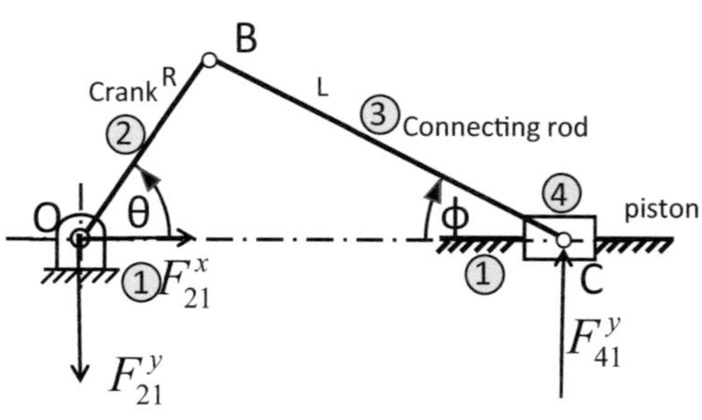

Fig. 9.20 Reaction forces of a reciprocating mechanism on the frame

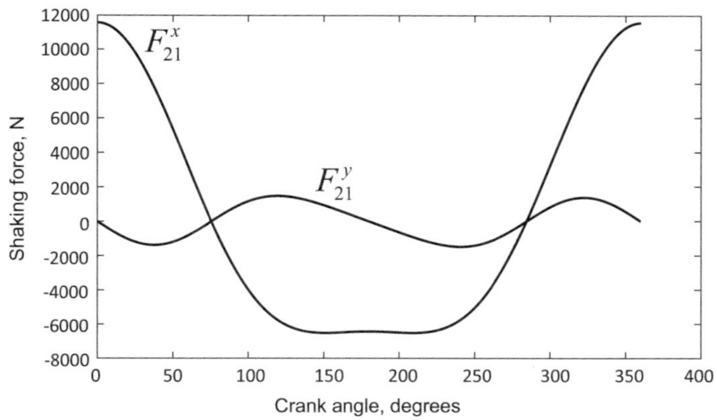

Fig. 9.21 Shaking forces of Example 9.5

Example 9.5

A reciprocating mechanism has a connecting rod of length 350 mm and a crank of length 100 mm. The total mass that has to be accounted for shaking forces and couples (mass of the piston plus the mass of the connecting rod assumed to be located at the piston) is 1 kg. The crank is balanced such that the crank mass and the connecting rod mass assumed to be at the crank pin are balanced such that the mass center of the crank is at the crank hinge. If the crank rotates counterclockwise at 300 rad/s, determine the shaking forces for one complete rotation of the crank. In addition, determine the torque output of the crankshaft. Neglect gas forces as the objective is to only determine the dominant shaking force and nature of variation of the torque at the crankshaft.

By using the above Eqs. (9.39) and (9.41), the shaking forces F_{21}^x and F_{21}^y are plotted as shown in Fig. 9.21. From the figure, it is clear that F_{21}^x is the dominant shaking force.

Using Eq. (9.42), the torque variation as a function of the crank angle is plotted in Fig. 9.22. Although accounting for the gas force can result in slight variation, since the inertial forces play a dominant role in the forces acting on various links of the system, it would be justified to use Eq. (9.42) for computing the torque variation in each cycle. As shown in Fig. 9.22, there will be significant torque variation in each cycle. Such a variation will not result in the smooth operation of the driven machine at a constant angular velocity. Hence, it would be necessary to have multi-cylinder engines that can have minimum torque variation at the crankshaft. With the twin objective of obtaining uniform torque and minimizing the effect of shaking forces, it is required to use multiple cylinders in internal combustion engines. There are many ways in which multiple cylinders can be configured. We shall discuss typical designs that would minimize the effect of shaking forces in this section. Torque variations despite the use of multiple cylinders can be reduced using flywheels.

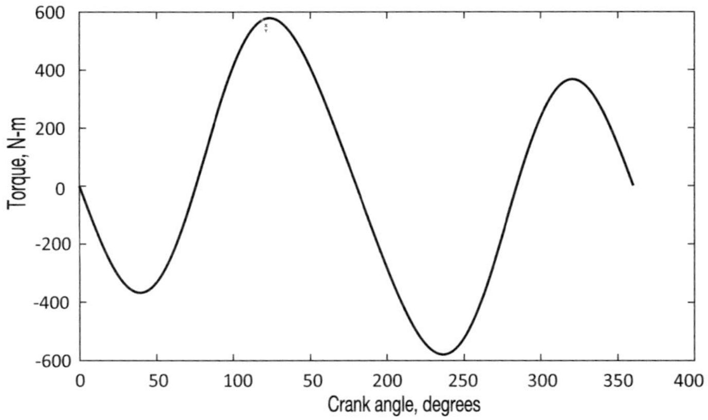

Fig. 9.22 Torque variation of Example 9.5

9.4 Radial Multi-cylinder Engines

Multiple cylinders are used to smoothen the torque output and to minimize the effect of shaking forces that result in the vibration of a foundation on which a reciprocating machine is placed. Multiple cylinders can be obtained in various configurations. In any configuration, the cranks of the several cylinders are on a common crankshaft and they make a definite angle between them. Before we go into the details of such configurations, we should try to obtain necessary equations that can resolve the shaking forces in the coordinate directions.

From Eq. (9.39), the equation for the dominant shaking force that acts along the direction of piston movement is given by

$$F_s = mr\omega^2\left[\cos\theta + \left(\frac{r}{\ell}\right)\cos 2\theta\right] \tag{9.43}$$

The angle θ in Eq. (9.43) is with respect to the angle of the crank measured from the line of action of the shaking force, which is also incidentally in the horizontal direction. But in the case of multiple cylinders, since the orientation of the cylinders is at different angles and the cranks have a definite angular relationship between them, necessary changes will have to be made regarding this angle θ. Then the shaking forces in the coordinate directions will have to be separately equal to zero, if there were to be no shaking force transmitted to the foundation. In addition, the moments due to the shaking forces must also be equal to zero.

For example, consider a radial arrangement of three engines connected to a common crankshaft in the same plane. Let ϕ_1, ϕ_2, and ϕ_3 are respectively the angles made by the lines of movement of the pistons of each cylinder with respect to the horizontal direction; they are also known as banking angles. It is assumed that the crank angles are measured with respect to the first crank, as various cranks of a multiple cylinder engine will have a constant phase difference; Hence, $\psi_1 = 0$, ψ_2 is

Fig. 9.23 Radial view of multiple cylinder engines in the same plane

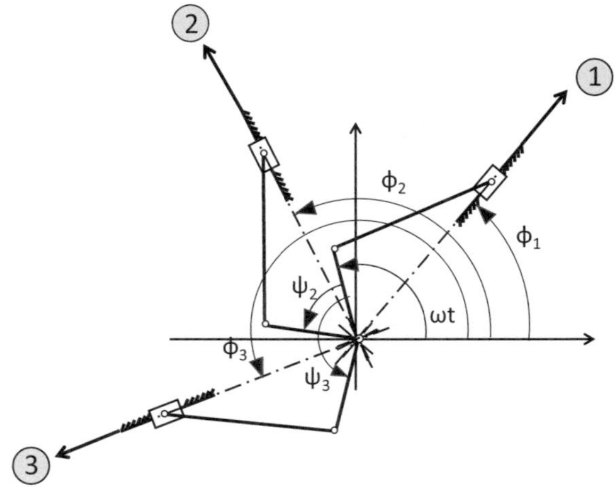

the angle between the second and first crank, ψ_3 is the angle between the third and the first crank and so on. It is assumed that the first crank makes angle of $\theta_1 = \omega t$ with respect to the horizontal direction. However, the direction of the first shaking force is at angle of $\omega t\text{-}\phi_1$ with respect to the first crank. Similarly, the second shaking force is at an angle of $\omega t + \psi_2 - \phi_2$ with respect to the second crank and the third shaking force is at an angle of $\omega t + \psi_3 - \phi_3$. The same logic can be extended to any number of engines.

Extending Eq. (9.43) to a multiple cylinder system of n cylinders, similar to the one shown in Fig. 9.23, the shaking forces can be defined as follows.

$$
\begin{aligned}
F_{s_1} &= mr\omega^2\left[\cos(\omega t - \phi_1) + \left(\frac{r}{\ell}\right)\cos 2(\omega t - \phi_1)\right]\\
F_{s_2} &= mr\omega^2\left[\cos(\omega t + \psi_2 - \phi_2) + \left(\frac{r}{\ell}\right)\cos 2(\omega t + \psi_2 - \phi_2)\right]\\
F_{s_3} &= mr\omega^2\left[\cos(\omega t + \psi_3 - \phi_3) + \left(\frac{r}{\ell}\right)\cos 2(\omega t + \psi_3 - \phi_3)\right]\\
&\vdots\\
F_{s_n} &= mr\omega^2\left[\cos(\omega t + \psi_n - \phi) + \left(\frac{r}{\ell}\right)\cos 2(\omega t + \psi_n - \phi)\right]
\end{aligned}
\tag{9.44}
$$

9.5 Inline Engines

In this type of configuration, the cylinders are arranged in a row separated, by the same distance *a* and the axes of the cylinders are parallel to each other. Hence, the line of action of the shaking forces will be parallel to each other for all the cylinders, but their magnitude will depend on the position of their cranks with respect to each other; all the cranks will be connected to a common crankshaft.

Since all the cylinders have the line of action of the shaking forces parallel to each other, $\phi_1 = \phi_2 = \phi_3 = 0 \ldots \phi_n = \phi$. Hence the shaking forces of Eq. (9.44) become

$$
\begin{aligned}
F_{s_1} &= mr\omega^2\left[\cos(\omega t - \phi) + \left(\frac{r}{\ell}\right)\cos 2(\omega t - \phi)\right] \\
F_{s_2} &= mr\omega^2\left[\cos(\omega t + \psi_2 - \phi) + \left(\frac{r}{\ell}\right)\cos 2(\omega t + \psi_2 - \phi)\right] \\
F_{s_3} &= mr\omega^2\left[\cos(\omega t + \psi_3 - \phi) + \left(\frac{r}{\ell}\right)\cos 2(\omega t + \psi_3 - \phi)\right] \\
&\vdots \\
F_{s_n} &= mr\omega^2\left[\cos(\omega t + \psi_n - \phi) + \left(\frac{r}{\ell}\right)\cos 2(\omega t + \psi_n - \phi)\right]
\end{aligned}
\tag{9.45}
$$

The net shaking force from all the cylinders in the coordinate directions will be

$$
\mathbf{F}_s = \sum_{n=1}^{N}\left(F_{s_n}\cos\phi\hat{\boldsymbol{i}} + F_{s_n}\sin\phi\hat{\boldsymbol{j}}\right) \tag{9.46}
$$

From Eqs. (9.45) and (9.46)

$$
\begin{aligned}
\mathbf{F}_s &= \sum_{n=1}^{N}\left(F_{s_n}\cos\phi\hat{\boldsymbol{i}} + F_{s_n}\sin\phi\hat{\boldsymbol{j}}\right) \\
&= mr\omega^2\sum_{n=1}^{N}[\cos(\omega t + \psi_n - \phi)\cos\phi]\hat{\boldsymbol{i}} \\
&\quad + mr\omega^2\left(\frac{r}{\ell}\right)\sum_{n=1}^{N}\{\cos[2(\omega t + \psi_n - \phi)]\cos\phi\}\hat{\boldsymbol{i}} \\
&\quad + mr\omega^2\sum_{n=1}^{N}[\cos(\omega t + \psi_n - \phi)\sin\phi]\hat{\boldsymbol{j}} \\
&\quad + mr\omega^2\left(\frac{r}{\ell}\right)\sum_{n=1}^{N}\{\cos[2(\omega t + \psi_n - \phi)]\sin\phi\}\hat{\boldsymbol{j}}
\end{aligned}
\tag{9.47}
$$

If the net shaking force has to be zero, each of the terms of Eq. (9.47) must be equal to zero. By using $\cos(A + B) = \cos A\cos B - \sin A\sin B$, the following terms of Eq. (9.47), by grouping ωt and $\psi_n - \phi$ separately, can be simplified as

$$
\begin{aligned}
\cos(\omega t + \psi_n - \phi) &= \cos\omega t\cos(\psi_n - \phi) - \sin\omega t\sin(\psi_n - \phi) \\
\cos 2(\omega t + \psi_n - \phi) &= \cos 2\omega t\cos 2(\psi_n - \phi) - \sin 2\omega t\sin 2(\psi_n - \phi)
\end{aligned}
\tag{9.48}
$$

Using Eq. (9.48), Eq. (9.47) becomes

$$
\begin{aligned}
\mathbf{F}_s &= \sum_{n=1}^{N} \left(F_{s_n} \cos\phi \hat{\boldsymbol{i}} + F_{s_n} \sin\phi \hat{\boldsymbol{j}}\right) \\
&= mr\omega^2 \sum_{n=1}^{N} [\cos\omega t \cos(\psi_n - \phi)\ \cos\phi - \sin\omega t \sin(\psi_n - \phi) \cos\phi]\hat{\boldsymbol{i}} + \\
&\quad + mr\omega^2 \left(\frac{r}{\ell}\right) \sum_{n=1}^{N} \{\cos 2\omega t \cos 2(\psi_n - \phi)\ \cos\phi - \sin 2\omega t \sin 2(\psi_n \ - \phi) \cos\phi\}\hat{\boldsymbol{i}} \\
&\quad + mr\omega^2 \sum_{n=1}^{N} [\cos\omega t \cos(\psi_n - \phi) \sin\phi\ - \sin\omega t \sin(\psi_n - \phi) \sin\phi]\hat{\boldsymbol{j}} \\
&\quad + mr\omega^2 \left(\frac{r}{\ell}\right) \sum_{n=1}^{N} \{\cos 2\omega t \cos 2(\psi_n - \phi)\ \sin\phi - \sin 2\omega t \sin 2(\psi_n \ - \phi) \sin\phi\}\hat{\boldsymbol{j}}
\end{aligned}
\tag{9.49}
$$

If the shaking forces are to be zero, all the individual terms of Eq. (9.49) must be equal to zero. Taking out term $mr\omega^2$ out of the summation of Eq. (9.49) results in the following equations.

From the terms (1) and (3) of Eq. (9.49)

$$
\begin{aligned}
&\sum_{n=1}^{N} [\cos(\psi_n - \phi) \cos\phi] = 0 \\
&\sum_{n=1}^{N} \{\cos[2(\psi_n - \phi)] \cos\phi\} = 0
\end{aligned}
\tag{9.50}
$$

From terms (5) and (7) of Eq. (9.49)

$$
\begin{aligned}
&\sum_{n=1}^{N} [\cos(\psi_n - \phi) \sin\phi] = 0 \\
&\sum_{n=1}^{N} \{\cos[2(\psi_n - \phi)] \sin\phi\} = 0
\end{aligned}
\tag{9.51}
$$

From terms (2) and (4) of Eq. (9.49)

$$\sum_{n=1}^{N} [\sin(\psi_n - \phi) \cos\phi] = 0$$
$$\sum_{n=1}^{N} \{\sin[2(\psi_n - \phi)] \cos\phi\} = 0 \tag{9.52}$$

From terms (6) and (8) of Eq. (9.49)

$$\sum_{n=1}^{N} [\sin(\psi_n - \phi) \sin\phi] = 0$$
$$\sum_{n=1}^{N} \{\sin[2(\psi_n - \phi)] \sin\phi\} = 0 \tag{9.53}$$

If the moments of the shaking forces are to be equal to zero, the above equations can be easily extended to the following equations by accounting the distance a_n of each cylinder from a reference plane.

From Eq. (9.50)

$$\sum_{n=1}^{N} [a_n \cos(\psi_n - \phi) \cos\phi] = 0$$
$$\sum_{n=1}^{N} \{a_n \cos[2(\psi_n - \phi)] \cos\phi\} = 0 \tag{9.54}$$

From Eq. (9.51)

$$\sum_{n=1}^{N} a_n[\cos(\psi_n - \phi) \sin\phi] = 0$$
$$\sum_{n=1}^{N} a_n\{\cos[2(\psi_n - \phi)] \sin\phi\} = 0 \tag{9.55}$$

From Eq. (9.52)

$$\sum_{n=1}^{N} a_n[\sin(\psi_n - \phi) \cos\phi] = 0$$
$$\sum_{n=1}^{N} a_n\{\sin[2(\psi_n - \phi)] \cos\phi\} = 0 \tag{9.56}$$

From Eq. (9.53)

$$\sum_{n=1}^{N} a_n[\sin(\psi_n - \phi)\sin\phi] = 0$$
$$\sum_{n=1}^{N} a_n\{\sin[2(\psi_n - \phi)]\sin\phi\} = 0 \quad (9.57)$$

The above equations can be used to check the shaking forces from multi-cylinder inline engines.

9.5.1 Four-Cylinder Inline Engine

A four-cylinder inline engine is shown in Fig. 9.24. It consists of four cylinders whose axes are aligned with the vertical direction; they are equally spaced at a distance of *a* from each other. The cranks of cylinders 1 and 4 are in phase with each other and those of cylinders 2 and 3 are out of phase by 180°, with respect to the first and fourth cylinders.

The angle of the axes of all the cylinders is aligned to the vertical directions and hence it can be assumed that the line of action of the shaking force is in the vertical direction and the angle ϕ can be taken as zero for all the cylinders. From the given

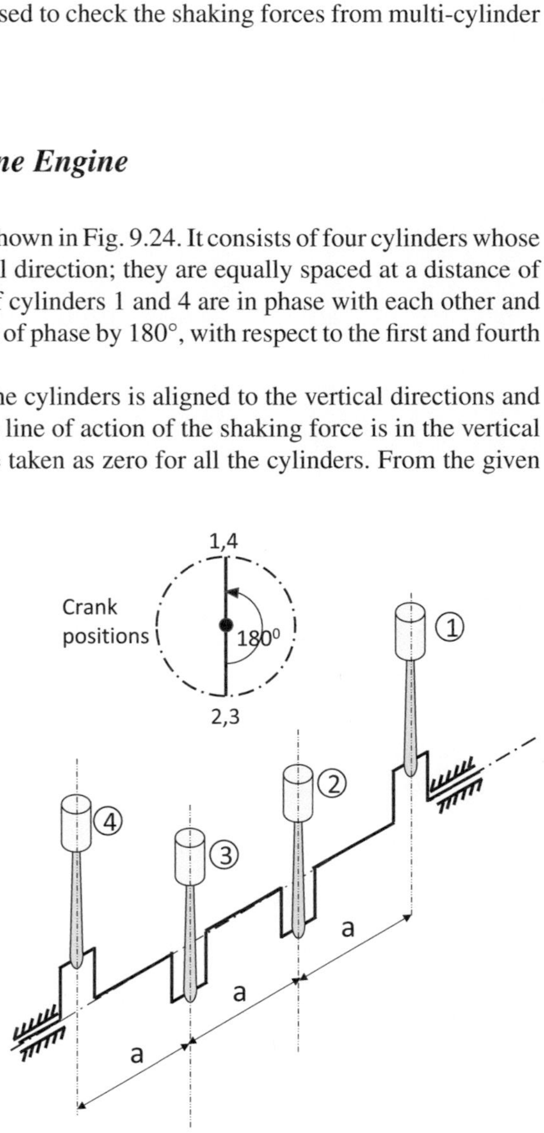

Fig. 9.24 Four-cylinder inline engine

configuration, it can be assumed that $\psi_1 = \psi_4 = 0$ and $\psi_2 = \psi_3 = 180°$. Assuming that the moments of the shaking force are taken around the plane of the first cylinder, $a_1 = 0$, $a_2 = a$, $a_3 = 2a$, and $a_4 = 3a$. It is assumed that the cylinders are symmetrical in construction and hence their reciprocating mass, dimensions of the crank, and connecting rod are the same.

Shaking forces

From Eq. (9.50), sum of the trigonometric entities of primary forces is given by

$$\begin{aligned}&\sum_{n=1}^{N}[\cos(\psi_n - \phi)\cos\phi] = 0\\&\Rightarrow \cos(\psi_1) + \cos(\psi_2) + \cos(\psi_3) + \cos(\psi_4)\\&= 1 - 1 - 1 + 1 = 0\end{aligned} \tag{9.58}$$

From Eq. (9.50), sum of the trigonometric entities of secondary forces is given by

$$\begin{aligned}&\sum_{n=1}^{N}\{\cos[2(\psi_n - \phi)]\cos\phi\} = 0\\&\Rightarrow \cos[2(\psi_1)] + \cos[2(\psi_2)] + \cos[2(\psi_3)] + \cos[2(\psi_4)]\\&= 1 + 1 + 1 + 1 = 4\end{aligned} \tag{9.59}$$

Equation (9.51) is zero for primary forces, but non-zero secondary forces for all four-cylinder inline engines.

From Eq. (9.52), sum of the trigonometric entities of primary forces is given by

$$\begin{aligned}&\sum_{n=1}^{N}[\sin(\psi_n)] = 0\\&\Rightarrow \sin(\psi_1) + \sin(\psi_2) + \sin(\psi_3) + \sin(\psi_4)\\&= 0 + 0 + 0 + 0 = 0\end{aligned} \tag{9.60}$$

From Eq. (9.52), sum of the trigonometric entities of the secondary forces is given by

$$\begin{aligned}&\sum_{n=1}^{N}\{\sin[2(\psi_n)]\} = 0\\&\sin 2\psi_1 + \sin 2\psi_2 + \sin 2\psi_3 + \sin 2\psi_4 = 0\end{aligned} \tag{9.61}$$

From Eq. (9.53), sum of the trigonometric entities of primary and secondary forces is zero for all inline engines.

Shaking moments

From Eq. (9.54), sum of the trigonometric entities of the primary couples is given by

$$\sum_{n=1}^{N} a_n \cos(\psi_n) = 0$$
$$\Rightarrow a_1 \cos \psi_1 + a_2 \cos \psi_2 + a_3 \cos \psi_3 + a_4 \cos \psi_4$$
$$= 0 - a - 2a + 3a = 0 \tag{9.62}$$

From Eq. (9.54), sum of the trigonometric entities of the secondary couples is given by

$$\sum_{n=1}^{N} \{a_n \cos[2(\psi_n)]\}$$
$$a_1 \cos 2\psi_1 + a_2 \cos 2\psi_2 + a_3 \cos 2\psi_3 + a_4 \cos 2\psi_4$$
$$= 0 + a + 2a + 3a = 6a \tag{9.63}$$

If the center vertical plane of the four-cylinder inline engine is taken as the reference, the above equation will be zero.

From Eq. (9.55), sum of the trigonometric entities of primary and secondary couples is zero for all inline engines.

From Eq. (9.56), sum of the trigonometric entities of the primary couples is given by

$$\sum_{n=1}^{N} a_n \sin(\psi_n) = 0$$
$$\Rightarrow a_1 \sin \psi_1 + a_2 \sin \psi_2 + a_3 \sin \psi_3 + a_4 \sin \psi_4 = 0 \tag{9.64}$$

From Eq. (9.56), sum of the trigonometric entities of the secondary couples is given by

$$\sum_{n=1}^{N} a_n \sin(2\psi_n) = 0$$
$$\Rightarrow a_1 \sin 2\psi_1 + a_2 \sin 2\psi_2 + a_3 \sin 2\psi_3 + a_4 \sin 2\psi_4 = 0 \tag{9.65}$$

There is only one equation corresponding to the secondary forces of Eq. (9.59) that is non-zero. Hence, a four-cylinder inline engine does minimize the action of shaking forces and moments, except for one set of secondary forces of Eq. (9.59). Therefore, it is not a perfect system for minimizing shaking forces. But since secondary forces and moments have a multiplication factor of r/ℓ, their contribution is much less than the primary forces and moments. Despite that it does not balance all the secondary forces, a four-cylinder inline configuration is one of the most popular configurations used in automobile engines.

9.5.2 Six-Cylinder Inline Engine

A six-cylinder engine is shown in Fig. 9.25. It consists of six cylinders whose axes are aligned with the vertical direction; they are equally spaced at a distance of a from each other.

The angle of the axes is aligned to the vertical direction and hence it can be assumed that the line of action of the shaking force is in the vertical direction and the angle ϕ can be taken as zero for all the cylinders. From the given configuration, it can be assumed that $\psi_1 = \psi_6 = 0$ and $\psi_3 = \psi_4 = -120°$, and $\psi_2 = \psi_5 = 120°$ Assuming that the moments of the shaking force are taken around the plane of the first cylinder, $a_1 = 0$, $a_2 = a$, $a_3 = 2a$, $a_4 = 3a$, $a_5 = 4a$, and $a_6 = 5a$. It is assumed that the cylinders are symmetrical in construction and hence their reciprocating mass, dimensions of the crank, and connecting rod are the same.

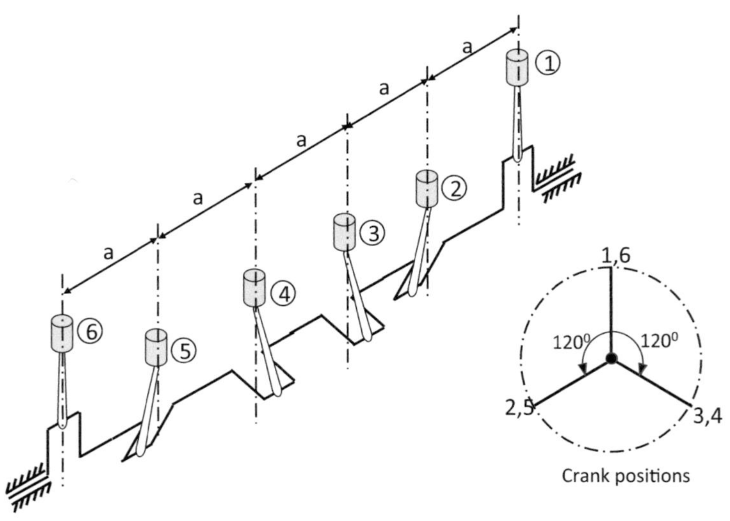

Fig. 9.25 Six-cylinder inline engine

Shaking forces

From Eq. (9.50), sum of the trigonometric entities of primary forces is given by

$$\sum_{n=1}^{N}[\cos(\psi_n)] = 0$$
$$\Rightarrow \cos\psi_1 + \cos\psi_2 + \cos\psi_3 + \cos\psi_4 + \cos\psi_5 + \cos\psi_6$$
$$= 1 - 0.5 - 0.5 - 0.5 - 0.5 + 1 = 0 \quad (9.66)$$

From Eq. (9.50), sum of the trigonometric entities of secondary forces is given by

$$\sum_{n=1}^{N}\{\cos[2(\psi_n)]\} = 0$$
$$\Rightarrow \cos 2\psi_1 + \cos 2\psi_2 + \cos 2\psi_3 + \cos 2\psi_4 + \cos 2\psi_5 + \cos 2\psi_6$$
$$= 1 - 0.5 - 0.5 - 0.5 - 0.5 + 1 = 0 \quad (9.67)$$

Equation (9.51) is zero for primary and secondary forces for all inline engines.
From Eq. (9.52), sum of the trigonometric entities of primary forces is given by

$$\sum_{n=1}^{N}[\sin(\psi_n)] = 0$$
$$\sin\psi_1 + \sin\psi_2 + \sin\psi_3 + \sin\psi_4 + \sin\psi_5 + \sin\psi_6$$
$$= 0 + 0.87 - 0.87 - 0.87 + 0.87 + 0 = 0 \quad (9.68)$$

From Eq. (9.52), sum of the trigonometric entities of the secondary forces is given by

$$\sum_{n=1}^{N}[\sin(2\psi_n)] = 0$$
$$\sin 2\psi_1 + \sin 2\psi_2 + \sin 2\psi_3 + \sin 2\psi_4 + \sin 2\psi_5 + \sin 2\psi_6$$
$$= 0 - 0.87 + 0.87 + 0.87 - 0.87 + 0 = 0 \quad (9.69)$$

From Eq. (9.53), sum of the trigonometric entities of primary and secondary forces is zero for all inline engines.

Shaking moments

From Eq. (9.54), sum of the trigonometric entities of the primary couples is given by

$$\sum_{n=1}^{N} [a_n \cos(\psi_n)] = 0$$

$$a_1 \cos \psi_1 + a_2 \cos \psi_2 + a_3 \cos \psi_3 + a_4 \cos \psi_4 + a_5 \cos \psi_5 + a_6 \cos \psi_6$$
$$= 0 - 0.5a - a - 1.5a - 2a + 5a = 0 \quad (9.70)$$

From Eq. (9.54), sum of the trigonometric entities of the secondary couples is given by

$$\sum_{n=1}^{N} [a_n \cos(2\psi_n)] = 0$$

$$a_1 \cos 2\psi_1 + a_2 \cos 2\psi_2 + a_3 \cos 2\psi_3 + a_4 \cos 2\psi_4 + a_5 \cos 2\psi_5 + a_6 \cos 2\psi_6$$
$$= 0 - 0.5a - a - 1.5a - 2a + 5a = 0 \quad (9.71)$$

From Eq. (9.55), sum of the trigonometric entities of primary and secondary couples is zero for all inline engines.

From Eq. (9.56), sum of the trigonometric entities of the primary couples is given by

$$\sum_{n=1}^{N} [a_n \sin(\psi_n)] = 0$$

$$a_1 \sin \psi_1 + a_2 \sin \psi_2 + a_3 \sin \psi_3 + a_4 \sin \psi_4 + a_5 \sin \psi_5 + a_6 \sin \psi_6$$
$$= 0 + a0.87 - 2a \times 0.87 - 3a \times 0.87 + 4a \times 0.87 + 0 = 0 \quad (9.72)$$

From Eq. (9.56), sum of the trigonometric entities of the secondary couples is given by

$$\sum_{n=1}^{N} [a_n \sin(2\psi_n)] = 0$$

$$a_1 \sin 2\psi_1 + a_2 \sin 2\psi_2 + a_3 \sin 2\psi_3 + a_4 \sin 2\psi_4 + a_5 \sin 2\psi_5 + a_6 \sin 2\psi_6$$
$$= 0 - 0.87a + 2 \times 0.87a + 3 \times 0.87a - 4 \times 0.87a + 0 = 0 \quad (9.73)$$

From Eqs. (9.66) to (9.73), it has been proved that a six-cylinder inline engine is perfectly balanced for primary and secondary shaking forces and moments. Perhaps that might have been the reason why some small propeller aircrafts had six-cylinder inline engines.

9.5.3 Eight-Cylinder Inline Engine

An eight-cylinder inline engine is shown in Fig. 9.26. It consists of eight cylinders whose axes are aligned with the vertical direction; they are equally spaced at a distance of *a* from each other. The above eight-cylinder combination can be presumed as a combination of two four-cylinder inline engines with 1–2–7–8 forming the first set and 3–4–5–6 the second set of four-cylinder inline engines. The angle of the axes is aligned to the vertical directions and hence it can be assumed that the line of action of the shaking force is in the vertical direction and the angle ϕ can be taken as zero for all the cylinders. From the given configuration, it can be assumed that $\psi_1 = \psi_8 = 0$ and $\psi_3 = \psi_6 = -90°$, $\psi_2 = \psi_7 = -180°$, and $\psi_4 = \psi_5 = -270°$. Assuming that the moments of the shaking force are taken around the plane of the center of the cylinders as shown in Fig. 9.26, a1 = −3.5a, $a_2 = -2.5a$, $a_3 = -1.5a$, $a_4 = -0.5a$, $a_5 = 0.5a$, $a_6 = 1.5a$, $a_7 = 2.5a$, and $a_8 = 3.5a$. It is assumed that the cylinders are symmetrical in construction and hence the reciprocating mass, dimensions of the crank, and connecting rod are the same for all cylinders.

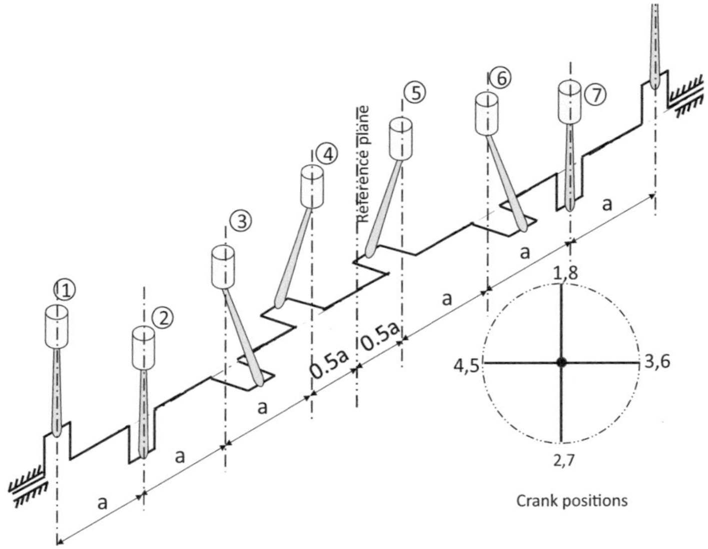

Fig. 9.26 Eight-cylinder inline engine

Shaking forces

From Eq. (9.50), sum of the trigonometric entities of primary forces is given by

$$\sum_{n=1}^{8}[\cos(\psi_n)]$$
$$\Rightarrow \cos\psi_1 + \cos\psi_2 + \cos\psi_3 + \cos\psi_4 + \cos\psi_5 + \cos\psi_6 + \cos\psi_7 + \cos\psi_8$$
$$= 1 - 1 + 0 + 0 + 0 + 0 - 1 + 1 = 0 \tag{9.74}$$

From Eq. (9.50), sum of the trigonometric entities of secondary forces is given by

$$\sum_{n=1}^{8}\{\cos[2(\psi_n)]\} = 0$$
$$\Rightarrow \cos 2\psi_1 + \cos 2\psi_2 + \cos 2\psi_3 + \cos 2\psi_4 + \cos 2\psi_5 + \cos 2\psi_6$$
$$= 1 + 1 - 1 - 1 - 1 - 1 + 1 + 1 = 0 \tag{9.75}$$

Equation (9.51) is zero for primary and secondary forces for all inline engines. From Eq. (9.52), sum of the trigonometric entities of primary forces is given by

$$\sum_{n=1}^{8}[\sin(\psi_n)] = 0$$
$$\sin\psi_1 + \sin\psi_2 + \sin\psi_3 + \sin\psi_4 + \sin\psi_5 + \sin\psi_6$$
$$= 0 + 0 - 1 + 1 + 1 - 1 + 0 + 0 = 0 \tag{9.76}$$

From Eq. (9.52), sum of the trigonometric entities of the secondary forces is given by

$$\sum_{n=1}^{N}[\sin(2\psi_n)] = 0$$
$$\sin 2\psi_1 + \sin 2\psi_2 + \sin 2\psi_3 + \sin 2\psi_4 + \sin 2\psi_5 + \sin 2\psi_6$$
$$= 0 + 0 + 0 + 0 + 0 + 0 + 0 + 0 = 0 \tag{9.77}$$

From Eq. (9.53), sum of the trigonometric entities of primary and secondary forces is zero.

From Eqs. (9.74) to (9.77), primary and secondary forces for all eight-cylinder inline engines is zero.

Shaking moments

From Eq. (9.54), sum of the trigonometric entities of the primary couples is given by

$$\sum_{n=1}^{N}[a_n \cos(\psi_n)] = 0$$

$$a_1 \cos\psi_1 + a_2 \cos\psi_2 + a_3 \cos\psi_3 + a_4 \cos\psi_4 + a_5 \cos\psi_5 + a_6 \cos\psi_6$$
$$= -\frac{7a}{2} + \frac{5a}{2} + 0 + 0 + 0 + 0 - \frac{5a}{2} + \frac{7a}{2} = 0 \quad (9.78)$$

From Eq. (9.54), sum of the trigonometric entities of the secondary couples is given by

$$\sum_{n=1}^{N}[a_n \cos(2\psi_n)] = 0$$

$$a_1 \cos 2\psi_1 + a_2 \cos 2\psi_2 + a_3 \cos 2\psi_3 + a_4 \cos 2\psi_4 + a_5 \cos 2\psi_5 + a_6 \cos 2\psi_6$$
$$= -\frac{7a}{2} - \frac{5a}{2} + \frac{3a}{2} + \frac{a}{2} - \frac{a}{2} - \frac{3a}{2} + \frac{5a}{2} + \frac{7a}{2} = 0 \quad (9.79)$$

From Eq. (9.55), sum of the trigonometric entities of primary and secondary couples is zero for all inline engines.

From Eq. (9.56), sum of the trigonometric entities of the primary couples is given by

$$\sum_{n=1}^{8}[a_n \sin(\psi_n)] = 0$$

$$a_1 \sin\psi_1 + a_2 \sin\psi_2 + a_3 \sin\psi_3 + a_4 \sin\psi_4 + a_5 \sin\psi_5 + a_6 \sin\psi_6$$
$$= 0 + 0 + \frac{3a}{2} - \frac{a}{2} + \frac{a}{2} - \frac{3a}{2} + 0 + 0 = 0 \quad (9.80)$$

From Eq. (9.56), sum of the trigonometric entities of the secondary couples is given by

$$\sum_{n=1}^{8}[a_n \sin(2\psi_n)] = 0$$

$$a_1 \sin 2\psi_1 + a_2 \sin 2\psi_2 + a_3 \sin 2\psi_3 + a_4 \sin 2\psi_4 + a_5 \sin 2\psi_5 + a_6 \sin 2\psi_6$$
$$= 0 + 0 + 0 + 0 + 0 + 0 + 0 + 0 \quad (9.81)$$

From the above equations, it is proved that an eight-cylinder inline engine is balanced for primary and secondary forces and couples.

9.6 Conclusions

Balancing is one of the most practical and cost-effective possibilities for reducing vibration from machinery, especially in rotating machinery. Hence, most practical engineers ensure that rotary balancing is carried out to the extent possible. There could be other reasons for machinery vibration that may not be simple to solve, but balancing is considered as a low-hanging fruit. Apart from rotary machines, reciprocating machines can also be balanced by choosing a proper combination of multiple cylinder engines; in addition to balancing, that will also ensure minimum torque variation.

9.7 Exercise Problems

9.1 A thin rotor has the following unbalanced masses in a single plane, at various angles and radii as shown in Table P9.1. Determine the mass, m_4, that has to be placed at a radius of 100 mm and the corresponding angle. Angles are measured with respect to the radius of mass, m_1.

9.2 A thin rotor has the following unbalanced masses in a single plane, at various angles and radii as shown in Table P9.2. Determine the mass, m_5, that has to be placed at a radius of 140 mm and the corresponding angle. Angles are measured with respect to the radius of mass, m_1.

9.3 A thin rotor has the following unbalanced masses in a single plane, at various angles and radii as shown in Table P9.3. Determine the masses, m_4 and m_5, that have to be placed at a radius of 100 mm and the given angles. Angles are measured with respect to the radius of mass, m_1.

Table P9.1 Details of unbalance masses, and their locations of exercise problem 9.1

No	Mass (kg)	Radius (mm)	Angle (°)
1	10	120	0
2	12	135	75
3	25	80	140
4	?	100	?

Table P9.2 Details of unbalance masses, and their locations of exercise problem 9.2

No	Mass (kg)	Radius (mm)	Angle (°)
1	5	120	0
2	3	135	75
3	12	80	140
4	9	100	265
5	?	140	?

Table P9.3 Details of unbalance masses, and their locations of exercise problem 9.3

No	Mass (kg)	Radius (mm)	Angle (°)
1	6	200	0
2	4	200	70
3	5	200	135
4	?	200	220
5	?	200	330

9.4 Three unbalanced masses m_1, m_2, and m_3 are rotating along different vertical planes of a horizontal shaft at different radii and angular positions as shown in Table P9.4. Two balancing masses, m_4 and m_5 are placed along two different vertical planes to balance the above masses. The axial locations of the planes of all the masses are shown with respect to the mass m_4 and the angular positions are measured with respect to the mass m_1. Determine the unbalanced masses and their angular positions.

9.5 Three unbalanced masses m_1, m_2, and m_3 are rotating along different vertical planes of a horizontal shaft at different radii and angular positions as shown in Table P9.5. Two balancing masses, m_4 and m_5 are placed along two different vertical planes to balance the above masses. The axial locations of the planes of all the masses are shown with respect to the mass m_4 and the angular positions are measured with respect to the mass m_1. Determine the unbalanced masses and their angular positions.

Table P9.4 Details of unbalance masses, and their locations of exercise problem 9.4

No	Mass (kg)	Radius (mm)	Angle (°)	Axial location (mm)
1	8	80	0	−450
2	6	100	65	450
3	14	65	140	2000
4	?	75	?	0
5	?	75	?	1000

Table P9.5 Details of unbalance masses, and their locations of exercise problem 9.5

No	Mass (kg)	Radius (mm)	Angle (°)	Axial location (mm)
1	9	50	0	−400
2	8	50	70	400
3	10	50	135	1800
4	?	150	?	0
5	?	150	?	900

9.6 Four masses , of equivalent unbalance ,m_1, m_2, m_3, and m_4 are rotating along different vertical planes of a horizontal shaft at different radii and angular positions as shown in Table P9.6. Determine the unknown values so there is a complete balance. The angular and axial values are with respect to mass m_1.

9.7 A three-cylinder engine has three cranks that are 120° apart and are in three different vertical planes on a horizontal shaft that are 0.5 m apart as shown in Fig. P9.7. Each of these cranks has an unbalanced mass of 150 kg at a radius of 200 mm. The unbalance action of these cranks in different planes is balanced by placing balancing masses in two vertical planes, whose locations with respect to the crank are as shown; the balancing masses are also placed at a radius of 200 mm. The above values are summarized in Table P9.7. Determine the magnitude of the balancing masses and their angular positions. The right plane is considered as the reference plane.

9.8 Four masses, m_1, m_2, m_3, and m_4 are rotating in four vertical planes of a horizontal shaft that are equally spaced at 100 mm and all the masses are located at the same radius of 75 mm. Only the magnitude of the second mass, m_2, is known and the angular positions of the third and fourth masses, m_3 and m_4, with respect to the second mass, m_2, are given. The above values are summarized in Table P9.8. Determine the angular position of m_1 with respect to m_2 and the magnitude of m_1, m_3, and m_4 so there is complete dynamic balance.

Table P9.6 Details of unbalance masses, and their locations of exercise problem 9.6

No	Mass (kg)	Radius (mm)	Angle (°)	Axial location (mm)
1	85	50	0	0
2	?	50	?	250
3	180	50	?	900
4	100	50	?	1250

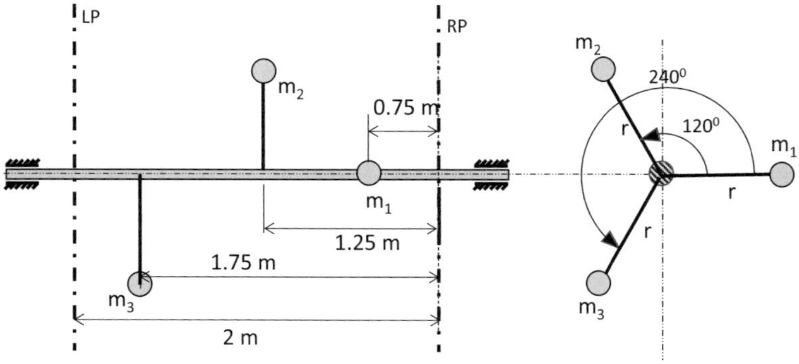

Fig. P9.7 Balancing masses in two planes

Table P9.7 Details of unbalance masses, and their locations of exercise problem 9.7

No	Mass (kg)	Radius (mm)	Angle (°)	Axial location (m)
1	150	200	0	−0.75
2	150	200	120	−1.25
3	150	200	240	−1.75
4	?	200	?	−2.00
5	?	200	?	0

Table P9.8 Details of unbalance masses, and their locations of exercise problem 9.8

No	Mass (kg)	Radius (mm)	Angle (°)	Axial location (mm)
1	?	75	?	0
2	5	75	0	100
3	?	75	85	200
4	?	75	220	300

9.9 Three massess, m1, m2, and m3 are rotating in three vertical planes of a horizontal shaft that are equally spaced at 250 mm and all the masses are located at the same radius of 75 mm and at the angular positions as shown in Fig. P9.9. The above values are summarized in Table P9.9. The shaft is rotating in two bearings A and B at an angular speed of 500 rpm. Determine the reactions at the bearings A and B due to the unbalance

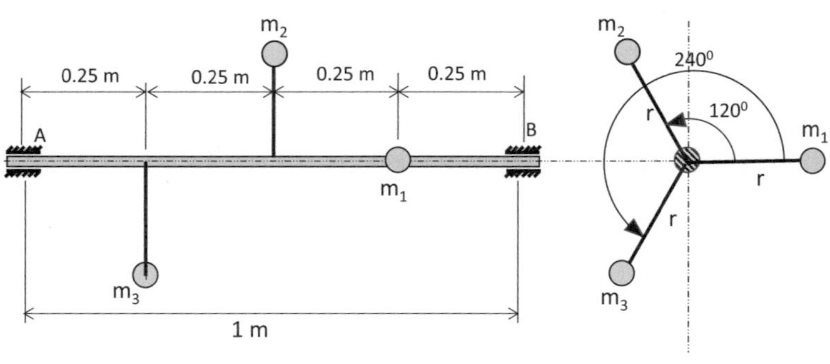

Fig. P9.9 Bearing reactions due to unbalance masses

Table P9.9 Details of unbalance masses, and their locations of exercise problem 9.9

No	Mass (kg)	Radius (mm)	Angle (°)	Axial location (m)
1	6	75	0	−0.25
2	10	75	120	0.50
3	6	75	240	1.25

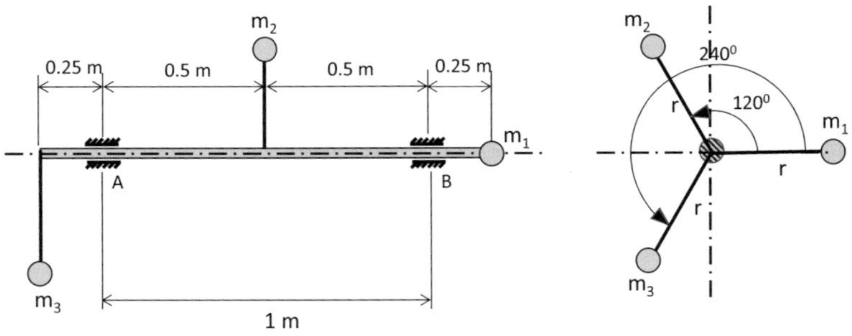

Fig. P9.10 Bearing reactions due to unbalance masses

9.10 Three masses, m_1, m_2, and m_3 are rotating in three vertical planes as shown in the figure and all the masses are located at the same radius of 75 mm and at the angular positions as shown in Fig. P9.10. The above values are summarized in Table P9.10. The shaft is rotating in two bearings A and B at an angular speed of 500 rpm. Determine the reactions at the bearings A and B. The axial locations are with respect to the left bearing A.

9.11 A horizontal reciprocating engine has a crank of 200 mm and connecting rod of 800 length. The crank pin has an equivalent rotational unbalance of 50 kg and the piston pin has a reciprocating mass of 50 kg; the connecting rod weighs 50 kg and the center of gravity of the connecting rod is at a distance of 250 mm from the crank pin end. If all the rotational unbalance and 50% of the reciprocating unbalance is balanced using a balancing mass at a distance of 250 mm opposite to the crank, determine the balancing mass. If the engine runs at 500 rpm, determine the residual unbalance load on the crankshaft.

9.12 Consider the three-cylinder radial engine that is equally spaced shown in Fig. P9.12. in which all the connecting rods are connected to a common crank. The mass of the crank is balanced. The total equivalent mass of the connecting rod at the crank pin is 5 kg and the mass of each piston is 1.5 kg and the equivalent mass of each connecting rod at the corresponding piston pin is 0.75 kg. The crank radius is 150 mm and the length of each connecting rod is 550 mm. Calculate the resultant shaking force on the engine frame for the phase shown, when the crank speed is 500 rpm.

Table P9.10 Details of unbalance masses, and their locations of exercise problem 9.10

No	Mass (kg)	Radius (mm)	Angle (°)	Axial location (m)
1	6	75	0	−0.25
2	10	75	120	0.50
3	6	75	240	1.25

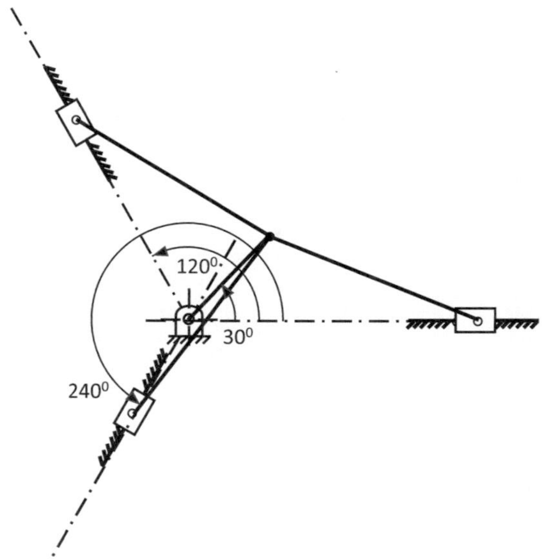

Fig. P9.12 Three cylinder radial engine

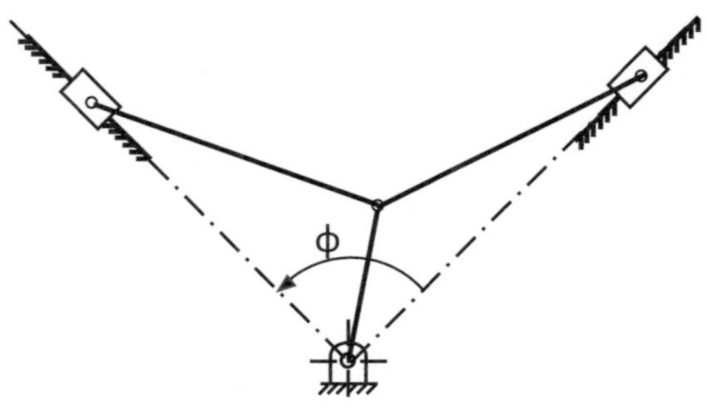

Fig. P9.13 Motor cycle V engine

9.13 Two exactly similar reciprocating engines of a motor cycle are joined to a common crank in the V-engine configuration as shown in Fig. P9.13. Determine the optimum angle ϕ between the cylinders that will result in the transfer of minimum shaking force to the rest of the motor cycle.

9.14 The cylinders of a six-cylinder inline arrangement are placed at intervals of 250 mm in the axial direction as shown in Fig. 9.14. The massless cranks have the same lengths—100 mm and their angular positions are as shown. The length of all connecting rods is 350 mm, which have a mass of 1.25 kg with their center of mass 70 mm from the crank pin. A mass equivalent to the connecting rod mass at the crank pin is placed (180 degrees from each crank pin) as a balancing mass on each of the cranks at a radius equal to the crank length. The piston mass is 1.25 kg for each cylinder. The left and the right bearings are at a distance of 150 mm from the center line of the nearest

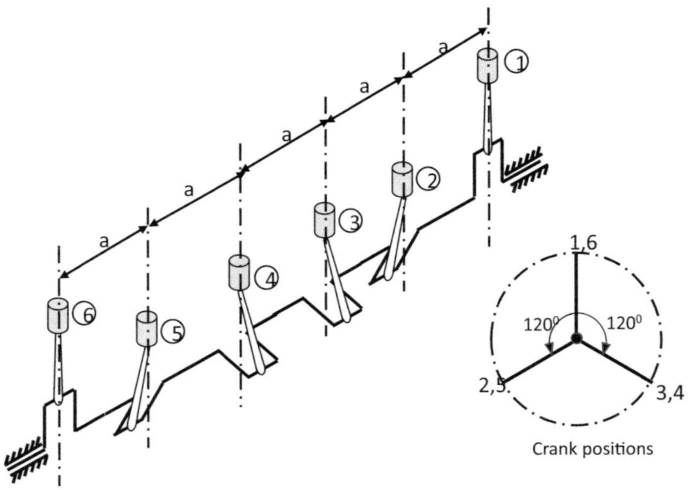

Fig. P9.14 Six-cylinder inline engine

cylinders. Find the reactions at the bearing when the first and sixth cylinders have maximum acceleration and the engine speed is 2500 rpm (Fig. P9.14).

9.15 The cylinders of an eight-cylinder inline arrangement are placed at intervals of 250 mm in the axial direction as shown in Fig. P9.15. The massless cranks have the same lengths—100 mm and their angular positions are as shown. The length of all connecting rods is 350 mm, which have a mass of 1.25 kg with a center of mass 70 mm from the crank pin. A mass equivalent to the connecting rod mass at the crank pin is placed (180° from each crank pin) as a balancing mass on each of the cranks at a radius equal to the crank length. The piston mass is 1.25 kg for each cylinder. The left and the right bearings are at a distance of 150 mm from the center line of the nearest cylinders. Find the reactions at the bearing when the first and eighth cylinders have maximum acceleration and the engine speed is 2500 rpm.

9.16 The cylinders of a four-cylinder inline arrangement are placed at intervals of 250 mm in the axial direction as shown in Fig. P9.16. The massless cranks have the same lengths—100 mm and their angular positions are as shown. The length of all connecting rods is 350 mm, which have a mass of 1.25 kg with a center of mass 70 mm from the crank pin. A mass equivalent to the connecting rod mass at the crank pin is placed (180° from each crank pin) as a balancing mass on each of the cranks at a radius equal to the crank length. The piston mass is 1.25 kg for each cylinder. The left and the right bearings are at a distance of 150 mm from the center line of the nearest cylinders. Find the reactions at the bearing when the first and fourth cylinders have maximum acceleration and the engine speed is 2500 rpm.

9.17 A V-8 engine consists of two banks of four-cylinder inline engines, connected at a banking angle of 90^0 with the crank positions is shown in Fig. P9.17.

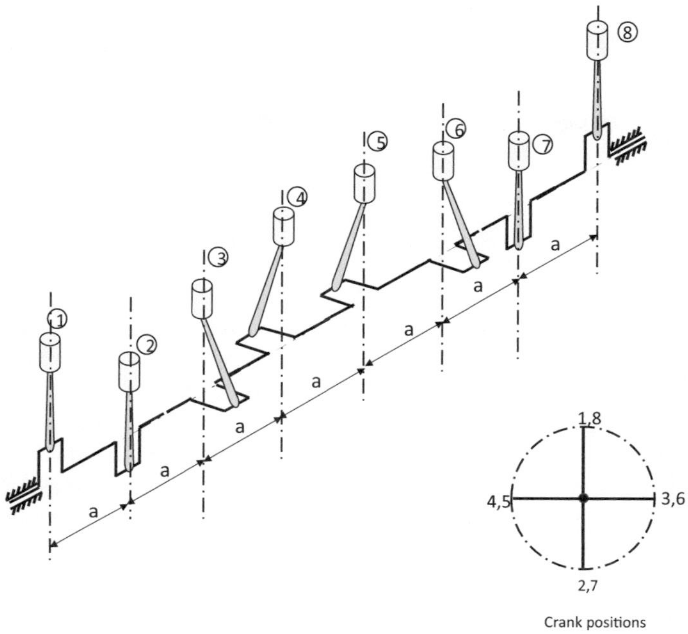

Fig. P9.15 Eight-cylinder inline engine

Fig. P9.16 Four-cylinder inline engine

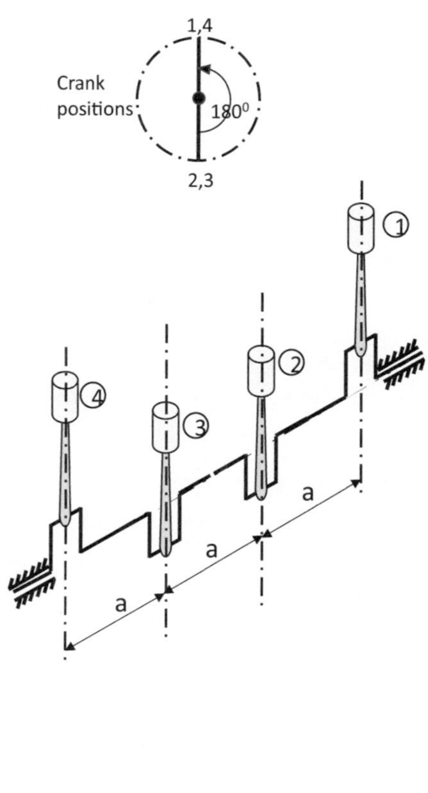

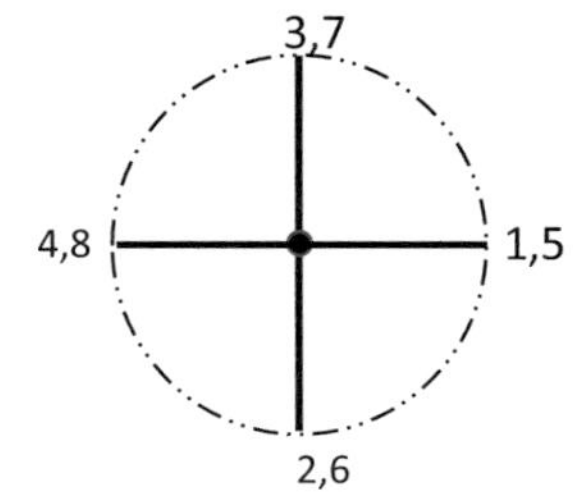

Fig. P9.17 Crank positions of an eight-cylinder V engine

The engines in each bank are placed at intervals of 250 mm in the axial direction. The massless cranks have the same lengths—100 mm and their angular positions are as shown. The length of all connecting rods is 350 mm, which have a mass of 1.25 kg with a center of mass 70 mm from the crank pin. A mass equivalent to the connecting rod mass at the crank pin is placed (180° from each crank pin) as a balancing mass on each of the cranks at a radius equal to the crank length. The piston mass is 1.25 kg for each cylinder. The left and the right bearings are at a distance of 150 mm from the center line of the nearest cylinders. Find the reactions at the bearings when the first- and fifth-cylinders' cranks are along the vertical and the engine speed is 2500 rpm.

9.18 Consider the three-cylinder engine of problem 9.12. If the crank mass is 0.5 kg that has its center of gravity at a distance of 75 mm and is not balanced, determine to what order the shaking forces need to be accounted for.

9.19 Consider the V-engine of 9.17. If each crank has a mass of 0.6 kg, determine the force and moment reactions on the engine frame.

9.20 A horizontal unbalanced rotor is mounted in between two bearings. A reference plane, termed as the left reference plane, is located close to the left bearing; the other reference plane close to the right bearing is termed as the right reference plane. The distance between the reference planes and the bearings is not given, but based on the vibration measurements made with and without correction masses and the knowledge of trial masses and their angular locations, the unknown unbalance in the reference plane is determined. Table P9.20a shows the location of trial masses and Table P9.20b shows the details of vibration measurements. Determine the unknown mass unbalance and its angular location in both the reference planes using the above data.

Table P9.20a Location of trial masses

Location of unbalance and correction masses				
	Left balancing plane, A		Right balancing plane, B	
	Mass (g)	Angle (°)	Mass (g)	Angle (°)
Original unbalance mass	M_L	θ_L	M_R	θ_R
Correction masses	15	225°	8	80°

Table P9.20b Details of vibration measurement

Vibration at the bearings	Left bearing		Right bearing	
	Displacement units/s	Angle (°)	Displacement units/s	Angle (°)
Without correction masses	6	130°	2	−30°
With correction masses, left plane	10	150°	6	240°
With correction masses, right plane	4	65°	2	−95°

9.21 Consider the shaking forces of Eqs. (9.40) and (9.41). Prove that by placing 50% of the reciprocating mass and all the rotational unbalanced is balanced using a balancing mass on the crank, the above forces can be minimized.

Chapter 10
Governors

10.1 Introduction

Governing in a broad sense means controlling any parameter of a machine, but in this chapter, it will be concerned only with its speed. Controlling speed has been a longstanding requirement since the beginning of steam engines and the simplest of the speed control devices were devised for this purpose. Requirement of constant speed is also a requirement in many electricity generators and internal combustion engines. Hence, a suitable modification of the devices discussed in this chapter can be used. All rotating components experience centripetal force that counteracts the inertial centrifugal force. The equilibrium between these forces can be used to devise a mechanism that can ensure that a machine can operate at a constant speed. The Watt governor is the simplest among the governors used for speed control is discussed first, followed by further improvements of the Watt governor in Porter and Proell governors. The main disadvantage of the above governors is their large size that may not be possible for application in all types of machines. Hence, a spring-loaded Hartnell governor is discussed at the end.

10.2 Watt Governors

The schematic of a Watt governor is shown in Fig. 10.1. It was one of the earliest mechanisms that was used for controlling the speed of steam engines. The interface to this governor and the steam engine is the control valve that allows steam into the engine and the drive from the engine that rotates the Watt governor at a speed proportional to the speed of the engine. The configuration consists of a central shaft driven by the engine. Two upper arms are hinged at the top of the shaft at A and are attached with a ball. When the upper arms rotate with the balls, they fly apart due to the centrifugal force. The balls are connected to two lower arms that are hinged to a

D. N. Manik, *Fundamentals of Mechanisms and Machines*,
https://doi.org/10.1007/978-981-97-1810-8_10

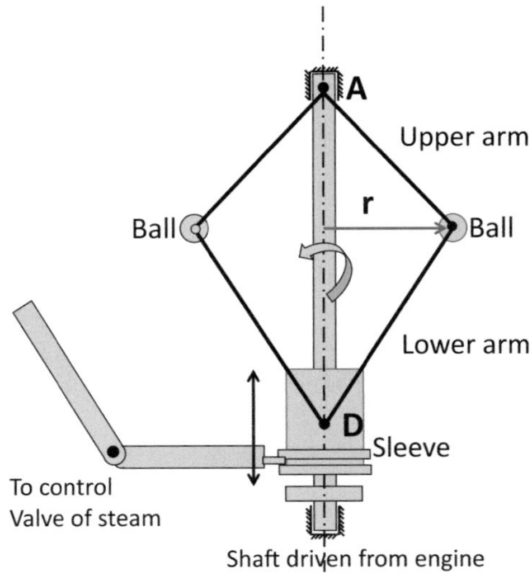

Fig. 10.1 Watt governor

sleeve that is free to move up and down. The control valve lever that is attached to the groove of the sleeve moves up and down with the sleeve and controls the amount of steam let into the engine. It is designed to move in such a way that when the speed increases, the balls fly apart and move up the sleeve and then move the control valve lever that controls the flow of steam to decrease the steam flow rate. If the speed of the engine decreases, the balls move inwards, thus moving the sleeve downwards that enables the control valve to admit more steam. Hence, there is an automatic regulation of the steam engine speed.

An engine speed control feedback system using a governor is shown in Fig. 10.2. This typical feedback system shows that the governor plays the role of a controller, boiler is the actuator and engine is the plant. The steam engine shaft cannot be directly connected to the Watt governor as the governor has to run at a higher speed so its arms can apply appropriate force to move the balls. Hence, the steam engine shaft is connected to a worm gearing in the feedback loop at governor shaft. The desired speed of the steam engine is the reference input to the feedback system. The reference speed can be set on the governor by adjusting the position of the sleeve, and the range of reference speeds can be obtained by appropriately designing the dimensions of the upper and lower arms and the mass of the balls.

The force analysis of the simplified force analysis of a Watt governor is shown in Fig. 10.3. Mass of the sleeve and force in the lower arms are neglected and the central shaft is assumed to be rotating at ω rad/s. The balls are assumed to be at a radius of r with the upper arm making an angle θ with respect to the vertical. m_B is the mass of each ball, h is the vertical distance from the hinge to the plane of rotation of the balls and T is the force on the upper arms; h is known as the height of the governor. The centrifugal force in the horizontal direction and the weight in the vertical direction are the external forces acting on the above system.

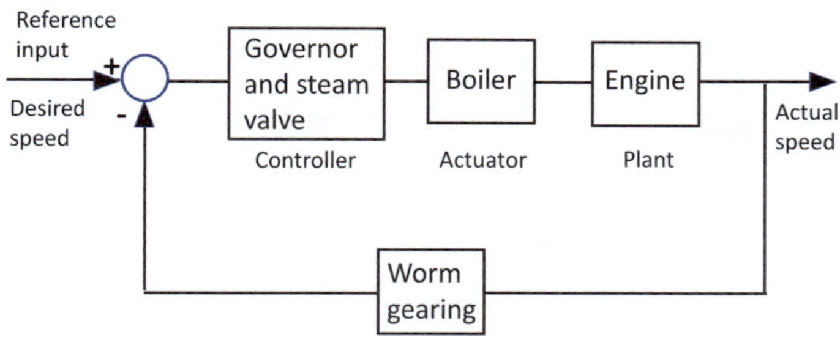

Fig. 10.2 Steam engine speed control system

Fig. 10.3 Watt governor: force analysis

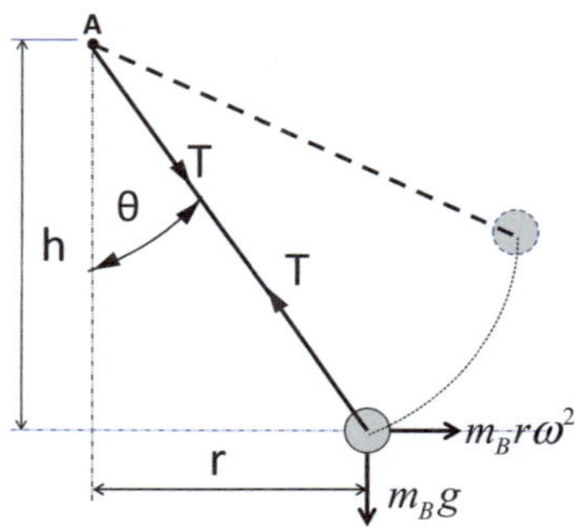

Taking moments of external forces about the hinge A,

$$m_B r\omega^2 h = m_B g r \tag{10.1}$$

Equation (10.1) simplifies to

$$h = \frac{g}{\omega^2} \tag{10.2}$$

The height and radius of the governor are related by

$$h = r\cot\theta = \sqrt{\ell^2 - r^2} \tag{10.3}$$

Equations (10.2) and (10.3) are used together to design a Watt governor.

Example 10.1

The arms of a Watt governor are 300 mm long and are pivoted on the axis of rotation. There are two balls, each ball has a mass of 10 kg. Find the equilibrium speeds corresponding to the radii of the ball positions at 200 and 250 mm (Fig. 10.4).

$\ell = 300$ mm, $m_B = 10$, $r_{1,2} = 200, 250$ mm. From Eq. (10.3), the Watt governor heights for the given radii are given by

Fig. 10.4 Watt governor of Example 10.1

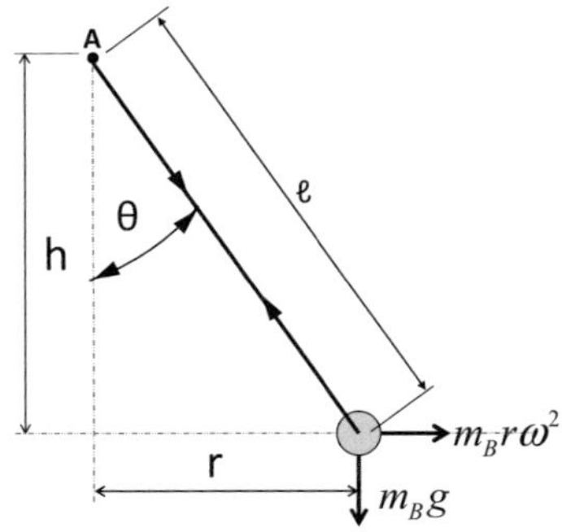

$$h_1 = \sqrt{\ell^2 - r_1^2} = \sqrt{0.3^2 - 0.2^2} = 0.224\,\text{m}$$
$$h_2 = \sqrt{\ell^2 - r_2^2} = \sqrt{0.3^2 - 0.250^2} = 0.166\,\text{m} \quad (1)$$

From Eq. (10.2), using values of the above governor heights from Eq. (1), the equilibrium speeds corresponding to the given radii are given by

$$\omega_1 = \sqrt{\frac{g}{h_1}} = \sqrt{\frac{9.81}{0.224}} = 6.62\,\text{rad/s}\ (63\,\text{RPM})$$
$$\omega_2 = \sqrt{\frac{g}{h_2}} = \sqrt{\frac{9.81}{0.166}} = 7.69\,\text{rad/s}\ (73\,\text{RPM}) \quad (2)$$

10.3 Porter Governor

The Porter governor, designed by Charles T. Porter, shown in Fig. 10.5 is similar to a Watt governor discussed in the previous section, except that the sleeve mass is significant in comparison to the mass of the balls. Hence, internal forces in the lower arms are not negligible as in a Watt governor. In addition, the upper and lower arms are assumed equal. Since the lower arms are offset by a certain distance on the sleeve at D, the angles made by the upper and lower arms are different. The masses of the arms (upper and lower) are considered negligible.

Consider the force analysis of a Porter governor presented in Fig. 10.6. The upper and lower arms make angles α and β respectively with respect to the vertical. The weight of the sleeves is assumed to be equally distributed at the hinge point D. T_1 is the force in each of the upper arms and T_2 is the force in each of the lower arms. Taking moments of forces on the left half symmetry of the governor about the point E. Note that T_1 and T_2 do not contribute toward the moments, since they generate equal and opposite moments. Taking moments about E

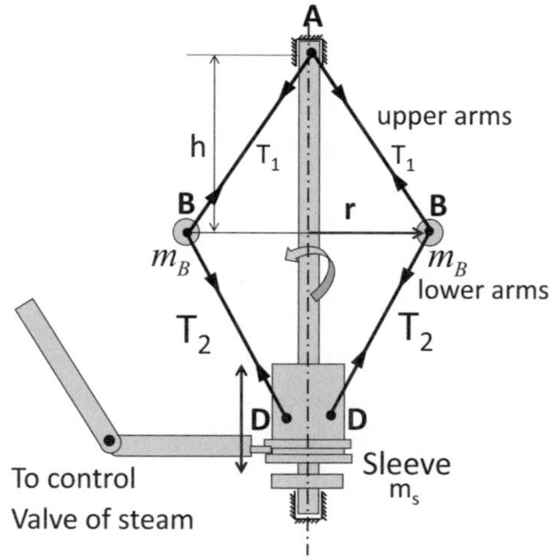

Fig. 10.5 Porter governor

$$m_B r\omega^2 BF = m_B g EF + \frac{m_s g}{2} ED \tag{10.4}$$

Simplifying Eq. (10.4)

$$m_B r\omega^2 = m_B g \frac{EF}{BF} + \frac{m_s g}{2} \frac{ED}{BF} \tag{10.5}$$

Equation (10.5) simplifies to

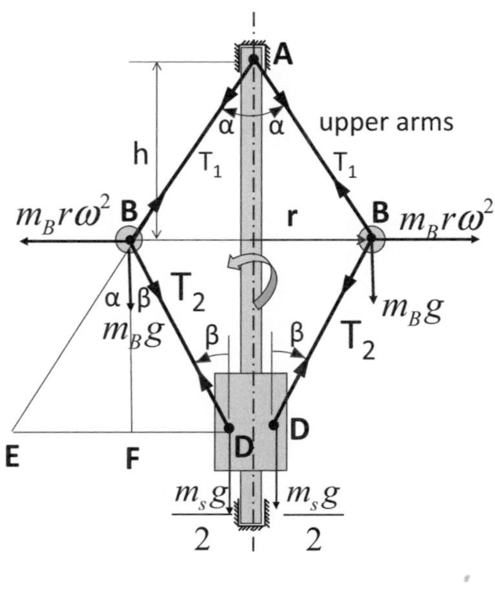

Fig. 10.6 Force analysis of a Porter governor

$$m_B r\omega^2 = \tan\alpha\left[m_B g + \frac{m_s g}{2}\left(1 + \frac{\tan\beta}{\tan\alpha}\right)\right] \tag{10.6}$$

Since $\tan\alpha = \frac{r}{h}$, Eq. (10.6) becomes

$$h = \frac{g}{\omega^2}\left[1 + \frac{m_s}{2m_B}\left(1 + \frac{\tan\beta}{\tan\alpha}\right)\right] \tag{10.7}$$

The above equations are derived assuming that there is no friction at the sleeve interface. Since the sleeve friction is in the vertical direction, depending on whether it is acting upwards or downwards, it can be accounted by modifying Eq. (10.7) as follows:

$$h = \frac{g}{\omega^2}\left[1 + \frac{m_s}{2m_B}\left(1 \pm \frac{F_f}{g}\right)\left(1 + \frac{\tan\beta}{\tan\alpha}\right)\right] \tag{10.8}$$

Hence, since downward forces are considered positive, a positive sign should be used when the sleeve is moving upwards and a negative sign when it is moving downwards.

An alternate derivation of Porter Governor can be obtained as follows.

Figure 10.7 shows the force analysis of a Porter governor. Figure 10.7a shows the forces acting at each flyball, consisting of tension in the upper and lower members, the weight of each ball, and the centrifugal force. Figure 10.7b shows the forces acting on the sleeve. The left side of Fig. 10.7b is easier to visualize in which the tension in the lower arms is balancing the weight of the sleeve; the figure on the right is the force triangle.

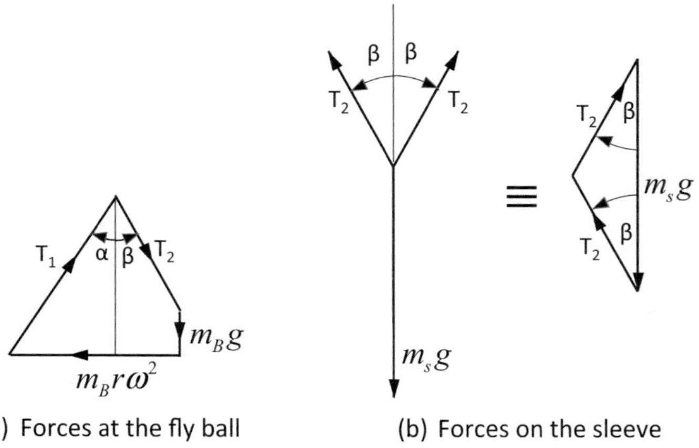

Fig. 10.7 Force analysis of a Porter governor

From Fig. 10.7b, the tension in the lower arm is given by

$$T_2 = \frac{m_s g}{2\cos\beta} \tag{10.9}$$

Considering the vertical forces on each ball from Fig. 10.7a

$$T_1 \cos\alpha = T_2 \cos\beta + m_B g \tag{10.10}$$

From Eqs. 10.9 and 10.10

$$T_1 = \frac{\frac{m_s g}{2} + m_B g}{\cos\alpha} \tag{10.11}$$

Considering the horizontal forces on each ball from Fig. 10.7a

$$m_B r\omega^2 = T_1 \sin\alpha + T_2 \sin\beta \tag{10.12}$$

From Eqs. 10.11 and 10.12

$$\begin{aligned} m_B r\omega^2 &= \left(\frac{\frac{m_s g}{2} + m_B g}{\cos\alpha}\right)\sin\alpha + \left(\frac{m_s g}{2\cos\beta}\right)\sin\beta \\ &= \tan\alpha\left[m_B g + \frac{m_s g}{2}\left(1 + \frac{\tan\beta}{\tan\alpha}\right)\right] \end{aligned} \tag{10.13}$$

The above equation is same as Eq. 10.6 derived earlier, but the above derivation will help compute tension in the members as well.

Example 10.2

The upper and lower arms of a Porter governor are 300 mm. long and the lower arms are offset by a small distance from the axis that can be considered negligible. There are two balls and each ball has a mass of 10 kg and a sleeve of mass 60 kg. (a) Find the equilibrium speeds corresponding to the radii of the ball positions at 200 mm. and 250 mm (b) If there is a friction force of 50 N due to the movement of the sleeve, determine the equilibrium speed at a radius of 200 mm, when the sleeve is moving up and down (c) Determine the tension in the upper and lower arms when there is no friction in the sleeves.

Since the offset is negligible, upper and lower arms make the same angle with the vertical, that is $\alpha = \beta = \tan^{-1}\left(\frac{r}{h}\right)$.

The Porter governor heights for the given radii are given by

$$\begin{aligned} h_1 &= \sqrt{\ell^2 - r_1^2} = \sqrt{0.3^2 - 0.2^2} = 0.224\,\text{m} \\ h_2 &= \sqrt{\ell^2 - r_2^2} = \sqrt{0.3^2 - 0.250^2} = 0.166\,\text{m} \end{aligned} \tag{1}$$

(a) When there is no friction at the sleeve

From Eq. (10.7), the angular velocity of the Porter governor for $r_1 = 200$ mm is given by

$$h = \frac{g}{\omega^2}\left(1 + \frac{m_s(1+1)}{2m_B}\right) \Rightarrow 0.224 = \frac{9.81}{\omega_1^2}\left(1 + \frac{2\times 60}{2\times 10}\right)$$

$$\omega_1 = 17.52\,\text{rad/s}\,(167\,\text{RPM}) \tag{2}$$

Similarly, the angular velocity of the Porter governor for $r_1 = 250$ mm is given by

$$h = \frac{g}{\omega^2}\left(1 + \frac{2m_s}{nm_B}\right) \Rightarrow 0.1658 = \frac{9.81}{\omega_2^2}\left(1 + \frac{2\times 60}{2\times 10}\right)$$

$$\omega_2 = 20.34\,\text{rad/s}\,(194\,\text{RPM}) \tag{3}$$

(b) When there is friction at the sleeve

From Eq. (10.8), the equilibrium speeds at $r_1 = 200$ mm due to friction force are given by

$$h = \frac{g}{\omega^2}\left[1 + \frac{2\left(m_s \pm \frac{F_f}{g}\right)}{nm_B}\right] \Rightarrow 0.2236 = \frac{9.81}{\omega^2}\left[1 + \frac{2\left(60 \pm \frac{50}{9.81}\right)}{2\times 10}\right] \tag{4}$$

$\omega = 18.15\,\text{rad/s}\,(173\,\text{RPM})$ when the sleeve is moving up
$\omega = 16.87\,\text{rad/s}\,(161\,\text{RPM})$ when the sleeve is moving down.

(c) Tension in the arms when there is no friction

Corresponding to r_1, the angle of the arms with respect to the vertical is given by

$$\alpha_1 = \beta_1 = \tan^{-1}\left(\frac{r_1}{h_1}\right) = \tan^{-1}\left(\frac{0.2}{0.2236}\right) = 42^\circ \tag{5}$$

Corresponding to r_2, the angle of the arms with respect to the vertical is given by

$$\alpha_2 = \beta_2 = \tan^{-1}\left(\frac{r_2}{h_2}\right) = \tan^{-1}\left(\frac{0.25}{0.1658}\right) = 56^\circ \tag{6}$$

Corresponding to r_1, the tensions in the arms are given by

$$T_1 = \frac{\frac{m_s g}{2} + m_B g}{\cos\alpha_1} = \frac{\frac{60\times 9.81}{2} + 10\times 9.81}{\cos 42} = 526\,\text{N}$$

$$T_2 = \frac{m_s g}{2\cos\beta} = \frac{60\times 9.81}{2\cos 42} = 395\,\text{N} \tag{7}$$

Corresponding to r_2, the tensions in the arms are given by

$$T_1 = \frac{\frac{m_s g}{2} + m_B g}{\cos \alpha_1} = \frac{\frac{60 \times 9.81}{2} + 10 \times 9.81}{\cos 56} = 710\,\text{N}$$

$$T_2 = \frac{m_s g}{2 \cos \beta} = \frac{60 \times 9.81}{2 \cos 42} = 532\,\text{N} \tag{8}$$

The following example illustrates an offset at the lower hinge point of a Porter governor.

Example 10.3
A Porter governor has equal upper and lower arms of length 300 mm. The upper arms are pivoted to the axis of rotation of the shaft and lower arms are attached to the sleeve with an offset distance of 40 mm from the shaft axis. The mass of each flyball is 10 kg and the sleeve mass is 60 kg. If the minimum radius of rotation of flyballs is 200 mm and the maximum radius is 250 mm, determine the equilibrium speeds of the governor at these extreme positions (Fig. 10.8).

Given, $\ell = 300$ mm, $r_1 = 200$ mm, $r_2 = 250$ mm, offset $\delta = 40$ mm, $m_B = 10$ kg, $m_s = 50$ kg.

The height of the upper arms (also the governor height) of the governor corresponding to the minimum speed is given by

$$h_1 = \sqrt{\ell^2 - r_1^2} = \sqrt{300^2 - 200^2} = 223.6\,\text{mm} \tag{1}$$

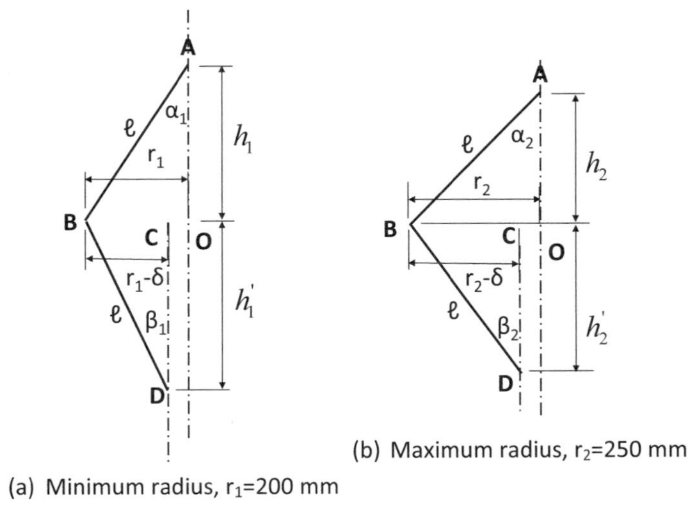

Fig. 10.8 Porter arm positions of Example 10.3

The height of the lower arms of the governor corresponding to the minimum speed is given by

$$h_1' = \sqrt{\ell^2 - (r_1 - \delta)^2} = \sqrt{300^2 - (200 - 40)^2} = 253.8\,\text{mm} \tag{2}$$

The height of the upper arms (also the governor height) of the governor corresponding to the maximum speed is given by

$$h_2 = \sqrt{\ell^2 - r_2^2} = \sqrt{300^2 - 250^2} = 165.8\,\text{mm} \tag{3}$$

The height of the lower arms of the governor corresponding to the maximum speed is given by

$$h_2' = \sqrt{\ell^2 - (r_2 - \delta)^2} = \sqrt{300^2 - (250 - 40)^2} = 214.2\,\text{mm} \tag{4}$$

The angles of the arms corresponding to the minimum speed are given by

$$\begin{aligned}\alpha_1 &= \tan^{-1}\left(\frac{r_1}{h_1}\right) = \tan^{-1}\left(\frac{200}{223.6}\right) = 41.8^\circ \\ \beta_1 &= \tan^{-1}\left(\frac{r_1 - \delta}{h_1}\right) = \tan^{-1}\left(\frac{200 - 40}{223.6}\right) = 32.2^\circ\end{aligned} \tag{5}$$

The angles of the arms corresponding to the maximum speed are given by

$$\begin{aligned}\alpha_2 &= \tan^{-1}\left(\frac{r_2}{h_2}\right) = \tan^{-1}\left(\frac{250}{253.8}\right) = 56.44^\circ \\ \beta_2 &= \tan^{-1}\left(\frac{r_2 - \delta}{h_2'}\right) = \tan^{-1}\left(\frac{250 - 40}{214.2}\right) = 44.42^\circ\end{aligned} \tag{6}$$

From Eq. 10.13, the minimum speed is given by

$$\begin{aligned}\omega_1 &= \sqrt{\frac{g}{h_1}\left[1 + \frac{m_s}{2m_B}\left(1 + \frac{\tan\beta_1}{\tan\alpha_1}\right)\right]} \\ &= \sqrt{\frac{9.81}{0.2236}\left[1 + \frac{60}{2 \times 10}\left(1 + \frac{\tan 32.9^\circ}{\tan 41.81^\circ}\right)\right]} = 16.45\,\text{rad/s}\,(157\,\text{RPM})\end{aligned} \tag{7}$$

Similarly, the maximum speed is given by

$$\omega_2 = \sqrt{\frac{g}{h_2}\left[1 + \frac{m_s}{2m_B}\left(1 + \frac{\tan\beta_2}{\tan\alpha_2}\right)\right]}$$
$$= \sqrt{\frac{9.81}{0.1658}\left[1 + \frac{60}{2\times 10}\left(1 + \frac{\tan 45.23^\circ}{\tan 56.44^\circ}\right)\right]} = 18.84\,\text{rad/s}\,(180\,\text{RPM}) \quad (8)$$

10.4 Proell Governor

A Proell governor is shown in Fig. 10.9. Similar to a Porter Governor, it has upper and lower arms, with the lower arms hinged at an offset at the sleeve. However, the balls, each of them of mass m_B, are placed at a rigid extension of the lower arm. At the minimum speed, the balls are vertically above the hinge point of the upper and lower arms, as the speed increases, the balls fly apart, but the angle ψ remains the same, due to rigid extension nature of the lower arm that is clear in Fig. 10.10. The lower arms are attached to a sleeve of mass m_s.

Force analysis of the Proell governor is shown in Fig. 10.10. T_1 is the tension in the upper arms and T_2 is the tension in the lower arms. The upper arms are symmetrically extended on both sides. The sleeve and ball weight are acting vertically downwards and the centrifugal forces acting are in the horizontal direction. Extension of the

Fig. 10.9 Proell governor

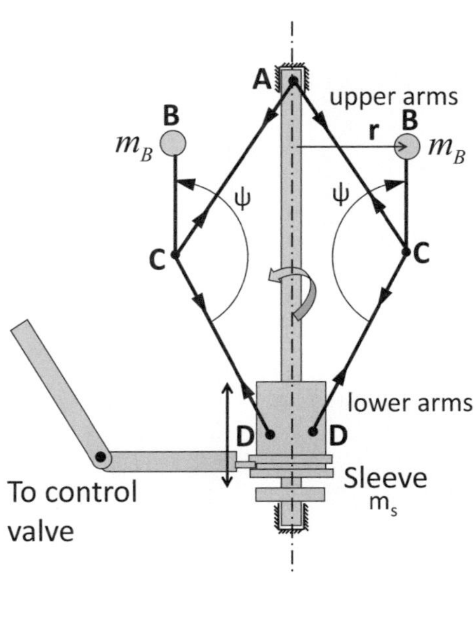

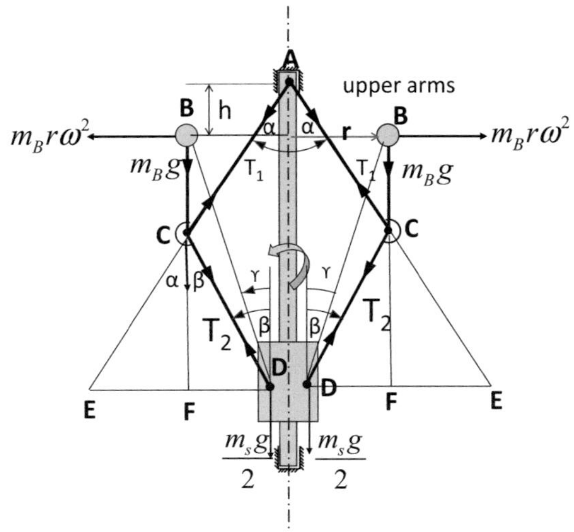

Fig. 10.10 Force analysis of the Proell governor

upper arm and the horizontal line from the sleeve hinge D meet at point E, which is the instantaneous center of rotation of the lower arm.

Taking moments about the point E

$$m_B r\omega^2 BF = m_B g.EF + \frac{m_s g}{2}.ED \tag{10.14}$$

Dividing Eq. 10.14 by CF throughout

$$m_B r\omega^2.\frac{BF}{CF} = m_B g.\frac{EF}{CF} + \frac{m_s g}{2}.\frac{ED}{CF} \tag{10.15}$$

Equation 10.15 can be written as

$$m_B r\omega^2.\frac{BF}{CF} = m_B g.\frac{EF}{CF} + \frac{m_s g}{2}.\frac{EF + FD}{CF} \tag{10.16}$$

Noting that $\tan\alpha = \frac{EF}{CF}$ and $\tan\beta = \frac{FD}{CF}$, Eq. (10.16) becomes

$$m_B r\omega^2 = \frac{CF}{BF}\left[m_B g \tan\alpha + \frac{m_s g}{2}(\tan\alpha + \tan\beta)\right] \tag{10.17}$$

Substituting $\tan\alpha = \frac{r}{h}$ in Eq. 10.17

$$m_B r\omega^2 = \frac{CF}{BF}.\frac{gr}{h}\left[m_B + \frac{m_s}{2}\left(1 + \frac{\tan\beta}{\tan\alpha}\right)\right] \tag{10.18}$$

From Eq. 10.18, the angular speed of the governor becomes

$$\omega^2 = \frac{CF}{BF} \cdot \frac{g}{h}\left[1 + \frac{m_s}{2m_B}\left(1 + \frac{\tan\beta}{\tan\alpha}\right)\right] \tag{10.19}$$

If the offset of the lower arm hinge at the sleeve is zero, $\alpha = \beta$. Then, Eq. 10.19 becomes

$$\omega^2 = \frac{CF}{BF} \cdot \frac{g}{h}\left[1 + \frac{m_s}{m_B}\right] \tag{10.20}$$

From Eq. 10.20 it is clear that a Proell governor can achieve lower equilibrium angular velocity than a Porter governor of the same mass of the flyballs and the sleeve.

Example 10.4

A Proell governor has equal upper and lower arms of length 300 mm. The upper arms are pivoted to the axis of rotation of the shaft and lower arms are attached to the sleeve with an offset distance of 40 mm from the shaft axis and to a ball at B. The mass of each flyball is 10 kg and it is fixed vertical 100 mm above the hinge of the upper and lower arms at the minimum radius (speed) position and the sleeve mass is 60 kg. If the minimum radius of rotation of flyballs is 200 mm and the maximum is 250 mm, determine the equilibrium speeds of the governor at these extreme positions.

Minimum speed

Configuration of the Proell governor of Example 10.4 is shown in Fig. 10.11a and only the lower arm is shown in Fig. 10.11b.

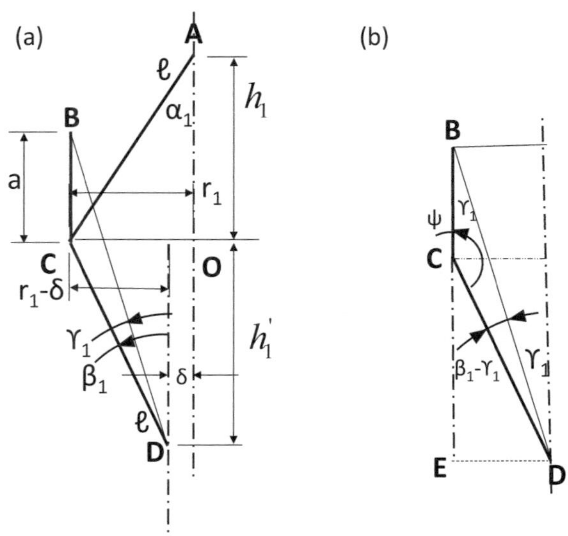

Fig. 10.11 Proell governor at the minimum radius of Example 10.4 for minimum speed

Given, $\ell = 300$ mm, $r_1 = 200$ mm, $r_2 = 250$ mm, offset $\delta = 40$ mm, $m_B = 10$ kg, $m_s = 50$ kg, a = 100 mm.

The height of the upper arms (also the governor height) of the governor corresponding to the minimum speed is given by

$$h_1 = \sqrt{\ell^2 - r_1^2} = \sqrt{300^2 - 200^2} = 223.6\,\text{mm} \quad (1)$$

The height of the lower arms of the governor corresponding to the minimum speed is given by

$$h_1' = \sqrt{\ell^2 - (r_1 - \delta)^2} = \sqrt{300^2 - (200 - 40)^2} = 253.8\,\text{mm} \quad (2)$$

The angles of the arms corresponding to the minimum speed are given by

$$\alpha_1 = \tan^{-1}\left(\frac{r_1}{h_1}\right) = \tan^{-1}\left(\frac{200}{223.6}\right) = 41.8^\circ$$
$$\beta_1 = \tan^{-1}\left(\frac{r_1 - \delta}{h_1}\right) = \tan^{-1}\left(\frac{200 - 40}{223.6}\right) = 32.2^\circ \quad (3)$$

From Eq. 10.19, the minimum angular speed of the governor becomes

$$\begin{aligned}\omega_1 &= \sqrt{\frac{h_1'}{(h_1' + a)}\cdot\frac{g}{h_1}\left[1 + \frac{m_s}{2m_B}\left(1 + \frac{\tan\beta_1}{\tan\alpha_1}\right)\right]} \\ &= \sqrt{\frac{0.2538}{(0.2538 + 0.1)}\cdot\frac{9.81}{0.2236}\left[1 + \frac{10}{2\times 50}\left(1 + \frac{\tan 32.23}{\tan 41.81}\right)\right]} \\ &= 12.87\,\text{rad/s}\,(123\,\text{RPM})\end{aligned} \quad (4)$$

Now, before computing the maximum speed, some parameters of the lower arms will have to be computed as follows:

The angle made by the line joining the sleeve hinge and the flyball location is given by

$$\gamma_1 = \tan^{-1}\left(\frac{r_1 - \delta}{h_1'}\right) = \tan^{-1}\left(\frac{200 - 40}{253.8}\right) = 24.34^\circ \quad (5)$$

Referring to Fig. 10.11b, the angle between the slant and straight portion of the lower arms is given by

$$\psi = 180^\circ - \beta_1 = 180^\circ - 32.23^\circ = 147.8^\circ \quad (6)$$

Length of the line joining the sleeve hinge and flyball location is given by

$$BD = \sqrt{\left(BC + h_1'\right)^2 + (r_1 - \delta)^2} = \sqrt{(40 + 253.8)^2 + (200 - 40)^2} = 388.3\,\text{mm} \tag{7}$$

Since the balls are mounted on the rigid extension of the lower arms, the above angle ψ and the distance BD remain constant for all positions taken by the balls. However, the only difference between the minimum and maximum speed positions is that at the minimum speed the balls are vertically up on the lower arms; whereas at the other speeds greater than the minimum speed, they are slightly tilted that has to be accounted for in computing the equilibrium maximum speed.

Maximum speed

The height of the upper arms (also the governor height) of the governor corresponding to the maximum speed is given by

$$h_2 = \sqrt{\ell^2 - r_2^2} = \sqrt{300^2 - 250^2} = 165.8\,\text{mm} \tag{8}$$

The height of the lower arms of the governor corresponding to the maximum speed is given by

$$h_2' = \sqrt{\ell^2 - \left(r_2 - \delta\right)^2} = \sqrt{300^2 - (250 - 40)^2} = 214.2\,\text{mm} \tag{9}$$

The angles of the arms corresponding to the maximum speed are given by

$$\begin{aligned} \alpha_2 &= \tan^{-1}\left(\frac{r_2}{h_2}\right) = \tan^{-1}\left(\frac{250}{253.8}\right) = 56.44^\circ \\ \beta_2 &= \tan^{-1}\left(\frac{r_2 - \delta}{h_2'}\right) = \tan^{-1}\left(\frac{250 - 40}{214.2}\right) = 44.42^\circ \end{aligned} \tag{10}$$

From Fig. 10.12b, the angle made by the line joining the sleeve hinge and the flyball location is given by

$$\gamma_2 = \beta_2 - \sin^{-1}\left(\frac{BC \sin\psi}{BD}\right) = 44.43^\circ - \sin^{-1}\left(\frac{100 \sin 147.8^\circ}{388.3}\right) = 36.53^\circ \tag{11}$$

Referring to Fig. 10.12b, ϕ is given by

$$\phi = 180^\circ - (\psi + \beta_2 - \gamma_2) = 180^\circ - (147.8^\circ + 44.43^\circ - 36.53^\circ) = 24.3^\circ \tag{12}$$

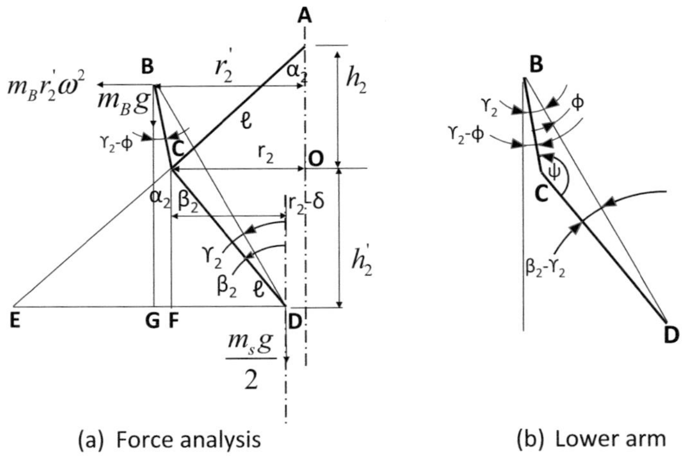

Fig. 10.12 Proell governor of Example 10.4 for maximum speed

The distance BG of Fig. 10.12a is given by

$$BG = h_2' + a.\cos(\gamma_2 - \phi) = 214.2 + 40.\cos(36.53 - 24.34) = 312\,\text{mm} \quad (13)$$

The distance GF of Fig. 10.12a is given by

$$GF = BC.\sin(\gamma_2 - \phi) = 40.\sin(36.53 - 24.34) = 21.1\,\text{mm} \quad (14)$$

The distance EF of Fig. 10.12a is given by

$$EF = h_2' \tan(\alpha_2) = 214.2.\tan(56.44) = 323\,\text{mm} \quad (15)$$

The actual radius of rotation of flyballs is given by

$$r_2' = r_2 + BC.\sin(\gamma_2 - \phi) = 250 + 40.\sin(36.53 - 24.34) = 271.1\,\text{mm} \quad (16)$$

Taking moments about the point E of Fig. 10.12a, the maximum equilibrium speed is given by

$$\begin{aligned}\omega_2 &= \sqrt{\frac{m_B g(EF - GF) + \frac{m_s g h_2'}{2}[\tan(\alpha_2) + \tan(\beta_2)]}{m_B r_2' BG}} \\ &= \sqrt{\frac{10 \times 9.81(0.323 - 0.0211) + \frac{50\times 9.81\times 0.2142}{2}[\tan(56.44) + \tan(44.43)]}{10 \times 0.2711 \times 0.312}} \\ &= 13.77\,\text{rad/s}\,(131\,\text{RPM}) \quad (17)\end{aligned}$$

10.5 Spring-Loaded Governors

Thus far we have studied many variations of governors that are related to centrifugal and gravity forces. But if the applications require governors that should operate at much higher speeds in a compact space, spring-loaded governors are preferred since they offer much more flexibility in their design. Among the various types of spring-loaded governors, Hartnell governor is widely used, that is discussed in the following section.

10.5.1 Hartnell Governor

The schematic of a Hartnell governor is shown in Fig. 10.13. Figure 10.13a shows a general position and Fig. 10.13b shows a position in which the bell crank is aligned along the horizontal and vertical directions. It consists of a bracket B that revolves with the governor shaft. Two bell crank levers of dimensions a and b are attached to this bracket and one end of the bell crank is attached with a flyball. There is a central spring-loaded spindle that presses on to the bell crank and the spring force is adjustable by means of a screw. The spring force essentially works similar to the heavy sleeve of Porter and Proell governors. Each ball is of mass m_B, sleeve is of mass m_s, and spring stiffness is k.

To derive the equation for spring stiffness, assume a position shown in Fig. 10.13a. Although the position of the bell crank can change between minimum and maximum speeds, we can assume that the bell crank is aligned with vertical and horizontal directions at the minimum speed; but of course, with the flyballs at different radii and the sleeve pressed upon by a spring force.

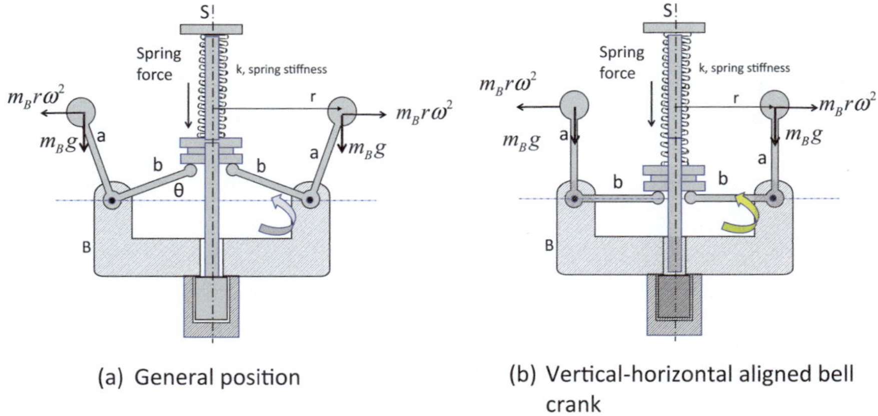

Fig. 10.13 Hartnell governor

Taking moments about the bell crank lever and neglecting the gravitational force of the balls and the arms, the centrifugal force corresponding to the minimum speed

$$m_B r_1 \omega_1^2 . a = \left(\frac{m_s g + F_1}{2} \right) b \tag{10.21}$$

where r_1 is the radial position of the ball and F_1 is the force exerted by the spring on the sleeve.

Similarly, taking moments about the bell crank lever and neglecting the gravitational force of the balls and the arms, the centrifugal force corresponding to the maximum speed is given by

$$m_B r_2 \omega_2^2 . a = \left(\frac{m_s g + F_2}{2} \right) b \tag{10.22}$$

where r_2 is the radial position of the ball and F_2 is the force exerted by the spring on the sleeve.

Subtracting Eq. 10.21 from 10.22,

$$F_2 - F_1 = \frac{2a.m_B}{b} \left(r_2 \omega_2^2 - r_1 \omega_1^2 \right) \tag{10.23}$$

Refer to Fig. 10.14 in which two positions of the bell crank lever corresponding to the minimum and maximum speeds are shown; it is assumed that at the minimum speed, the bell crank lever is aligned respectively in the horizontal and vertical directions. Since the angle traversed by both the horizontal and vertical arms of the bell crank lever is the same, the vertical lift of the sleeve between minimum and maximum speeds is given by

$$y = (r_2 - r_1) \frac{b}{a} \tag{10.24}$$

Since the force in a spring is the product of stiffness and deflection, dividing Eqs. 10.23 by 10.24, the spring stiffness is given by

$$k = 2 \left(\frac{a}{b} \right)^2 \frac{m_B \left(r_2 \omega_2^2 - r_1 \omega_1^2 \right)}{r_2 - r_1} \tag{10.25}$$

Example 10.5

A Hartnell governor, shown in Fig. 10.14, has a = 100 mm, b = 50 mm with the mass of each flyball, m_B = 2 kg. Its minimum speed is 300 RPM and maximum speed 320 RPM; and its minimum radius, r_1 = 70 mm, and maximum radius r_2 = 110 mm. (a) What is the spring stiffness required for this governor (b) What are the spring forces in the above positions (c) For a speed of 310 RPM, what is the radius at which the flyballs rotate? Neglect the weight of the sleeve.

Fig. 10.14 Relationship between vertical shift and radii of rotation

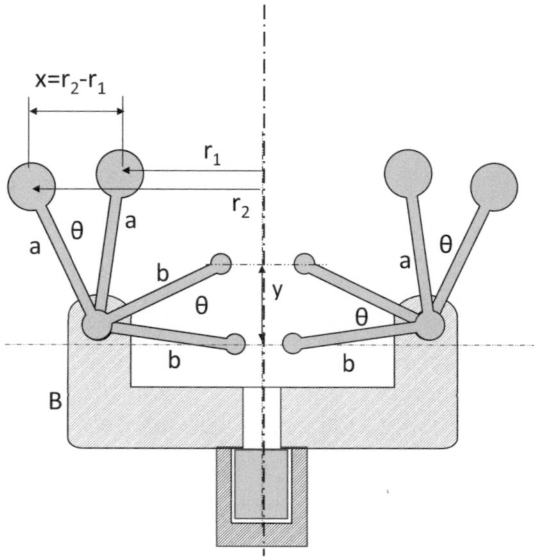

Minimum angular velocity, $\omega_1 = \frac{2\pi N_1}{60} = \frac{2\pi \times 300}{60} = 31.42\,\text{rad/s}$
Maximum angular velocity, $\omega_2 = \frac{2\pi N_2}{60} = \frac{2\pi \times 320}{60} = 33.51\,\text{rad/s}$
From Eq. (10.25), the spring stiffness of the Hartnell governor is given by

$$k = 2\left(\frac{a}{b}\right)^2 \frac{m_B\left(r_2\omega_2^2 - r_1\omega_1^2\right)}{r_2 - r_1}$$
$$= 2\left(\frac{0.1}{0.05}\right)^2 \frac{2\left(0.11 \times 33.51^2 - 0.07 \times 31.42^2\right)}{0.11 - 0.07} = 14.81\,\text{kN/m} \quad (1)$$

The spring forces corresponding to the minimum and maximum speeds are given by

$$F_1 = \frac{2am_Br_1\omega_1^2}{b} = \frac{2 \times 0.1 \times 2 \times 0.07 \times 31.42^2}{0.05} = 553\,\text{N}$$
$$F_2 = \frac{2am_Br_2\omega_2^2}{b} = \frac{2 \times 0.1 \times 2 \times 0.11 \times 33.51^2}{0.05} = 988\,\text{N} \quad (2)$$

The spring force F is linearly related to the radius of the flyballs r in the following equation

$$F = mr + C \quad (3)$$

where m is the slope and C is the intercept.

Based on Eq. (3), the coefficients of the linear equation can be obtained from

$$\begin{aligned}\begin{Bmatrix} m \\ C \end{Bmatrix} &= \begin{bmatrix} r_1 & 1 \\ r_2 & 1 \end{bmatrix}^{-1} \begin{Bmatrix} F_1 \\ F_2 \end{Bmatrix} \\ &= \begin{bmatrix} 0.07 & 1 \\ 0.11 & 1 \end{bmatrix}^{-1} \begin{Bmatrix} 553 \\ 988 \end{Bmatrix} \\ &= \begin{Bmatrix} 1.09 \times 10^4 \\ 0.05 \end{Bmatrix}\end{aligned} \tag{4}$$

The angular velocity at the unknown radius is given by $\omega_x = \frac{2\pi N_x}{60} = \frac{2\pi \times 310}{60} =$ 32.46 rad/s

From Eqs. (10.22) and (3), the unknown radius is given by

$$r_x = \frac{b}{\frac{2am_B\omega_x^2}{b} - m} = \frac{0.05}{\frac{2\times0.1\times2\times32.46^2}{0.05} - 1.09 \times 10^4} = 0.0852\,\text{m} \tag{5}$$

10.6 Controlling Force

The controlling force of a governor is an equal and opposite force to the centrifugal force that holds the balls in the equilibrium position. It is a result of the tensions in the arms and weight of the balls. A plot of the centrifugal and controlling forces as a function of the radial position of the balls is shown in Fig. 10.15. As shown above, they are equal at the equilibrium position of the governor at which all the balls are moving, and hence their vertical position remains stationary. The centrifugal force varies linearly with radius whereas the controlling force shows a nonlinear behavior as shown; the controlling force is less than the centrifugal force when the radial location of the balls is less than that of the equilibrium and greater than the centrifugal force when the radial location of the balls is greater than at the equilibrium. Hence the governor moves to an equilibrium position for a certain value of the centrifugal force that is related to its rotational speed. From the above discussion, it is clear that a governor operates in an intended manner only if the controller force and the centrifugal force have a definite relationship around the equilibrium position as described above. The relationship can be quantified as follows.

As shown in Fig. 10.16, consider two possible cases wherein the controlling force has different characteristics with respect to the centrifugal forces. In Fig. 10.16a, the slope of the controlling force at the equilibrium speed is more than the slope of the centrifugal force that ensures that if there is an increase in speed, the larger controlling force will bring back the governor into equilibrium speed. In Fig. 10.16b, the slope of the controlling force at the equilibrium speed is less than the slope of the centrifugal force. In such a case, if there is an increase speed of the governor, the

Fig. 10.15 Controlling and centrifugal forces of a governor

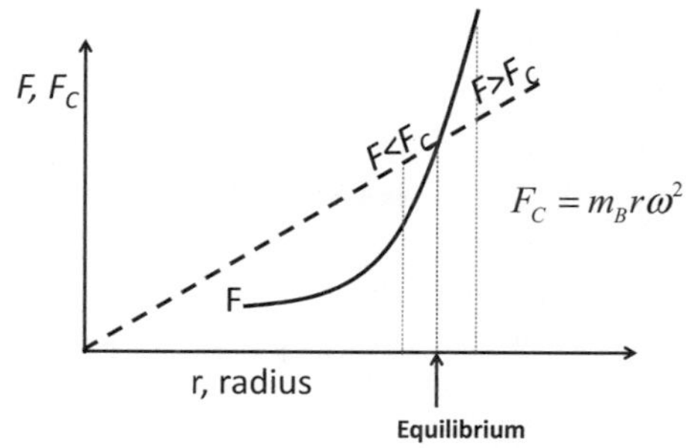

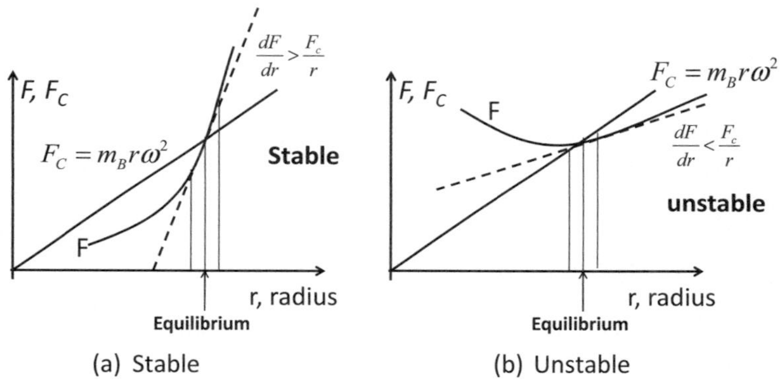

Fig. 10.16 Stability conditions for controlling force versus centrifugal force

controlling force keeps on decreasing with respect to the centrifugal force and the governor speed keeps on increasing, thus making it unstable.

10.7 Conclusions

We have covered various types of governors in this chapter. The main issues covered are the speed ranges in which these governors can be used and the reaction forces that result due to speed control. In addition, the friction in sliding surfaces that plays a role in determining the equilibrium speeds of various governors was also discussed. Based on the discussions in this chapter, new designs of governors can be obtained for future applications and some of them could include electronic components that have become very inexpensive in the recent past.

10.8 Exercise Problems

10.1 A Watt governor has arms that are 400 mm long and each of the two balls has a mass of 15 kg. Determine (a) the equilibrium speed corresponding to a ball radius of 300 mm (b) controlling force.

10.2 A Porter governor has equal arms that are 300 mm long. The upper arms and lower arms are fixed to a pivot on the axis of rotation. Mass of the sleeve is 45 kg and each ball of the two balls has a mass of 5 kg. Determine the equilibrium speeds when the radius of rotation of the balls are 150 and 175 mm.

10.3 A Porter governor has equal arms that are 350 mm long. The upper arms are fixed to a pivot on the axis of rotation and lower arms are attached to the sleeve at an offset of 40 mm from the central axis. Mass of the sleeve is 50 kg and each of the two balls has a mass of 10 kg. Determine the equilibrium speed when the radius of rotation of the balls is 200 mm.

10.4 The upper arms and lower arms of a Porter governor are equal and are fixed to a pivot on the axis of rotation. It has two balls and each ball has a mass of 5 kg. If the height of the governor corresponding to an equilibrium speed of 150 rpm is 300 mm, determine the mass of the sleeve.

10.5 The upper arms and lower arms of a Porter governor are equal and are fixed to a pivot on the axis of rotation; the length of each arm is 275 mm. It has two balls and each ball has a mass of 5 kg and the sleeve mass is 45 kg. Determine the equilibrium speeds when the radius of rotation of the balls are 125 and 175 mm and the speed regulation.

10.6 Consider the exercise problem 10.5. If there is a frictional force of 50 N at the sleeve interface, determine (a) the equilibrium speeds when the sleeve is going up and down (b) tension in the arms.

10.7 A Proell governor has equal upper and lower arms of length 300 mm. The upper arms are pivoted to the axis of rotation of the shaft and lower arms are attached to the sleeve at the axis of rotation. The mass of each flyball is 5 kg and it is fixed vertically 125 mm above the hinge of the upper and lower arms at the minimum radius (speed) position and the sleeve mass is 70 kg. If the minimum radius of rotation of flyballs is 200 mm determine the equilibrium speeds of the governor at minimum speeds.

10.8 A Proell governor has equal upper and lower arms of length 300 mm. The upper arms are pivoted at 15 mm from the axis of rotation of the shaft and lower arms are attached to the sleeve at 20 mm from the axis of rotation. The mass of each flyball is 5 kg and it is fixed vertically 100 mm above the hinge of the upper and lower arms at the minimum radius position and the sleeve mass is 50 kg. If the minimum radius of rotation of flyballs is 200 mm, determine the equilibrium speeds of the governor at the minimum speed.

10.9 Consider the exercise problem 10.7. Determine the minimum speed if the lower arms are hinged at a distance of 40 mm from the central axis.

10.10 Consider the exercise problem 10.8. If there is a frictional force of 50 N at the sleeve interface, determine (a) the equilibrium speeds when the sleeve is going up and down (b) tension in the arms.

10.11 A Hartnell governor has two balls of mass 1.5 kg each and are attached to a bell crank lever (a = 60 mm and b = 40 mm)and the mass of the sleeve is 5 kg and the sleeve is lifted against the spring force. If the minimum speed of the governor is 300 RPM at which the sleeve begins to lift up, and the radius is 150 mm, determine the initial load. If the maximum speed is limited to an increase of 10% of the minimum speed, corresponding to a radius of 200 mm, determine the required stiffness of the spring.

10.12 A Hartnell governor has two balls of mass 2 kg each and are attached to a bell crank lever (a = 125 mm and b = 80 mm) and the mass of the sleeve is 5 kg and the sleeve is lifted against the spring force. If the minimum speed of the governor is 250 RPM at which the sleeve begins to lift up, and the radius is 100 mm. The maximum speed is limited to an increase of 10% of the minimum speed, corresponding to a radius of 150 mm, determine the required stiffness of the spring and the speed when the radius is 125 mm.

10.13 A Hartnell governor has two balls of mass 2.5 kg each and are attached to a bell crank lever (a = 100 mm and b = 75 mm) and the mass of the sleeve is 6 kg and the sleeve is lifted against the spring force. If the minimum speed of the governor is 300 RPM at which the sleeve begins to lift up, and the radius is 70 mm. If the maximum speed is limited to an increase of 8% of the minimum speed, for a rise of 10 mm of the sleeve, determine the required stiffness of the spring and the initial load on the spring at the minimum speed.

10.14 A Hartnell governor has two balls of mass 2 kg each and are attached to a bell crank lever (a = 115 mm and b = 90 mm) and the mass of the sleeve is 5.5 kg and the sleeve is lifted against the spring force of 20 kN/m. The minimum speed of the governor is 300 RPM at which the sleeve begins to lift and the maximum rise is 10 mm of the sleeve. Determine the maximum equilibrium speed and the initial load on the spring at the minimum speed.

10.15 A Hartnell governor has two balls of mass 2 kg each and are attached to a bell crank lever (a = 125 mm and b = 50 mm) and the mass of the sleeve is 6 kg and the sleeve is lifted against the spring force of 30 kN/m. The minimum speed of the governor is 275 RPM corresponding to a ball radius of 50 mm. If the allowable increase in speed is limited to 9%, Determine the total lift of the sleeve and the initial load on the spring at the minimum speed.

10.16 A Hartnell governor has two balls weighing 2 kg each and are attached to a bell crank lever (a = 125 mm and b = 50 mm) and the mass of the sleeve is 7 kg. The minimum speed of the governor is 275 RPM corresponding to a ball radius of 60 mm. If the allowable increase in speed is limited to 8%, If the total lift of the sleeve is 15 mm, determine the spring stiffness and the initial load on the spring at the minimum speed.

10.17 A spring-loaded governor of the Hartnell type has arms of equal length. The masses rotate in a circle of 175 mm diameter when the sleeve is in its mid-position **and the arms attached with the mass** are vertical. The equilibrium speed for this position is 500 rpm. The maximum sleeve movement between maximum and minimum speeds is 45 mm, and the maximum variation is to be $\pm$ 4 percent of the mid-position speed. The mass of the sleeve is 5 kg and the equivalent friction at the sleeve is 35 N. Determine the rotating mass, the spring stiffness, and the initial compression of the spring.

10.18 In a Hartnell governor, the two balls of mass 1.5 kg each revolve at a radius of 50 mm when running at a minimum operating speed of 500 rpm and at the maximum speed, the ball radius is 100 mm and the spring is compressed by 30 mm.

(a) If the spring stiffness is 40 kN/m in compression find the speed regulation.
(b) What must be the spring stiffness to make the governor theoretically isochronous?

10.19 A Hartnell governor has a minimum radius of rotation of 100 mm and a maximum radius of 150 mm and each governor ball has a mass of 4 kg. The controlling force curve for the governor varies linearly and corresponds to 200 and 350 N respectively at the above radii, determine the extreme equilibrium speeds and sensitivity at a radius of 125 mm.

10.20 Consider the exercise problem 10.19. If the friction in the mechanism corresponding to each ball is 4 N, determine the extreme equilibrium speeds and sensitivity at a radius of 135 mm.

Chapter 11
Vibrations in Machines

During our initial study of mechanisms, and in later chapters, we assumed that most of the links are rigid. Although this assumption simplified the respective analyses to some extent, there are machines in which the flexibility of its elements will have to be accounted for. Because, it is this flexibility, sometimes under certain conditions, that can result in large amplitude vibrations, even for small amplitude of force excitation due to unbalance, misalignment, etc. Therefore, vibration becomes very important in many machines and sometimes may result in excessive noise. Unfortunately, most machine elements are not designed from the viewpoint of vibration at the design stage itself and vibration is noticed only when there is a serious problem arising from it. The discussions in this chapter should, therefore, help in putting "a stitch in time."

This chapter begins with the discussion of a single degree of freedom (SDOF) constructed from a spring-mass-damper unit. Using this simple model, various aspects of dynamics systems are discussed: free vibration, forced vibration, impulse response function, Duhamel's integral, etc. The concept of an SDOF is then used to demonstrate its application in designing vibration isolators to prevent vibration from machines moving to other connected systems and to isolate sophisticated instruments that are to be mounted on vibrating platforms.

Although SDOF is a very simplistic model, it is very useful in explaining most characteristics of the general class of dynamic systems that may be more complex than it. These concepts of an SDOF are then extended to multi-degree of freedom (MDOF) systems, to show that the number of natural frequencies of a dynamic system is related to the number of degrees of freedom. Knowledge of an MDOF is very useful in relating the experimental modal analysis of many practical systems, from a turbine blade to an aircraft. In addition, using dynamic absorbers to deal with constant frequency vibration problems is presented using the concept of an MDOF system.

The most important and common aspect of machinery vibration is the design of shafts to prevent them from operating at critical speeds that will result in very high

D. N. Manik, *Fundamentals of Mechanisms and Machines*,
https://doi.org/10.1007/978-981-97-1810-8_11

vibration levels that may result in catastrophic failure. Although many analytical methods are available to solve such problems, they cannot be easily mastered within a limited space of this chapter. Therefore, approximate methods to compute the natural frequencies of shafts in flexural and torsional vibration are presented that are accurate enough to apply to many practical problems.

11.1 Single Degree of Freedom System (SDOF)

A single degree of freedom system shown in Fig. 11.1 is the basic building block of many dynamic systems that are analyzed for their vibration response. It consists of a mass (kg), spring (stiffness, N/m), and viscous damper (damping coefficient, N s/m); the mass stores kinetic energy, the spring stores potential energy, and the viscous damper represents an energy dissipating element. The above system is represented by a linear second-order differential equation, as will be derived later, which is also similar to the mathematical model of many dynamic systems in other areas like electrical, chemical engineering, control systems, etc. Hence, basic concepts of vibration analysis can be very easily understood by studying such systems. However, most dynamic systems are more complex than a single degree of freedom system. But they can be modeled as a combination of such single degree of freedom systems. Therefore, the study of a single degree of freedom system is considered first. Hence, the dynamic characteristics of an SDOF are studied in greater detail in this chapter.

The study of an SDOF has two important aspects that are related to each other. First, the response of an SDOF is due to initial conditions imposed on it, which is known as free vibration. During free vibration, an SDOF behaves in a characteristic way that actually influences the dynamic response of the SDOF during forced vibration, which is an equally important aspect of its dynamic response that can be related to vibration in practical machinery. During forced vibration, an external force is either momentarily or consistently applied, which is commonly encountered in many practical machines. Free vibration brings out unique characteristic features of a dynamic system, and the response due to force excitation depends on these characteristic features. Hence, free vibration is studied first before studying forced vibration.

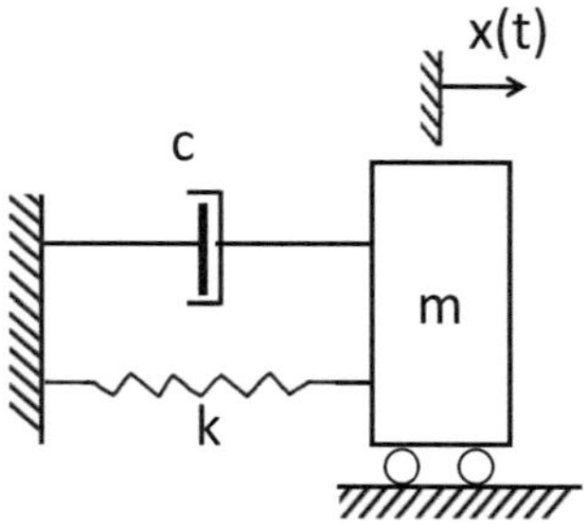

Fig. 11.1 Single degree of freedom system

11.1.1 Undamped SDOF

The mechanism of damping that actually brings a vibrating system under the action of initial conditions to rest, is actually small in many practical systems. In addition, damping becomes relevant only under certain conditions, as will be discussed later. Therefore, the characteristic features of dynamic systems can be studied by initially neglecting damping and then introducing it at an appropriate time, since it is not necessary to introduce all the complexities together. Although all practical systems do have a certain amount of damping, an undamped model nevertheless enables us to study the basic concept of exchange of kinetic and potential energy, which is the most important aspect of vibration in an SDOF. Hence, after the basic concepts are very clear, damping can be introduced later.

Since the absence of damping simplifies the discussion of dynamic systems, the free vibration of an undamped system is studied first. Once we understand the important dynamic characteristics of an undamped system, it can then be followed with the study of free vibration of a damped SDOF. Then, important concepts of logarithmic decrement and impulse response function are discussed, followed by convolution integral and force excitation.

11.1.2 Free Undamped Vibration

Consider a single degree of freedom (SDOF) system without damping as shown in Fig. 11.2, which is the same as shown in Fig. 11.1, but without the damping element; the mass is constrained to move horizontally on a frictionless surface. When the mass is initially pulled toward the right, a certain distance x_o and let go, it performs to and fro motion about its equilibrium position. This to and fro motion that can be termed as free vibration can be explained as follows.

When the mass is displaced toward right, the spring is elongated and hence stores a certain amount of potential energy. When the mass is let go, due to the restoring force of the spring, it moves left toward its equilibrium position. During the above action, the mass uses the potential energy that was initially stored in the spring by acquiring kinetic energy due to its movement. This results in a proportional decrease

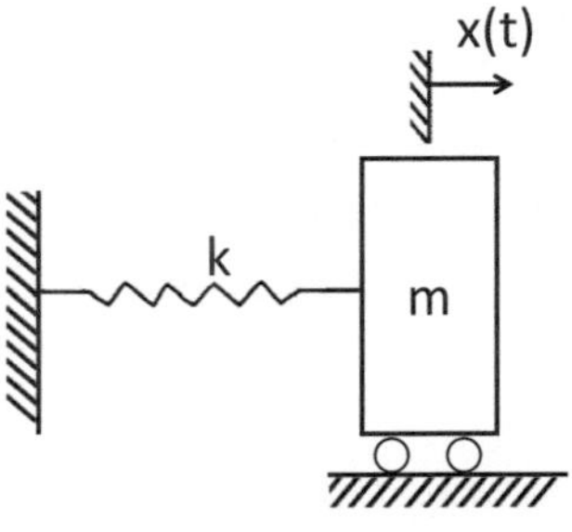

Fig. 11.2 Undamped single degree of freedom (SDOF) system

in the potential energy of the spring and a proportional increase in the kinetic energy of the mass. This conversion of potential energy of the spring to kinetic energy of the mass continues till the mass reaches its static equilibrium position. At this position, the entire potential energy of the spring would have been spent to increase the kinetic energy of the mass. Due to inertia, the mass that now has acquired maximum kinetic energy will move toward left that results in compressing the spring, due to which the kinetic energy of the mass is converted into potential energy of the spring till the mass moves to its left extreme position. This process continues till all of the kinetic energy of the mass is converted into potential energy of the spring. The position of the mass now would be at a distance of x_0 toward left from the equilibrium position, which was actually the initial displacement given to it. The spring that is now fully compressed and has stored maximum potential energy, tries to push the mass to right, reversing its direction. A similar conversion of potential energy to kinetic energy takes place as before, thus resulting in the to and fro motion of the mass. This vibratory motion should go on forever, at least theoretically, in the absence of damping. In practice, however, due to the damping provided by air or due to the structural damping of the spring, the motion comes to a complete stop after some time. That will be discussed when we introduce damping to the system.

The motion of an undamped system described above due to the exchange of potential and kinetic energy is known as free vibration. The frequency of free vibration is known as the undamped ***natural frequency*** or **resonance frequency** of the system. Obviously, it depends on mass and stiffness, which respectively store kinetic and potential energy. Free vibration can result not only due to displacement as the initial condition, but also due to velocity as the initial condition. Hence, a similar to and fro motion takes place if the mass were to be given an initial velocity, by striking it with an impulsive force. However, both displacement and velocity together can also be given as initial conditions, but that is rarely encountered in practice.

Since we have now understood the physical mechanism of free vibration, the differential equation of motion for the undamped system can be formulated. The differential equation will enable us to qualitatively and quantitatively assess the dynamic behavior of an SDOF.

11.1.3 Differential Equations of Undamped Free Vibration

Now, let us formulate a mathematical model of the undamped system presented in Fig. 11.2.

In the above experiment of the previous section, the restoring force, f_r, is proportional to the displacement of the mass from its initial position and is given by,

$$f_r = -kx \tag{11.1}$$

where k is the stiffness of the spring that is the proportionality constant between force and displacement and x is the displacement of the mass from its equilibrium position. The negative sign indicates that the restoring force is always in a direction opposite to the direction of motion of the mass.

From Eq. (11.1), applying Newton's second law of motion to the undamped system shown in Fig. 11.2, we have,

$$f_r = -kx = m\frac{d^2x}{dt^2} \tag{11.2}$$

From Eq. (11.2), the following differential equation can be written

$$m\ddot{x} + kx = 0, \tag{11.3}$$

where $\ddot{x}$ is the short form for the second derivative of time.

Dividing Eq. (11.3) by m throughout, we get the following equation

$$\ddot{x} + \frac{k}{m}x = 0 \tag{11.4}$$

The above differential equation can be solved using the method of differential operators as follows.

Equation (11.4) can be written as

$$\left(D^2 + \frac{k}{m}\right)x = 0 \tag{11.5}$$

where the differential operator $D = d/dt$. The characteristic equation of the differential equation represented by Eq. (11.5) in terms of the differential operator can be written as

$$D^2 + \frac{k}{m} = 0 \tag{11.6}$$

or

$$D = \pm j\sqrt{\frac{k}{m}} \tag{11.7}$$

From the theory of differential equations, we know that purely imaginary roots of the characteristic equation correspond to an oscillatory response. Therefore, the general solution for displacement is of the form

$$x(t) = A\sin\sqrt{\frac{k}{m}}t + B\cos\sqrt{\frac{k}{m}}t \tag{11.8}$$

Although the above equation contains sine and cosine terms, the periodic motion can only be either of them depending on the value of the constants. Displacement of the mass results in storing potential energy of the spring and velocity input due to the impulse applied to the mass results in the mass acquiring kinetic energy. Irrespective of whether the system acquires kinetic or potential energy first, it begins to vibrate as per the energy exchange mechanism explained earlier.

Equation (11.8) shows that the term $\sqrt{\frac{k}{m}}$ represents the undamped natural frequency of periodic motion of the system. Therefore, the undamped natural frequency (rad/s) is given by the following equation:

$$\omega_n = \sqrt{\frac{k}{m}} \text{rad/s} \tag{11.9}$$

In terms of undamped natural frequency, ω_n, Eq. (11.8) can be written as

$$x(t) = A \sin \omega_n t + B \cos \omega_n t \tag{11.10}$$

Which can also be written as

$$x(t) = C \cos(\omega_n t + \theta) \tag{11.11}$$

A and B in Eq. (11.10) and C and θ in Eq. (11.11) are constants, which can be obtained by using the initial conditions of displacement or velocity that were provided to initiate motion. For example, let us consider the case of an initial displacement x_0 given to an undamped SDOF system. The initial conditions for giving an initial displacement can be written in the following form:

$$x(0) = x_0 \tag{11.12}$$

and

$$\dot{x}(0) = 0 \tag{11.13}$$

It is important to note that before applying the initial conditions, the general equations of displacement and velocity must be obtained. The initial condition for displacement is used in Eq. (11.10) and the initial condition for velocity is applied after differentiating Eq. (11.10) with respect to time. From this, we get the following values of the constants A and B.

$$A = 0, B = x_0 \tag{11.14}$$

Substituting the values of A and B in Eq. (11.10) we get

$$x(t) = x_0 \cos \omega_n t \tag{11.15}$$

Therefore, the given initial condition results in the system performing a cosine periodic motion at its undamped natural frequency, with a peak amplitude equal to the initial displacement of the mass.

Similarly, the constants in Eq. (11.11), for the initial conditions given in Eqs. (11.12) and (11.13) can be obtained as:

$$C = x_0, \theta = 0. \tag{11.16}$$

The constants obtained in Eq. (11.16) will give the same equation obtained in Eq. (11.15).

Example 11.1

A single degree of freedom system with no damping consists of a mass of 0.1 kg attached to a spring of stiffness 1000 N/m attached to a fixed support. (a) Compute the undamped natural frequency in rad/s and Hz and period of free oscillation (b) If an initial displacement of 10 mm is given to the mass by pulling the mass down, determine the equation of motion of the mass (c) What is the maximum velocity of the mass and when does it occur (d) Draw the displacement and velocity response due to the above initial condition for four cycles

(a) The undamped natural frequency (rad/s) from Eq. (11.9) is given by

$$\omega_n = \sqrt{\frac{k}{m}} = \sqrt{\frac{1000}{0.1}} = 100 \text{ rad/s} \tag{1}$$

The undamped natural frequency (Hz) is given by

$$f_n = \frac{\omega_n}{2\pi} = \frac{100}{2\pi} = 15.92 \text{ Hz (cycles/s)} \tag{2}$$

Hertz[1] shortened as Hz is same as cycles/s and it is used in measurement systems. Therefore, one has to carefully use rad/s in mathematical analysis and Hz in experimental measurements.

The period of oscillation for undamped free vibration of Eq. (11.2) is given by the reciprocal of the frequency in Hertz and is given by

$$T = \frac{1}{f_n} = \frac{1}{15.92} = 63 \text{ ms} \tag{3}$$

(b) Assuming a solution of the following form to represent displacement response due to undamped free vibration

$$x(t) = x_o \cos(\omega_n t + \phi) \tag{4}$$

[1] Hertz-Armstrong-Marconi are the three scientists who have made significant contribution to inventions related to radio and the first letter of each of their last names is given the name HAM for amateur radio.

Differentiating Eq. (11.4) with respect to time, the velocity response due to undamped free vibration is given by

$$\dot{x}(t) = v(t) = -x_o\omega_n \sin(\omega_n t + \phi) \tag{5}$$

Using the given initial conditions of 10 mm displacement and zero velocity, $x_o = 10$ mm and $\phi = 0$. Using these constants in Eqs. (11.4) and (11.5), the equation for displacement response due to free vibration becomes

$$x(t) = 10\cos\omega_n t, \text{ mm} \tag{6}$$

And the equation for velocity response due to free vibration becomes

$$\dot{x}(t) = v(t) = -10\omega_n \sin\omega_n t, \text{ mm/s} \tag{7}$$

(c) From Eq. (11.7), the maximum amplitude of velocity is given by $10\omega_n = 1000$ mm/s, which is rather a large value of velocity vibration that is not normally not encountered in practice.

The maximum velocity occurs when $\omega_n t = \frac{\pi}{2} = \frac{3\pi}{2}$ or when $t = \dfrac{\pi}{2\omega_n} = \dfrac{\pi}{2 \times 1000} = 1.57$ ms and

$t = \dfrac{3\pi}{2\omega_n} = \dfrac{3\pi}{2 \times 1000} = 4.7$ ms

(d) The displacement and velocity response of the undamped system are shown in Fig. 11.3.

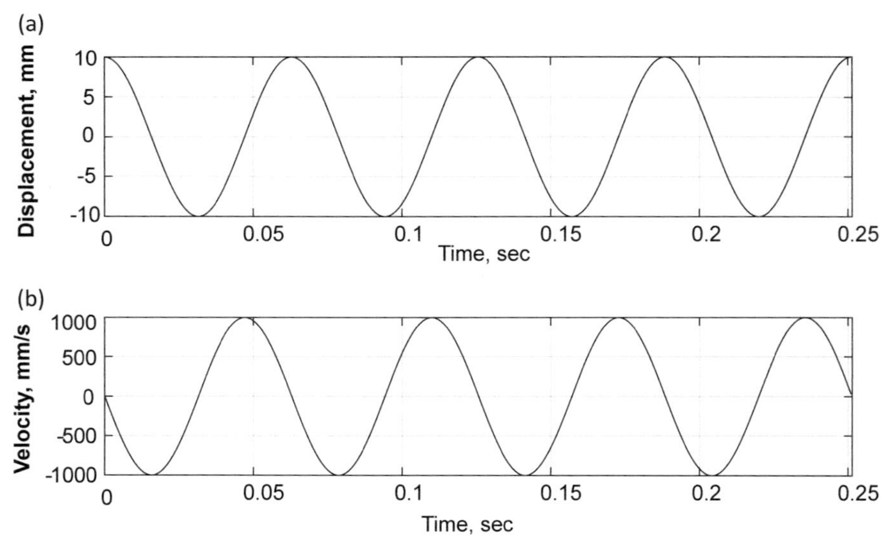

Fig. 11.3 **a** Displacement and **b** velocity response of an undamped system

11.2 Damped SDOF

Let us return to Fig. 11.1 that has all the three elements-mass, stiffness and damping. Although damping was neglected in the initial analysis, energy dissipation by damping, however small, is invariably present in all practical systems. This energy dissipating mechanism will ensure that any system excited into free vibration will eventually come to rest. In addition, damping will also restrict the amplitude of response due to a sinusoidal force excitation at the resonance frequency that will be discussed later.

There are many models of damping. The most common and convenient model is the one in which the damping force opposing the motion is assumed to be proportional to the velocity of the mass. This type of damping is known as viscous damping. The proportionality constant is defined as the damping coefficient c, expressed as N s/m.

11.2.1 Differential Equations of Damped Vibration

Consider the SDOF shown in Fig. 11.1.

Assuming viscous damping, the damping force is given by

$$f_d = -cv \tag{11.17}$$

The negative sign indicates that it is opposite to the direction of velocity of the mass.

Applying Newton's second law to the system shown in Fig. 11.1, and accounting for the restoring force of the spring from Eq. (11.1) and damping force from Eq. (11.17), we get the following differential equation of motion:

$$m\ddot{x} + c\dot{x} + kx = 0 \tag{11.18}$$

Dividing the above differential equation throughout by m, we get

$$\ddot{x} + \frac{c}{m}\dot{x} + \frac{k}{m}x = 0 \tag{11.19}$$

From our earlier discussion, undamped natural frequency, $\omega_n = \sqrt{\frac{k}{m}}$. Hence the last term of the above equation is given by

$$\omega_n^2 = \frac{k}{m} \tag{11.20}$$

A new term known as damping factor that relates mass, stiffness, and damping coefficient is defined by

$$\zeta = \frac{c}{2m\omega_n} = \frac{c}{2\sqrt{\mathrm{km}}} \tag{11.21}$$

The definition of damping factor is based on the limiting values of damping for exponential decay, oscillatory response, or a combination of both, of damped systems during free vibration. This could be known only after completely solving the differential equation. But since the definition of a damping factor through Eq. (11.21), is well known based on past experience, we can make such an assumption to save time; a zero damping factor results in an oscillatory response that has already been discussed; a damping factor less than one results in an exponential, oscillatory decay; and a damping factor of more than one results in exponential decay of the response.

Rewriting Eq. (11.21)

$$\frac{c}{m} = 2\zeta\omega_n \tag{11.22}$$

From Eqs. (11.20) and (11.22), (11.19) can be written as

$$\ddot{x} + 2\zeta\omega_n\dot{x} + \omega_n^2 x = 0 \tag{11.23}$$

Equation (11.23) in terms of undamped natural frequency and damping factor is generally used for analysis of many second-order systems. In terms of the differential operator D, Eq. (11.23) becomes

$$\left(D^2 + 2\zeta\omega_n D + \omega_n^2\right)x = 0 \tag{11.24}$$

From Eq. (11.24), the characteristic equation is obtained as

$$D^2 + 2\zeta\omega_n D + \omega_n^2 = 0 \tag{11.25}$$

The solution of the above characteristic Eq. (11.25) gives

$$D = -\zeta\omega_n \pm \omega_n\sqrt{\zeta^2 - 1} \tag{11.26}$$

Depending on the value of the damping factor ζ, the roots of the above characteristic equation take various types of values (like real, complex, repeated) and the response of the corresponding system will, therefore, be distinctively different for different values of damping factors.

Overdamping

When the damping factor $\zeta > 1$, $\omega_{d_1}, \omega_{d_2} = -\zeta\omega_n \pm \omega_n\sqrt{\zeta^2 - 1}$ is a real number, the condition is known as overdamping, hence the roots of the characteristic equation are negative and real, and therefore the time-domain response is an exponential decay.

The time-domain response due to initial conditions of an overdamped system is given by

$$x(t) = e^{-\zeta\omega_n t}\left(Ae^{\left(\omega_n\sqrt{\zeta^2-1}\right)t} + Be^{\left(-\omega_n\sqrt{\zeta^2-1}\right)t}\right) \tag{11.27}$$

where A and B are constants that can be evaluated from initial conditions and in addition, Eq. (11.27) can be expressed in terms of hyperbolic functions (sinh or cosh) also.

Critical damping

For the damping factor $\zeta = 1$, the condition is known as critical damping and the roots of the characteristic equation are $-\zeta\omega_n$ and repeated. Therefore, one of the terms will be an exponential decay and the other a product of exponential decay and time. However, the exponential decay will dominate the response of a critically damped system and hence will be similar to that of the overdamped case. It is given by

$$x(t) = \left(Ae^{-\zeta\omega_n t} + Bte^{-\zeta\omega_n t}\right) \tag{11.28}$$

where A and B are constants that can be evaluated based on initial conditions

Underdamping

For the damping factor $\zeta < 1$, the condition is known as underdamping. Underdamping is the most common damping type of many practical systems that have a very low value of damping. The roots of the characteristic equation for underdamping are complex with a negative real part that result in an exponential oscillatory decay. They are given by

$$D = -\zeta\omega_n \pm j\omega_n\sqrt{1-\zeta^2} \tag{11.29}$$

The imaginary part of Eq. (11.29) can be expressed in terms of the damped natural frequency ω_d as follows:

$$\omega_d = \omega_n\sqrt{1-\zeta^2} \tag{11.30}$$

Using Eq. (11.30), the roots of the characteristic Eq. (11.29) become

$$D = -\zeta\omega_n \pm j\omega_d \tag{11.31}$$

The time-domain response of an underdamped system ($\zeta < 1$) due to initial conditions is given by

$$x(t) = e^{-\zeta\omega_n t}(A \sin \omega_d t + B \cos \omega_d t) \tag{11.32}$$

where A and B are constants that can be evaluated from initial conditions.

Alternatively, Eq. (11.32) can also be expressed as follows

$$x(t) = Ae^{-\zeta\omega_n t} \cos(\omega_d t + \varphi) \tag{11.33}$$

where A and φ can be obtained from initial conditions.

From Eqs. (11.32) or (11.33) it is clear that the response is oscillatory at the damped natural frequency ω_d and has an exponentially decaying component $-\zeta\omega_n$.

The equation for velocity response can be obtained by taking the time-derivative of Eq. (11.33)

$$\dot{x}(t) = -Ae^{-\zeta\omega_n t}[\zeta\omega_n \cos(\omega_d t + \varphi) + \omega_d \sin(\omega_d t + \varphi)] \tag{11.34}$$

If the initial conditions are $x(0)$ and $\dot{x}(0)$, the constants of Eqs. (11.32) and (11.33) can be obtained as follows:

$$\phi = -\tan^{-1}\left(\frac{\zeta\omega_n x(0) + \dot{x}(0)}{\omega_d x(0)}\right) \tag{11.35}$$

and

$$A = \sqrt{[x(0)]^2 + \left[\frac{\zeta\omega_n x(0) + \dot{x}(0)}{\omega_d}\right]^2} \tag{11.36}$$

Example 11.2

A single degree of freedom system (SDOF) has a mass of 0.1 kg, stiffness of 1000 N/m, and a damping coefficient of 5 N s/m is subjected to an initial condition by moving it by 10 mm from its equilibrium position. (a) Determine the damping factor and damped natural frequency of oscillations (b) Determine the constants from initial conditions and obtain the equation for displacement and velocity response (c) Plot the displacement and velocity response for four cycles.

(a) The undamped natural frequency (rad/s) from Eq. (11.9) is given by

$$\omega_n = \sqrt{\frac{k}{m}} = \sqrt{\frac{1000}{0.1}} = 100 \text{ rad/s} \tag{1}$$

The damping factor from Eq. (11.21) is given by

$$\zeta = \frac{c}{2\sqrt{km}} = \frac{c}{2m\omega_n} = \frac{5}{2 \times 0.1 \times 100} = 0.25 \tag{2}$$

Therefore, this is an underdamped system.
The damped natural frequency from Eq. (11.30) is given by

$$\omega_d = \omega_n\sqrt{1-\zeta^2} = 100\sqrt{1-0.25^2} = 96.82\,\text{rad/s} \tag{3}$$

(b) Since the initial condition of velocity is not given, it can be assumed as zero. Therefore, the given initial conditions are $x(0) = 20\,\text{mm}$; $\dot{x}(0) = 0$. From Eqs. (11.35) and (11.36), the constants due to these initial conditions can be respectively obtained as

$$\begin{aligned}\phi &= -\tan^{-1}\left(\frac{\zeta\omega_n x(0) + \dot{x}(0)}{\omega_d x(0)}\right)\\ &= -\tan^{-1}\left(\frac{0.25 \times 100 \times 20)}{96.82 \times 20}\right) = -14.47^\circ\end{aligned} \tag{4}$$

and

$$\begin{aligned}A &= \sqrt{[x(0)]^2 + \left[\frac{\zeta\omega_n x(0) + \dot{x}(0)}{\omega_d}\right]^2}\\ &= \sqrt{20^2 + \left[\frac{0.25 \times 100 \times 20}{96.82}\right]^2} = 20.65\ \text{mm}\end{aligned} \tag{5}$$

Using the above constants from Eqs. (11.4) and (11.5) in (11.33), the equation for displacement becomes

$$x(t) = 20.65e^{-25t}\cos(\omega_d t - 14.14^\circ)\,\text{mm} \tag{6}$$

Differentiating Eq. (11.6), the equation for velocity is given by

$$\dot{x}(t) = -20.65e^{-25t}\begin{bmatrix} 25\cos(\omega_d t - 14.14^\circ) \\ +96.82\sin(\omega_d t - 14.14^\circ)\end{bmatrix}\text{mm/s} \tag{7}$$

(c) The displacement and velocity response due to initial conditions respectively from Eqs. (11.6) and (11.7) are shown in Fig. 11.4a, b.

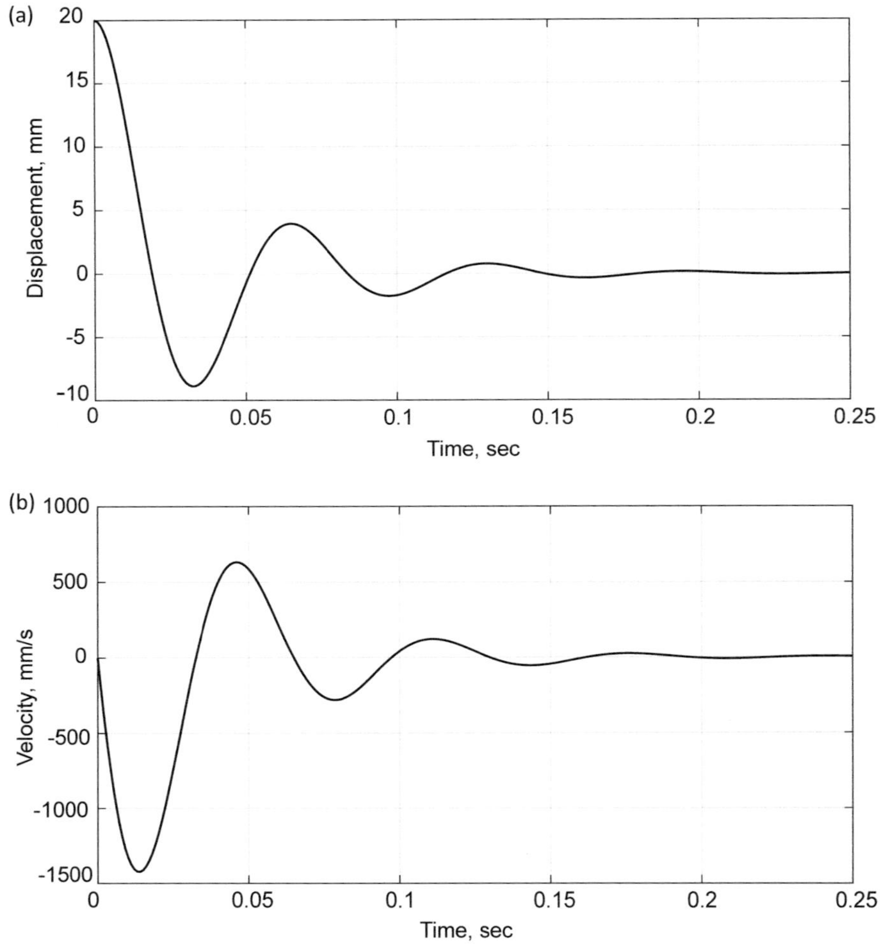

Fig. 11.4 **a** Displacement and **b** velocity response of a single degree of freedom system due to initial conditions

11.2.2 Logarithmic Decrement

For an underdamped system, the peak amplitudes of oscillation during free vibration are reduced by a definite amount (that still needs to be computed), after each period of oscillation, until the oscillations come to rest. This is shown in Fig. 11.5 in which the decay of vibration of an SDOF is traced for seven cycles after giving an initial displacement x_0.

Let the equation of motion of an underdamped system shown in Fig. 11.5 be given by

$$x = x_0 e^{-\zeta\omega_n t}\cos(\omega_d t) \tag{11.37}$$

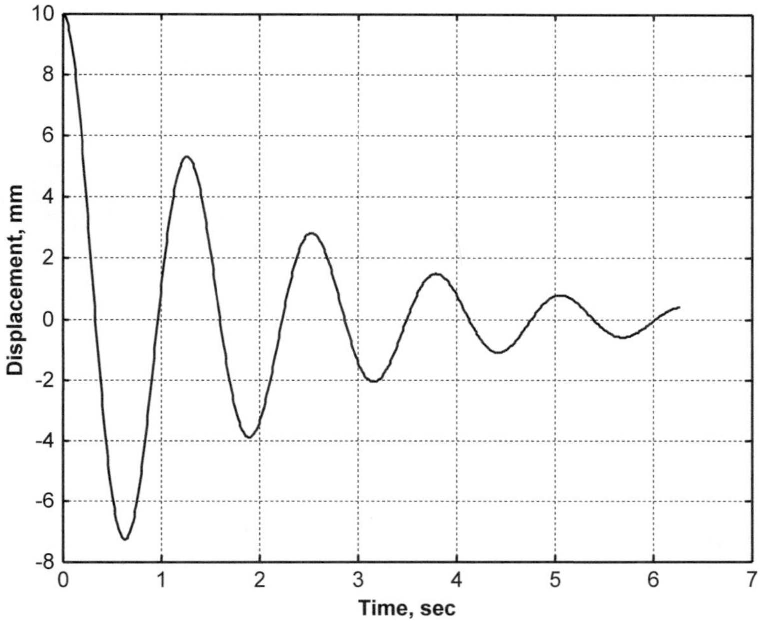

Fig. 11.5 Decay of damped vibration

After giving an initial displacement of x_o, the second peak of displacement response will occur after one period of oscillation, corresponding to the damped natural frequency ω_d and the period of this cycle is given by $2\pi/\omega_d$. The second peak therefore appears at $t_1 = 2\pi/\omega_d$ and is given by

$$x_1 = Ae^{-\zeta\omega_n(2\pi/\omega_d)}\cos 2\pi \tag{11.38}$$

Similarly, the third peak x_2, can be obtained as

$$x_2 = Ae^{-\zeta\omega_n(4\pi/\omega_d)}\cos 4\pi \tag{11.39}$$

Natural logarithm of the ratio of successive peak amplitudes of a decaying response that remains constant is defined as *logarithmic decrement.* From the above Eqs. (11.37)–(11.39), and the further ratio of peak amplitudes obtained as a logical extension, the logarithmic decrement is given by

$$\Delta = \ln\frac{x_0}{x_1} = \ln\frac{x_1}{x_2} = \ln\frac{x_2}{x_3}\cdots = \frac{2\pi\zeta\omega_n}{\omega_d} = \frac{2\pi\zeta}{\sqrt{1-\zeta^2}} \tag{11.40}$$

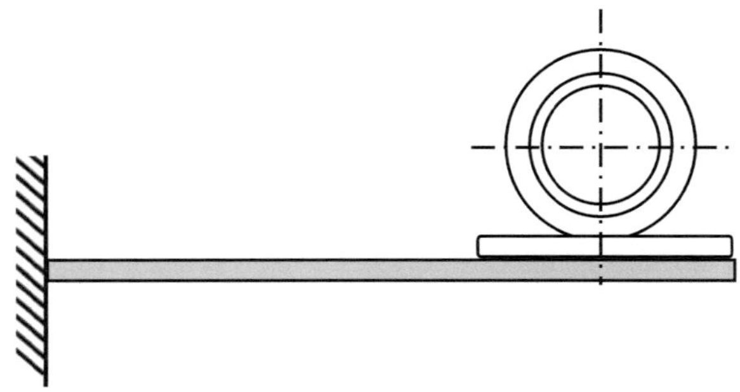

Fig. 11.6 Electric motor on a cantilever of Example 11.3

Determination of the successive peak amplitudes of an underdamped system performing free vibration is one of the methods of estimating the damping of a system using Eq. (11.40).

Example 11.3

An electric motor has a mass of 30 kg, which is mounted on a cantilever beam as shown in Fig. 11.6. When the motor is not running, it is pushed down by 10mm and allowed to vibrate along with the cantilever. It was observed that the vibration amplitude reduces to 1 mm in ten cycles within one second. Determine

(a) damping factor
(b) damped natural frequency
(c) Undamped natural frequency
(d) Stiffness of the equivalent system

(e) Equivalent viscous damping.

Mass of the motor m = 30 kg
Initial displacement $\delta_o = 10$ mm
Number of cycles of damped vibration, n = 10
Displacement at the end of 10 cycles, $\delta_{10} = 1$ mm
Extending Eq. (11.40) to obtain the logarithmic ratio of the initial to final amplitude at the nth cycle.

(a) $$\ln\left(\frac{\delta_0}{\delta_n}\right) = \frac{2\pi n\zeta}{\sqrt{1-\zeta^2}} \tag{1}$$

where ζ is the damping factor.

Equation (1) can be rearranged to obtain the damping factor as

$$\zeta = \sqrt{\frac{\left(\ln\left(\frac{\delta_0}{\delta_{10}}\right)\right)^2}{(2\pi n)^2 + \left(\ln\left(\frac{\delta_0}{\delta_{10}}\right)\right)^2}} \tag{2}$$

Natural logarithm of the ratio of the initial to final amplitude, $\ln\left(\frac{\delta_0}{\delta_{10}}\right) = \ln\left(\frac{10}{1}\right) = 2.3026$.

From Eq. (2), the damping factor is given by

$$\zeta = \sqrt{\frac{(2.3026)^2}{(2\pi \times 10)^2 + (2.3026)^2}} = 0.0133 \tag{3}$$

(b) The damped natural frequency is given by

$$\omega_d = \frac{2\pi \times n}{t_{10}} = \frac{2\pi \times 10}{1} = 62.83\,\text{rad/s} \tag{4}$$

From Eq. (11.30), using the damping factor obtained from Eq. (11.3), the undamped natural frequency can be obtained as

$$(c) \quad \omega_n = \frac{62.83}{\sqrt{1 - 0.0133^2}} = 62.8374\ \text{rad/s} \tag{5}$$

(d) Stiffness of the equivalent system, $km\omega_n^2 = 30 \times 62.8374^2 = 0.118$ MN/m
(e) Damping coefficient of the equivalent system,

$$c = 2\zeta\sqrt{km} = 2 \times 0.0133 \times \sqrt{0.118 \times 10^6 \times 30} = 50.04\,\text{N s/m} \tag{6}$$

11.2.3 Impulse Response Function

Impulse response function of a dynamic system is defined as the response of the system at rest for a unit impulse, applied as an initial condition. Hence, the equation for impulse response can be solved as an initial value problem with initial conditions of zero displacement and unit impulse at t = 0 that can be expressed as

$$x(0) = 0 \quad m\dot{x}(0) = 1 \tag{11.41}$$

From Eq. (11.32), the general equation for the free response of a damped SDOF is given by

$$x = Ae^{-\zeta\omega_n t}\cos(\omega_d t + \phi) \tag{11.42}$$

The first time-derivative of the above equation is given by

$$\dot{x} = -Ae^{-\zeta\omega_n t}[\zeta\omega_n\cos(\omega_d t + \varphi) + \omega_d\sin(\omega_d t + \varphi)] \tag{11.43}$$

Applying the initial conditions of Eqs. (11.41), (11.42) and (11.43)

$$\phi = \frac{\pi}{2} \quad A = -\frac{1}{m\omega_d \sin\phi} \tag{11.44}$$

Using the constants obtained in Eq. (11.44), (11.42) thus becomes

$$x(t) = \frac{e^{-\zeta\omega_n t}\sin(\omega_d t)}{m\omega_d} \tag{11.45}$$

Equation (11.45) is the free response of a damped SDOF system to a unit impulse, which is known as the impulse response function; it is given a special symbol $h(t)$. Thus the impulse response function of a damped SDOF is given by

$$h(t) = \frac{e^{-\zeta\omega_n t}\sin(\omega_d t)}{m\omega_d} \tag{11.46}$$

The impulse response function and its frequency-domain counterpart, the system transfer function H(ω), are related through Fourier integral, which is useful in measuring natural frequencies and mode shapes of dynamic systems using a spectrum analyzer. This is known as experimental modal analysis that plays a major role in identifying characteristics of many practical systems from turbine blades to aircrafts.

We experience impulse excitation in our everyday lives, but we are not aware of it. Knocking on the door, dropping a spoon, jumping on the floor, etc. are all impulse excitations. Therefore, there is a difference in the response of a system subjected to impulse excitation and steady-state force excitation (as we shall see in the next section). An important aspect of impulse excitation is that it excites the natural frequencies of a system. Hence, impulse excitation through an instrumented hammer that has a force transducer is commonly used in obtaining dynamic characteristics of systems (like system transfer function). The system transfer function is obtained after converting the response of the system and the force applied onto it into frequency domain using an FFT (Fast Fourier Transform) algorithm. The system transfer function can give information about natural frequencies and damping of a system. Even if it is not possible to measure the force input to a system, by giving an impulse input using a simple hammer of suitable weight, measuring the response, and converting it into a frequency domain can give a scaled system transfer function that will at least give information about the natural frequencies, if not mode shapes. Using a similar principle, a device known as Periotest is used by dentists to ascertain the natural frequencies of teeth to detect gum diseases.

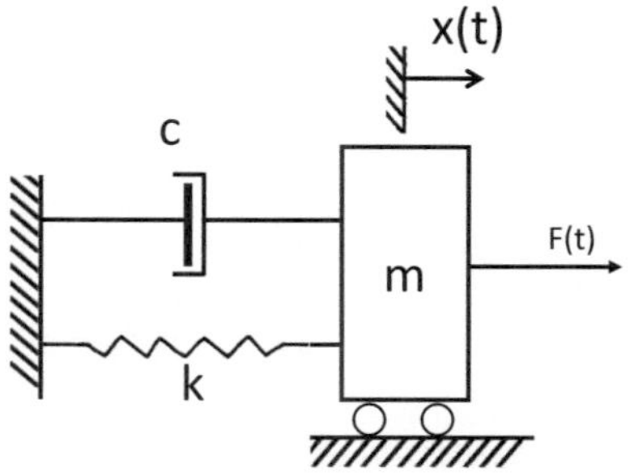

Fig. 11.7 SDOF subjected to force excitation

11.2.4 Force Excitation

A single degree of freedom system subjected to an external force excitation F(t) is shown in Fig. 11.7. From the free-body diagram of all the forces acting on the mass and equating it to the product of mass and acceleration, the differential equation of motion for an SDOF subjected to an external force excitation F(t) is given by

$$m\ddot{x} + c\dot{x} + kx = F(t) \tag{11.47}$$

The external force excitation F(t) forces the system to respond with a similar but scaled time history, depending on the nature of the force and parameters of the SDOF. Therefore, one method of computing the response of the above system is to assume a solution similar to the force excitation and then match the parameters that satisfy the Eq. (11.47). We shall use this method to determine the response of an SDOF to sinusoidal excitation in Sect. 11.2.5.1.

However, a general approach that would use a similar procedure to any type of excitation is the use of Duhamel's integral, which can compute the response for any arbitrary force excitation using the impulse response function. This integral is useful in determining the time-domain response to any arbitrary force excitation and has the additional benefit of using a numerical technique for integration. The derivation of a convolution integral is explained as follows.

11.2.5 Convolution Integral

Consider an arbitrary force excitation as a function of time as shown in Fig. 11.8. The force excitation can be divided into a number of small slices with an infinitesimal time interval $\Delta\eta$. Therefore, the force excitation is assumed to be a combination of several such sequential slices as shown in the figure.

Consider a time instant η and a corresponding slice of the force with a width of $\Delta\eta$. The impulse corresponding to this slice is given by $F(\eta)\Delta\eta$.

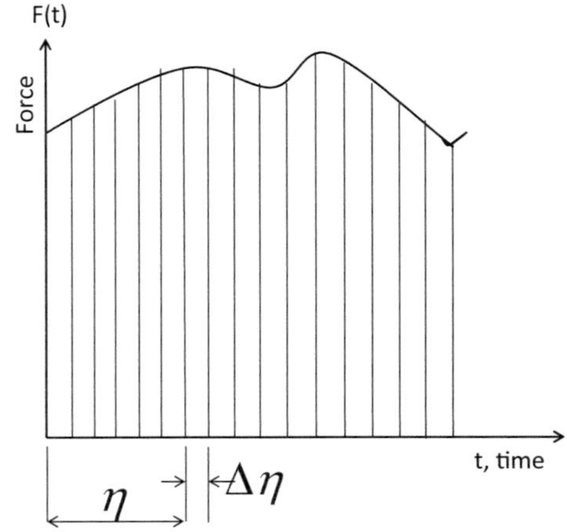

Fig. 11.8 Arbitrary force excitation

From Eq. (11.46), we know the response of a system to a unit impulse. We can use the same equation to determine the response due to the force excitation that is a combination of many impulses shown in Fig. 11.8. However, it is important to note that there can be a response only after applying the corresponding impulse. Hence, the corresponding response of the system that is applied at the instance η can only be counted after η. With respect to the force excitation that starts at t = 0, we can say that the response will start only at t- η sec after applying the corresponding slice of force.

Using the impulse response equation of (11.46), the response only due to the impulse force $F(\eta)\Delta\eta$ applied at the instant η is given by

$$x(t)|_{F(\eta)\Delta\eta} = \frac{F(\eta)\Delta\eta e^{-\zeta\omega_n\eta}\sin[\omega_d(t-\eta)]}{m\omega_d} \tag{11.48}$$

The response due to all the impulses will be an integral value of responses similar to Eq. (11.48), given by

$$x(t) = \int_0^t \frac{F(\eta)e^{-\zeta\omega_n\eta}\sin[\omega_d(t-\eta)]d\eta}{m\omega_d} \tag{11.49}$$

The short form of the above equation is

$$x(t) = \int_0^t F(\eta)h(t-\eta)d\eta = \int_0^t F(t-\eta)h(\eta)d\eta \tag{11.50}$$

The above equation is known as Duhamel's integral or convolution integral. If the time-history of any arbitrary force is known, Eq. (11.50) can be used to compute the response.

Example 11.4

A spring-mass system shown in Fig. 11.9 is initially at rest. It is subjected to an excitation force that linearly drops from 100 N to 0 within half of the time period of natural oscillation of the spring-mass system. Determine the displacement response of the system during and after removing the force. Assume the mass m = 10 kg and stiffness k = 100 kN.

The undamped **natural frequency** from Eq. (11.9) is given by $\omega_n = \sqrt{\frac{k}{m}} = \sqrt{\frac{100000}{10}} = 100\,\text{rad/s}$.

The **time period** of the undamped response is given by $\tau = \frac{2\pi}{\omega_n} = \frac{2\pi}{100} = 0.0628\,s$.

The given force can be represented by

$$F(t) = 100 - 200\frac{t}{\tau}\ \text{N}$$

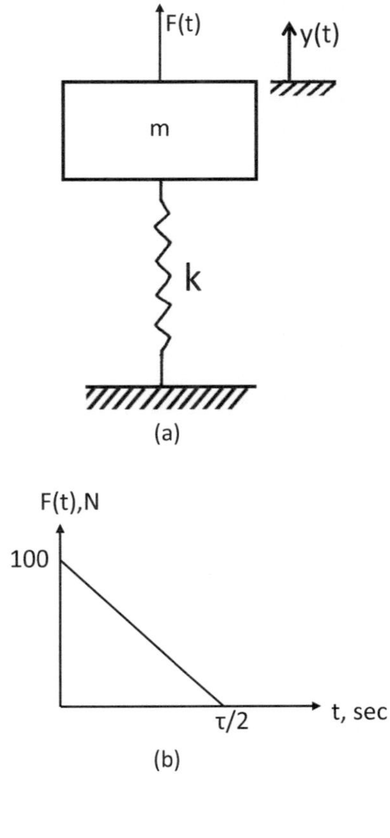

Fig. 11.9 **a** Spring-mass system of Example 11.4 **b** Time-history of force of Example 11.4

$$0 \leq t \leq \frac{\tau}{2} \tag{1}$$

where, τ is the time period of the undamped response, given by

$$\tau = \frac{2\pi}{\omega_n} \tag{2}$$

From Eq. (11.46), the impulse response function of the undamped system is given by

$$h(t) = \frac{1}{m\omega_n} \sin \omega_n t \tag{3}$$

From Eq. (11.50), the response of the SDOF for the given force can be expressed in terms of the Duhamel's integral as

$$y(t) = \frac{1}{m\omega_n} \int_0^t \sin \omega_n (t - \eta) \left(100 - 200 \frac{\eta}{\tau} \right) d\eta$$
$$0 \leq t \leq \frac{\tau}{2} \tag{4}$$

where, η is a dummy variable.

Integrating Eq. (4)

$$y(t) = \frac{100}{m\omega_n} \int_0^t \sin \omega_n (t - \eta) d\eta - \frac{200}{\tau m\omega_n} \int_0^t \eta \sin \omega_n (t - \eta) d\eta$$
$$= \frac{100}{m\omega_n} \left[\frac{\cos \omega_n (t - \eta)}{\omega_n} \Big|_0^t \right] - \frac{200}{\tau m\omega_n} \left[\frac{\eta \cos \omega_n (t - \eta)}{\omega_n} \Big|_0^t + \frac{1}{\omega_n^2} \sin \omega_n (t - \eta) \Big|_0^t \right]$$
$$0 \leq t \leq \frac{\tau}{2} \tag{5}$$

Simplifying Eq. (5)

$$y(t) = \frac{100}{k} (1 - \cos \omega_n t) - \frac{200}{k} \left(\frac{t}{\tau} - \frac{\sin \omega_n t}{2\pi} \right)$$
$$0 \leq t \leq \frac{\tau}{2} \tag{6}$$

Differentiating Eq. (6) will give the velocity response for the duration in which the force was applied

$$v(t) = \frac{100\omega_n}{k} \sin \omega_n t - \frac{200}{k\tau} (1 - \cos \omega_n t) \tag{7}$$

Since the force excitation is removed at t = τ/2, the response of the system will be in free vibration for t > τ/2 and the displacement and velocity at this instant will act as initial conditions for further motion.

From Eqs. (6) and (7)

$$\begin{aligned} y(t = \tau/2) &= 1\,\text{mm} \\ v(t = \tau/2) &= 68.2\,\text{mm/s} \end{aligned} \tag{8}$$

Time can be reset to zero at the end of forced response and the above values of displacement and velocity can be taken as initial conditions at t = 0 for free vibration. With the reset time, the new initial conditions are

$$\begin{aligned} y(t = 0) &= 1\,\text{mm} \\ v(t = 0) &= 68.2\,\text{mm/s} \end{aligned} \tag{9}$$

For free vibration, assume a solution of the following form

$$y(t) = A\sin(\omega_n t + \varphi) \tag{10}$$

where, A and ϕ are constants that can be determined based on initial conditions

From Eq. (10), the velocity of the mass during free vibration is given by

$$v(t) = A\omega_n\cos(\omega_n t + \varphi) \tag{11}$$

Applying the initial conditions of Eqs. (9), (10) and (11)

$$\begin{aligned} \varphi &= a\tan\left\{\omega_n\frac{y(0)}{v(0)}\right\} = 0.97\,\text{rad} \\ A &= \frac{y(0)}{\sin(\varphi)} = 1.2\,\text{mm} \end{aligned} \tag{12}$$

Equation (10) now becomes

$$\begin{aligned} &y(t) = 1.2\times 10^{-3}\sin(\omega_n t + 0.97)\ \text{mm} \\ &t \geq \frac{\tau}{2} \end{aligned} \tag{13}$$

11.2.5.1 Sinusoidal Excitation

Sinusoidal excitation is commonly encountered in machines with the excitation frequency generally related to multiples of its rpm. However, unbalance is the most common excitation that has an excitation frequency equal to its rpm and the amplitude of excitation is given by the product of unbalanced mass, eccentricity, and square of the speed in rad/s. Another relevant point regarding sinusoidal excitation is: In order to ensure satisfactory operation of dynamic systems to arbitrary excitation forces, from within or externally, that might be encountered in its field operation, they are subjected to sinusoidal excitation of a certain amplitude in a range of frequencies in a wide range. This is known as sweep sine testing that is one of the testing procedures of the seismic testing required for critical machinery. Therefore, the study of a single degree of freedom system subjected to sinusoidal excitation is very important.

From Eq. (11.47), a single degree of freedom system subjected to a sinusoidal excitation of peak amplitude F and frequency ω (rad/s) can be represented as

$$m\ddot{x} + c\dot{x} + kx = F\sin\omega t \tag{11.51}$$

Since the above sinusoidal excitation constantly acts on the system, the system has no choice but to respond at the same frequency ω as that of the excitation frequency. But the amplitude of response and its phase difference with the excitation force (**phase difference** is the **difference**, expressed in angular measurement, **between two waves** having the same frequency and referenced to the same point in time.) depends on all the parameters of the system-mass, damping, and stiffness- in addition to the excitation frequency ω. Therefore, the response of the system to sinusoidal excitation can be written as

$$x_p = A\sin(\omega t - \phi) \tag{11.52}$$

where A is the amplitude of response and ϕ is the phase difference of the response with the excitation force and it is assumed that the response lags behind the force excitation—hence the negative sign. The response is given a subscript p to indicate that it is part of the particular integral well known in mathematics and corresponds to the force excitation. This is done to clearly differentiate between the responses due to initial conditions. However, for most of the machines normally encountered in practice, it is the steady-state response due to force excitation that is more important than those introduced by initial conditions.

The first time-derivative of Eq. (11.52), velocity, is given by

$$\dot{x}_p = A\omega\cos(\omega t - \phi) \tag{11.53}$$

And the second time-derivative of Eq. (11.52), acceleration, is given by

$$\ddot{x}_p = -A\omega^2\sin(\omega t - \phi) \tag{11.54}$$

Using Eqs. (11.52) through (11.54) in (11.51), we have

$$\begin{aligned} &-mA\omega^2\sin(\omega t - \phi) + cA\omega\cos(\omega t - \phi) \\ &+ kA\sin(\omega t - \phi) = F\sin\omega t \end{aligned} \tag{11.55}$$

Expanding the sine and cosine functions on the left hand side of Eq. (11.55), we get

$$\begin{aligned} & -mA\omega^2[\sin\omega t\cos\phi - \cos\omega t\sin\phi] + cA\omega[\cos\omega t\cos\phi + \sin\omega t\sin\phi] \\ & + kA[\sin\omega t\cos\phi - \cos\omega t\sin\phi] = F\sin\omega t \end{aligned} \tag{11.56}$$

Regrouping terms in the above equation corresponding to sin ωt and cos ωt

$$\begin{aligned} & \sin\omega t\left[-mA\omega^2\cos\phi + kA\cos\phi + cA\omega\sin\phi\right] + \\ & \cos\omega t\left[mA\omega^2\sin\phi + cA\omega\cos\phi - kA\sin\phi\right] = F\sin\omega t \end{aligned} \tag{11.57}$$

Equating the coefficients of sin ωt on both sides of Eq. (11.57)

$$\left[\left(k - m\omega^2\right)\cos\phi + c\omega\sin\phi\right]A = F \tag{11.58}$$

Equating the coefficients of cos ωt on both sides of Eq. (11.57)

$$\left[\left(k - m\omega^2\right)\sin\phi - c\omega\cos\phi\right]A = 0 \tag{11.59}$$

From Eq. (11.59), the phase angle between displacement response and force excitation is given by

$$\tan\phi = \frac{c\omega}{k - m\omega^2} \tag{11.60}$$

From Eq. (11.60), using the trigonometric relations, the following equations are obtained

$$\sin\phi = \frac{c\omega}{\sqrt{\left(k - m\omega^2\right)^2 + (c\omega)^2}} \tag{11.61}$$

and

$$\cos\phi = \frac{k - m\omega^2}{\sqrt{\left(k - m\omega^2\right)^2 + (c\omega)^2}} \tag{11.62}$$

From Eqs. (11.58), (11.61), and (11.62) the peak amplitude of displacement response becomes

$$
\begin{aligned}
A &= \frac{F}{\left[\left(k - m\omega^2\right)\cos\phi + c\omega \sin\phi\right]} \\
&= \frac{F}{\left[\frac{\left(k-m\omega^2\right)^2}{\sqrt{\left(k-m\omega^2\right)^2+(c\omega)^2}} + \frac{(c\omega)^2}{\sqrt{\left(k-m\omega^2\right)^2+(c\omega)^2}}\right]} \\
&= \frac{F}{\sqrt{\left(k - m\omega^2\right)^2 + (c\omega)^2}}
\end{aligned}
\tag{11.63}
$$

A non-dimensional parameter known as forcing frequency ratio, that is the ratio of excitation frequency to undamped natural frequency, can be defined as follows

$$
r = \frac{\omega}{\omega_n} \tag{11.64}
$$

By using forcing frequency ratio and damping factor, the amplitude of response due to sinusoidal excitation is given by

$$
A = \frac{F}{k\sqrt{\left(1 - r^2\right)^2 + (2\zeta r)^2}} \tag{11.65}
$$

and the phase angle is given by

$$
\phi = \tan^{-1} \frac{2\zeta r}{1 - r^2} \tag{11.66}
$$

Using the amplitude and phase values from Eqs. (11.65) and (11.66) respectively, the response of the single degree of freedom system to sinusoidal excitation, represented by Eq. (11.52), is given by

$$
x_p = \frac{F}{k\sqrt{\left(1 - r^2\right)^2 + (2\zeta r)^2}} \sin(\omega t - \phi) \tag{11.67}
$$

The amplitude response is normalized by using the ratio of peak amplitude of force and stiffness, which is the equivalent static deflection due to the peak amplitude of force and is given by

$$
\frac{x_p}{F/k} = \frac{1}{\sqrt{\left(1 - r^2\right)^2 + (2\zeta r)^2}} \tag{11.68}
$$

The advantage of using normalized response is that it becomes easier to compare qualitative characteristics of similar systems. Because, the right hand side of Eq. (11.68) contains only frequency ratio and damping factor.

Example 11.5

Refer to Example 11.3. If the unbalanced mass of the rotating parts of the motor is 0.05 kg at a radius of 50 mm, determine the amplitude and phase of vertical displacement response when the motor is rotating at 4000 RPM.

Rotational speed of the motor, N = 4000 RPM.

Excitation frequency, $\omega = \frac{2\pi N}{60} = \frac{2\pi \times 4000}{60} = 418.87$ rad/s.

Amplitude of excitation due to the unbalanced mass,

$F = me\omega^2 = 0.05 \times 0.05 \times 418.87^2 = 438.65$ N.

Frequency ratio $r = \frac{\omega}{\omega_n} = \frac{418.87}{62.8374} = 6.66$.

From Eq. (11.65), the amplitude of response is given by

$$\begin{aligned} Y &= \frac{F}{k\sqrt{(1-r^2)^2 + (2\zeta r)^2}} \\ &= \frac{438.65}{0.118 \times 10^6\sqrt{(1-6.66^2)^2 + (2 \times 0.0133 \times 6.66)^2}} = 85.25\,\mu\text{m} \end{aligned} \tag{6}$$

From Eq. (11.66), phase of the response is given by

$$\phi = \tan^{-1}\frac{2\zeta r}{1-r^2} = \tan^{-1}\frac{2 \times 0.0133 \times 6.66}{1-6.66^2} \approx 180^\circ \tag{7}$$

The displacement response and the unbalanced force are out of phase with each other. Hence, the vibration response due to unbalance is very small. However, during starting and stopping of the motor, there will be an instantaneous large response when the motor speed momentarily passes through the resonance frequency of 62.83 rad/s (600 RPM).

Normalized response

The normalized amplitude responses of a single degree of freedom system, for various damping factors, due to sinusoidal excitation at various frequencies are shown in Fig. 11.10. The normalized response is unity at very low frequencies, and as the excitation frequency gets closer to the undamped natural frequency (also known as resonance frequency) near r = 1, the normalized response increases to a large value depending on the damping factor and then becomes negligible at higher frequencies. This is the typical response of most dynamic systems defined by a second-order differential equation. Since the excitation frequency is generally related to the machine rpm, when the machine starts from rest and slowly moves to its normal operating speed, the excitation frequency passes through the resonance frequency that results in a large response. If this phenomenon of the excitation frequency passing through the resonance frequency is for a very short period of time, it is not a matter of concern. Sometimes, in turbine rotors that operate at very high speeds of more than 10,000 rpm, there are many resonant frequencies for a multi-degree of freedom system and the excitation frequency may have to pass through the third resonant frequency before reaching the operating speed. We shall discuss such situations later in this chapter. It is important to note that the only way of reducing the response at frequencies around the resonance frequency is by increasing the amount of damping in the system.

Figure 11.11 shows the phase lag of the response due to sinusoidal excitation at various frequencies for different values of damping. At very low excitation frequencies, the phase lag is low and it increases with the excitation frequencies, and at a resonance frequency ($\omega = \omega_n$) the phase lag is 90° for all values of damping. In addition, there is a large shift in the phase around the resonance frequency.

Velocity response can be obtained by differentiating Eq. (11.67) with respect to time. The phase lag of the velocity response with respect to force excitation gives a better physical explanation to the phenomenon of resonance; the velocity response is in phase with the sinusoidal force excitation that is same as the resonant frequency. Hence, the maximum value of velocity response always occurs at the resonance frequency; whereas the maximum displacement occurs at an excitation frequency slightly less than the resonant frequency depending on the damping

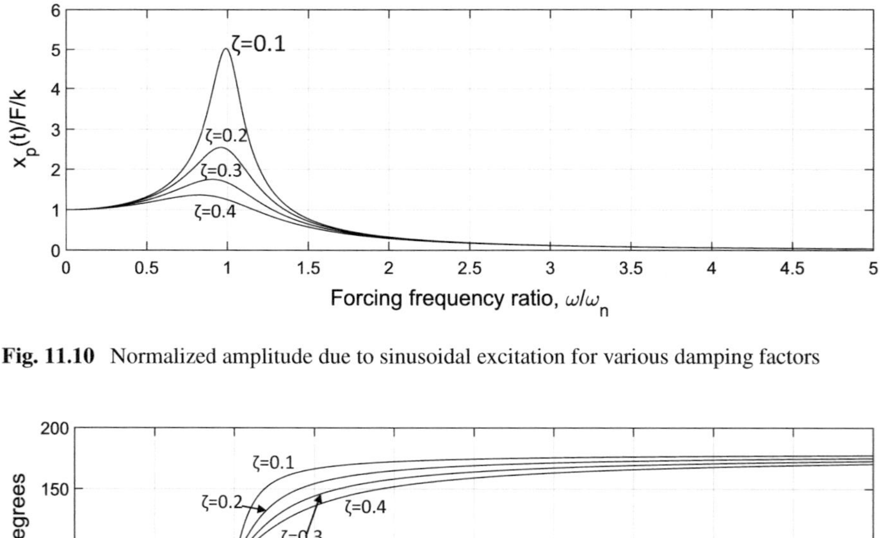

Fig. 11.10 Normalized amplitude due to sinusoidal excitation for various damping factors

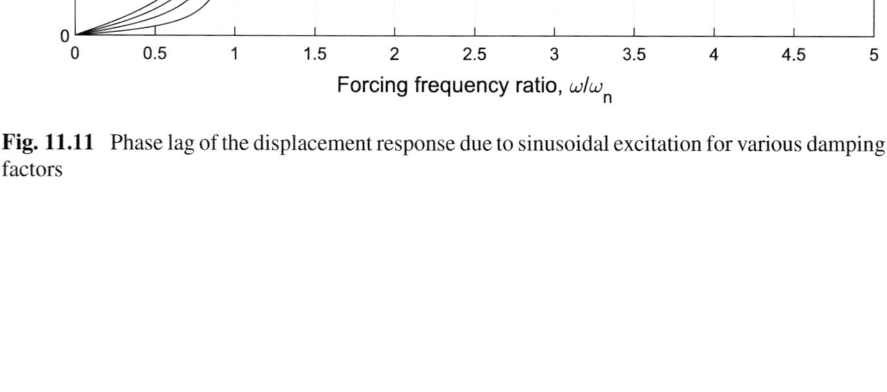

Fig. 11.11 Phase lag of the displacement response due to sinusoidal excitation for various damping factors

factor. It may be noted that the peak amplitude of velocity is a product of the peak amplitude of displacement and the excitation frequency. Hence, for imposing limits on general machinery vibration that has many frequencies, the average velocity response across frequencies (rms value) is used as an objective parameter.

Example 11.6

A spring-mass-damper system has a mass of 0.1 kg, a spring of stiffness constant 1000 N/m, and damping coefficient 0.15 N s/m. It is subjected to a driving force that has two excitation frequencies, represented by $F = 2\cos(85t) + 4\sin(120t)$ N. (a) Determine the time-domain displacement response of the above system (b) Why the phase angles are different by an order of magnitude?

Mass, m = 0.1 kg
Stiffness, k = 1000 N/m
Damping coefficient, c = 0.15 N s/m
From Eq. (9), the undamped natural frequency is given by

$$\omega_n = \sqrt{\frac{k}{m}} = \sqrt{\frac{1000}{0.1}} = 100\,\text{rad/s} \tag{1}$$

From Eq. (11.21), the damping factor is given by

$$\zeta = \frac{c}{2\sqrt{km}} = \frac{0.15}{2\sqrt{1000 \times 0.1}} = 0.0075 \tag{2}$$

Amplitude of the first force $F_1 = 2$ N
Amplitude of the second force $F_2 = 4$ N
Excitation frequency of the first force $\omega_1 = 85$ rad/s
Excitation frequency of the second force $\omega_2 = 120$ rad/s
Frequency ratio of the first force $r_1 = \frac{\omega_1}{\omega_n} = \frac{85}{100} = 0.85$
Frequency ratio of the second force $r_2 = \frac{\omega_2}{\omega_n} = \frac{120}{100} = 1.2$
From Eq. (11.67), the time-domain response of the SDOF is given by

$$\begin{aligned} y(t) = &\frac{F_1}{k\sqrt{(1-r_1^2)^2 + (2\zeta r_1)^2}} \cos(\omega_1 t - \phi_1) \\ &+ \frac{F_2}{k\sqrt{(1-r_2^2)^2 + (2\zeta r_2)^2}} \sin(\omega_2 t - \phi_2) \end{aligned} \tag{3}$$

where the respective phase angles for the two excitation frequencies are given by (from Eq. (11.66))

$$\phi_1 = \tan^{-1}\frac{2\zeta r_1}{1-r_1^2} = \tan^{-1}\frac{2 \times 0.0075 \times 0.85}{1 - 0.85^2} = 2.63^\circ \tag{4}$$

$$\phi_2 = \tan^{-1}\frac{2\zeta r_2}{1 - r_2^2} = \tan^{-1}\frac{2 \times 0.0075 \times 1.2}{1 - 1.2^2} = 177.65^\circ \quad (5)$$

From Eqs. (3), (4) and (5)

$$y(t) = \frac{2}{1000\sqrt{(1 - 0.85^2)^2 + (2 \times 0.0075 \times 0.85)^2}}\cos(85t - 2.63^\circ) + \frac{4}{1000\sqrt{(1 - 1.2^2)^2 + (2 \times 0.0075 \times 1.2)^2}}\sin(120t - 177.65^\circ)\ \text{m} \quad (6)$$

Simplifying Eq. (6)

$$y(t) = 0.0072\cos(85t - 2.63^\circ) + 0.0091\sin(120t - 177.65^\circ)\ \text{m} \quad (7)$$

(b) Excitation frequencies are on either side of the undamped resonance frequency and farther away from it. Hence, there is a distinct difference in phase between displacement and force at excitation frequencies on either side of the resonance.

11.3 Vibration Isolation and Transmissibility

Consider a single degree of freedom system of mass m, stiffness k, and viscous damping coefficient c, subjected to a force excitation F sinωt. The mass *m* and the force excitation can form part of a rigid machine and stiffness and damping can be part of the isolator on which the machine is mounted. The isolator elements need to be designed to prevent the machinery vibration from reaching its foundation.

From Eq. (11.67), the displacement response is given by

$$y(t) = \frac{F}{k\sqrt{(1 - r^2)^2 + (2\zeta r)^2}}\sin(\omega t - \phi) \quad (11.69)$$

And the velocity response is given by

$$v(t) = \dot{y}(t) = \frac{F\omega}{k\sqrt{(1 - r^2)^2 + (2\zeta r)^2}}\cos(\omega t - \phi) \quad (11.70)$$

From Eqs. (11.69) and (11.70), the force transmitted to the foundation is given by

$$F_T = ky(t) + cv(t) \quad (11.71)$$

From Eqs. (11.69) through (11.71), the above equation becomes

$$F_T = \frac{kF}{k\sqrt{(1 - r^2)^2 + (2\zeta r)^2}}\sin(\omega t - \phi)$$

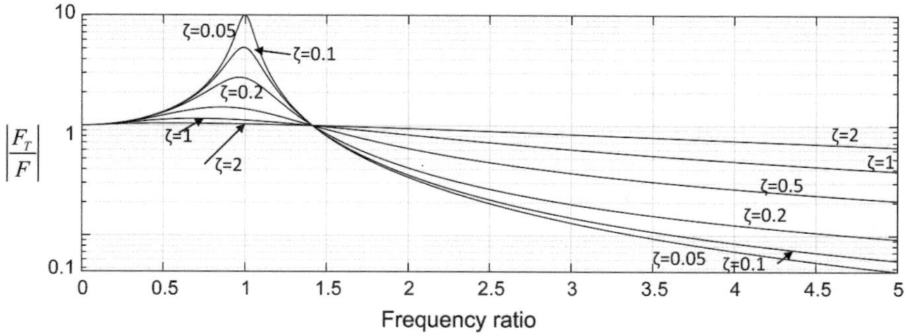

Fig. 11.12 Magnitude of force transmissibility ratio for various values of damping

$$+\frac{cF\omega}{k\sqrt{(1-r^2)^2+(2\zeta r)^2}}\cos(\omega t-\phi) \tag{11.72}$$

where $T=\frac{F_T}{F}$ is the force transmissibility ratio.

The above equation can be simplified as

$$T=\frac{F_T}{F}=\frac{\sqrt{1+(2\zeta r)^2}}{\sqrt{(1-r^2)^2+(2\zeta r)^2}}\sin(\omega t+\theta-\phi) \tag{11.73}$$

where the phase angles are given by

$$\theta=\tan^{-1}2\zeta r;\quad \phi=\tan^{-1}\frac{2\zeta r}{1-r^2} \tag{11.74}$$

Equation (11.73) is plotted in Fig. 11.12.

1. The transmissibility decreases with increase in damping at the resonance.
2. The family of curves for different values of damping intersect at $T=1$ and $r=\sqrt{2}$.
3. For frequency ratio r greater than $\sqrt{2}$, the transmissibility decreases.
4. The design of vibration isolators is based on the transmissibility curves for $r>\sqrt{2}$. The force transmissibility ratio curves will be lifted up from those presented in Fig. 11.12 if the excitation force has an amplitude related to operating frequency of the machine, as in the case of unbalance. Hence, vibration isolators will have to be carefully designed, especially at the normal machine speeds that are in the high-frequency ratio region of the above figure.

Many times, it may be necessary to isolate a sophisticated instrument from receiving vibration inputs from machines that are installed in the vicinity. A similar vibration isolator can be used for such a purpose. Since one may have to decide a location at which the instrument has to be mounted, the base vibration of the location is taken as the input to the isolator. The response of the instrument is taken as the output; it could be either displacement or velocity. For example, if Y_b is the peak amplitude of displacement of the base, the actual amplitude of displacement of the mass representing the instrument can be obtained on the same lines as force transmissibility and is given by

$$\frac{Y}{Y_b} = \frac{\sqrt{1 + (2\zeta r)^2}}{\sqrt{(1 - r^2)^2 + (2\zeta r)^2}} \sin(\omega t + \theta - \phi) \tag{11.75}$$

where the phase angles are given by

$$\theta = \tan^{-1} 2\zeta r; \quad \phi = \tan^{-1} \frac{2\zeta r}{1 - r^2} \tag{11.76}$$

Example 11.7

A measuring instrument of mass 10 kg has to be installed on a floor that is subject to vibration excitation by other machines that are installed on the same floor, which vibrate with an amplitude of 5 mm at a frequency of 3000 RPM. In order that the electronic components of the instrument satisfactorily operate, the instrument requires that the maximum displacement amplitude is limited to 100 μm.

(a) What is the equivalent stiffness of an isolator with a damping ratio of 0.01 to limit the transmitted displacement to an acceptable level?
(b) What is the maximum acceleration of the measuring instrument?
(c) What is the maximum deformation of the isolator?

The single degree system model of displacement transmissibility is shown in Fig. 11.13. The base vibration y_b represents the plant floor and the mass m representing the instrument has a displacement y. Stiffness and damping are provided by the isolators.

Mass m = 10 kg.

Maximum displacement amplitude of the floor, $Y_b = 5$ mm.

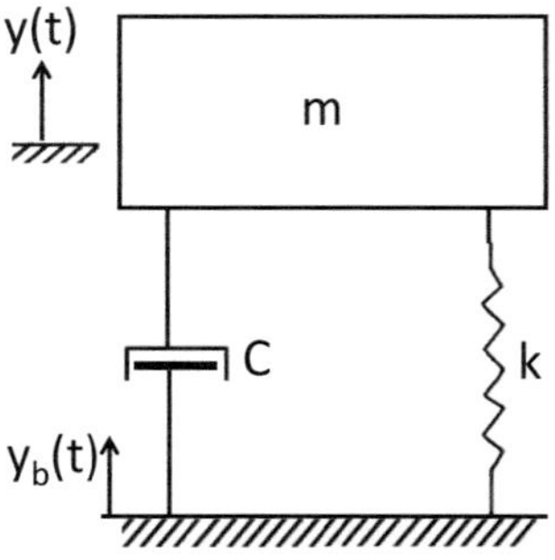

Fig. 11.13 Vibration isolator

Speed N = 3000 ω = 2πN/60 = 314.16 rad/s.
Maximum displacement amplitude of the instrument, Y = 100 μm.
Damping factor ζ = 0.01.

(a) Transmissibility of the isolation system at the given rotating frequency is given by

$$T = \frac{Y}{Y_b} = \frac{100 \times 10^{-6}}{5 \times 10^{-3}} = 0.02 \tag{1}$$

From Eq. (11.75), the transmissibility equation is given by

$$T = \frac{\sqrt{1 + (2\zeta r)^2}}{\sqrt{\left(1 - r^2\right)^2 + (2\zeta r)^2}} \tag{2}$$

where, r is the forcing frequency ratio.
Rearranging Eq. (2)

$$T^2\left[\left(1 - r^2\right)^2 + (2\zeta r)^2\right] = 1 + (2\zeta r)^2 \tag{3}$$

$$T^2 r^4 + \left[(2\zeta)^2\left(T^2 - 1\right) - 2T^2\right]r^2 + T^2 - 1 = 0 \tag{4}$$

Using the transmissibility value of T = 0.02 from Eqs. (1), (4) becomes

$$4 \times 10^{-4} r^4 - 1.199 \times 10^{-3} r^2 - 0.9996 = 0 \tag{5}$$

Equation (5) is a quadratic equation in r^2 and its roots are given by

$$r = \sqrt{\frac{0.001199 \pm \sqrt{0.001199^2 + 4 \times 4 \times 10^{-4} \times 0.9996}}{2 \times 4 \times 10^{-4}}} = 7.18 \tag{6}$$

The undamped natural frequency formed by the instrument and isolator system is given by

$$\omega_n = \frac{\omega}{r} = \frac{314.16}{7.18} = 43.75 \text{ rad/s} \tag{7}$$

The spring stiffness is given by

$$k = m\omega_n^2 = 10 \times 43.75^2 = 19.14 \text{ kN/m} \tag{8}$$

(b) The acceleration of the measuring instrument is given by

$$A = Y\omega^2$$

$$= 100 \times 10^{-6} \times 314.16^2 \text{m/s}^2$$
$$= 9.86 \text{ m/s}^2 \quad (9)$$

(c) Maximum deformation of the isolator is given by

$$\delta = (Y_b - Y)$$
$$= 5 - 0.1 = 4.9 \text{ mm} \quad (10)$$

11.4 Multi-degree of Freedom Discrete System

Most practical systems can be modeled in terms of a number of spring-mass-damper systems connected to each other. They have many degrees of freedom and are, therefore, known as multi-degree of freedom (MDOF) systems. Since this is an introductory chapter on vibrations, only the analysis corresponding to determining the natural frequencies and mode shapes of spring-mass systems are covered. Further discussions on force excitation of MDOF systems with damping, etc. are covered in specialized textbooks on vibration.

Consider a two-degree of freedom system shown in Fig. 11.14, formed by springs k_1, k_2, k_3, and masses m_1 and m_2. The system has two degrees of freedom y_1 and y_2, measured in the vertical direction from their respective static equilibrium positions.

A small displacement is given to both degrees of freedom and by considering the forces and acceleration of each of the masses and applying Newton's second law, the equations of motion are derived for free vibration. In the free-body diagrams of

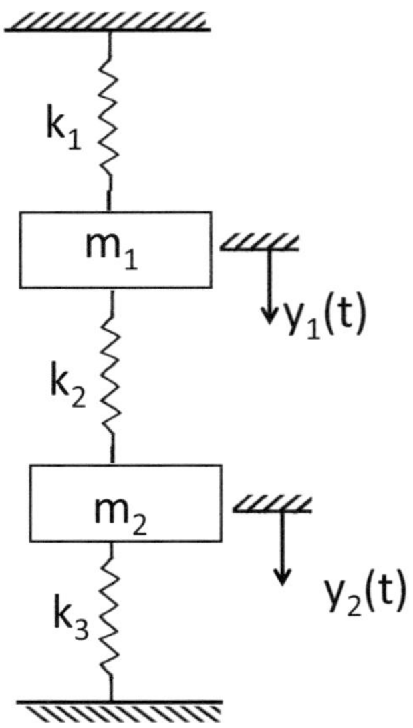

Fig. 11.14 Two-degree of freedom discrete system

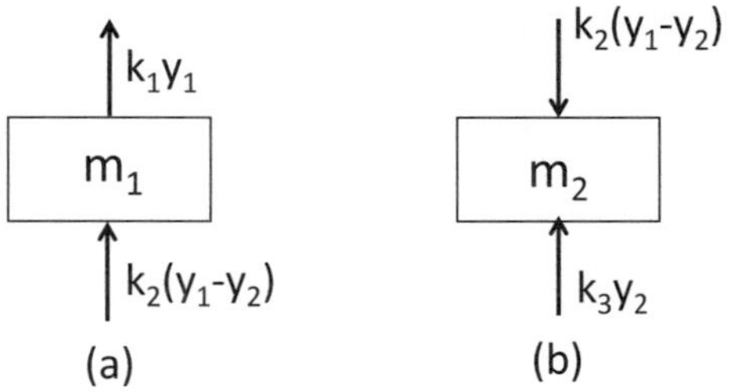

Fig. 11.15 Free-body diagram of masses shown in Fig. 11.14 due to initial deflection in the downward direction (assume $y_1 > y_2$)

both masses in Fig. 11.15, downward direction is taken as positive and y_1 is assumed greater than y_2. However, even if it is assumed that y_2 is greater than y_1, the same equations result.

From Fig. 11.15a, applying Newton's second law to the first mass

$$m_1\ddot{y}_1 = -k_1y_1 - k_2(y_1 - y_2) \tag{11.77}$$

From Fig. 11.15b, applying Newton's second laws to the second mass

$$m_2\ddot{y}_2 = k_2(y_1 - y_2) - k_3y_2 \tag{11.78}$$

Rearranging Eqs. (11.77) and (11.78)

$$m_1\ddot{y}_1 + k_1y_1 + k_2(y_1 - y_2) = 0 \tag{11.79}$$

$$m_2\ddot{y}_2 - k_2(y_1 - y_2) + k_3y_2 = 0 \tag{11.80}$$

Regrouping terms corresponding to y_1 and y_2 of Eqs. (11.79) and (11.80)

$$m_1\ddot{y}_1 + (k_1 + k_2)y_1 - k_2y_2 = 0 \tag{11.81}$$

$$m_2\ddot{y}_2 - k_2y_1 + (k_2 + k_3)y_2 = 0 \tag{11.82}$$

Equations (11.81) and (11.82) can be put in a matrix form as

$$\begin{bmatrix} m_1 & 0 \\ 0 & m_2 \end{bmatrix}\begin{Bmatrix} \ddot{y}_1 \\ \ddot{y}_2 \end{Bmatrix} + \begin{bmatrix} k_1 + k_2 & -k_2 \\ -k_2 & k_2 + k_3 \end{bmatrix}\begin{Bmatrix} y_1 \\ y_2 \end{Bmatrix} = 0$$
$$\Rightarrow \boldsymbol{M}\ddot{\boldsymbol{y}} + \boldsymbol{K}\boldsymbol{y} = 0 \tag{11.83}$$

The mass matrix $\boldsymbol{M}$ is a diagonal matrix that has value of masses corresponding to each degree of freedom along the diagonal and hence the off-diagonal elements are zero.

The stiffness matrix $\boldsymbol{K}$ can also be formed for any general MDOF based on the following rule, since it is difficult to write free-body diagrams for a very large interconnected MDOF system of springs and masses:

(1) Sum of all the stiffnesses connected to a degree of freedom form the diagonal terms
(2) Negative of the stiffness connecting one degree of freedom to the other degrees of freedom form the off-diagonal elements

To determine the natural frequencies of the above system, it is assumed that both degrees of freedom vibrate at a frequency ω, but with different peak amplitudes respectively Y_1 and Y_2. The time-domain response corresponding to each degree of freedom is given by

$$\begin{aligned} y_1 &= Y_1 \cos \omega t \\ y_2 &= Y_2 \cos \omega t \end{aligned} \tag{11.84}$$

The second derivative of Eq. (11.89) with respect to time is given by

$$\begin{aligned} \ddot{y}_1 &= -\omega^2 Y_1 \cos \omega t \\ \ddot{y}_2 &= -\omega^2 Y_2 \cos \omega t \end{aligned} \tag{11.85}$$

From Eqs. (11.83) through (11.85)

$$\begin{bmatrix} k_1 + k_2 - m_1\omega^2 & -k_2 \\ -k_2 & k_2 + k_3 - m_2\omega^2 \end{bmatrix} \begin{Bmatrix} Y_1 \\ Y_2 \end{Bmatrix} = 0 \tag{11.86}$$

Since the peak amplitudes cannot be zero in Eq. (11.86), the condition for both masses to vibrate at a frequency is given by the determinant of the matrix should be equal to zero is given by

$$\begin{vmatrix} k_1 + k_2 - m_1\omega^2 & -k_2 \\ -k_2 & k_2 + k_3 - m_2\omega^2 \end{vmatrix} = 0 \tag{11.87}$$

Expanding the above determinant of Eq. (11.87) and collecting terms corresponding to ω^4 and ω^2

$$m_1 m_2 \omega^4 - [m_1(k_2 + k_3) + m_2(k_1 + k_2)]\omega^2 + k_1 k_2 + k_2 k_3 + k_1 k_3 = 0 \tag{11.88}$$

Equation (11.88) is a quadratic in ω^2, which can be solved to obtain the resonant frequencies ω_1 and ω_2 that correspond to the positive roots of this equation. The same procedure can be applied to any MDOF system with much larger degrees of freedom. Equation (11.86) can also be cast as an eigenvalue problem; the eigenvalues and the corresponding eigen vectors respectively correspond to the natural frequencies and mode shapes of the MDOF.

If both masses have to vibrate at a single frequency, their amplitudes must have constant relative amplitude at this frequency. These are known as mode shapes. Once the resonant frequencies are computed from Eq. (11.88), each of them can be substituted in Eq. (11.86) to obtain the relative amplitudes (mode shapes) corresponding to each resonant frequency. Since Eq. (11.86) represents a system of homogeneous equations, only ratios of the peak amplitude corresponding to various degrees of freedom for an undamped natural frequency can be obtained. Assuming one of them to be unity, other relative displacements can be obtained. These relative displacements are determined for each of the undamped natural frequencies and represented as a column vector and the column vectors corresponding to all the undamped natural frequencies are assembled into a matrix called modal matrix, represented by $\boldsymbol{\psi}$. This is known as modal matrix and its elements depend on the reference degree of freedom chosen and its value. However, the ratios remain the same. In order to make a modal matrix unique, a mass-normalized modal matrix is defined in terms of the above modal matrix as $\boldsymbol{\Phi} = \boldsymbol{\psi}\left(\boldsymbol{\psi}^T \boldsymbol{M} \boldsymbol{\psi}\right)^{-0.5}$ (see exercise problem 11). Both the modal and mass-normalized modal matrices can be used for decoupling the system of differential equations for solving both initial value and force excitation problems.

Example 11.8

6 Consider the two-degree of freedom spring-mass-damper system shown in the Fig. 11.16. Assume $k_1 = 10$ kN/m, $k_2 = 30$ kN/m, $k_3 = 20$ kN/m and $m_1 = 2$ kg, $m_2 = 1$ kg.

(a) Determine the undamped natural frequencies of the system, (b) the corresponding mode shapes (assuming that the first degree of freedom moves unit distance), and (c) the mass-normalized modal matrix.

(b) The mass matrix is given by

$$\boldsymbol{M} = \begin{bmatrix} 2 & 0 \\ 0 & 1 \end{bmatrix} \tag{1}$$

Stiffness matrix is given by

$$\boldsymbol{K} = \begin{bmatrix} k_1 + k_2 & -k_2 \\ -k_2 & k_2 + k_3 \end{bmatrix} = 10^4 \begin{bmatrix} 4 & -3 \\ -3 & 5 \end{bmatrix} \tag{2}$$

(c) Thus, the equations of motion become

$$\begin{bmatrix} 2 & 0 \\ 0 & 1 \end{bmatrix} \begin{Bmatrix} \ddot{y}_1 \\ \ddot{y}_2 \end{Bmatrix} + 10^4 \begin{bmatrix} 4 & -3 \\ -3 & 5 \end{bmatrix} \begin{Bmatrix} y_1 \\ y_2 \end{Bmatrix} = 0 \tag{3}$$

Assuming the following solutions.

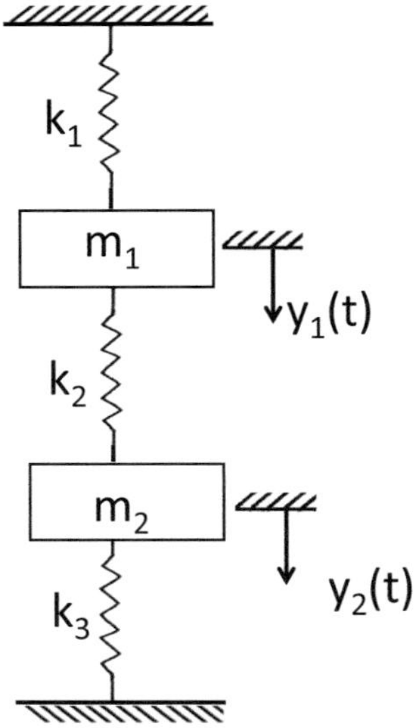

Fig. 11.16 Example 11.8

$$\begin{aligned} y_1 &= Y_1 \cos \omega t\ \ddot{y}_1 = -\omega^2 Y_1 \cos \omega t \\ y_2 &= Y_2 \cos \omega t\ \ddot{y}_2 = -\omega^2 Y_2 \cos \omega t \end{aligned} \tag{4}$$

Equation (3) becomes

$$\begin{bmatrix} 4 \times 10^4 - 2\omega^2 & -3 \times 10^4 \\ -3 \times 10^4 & 5 \times 10^4 - \omega^2 \end{bmatrix} \begin{Bmatrix} Y_1 \\ Y_2 \end{Bmatrix} = 0 \tag{5}$$

The determinant of Eq. (5) can be expanded as the following polynomial

$$\omega^4 - 7 \times 10^4 \omega^2 + 5.5 \times 10^8 = 0 \tag{6}$$

Solving the quadratic equation (6), the resonant frequencies of the system are given by

$$\omega_{1,2} = \sqrt{\frac{7 \times 10^4 \pm \sqrt{\left(7 \times 10^4\right)^2 - 4 \times 5.5 \times 10^8}}{2}} = 94.96,\ 246.94\,\text{rad/s} \tag{7}$$

(Natural frequencies are always arranged in the increasing order.)

For the first natural frequency, Eq. (5) becomes

$$\begin{bmatrix} 4 \times 10^4 - 2\omega_1^2 & -3 \times 10^4 \\ -3 \times 10^4 & 5 \times 10^4 - \omega_1^2 \end{bmatrix} \begin{Bmatrix} Y_{11} \\ Y_{21} \end{Bmatrix} = 0 \tag{8}$$

where, Y_{11} and Y_{21} are the peak amplitudes of free vibration of the degrees of freedom at the first resonant frequency.

From Eqs. (7) and (8)

$$\begin{aligned} Y_{21} &= \frac{4\times10^4 - 2\omega_1^2}{3\times10^4}Y_{11} = \frac{4\times10^4 - 2\times94.96^2}{3\times10^4}Y_{11}\ \text{(first mode)} \\ &= 0.732Y_{11} \end{aligned} \tag{9}$$

Similarly, from Eqs. (7) and (8), corresponding to the second mode

$$\left[k\begin{bmatrix}2 & -1\\ -1 & 2\end{bmatrix} - \omega_2^2 m\begin{bmatrix}1 & 0\\ 0 & 1\end{bmatrix}\right]\begin{Bmatrix}Y_{12}\\ Y_{22}\end{Bmatrix} = 0$$

$$\begin{aligned} Y_{22} &= \frac{4\times10^4 - 2\omega_2^2}{3\times10^4}Y_{12} = \frac{4\times10^4 - 2\times246.94^2}{3\times10^4}Y_{12} \\ &= -2.732Y_{12} \end{aligned} \tag{10}$$

where, Y_{12} and Y_{22} are the peak amplitudes of free vibration of the degrees of freedom at the second resonant frequency (Fig. 11.17).

From Eqs. (9) and (10), by assuming $Y_{11} = Y_{12} = 1$, the above modal vectors corresponding to the first and second natural frequency can be combined into a modal matrix, given by

$$\boldsymbol{\Psi} = \begin{bmatrix}1 & 1\\ 0.732 & -2.732\end{bmatrix} \tag{11}$$

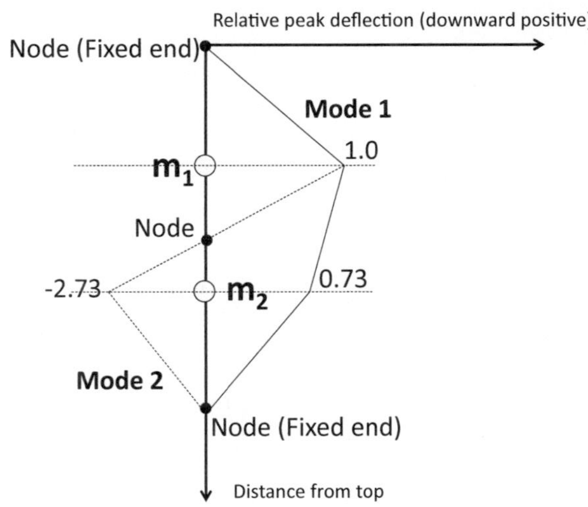

Fig. 11.17 Mode shapes

The mass-normalized modal matrix is given by

$$\Phi = \psi\{\psi^T M \psi\}^{-0.5}$$

$$= \begin{bmatrix} 1 & 1 \\ 0.732 & -2.732 \end{bmatrix} \left[\begin{bmatrix} 1 & 0.732 \\ 1 & -2.732 \end{bmatrix} \begin{bmatrix} 2 & 0 \\ 0 & 1 \end{bmatrix} \begin{bmatrix} 1 & 1 \\ 0.732 & -2.732 \end{bmatrix} \right]^{-0.5}$$

$$= \begin{bmatrix} 0.6280 & 0.3250 \\ 0.4597 & -0.8881 \end{bmatrix} \tag{12}$$

Since the modal matrix depends on the degree of freedom that has been kept as the reference and the reference value assigned to it, the modal matrix cannot be unique although the ratio of peak displacements at each undamped natural frequency remains the same. However, the mass-normalized modal matrix is unique irrespective of the reference degree of freedom and the reference value. Both these matrices can be further used to decouple the coupled differential equations to obtain the forced response or response to initial values of MDOF systems. Interested readers can refer to advanced vibration textbooks.

11.5 Dynamic Vibration Absorbers

Dynamic vibration absorbers, also called as tuned absorbers, are useful in controlling large vibrations occurring at a single frequency. Most machines used in practice are driven by electric induction motors and if the line frequency of the electric supply remains the same, their rotational speeds are also the same. Hence, any component of vibration generated due to the combined operation of the motor and the driven machine will generate vibration at a frequency that remains constant. For example, in transformers, gearboxes, etc., do have constant single frequency vibration in them. If the entire system could be modeled as a single degree of freedom system (referred to as the primary system), by attaching a suitably designed single degree of freedom system (much smaller in size, referred to as the secondary system) can actually bring down the vibration of the primary system. Its broad principle can be explained as follows.

Consider a primary system as shown in Fig. 11.18, constructed from stiffness k_1 and mass m_1 that is acted upon by a single frequency excitation force $me\omega^2\cos\omega t$, where the excitation is due to an unbalanced mass m at an eccentricity of *e*. However, the source of excitation could be different and the tuned vibration design remains the same as long as the excitation frequency is constant and is closer to the undamped resonant frequency of the primary system. This primary system is attached with a secondary system formed by stiffness k_2 and mass m_2, which forms the tuned absorber that has to be designed.

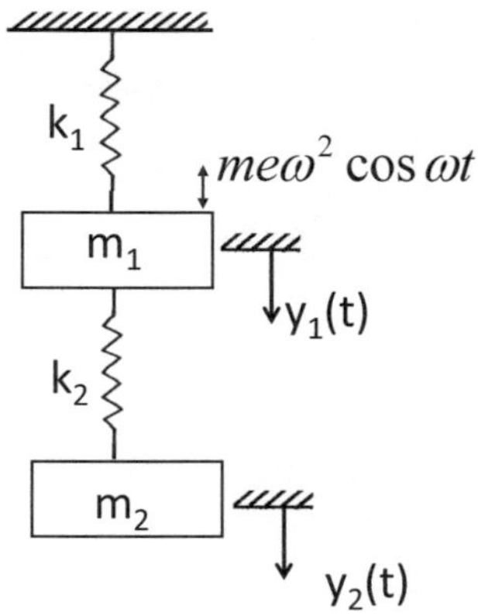

Fig. 11.18 Tuned vibration absorber

The dynamic equations of motion for a tuned absorber (secondary system) attached to a primary system, as shown in Fig. 11.18, are given by

$$\begin{bmatrix} m_1 & 0 \\ 0 & m_2 \end{bmatrix} \begin{Bmatrix} \ddot{y}_1 \\ \ddot{y}_2 \end{Bmatrix} + \begin{bmatrix} k_1 + k_2 & -k_2 \\ -k_2 & k_2 \end{bmatrix} \begin{Bmatrix} y_1 \\ y_2 \end{Bmatrix} = \begin{Bmatrix} F_o \cos \omega t \\ 0 \end{Bmatrix} \tag{11.89}$$

Since there is a single frequency excitation, the response of both the masses will be at this excitation frequency, given by

$$\begin{aligned} y_1 &= Y_1 \cos \omega t \\ y_2 &= Y_2 \cos \omega t \end{aligned} \tag{11.90}$$

where, Y_1 and Y_2 are the respective peak amplitudes of the two masses.

From Eqs. (11.89) and (11.90)

$$\begin{bmatrix} k_1 + k_2 - m_1\omega^2 & -k_2 \\ -k_2 & k_2 - m_2\omega^2 \end{bmatrix} \begin{Bmatrix} Y_1 \\ Y_2 \end{Bmatrix} = \begin{Bmatrix} F_o \\ 0 \end{Bmatrix} \tag{11.91}$$

Solving Eq. (11.91)

$$\begin{Bmatrix} Y_1 \\ Y_2 \end{Bmatrix} = \frac{1}{D} \begin{bmatrix} k_2 - m_2\omega^2 & k_2 \\ k_2 & k_1 + k_2 - m_1\omega^2 \end{bmatrix} = \begin{Bmatrix} F_o \\ 0 \end{Bmatrix} \tag{11.92}$$

where, D is given by

$$D = m_1 m_2 \omega^4 - (k_1 m_2 + k_2 m_2 + k_2 m_1)\omega^2 + k_1 k_2 \tag{11.93}$$

From Eqs. (11.92) and (11.93), the amplitude of vibration of the masses is given by

$$Y_1 = \frac{1}{D}(k_2 - m_2\omega^2)F_0$$
$$Y_2 = \frac{k_2}{D}F_0 \tag{11.94}$$

To make the amplitude of response of the main system, $Y_1 = 0$, which is the main objective, the condition is $(k_2 - m_2\omega^2) = 0$.

Therefore, when the excitation frequency is equal to the individual natural frequency of the absorber, the tuned absorber can reduce vibration of the primary mass to zero.

The above frequency relations can be expressed as

$$\omega = \omega_{n2} = \sqrt{\frac{k_2}{m_2}} \tag{11.95}$$

Using Eq. (11.95), (11.93) becomes

$$D = m_1 m_2\left(\frac{k_2}{m_2}\right)^2 - (k_1 m_2 + k_2 m_2 + k_2 m_1)\left(\frac{k_2}{m_2}\right) + k_1 k_2$$
$$= -k_2^2 \tag{11.96}$$

From Eqs. (11.94) and (11.96)

$$k_2 = \frac{F_0}{Y_2} = \frac{me\omega^2}{Y_2} \tag{11.97}$$

The tuned vibration absorber mass is given by

$$m_2 = \frac{k_2}{\omega_{n2}^2}\ \text{kg} \tag{11.98}$$

Now that we know the mass and stiffness of the absorber, they can be used to determine the natural frequencies of the combined system using Eq. (11.93).

$$m_1 m_2 \Omega^4 - (k_1 m_2 + k_2 m_2 + k_2 m_1)\Omega^2 + k_1 k_2 = 0 \tag{11.99}$$

Simplifying Eq. (11.99)

$$\Omega^4 - (\omega_{n_1}^2 + \mu\omega_{n_2}^2 + \omega_{n_2}^2)\Omega^2 + \omega_{n_1}^2\omega_{n_2}^2 = 0 \tag{11.100}$$

Example 11.9

A machine of mass 200 kg with a rotating unbalance of 0.1 kg-m operates at a speed of 1200 RPM. It is mounted on a spring of stiffness 700 kN/m.

(a) Determine the vibration due to the unbalanced force when no absorber is attached when the machine is running at 1200 RPM.
(b) Design a tuned vibration absorber such that when attached to the above system, the vibrations of the machine will cease at 1200 RPM and the steady state absorber amplitude will be less than 10 mm.
(c) What are the system's natural frequencies with the absorber in place?
(d) After installing the tuned absorber, if the machine is operated between 900 and 1400 RPM, what is the deflection at these speeds?

Mass of the machine $\mathrm{m}_1 = 200$ kg
Unbalance = me = 0.1 kg-m
RPM = 1200

(a) $\omega = \frac{2\pi N}{60} = \frac{2\pi \times 1200}{60} = 125.66$ rad/s (excitation frequency)

The natural frequency of the above machine mounted on the spring is given by

$$\omega_{n1} = \sqrt{\frac{k_1}{m_1}} = \sqrt{\frac{700 \times 10^3}{200}} = 59.16 \text{ rad/s} \quad (1)$$

From Eq. (11.63), the deflection due to unbalance when the shaft is rotating at 1200 RPM is given by $y = \frac{me\omega^2}{k - m\omega^2} = \frac{0.1 \times 125.66^2}{700 \times 10^3 - 200 \times 125.66^2} = 0.64$; mm

$$\omega_{n2} = \omega = 1200\,\text{RPM} = 125.66 \text{ rad/s}$$

(b) From Eq. (11.97)

$$\begin{aligned} k_2 &= \frac{F_0}{Y_2} = \frac{me\omega^2}{Y_2} = \frac{0.1 \times 125.66^2}{10 \times 10^{-3}} \\ &= 158\,\text{kN/m} \end{aligned} \quad (2)$$

From Eq. (11.98), the tuned vibration absorber mass is given by

$$m_2 = \frac{k_2}{\omega_{n2}^2} = \frac{158,000}{125.66^2} = 10 \text{ kg} \quad (3)$$

Using the above design of the tuned absorber, the machine response is zero at the operational speed of 1200 RPM from Eq. (11.94).

(c) Now that we know the mass and stiffness of the absorber; they can be used to determine the natural frequencies of the combined system using Eq. (11.99).

$$m_1m_2\Omega^4 - (k_1m_2 + k_2m_2 + k_2m_1)\Omega^2 + k_1k_2 = 0$$

$$200 \times 10\Omega^4 - (713 \times 10 + 158 \times 10 + 158 \times 200) \times 10^3\Omega^2 + 713 \times 158 \times 10^6 = 0$$

$$2000\Omega^4 - 40310 \times 10^3\Omega^2 + 112654 \times 10^6 = 0 \quad (4)$$

Solving Eq. (4) results in the following natural frequencies of the combined system: $\Omega_1 = 129.63$ rad/s and $\Omega_2 = 58$ rad/s.

(d) Minimum speed is 900 RPM (94.25 rad/s).
Maximum speed is 1400 RPM (146.61 ra/s).

Using Eq. (11.93), the value of D at 900 RPM is given by

$$\begin{aligned} D(900\text{ RPM}) &= m_1m_2\omega^4 - (k_1m_2 + k_2m_2 + k_2m_1)\omega^2 + k_1k_2 \\ &= 200 \times 10 \times 94.25^4 - (700 \times 10 + 158 \times 10 \\ &\quad + 158 \times 200) \times 10^3 \times 94.25^2 + 700 \times 158 \times 10^6 \\ &= -8.84 \times 10^{10} \end{aligned} \quad (5)$$

From Eq. (11.94), response of the machine is given by

$$\begin{aligned} Y_1(900\text{ RPM}) &= \frac{1}{D}(k_2 - m_2\omega^2)F_0 \\ &= \frac{1}{-8.84 \times 10^{10}}(158 \times 10^3 - 10 \times 94.25^2) \times 0.1 \times 94.25^2 \\ &= 694\,\mu\text{m} \end{aligned} \quad (6)$$

Using Eq. (11.93), the value of D at 1400 RPM is given by

$$\begin{aligned} D(1400\text{ RPM}) &= m_1m_2\omega^4 - (k_1m_2 + k_2m_2 + k_2m_1)\omega^2 + k_1k_2 \\ &= 200 \times 10 \times 146.61^4 - (700 \times 10 + 158 \times 10 + 158 \times 200) \times 10^3 \\ &\quad \times 146.61^2 + 700 \times 158 \times 10^6 \\ &= 1.7128 \times 10^{11} \end{aligned} \quad (7)$$

From Eq. (11.94), the response of the machine is given by

$$\begin{aligned} Y_1(1400\text{ RPM}) &= \frac{1}{D}(k_2 - m_2\omega^2)F_0 \\ &= \frac{1}{1.7128 \times 10^{11}}(158 \times 10^3 - 10 \times 146.61^2) \times 0.1 \times 146.61^2 \\ &= 715\,\mu\text{m} \end{aligned} \quad (8)$$

Therefore, from Eqs. (6) and (8), if the operation speed of the machine deviates from 1200 RPM, the tuned vibration absorber may not be very effective.

11.6 Critical Speeds of Shafts

Shafts are commonly used in machinery to transfer power between various elements. They are generally cylindrical elements with some masses representing gears, pulleys, turbines, etc., mounted on them. They can be subjected to both torsional oscillations and flexural vibration. However, flexural vibrations are generally related to the shaft speed and when the natural frequencies of flexural vibration coincide with the operational speed of the shaft, there will be significant vibration that can lead to its failure. Therefore, it is necessary to compute such speeds of the shaft that coincide with the natural frequencies of flexural vibration. They are known as critical speeds of shafts. Although there are detailed analytical methods to determine the natural frequencies of shafts using partial differential equations, such a high level of accuracy is not required in many practical machinery vibrations. Hence, the following approximate methods are discussed.

Torsional oscillations are also important, but they are not generally related to shaft speeds but on the torque variations of the driver like an electric motor or an IC engine. Therefore, they are discussed separately in Sect. 11.7.

11.6.1 Rayleigh's Method

This method of determining natural frequencies is applicable to systems without damping. The basis for this method is that maximum potential energy is equal to maximum kinetic energy for such systems. Although the application of this method is meant for continuous systems with masses that represent typical shaft systems in machines, we shall first apply this method to a single degree of freedom spring-mass system.

SDOF

Consider a spring-mass system of stiffness k and mass m. Let the system perform oscillations at its natural frequency ω_n and x_0 be its maximum displacement amplitude.

Potential energy is stored by the spring during vibration and the maximum potential energy is given by

$$U_{\max} = \frac{1}{2}kx_0^2 \tag{11.101}$$

Kinetic energy is stored by the mass during vibration and the maximum kinetic energy is given by

$$T_{\max} = \frac{1}{2}mv_{\max}^2 = \frac{1}{2}m\omega_n^2 x_0^2 \tag{11.102}$$

Based on the Rayleigh's method of determining natural frequency, equating the maximum potential and kinetic energies in the above equations gives the natural frequency as

$$\omega_n = \sqrt{\frac{k}{m}} \tag{11.103}$$

Before we can apply Rayleigh's method to a shaft model, we need an additional equation for natural frequency in terms of static deflection. For the spring-mass system considered above, when the mass is suspended from the spring, there will be an initial deflection δ before the mass reaches its static equilibrium position. The force in the spring due to this initial deflection is actually balanced by the gravitational force due to the mass. It can be expressed as

$$mg = k\delta \tag{11.104}$$

From Eqs. (11.103) and (11.104), the natural frequency can be obtained as

$$\omega_n = \sqrt{\frac{k}{m}} = \sqrt{\frac{g}{\delta}} \tag{11.105}$$

Therefore, static deflection can also be used to determine the natural frequency of a dynamic system. Hence, even for a more complex system than a simple spring-mass system, static deflection can be used to determine its natural frequency. Strength of materials equations can be used to compute the static deflection.

Cantilever beam

Consider a cantilever beam of uniform cross-section whose flexural natural frequencies (corresponding to vibration perpendicular to the beam length) will have to be determined. Since this is a continuous system, it will have many natural frequencies, which can be computed from its dynamic model represented by a partial differential equation. But computing all the natural frequencies may not be required in many practical cases and the Rayleigh's method can be used to compute the lowest or fundamental resonance frequency, without actually solving the partial differential equation. The objective of determining the fundamental frequency is to ensure that the excitation frequencies are much below the fundamental frequency of the shafts used in machines.

The static deflection at the free end of a cantilever, by considering its weight to be a uniformly distributed load, is given by

$$\delta = \frac{\rho A g \ell^4}{8EI} \tag{11.106}$$

where ρ is the mass density, A area of cross-section, ℓ its length, E Young's modulus, and I moment of inertia of the cross-section.

Since the deflection due to the weight of the cantilever closely resembles its mode shape during its vibration at the first natural frequency, the above equation for computing natural frequency using static deflection at the cantilever tip can be used to compute the first natural frequency of the cantilever and from Eqs. (11.105) and (11.106), it is given by

$$\omega_n(\text{approx}) = \sqrt{\frac{g}{\delta}} = \sqrt{\frac{8EI}{\rho A \ell^4}} = 2.83\sqrt{\frac{EI}{\rho A \ell^4}}\ \text{rad/s} \tag{11.107}$$

The exact value of first natural frequency for the above cantilever beam by solving the corresponding partial differential equation is given by

$$\omega_n(\text{exact}) = 3.5\sqrt{\frac{EI}{\rho A \ell^4}}\ \text{rad/s} \tag{11.108}$$

Although the approximate method underestimates the natural frequency by 20%, its simple calculation is an advantage for many practical situations. In addition, we can afford a slight increase in complexity of the dynamic system. For example, if we add a concentrated mass at the free end of the cantilever, the exact method has to be worked out all over again to accommodate the new change. Whereas the Rayleigh's method, by noting the deflection due to the weight of the concentrated mass at the cantilever tip as $\frac{Mg\ell^3}{3EI}$, can easily compute the fundamental flexural natural frequency of a cantilever beam with a concentrated mass at the free end as

$$\omega_n(\text{approx}) = \sqrt{\frac{g}{\delta}} = \sqrt{\frac{g}{\frac{g\ell^3}{EI}\left(\frac{\rho A \ell}{8} + \frac{M}{3}\right)}}$$
$$= \sqrt{\frac{EI}{\ell^3\left(\frac{m_b}{8} + \frac{M}{3}\right)}}\ \text{rad/s} \tag{11.109}$$

where m_b is the mass of the uniform beam alone. Therefore, if the concentrated mass M is heavier than the beam by an order of magnitude, natural frequency will depend on the second term only. This aspect is used to calculate the natural frequency of shaft systems by using only the masses at various locations of the shaft.

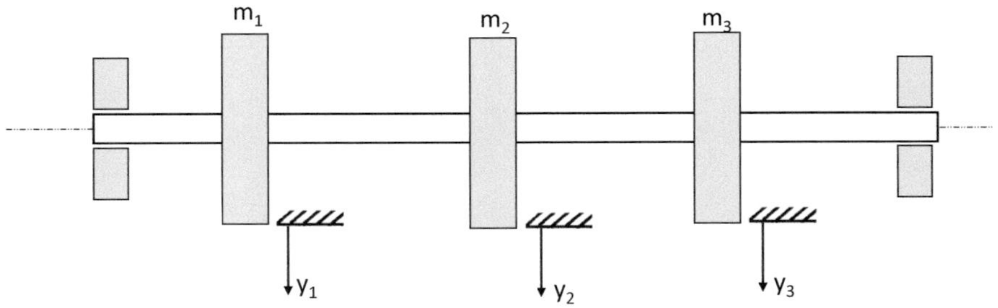

Fig. 11.19 Shaft carrying a number of masses

The main reason for the deviation of the natural frequency using static deflection at the tip of the cantilever is that deflections at other points of the beam were not considered. Therefore, the above method is modified to account for deflections at other points of a typical shaft that is simply supported with many masses mounted on it at various locations. These masses can represent typical machine elements such as pulleys, gears, rotors, etc. that may be mounted on a shaft. This is discussed in the next section.

Shaft carrying a number of masses

Consider a shaft of uniform cross-section with a number of masses along the length of the shaft as shown in Fig. 11.19. Assume that the entire shaft is vibrating at its first natural frequency, ω_n and let there be masses m_1, m_2,..., m_n mounted at various locations along the shaft and let y_1, y_2,...y_n be the peak amplitude of deflections at the locations of these masses.

By noting that the work done for moving each mass to its static deflection as $\frac{1}{2}m_i g y_i$, the maximum total potential energy of the system, considering only those associated with the masses, is given by

$$U_{\max} = \frac{g}{2}\sum_{i=1}^{n} m_i y_i \tag{11.110}$$

And the maximum kinetic energy is given by

$$T_{\max} = \frac{1}{2}\omega_n^2 \sum_{i=1}^{n} m_i y_i^2 \tag{11.111}$$

From Eqs. (11.110) and (11.111), equating maximum potential and kinetic energy, the fundamental natural frequency is given by

$$\omega_n = \sqrt{\frac{g\sum m_i y_i}{\sum m_i y_i^2}}\ \text{rad/s} \tag{11.112}$$

The peak amplitude of the above deflections is assumed to be proportional to the static deflections at the locations of masses based on Rayleigh's method.

11.6.2 Influence Coefficients' Method

To easily determine higher critical speeds of a shaft, influence coefficients' method can be used. It is much easier than to directly use the partial differential equation formulation. The basis for this method is the concept of an influence coefficient. For a general dynamic system, an influence coefficient α_{ij} is defined as the deflection at a point *i* due to a unit load at a point *j*. Since the influence coefficient has units of m/N, it is reciprocal to the units of stiffness that have units m/N.

Consider a shaft of negligible mass that is carrying a number of masses m_1, $m_2,\ldots,m_n$ similar to Fig. 11.19 located at various positions of the shaft. Influence coefficients can be defined at every location of these masses. For example, α_{12} is the deflection at the location of m_1 due to a unit load at the location of m_2 and so on. These influence coefficients will then fill a $n \times n$ matrix. Those along the diagonal will represent deflection at every mass due to a unit load at the same mass; the off-diagonal terms represent deflection at a certain mass due to a unit load at another. Due to the reciprocity principle of Maxwell, deflection at point 1 due to a unit load at point 2 is the same as deflection at point 2 due to a unit load at point 1. Hence, these off-diagonal terms will be the same and hence the matrix of influence coefficients is a symmetric matrix.

The deflection at the location of each of the masses can be expressed as

$$\begin{aligned} y_1 &= \alpha_{11}m_1g + \alpha_{12}m_2g + \cdots \alpha_{1n}m_ng \\ y_2 &= \alpha_{21}m_1g + \alpha_{22}m_2g + \cdots \alpha_{2n}m_ng \\ &\vdots \\ y_n &= \alpha_{n1}m_1g + \alpha_{n2}m_2g + \cdots \alpha_{nn}m_ng \end{aligned} \tag{11.113}$$

In terms of matrices and using the reciprocal relationship of Maxwell, the above equation can be written as

$$\begin{Bmatrix} y_1 \\ y_2 \\ \vdots \\ y_n \end{Bmatrix} = \begin{bmatrix} \alpha_{11} & \alpha_{12} & \cdots & \alpha_{1n} \\ \alpha_{12} & \alpha_{22} & \cdots & \alpha_{2n} \\ \vdots & \vdots & \ddots & \vdots \\ \alpha_{1n} & \alpha_{2n} & \cdots & \alpha_{nn} \end{bmatrix} \begin{Bmatrix} m_1g \\ m_2g \\ \vdots \\ m_ng \end{Bmatrix} \tag{11.114}$$

The force due to the weight of the masses mounted along the shaft must balance the product of mass and acceleration at these points. Assuming that every point on the shaft is vibrating at a common frequency ω, since $\ddot{y}_i = \omega^2 y_i$, the above equation becomes

$$\begin{Bmatrix} y_1 \\ y_2 \\ \vdots \\ y_n \end{Bmatrix} = \begin{bmatrix} \alpha_{11} & \alpha_{12} & \cdots & \alpha_{1n} \\ \alpha_{12} & \alpha_{22} & \cdots & \alpha_{2n} \\ \vdots & \vdots & \ddots & \vdots \\ \alpha_{1n} & \alpha_{2n} & \cdots & \alpha_{nn} \end{bmatrix} \begin{Bmatrix} m_1 y_1 \omega^2 \\ m_2 y_2 \omega^2 \\ \vdots \\ m_n y_n \omega^2 \end{Bmatrix} \quad (11.115)$$

Rearranging the above equation

$$\begin{bmatrix} \alpha_{11} m_1 - \frac{1}{\omega^2} & \alpha_{12} & \cdots & \alpha_{1n} \\ \alpha_{12} & \alpha_{22} m_2 - \frac{1}{\omega^2} & \cdots & \alpha_{2n} \\ \vdots & \vdots & \ddots & \vdots \\ \alpha_{1n} & \alpha_{2n} & \cdots & \alpha_{nn} m_n - \frac{1}{\omega^2} \end{bmatrix} \begin{Bmatrix} y_1 \\ y_2 \\ \vdots \\ y_n \end{Bmatrix} = 0 \quad (11.116)$$

The above Eq. (11.116) is a standard eigenvalue problem in which the eigenvalue is the reciprocal of the square of the natural frequency. Expanding the determinant and equating it to zero can also solve the above equation.

When the matrix of Eq. (11.116) is expanded as a determinant, the coefficient of the second highest term $1/\omega^{n-1}$ of the corresponding polynomial is given by the following term

$$-(\alpha_{11} m_1 + \alpha_{22} m_2 + \cdots \alpha_{nn} m_n) \quad (11.117)$$

Based on the properties of polynomials, we know that the negative of the coefficient of the second highest term is equal to the sum of all the roots. This condition leads to the following equation

$$\frac{1}{\omega_1^2} + \frac{1}{\omega_2^2} + \cdots \frac{1}{\omega_n^2} = (\alpha_{11} m_1 + \alpha_{22} m_2 + \cdots \alpha_{nn} m_n) \quad (11.118)$$

If the higher frequencies are much larger than the fundamental frequency, the above equation can be approximately used to compute the first natural frequency, given by

$$\frac{1}{\omega_1^2} \approx (\alpha_{11} m_1 + \alpha_{22} m_2 + \cdots \alpha_{nn} m_n) \quad (11.119)$$

This is known as Dunkerley's approximation.

Example 11.10

A steel shaft (E = 200 GPa and density 7800 kg/m^3) with two rotors respectively weighing 30 and 45 kg is as shown in Fig. 11.20. The shaft is stepped and its weight has to be accounted for. Considering the shaft as simply supported, determine the first critical frequency of the shaft using Rayleigh's method. All dimensions are in mm. The weight of each stepped portion of the shaft can be assumed to be concentrated at the middle of the respective length. The shaft has a circular cross-section of various

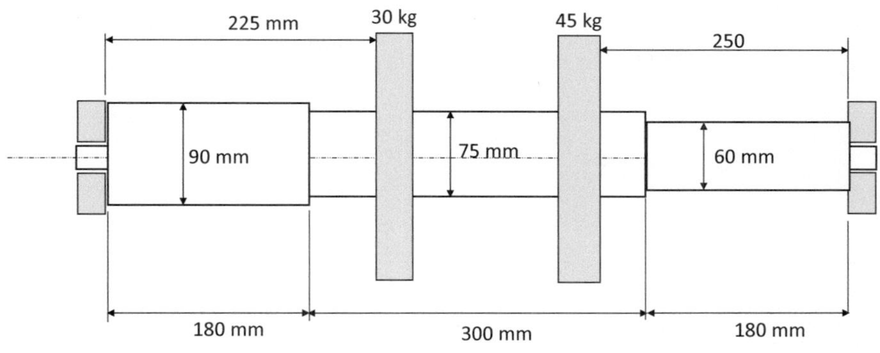

Fig. 11.20 Example 11.10

diameters as shown. (Assume average moment of inertia of the three area cross sections and use stiffness of the average diameter of the stepped shaft.)

Mass of the 90 dia, 180 length shaft, m_1 is given by

$$m_1 = \frac{\pi \times 0.090^2}{4} \times 0.180 \times 7800 = 8.93 \text{ kg} \tag{1}$$

$$\text{Mass of the rotor } m_2 = 30 \text{ kg} \tag{2}$$

Mass of the 75 dia, 300 length shaft, m_3 is given by

$$m_3 = \frac{\pi \times 0.075^2}{4} \times 0.3 \times 7800 = 10.33 \text{ kg} \tag{3}$$

$$\text{Mass of the rotor } m_4 = 45 \text{ kg} \tag{4}$$

Mass of the 60 dia, 180 length shaft, m_5 is given by

$$m_5 = \frac{\pi \times 0.060^2}{4} \times 0.180 \times 7800 = 3.96 \text{ kg} \tag{5}$$

Total length of the shaft is given by

$$\ell = 180 + 300 + 180 = 660 \text{ mm} \tag{6}$$

The above five masses and their locations can be summarized in Table 11.1.

The influence of varying diameters on stiffness is neglected by assuming an average moment of inertia for the entire shaft.

Table 11.1 Location of masses

No	Mass	kg	Location from the left end (mm)	Location from the right end (mm)
1	m_1	8.93	90	570
2	m_2	30	225	435
3	m_3	10.33	330	330
4	m_4	45	410	250
5	m_5	3.96	570	90

The influence coefficients are computed as follows

$$\alpha_{11} = \frac{b_1 a_1 \left(\ell^2 - b_1^2 - a_1^2\right)}{6EI\ell} = \frac{0.570 \times 0.09\left(0.66^2 - 0.57^2 - 0.09^2\right)}{EI(6 \times 0.66)}$$
$$= \frac{13 \times 10^{-4}}{EI} \text{m/N} \tag{7}$$

$$\alpha_{12} = \frac{b_2 a_1 \left(\ell^2 - b_2^2 - a_1^2\right)}{6EI\ell} = \frac{0.435 \times 0.090\left(0.660^2 - 0.435^2 - 0.090^2\right)}{6EI \times 0.660}$$
$$= \frac{24 \times 10^{-4}}{EI} \text{m/N} \tag{8}$$

$$\alpha_{13} = \frac{b_3 a_1 \left(\ell^2 - b_3^2 - a_1^2\right)}{6EI\ell} = \frac{0.330 \times 0.090\left(0.660^2 - 0.330^2 - 0.090^2\right)}{6EI \times 0.660}$$
$$= \frac{24 \times 10^{-4}}{EI} \text{m/N} \tag{9}$$

$$\alpha_{14} = \frac{b_4 a_1 \left(\ell^2 - b_4^2 - a_1^2\right)}{6EI\ell} = \frac{0.250 \times 0.090\left(0.660^2 - 0.250^2 - 0.090^2\right)}{6EI \times 0.660}$$
$$= \frac{21 \times 10^{-4}}{EI} \text{m/N} \tag{10}$$

$$\alpha_{15} = \frac{b_5 a_1 \left(\ell^2 - b_5^2 - a_1^2\right)}{6EI\ell} = \frac{0.090 \times 0.090\left(0.660^2 - 0.090^2 - 0.090^2\right)}{6EI \times 0.660}$$
$$= \frac{9 \times 10^{-4}}{EI} \text{m/N} \tag{11}$$

$$\alpha_{22} = \frac{b_2 a_2 \left(\ell^2 - b_2^2 - a_2^2\right)}{6EI\ell} = \frac{0.435 \times 0.225\left(0.66^2 - 0.435^2 - 0.225^2\right)}{6EI0.66}$$
$$= \frac{48 \times 10^{-4}}{EI} \text{m/N} \tag{12}$$

$$\alpha_{23} = \frac{b_3 a_2\left(\ell^2 - b_3^2 - a_2^2\right)}{6EI\ell} = \frac{0.330 \times 0.225\left(0.66^2 - 0.330^2 - 0.225^2\right)}{6EI0.66}$$
$$= \frac{52 \times 10^{-4}}{EI}\text{m/N} \tag{13}$$

$$\alpha_{24} = \frac{b_4 a_2\left(\ell^2 - b_4^2 - a_2^2\right)}{6EI\ell} = \frac{0.250 \times 0.225\left(0.66^2 - 0.250^2 - 0.225^2\right)}{6EI0.66}$$
$$= \frac{46 \times 10^{-4}}{EI}\text{m/N} \tag{14}$$

$$\alpha_{25} = \frac{b_5 a_2\left(\ell^2 - b_5^2 - a_2^2\right)}{6EI\ell} = \frac{0.090 \times 0.225\left(0.66^2 - 0.090^2 - 0.225^2\right)}{6EI0.66}$$
$$= \frac{19 \times 10^{-4}}{EI}\text{m/N} \tag{15}$$

$$\alpha_{33} = \frac{b_3 a_3\left(\ell^2 - b_3^2 - a_3^2\right)}{6EI\ell} = \frac{0.330 \times 0.330\left(0.66^2 - 0.330^2 - 0.330^2\right)}{6EI0.66}$$
$$= \frac{60 \times 10^{-4}}{EI}\text{m/N} \tag{16}$$

$$\alpha_{34} = \frac{b_4 a_3\left(\ell^2 - b_4^2 - a_3^2\right)}{6EI\ell} = \frac{0.250 \times 0.330\left(0.66^2 - 0.25^2 - 0.33^2\right)}{6EI0.66}$$
$$= \frac{55 \times 10^{-4}}{EI}\text{m/N} \tag{17}$$

$$\alpha_{35} = \frac{b_5 a_3\left(\ell^2 - b_5^2 - a_3^2\right)}{6EI\ell} = \frac{0.090 \times 0.330\left(0.66^2 - 0.090^2 - 0.330^2\right)}{6EI0.66}$$
$$= \frac{24 \times 10^{-4}}{EI}\text{m/N} \tag{18}$$

$$\alpha_{44} = \frac{b_4 a_4\left(\ell^2 - b_4^2 - a_4^2\right)}{6EI\ell} = \frac{0.250 \times 0.410\left(0.66^2 - 0.250^2 - 0.410^2\right)}{6EI0.66}$$
$$= \frac{53 \times 10^{-4}}{EI}\text{m/N} \tag{19}$$

$$\alpha_{45} = \frac{b_5 a_4\left(\ell^2 - b_5^2 - a_4^2\right)}{6EI\ell} = \frac{0.090 \times 0.410\left(0.66^2 - 0.09^2 - 0.410^2\right)}{6EI0.66}$$
$$= \frac{24 \times 10^{-4}}{EI}\text{m/N} \tag{20}$$

$$\alpha_{55} = \frac{b_5 a_5\left(\ell^2 - b_5^2 - a_5^2\right)}{6EI\ell} = \frac{0.090 \times 0.570\left(0.66^2 - 0.09^2 - 0.570^2\right)}{6EI0.66}$$
$$= \frac{13 \times 10^{-4}}{EI}\text{m/N} \tag{21}$$

An average diameter of the shaft $d_{avg} = (90 + 75 + 60)/3 = 75$ mm is used for calculating the moment of inertia of the area of cross-section of the shaft as

$$I = \frac{\pi d_{avg}^4}{64} = \frac{\pi \times 0.075^4}{64} = 1.5532 \times 10^{-6}\text{m}^4 \tag{22}$$

The static deflection at various locations of the masses is given by

$$\begin{Bmatrix} y_1 \\ y_2 \\ \vdots \\ y_5 \end{Bmatrix} = g\begin{bmatrix} \alpha_{11} & \alpha_{12} & \cdots & \alpha_{15} \\ \alpha_{12} & \alpha_{22} & \cdots & \alpha_{25} \\ \vdots & \vdots & \ddots & \vdots \\ \alpha_{15} & \alpha_{25} & \cdots & \alpha_{55} \end{bmatrix}\begin{Bmatrix} m_1 \\ m_2 \\ \vdots \\ m_5 \end{Bmatrix} \tag{23}$$

(please note the symmetry of influence coefficients)

$$\begin{Bmatrix} y_1 \\ y_2 \\ y_3 \\ y_4 \\ y_5 \end{Bmatrix} = \begin{Bmatrix} 6.44 \\ 13.69 \\ 15.66 \\ 14.57 \\ 6.45 \end{Bmatrix}\mu\text{m} \tag{24}$$

$$\omega_n = \sqrt{\frac{g\sum m_i y_i}{\sum m_i y_i^2}}$$
$$= \sqrt{\frac{9.81(8.93 \times 6.44 + 30 \times 13.69 + 10.33 \times 15.66 + 45 \times 14.57 + 3.96 \times 6.45) \times 10^{-6}}{\left(8.93 \times 6.44^2 + 30 \times 13.69^2 + 10.33 \times 15.66^2 + 45 \times 14.57^2 + 3.96 \times 6.45^2\right) \times 10^{-12}}}$$
$$= 839.77\ \text{rad/s}(8019\ \text{RPM}) \tag{25}$$

11.7 Torsional Vibrations

11.7.1 Torsional Shaft System

Torsional oscillations are common in many rotating machines. They can occur in such shaft systems due to variability of torque by driving systems like motors, gearbox, etc.; this is contrary to the popular notion that rotating machines receive constant torque. These torsional oscillations occur in either direction opposite to the direction

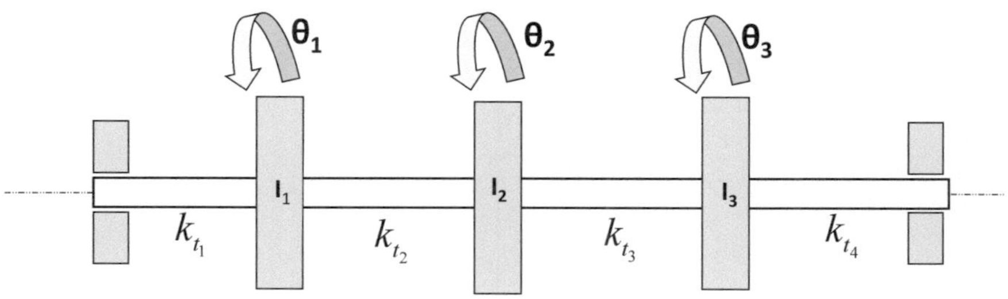

Fig. 11.21 Torsional shaft system

of rigid body rotation of the shaft and can also cause serious damage. Since the shaft is free to rotate in the bearings, the above systems are called semi-definite systems that have rigid body rotation as the first natural frequency. Since torsional systems do give rise to flexural vibrations, it is important to study them for practical machines.

Consider a shaft shown in Fig. 11.21 that has three rotors of moment of inertia respectively, I_1, I_2, and I_3. The torsional stiffness of the shaft in between the rotors respectively are k_{t_1}, k_{t_2}, k_{t_3} and k_{t_4}. These torsional oscillations are denoted as dynamic variables as θ_1, θ_2, θ_3. Torsional systems such as above can be easily modeled in terms of discrete systems using torsional stiffness and moment of inertia. Hence, the earlier study on multi-degree discrete systems can be extended to study torsional oscillations. In a later section, a numerical technique called the Holzer's method is presented for a linear semi-definite system that can be applied to torsional systems discussed above.

11.7.2 Geared Systems

Figure 11.22 shows a torsional geared system that has a gear ratio of n, which is equal to the ratio of diameter of the gear to pinion; it is assumed that the gear (larger diameter) is mounted on the generator (second) shaft and the pinion (smaller diameter) is mounted on the turbine (first) shaft. Since these shafts operate at different speeds, it is appropriate to model them as a single shaft system shown in Fig. 11.23. Therefore, if the system of Fig. 11.22 can be replaced by a single shaft equivalent system shown in Fig. 11.23, it can be easily analyzed for torsional oscillations. The moment of inertia of the gears is assumed to be negligible to those of the rotors in deriving the equivalent system. The equivalent system has a single shaft of equivalent stiffness $k_{t_{eq}}$ and equivalent moment of inertia I'_2.

The equivalent stiffness and moment of inertia can be derived as follows.

Let ω_1 be the angular velocity of shaft 1 and ω_2 that of shaft 2. We can find an equivalent system of rotors mounted on a single shaft of angular velocity ω_1 by equating the kinetic energy of the shaft shown in Fig. 11.22 to that shown in Fig. 11.23. It can be expressed in the following equation.

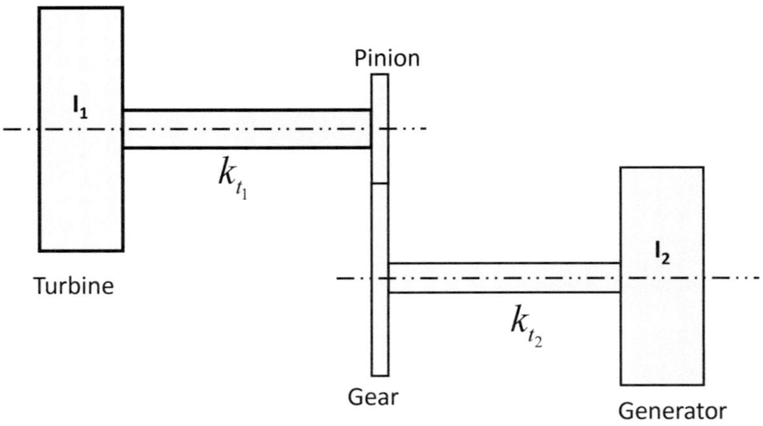

Fig. 11.22 Torsional geared system

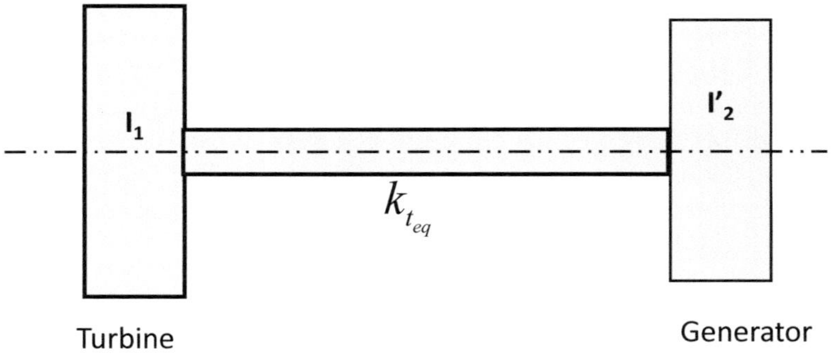

Fig. 11.23 Equivalent torsional system of Fig. 11.22

$$\frac{1}{2}I_1\omega_1^2 + \frac{1}{2}I_2\omega_2^2 = \frac{1}{2}(I_1 + I_2')\omega_1^2 \tag{11.120}$$

Since the angular velocity $\omega_2 = n\ \omega_1$, Eq. (11.120) becomes

$$\frac{1}{2}I_1\omega_1^2 + \frac{1}{2}I_2n^2\omega_1^2 = \frac{1}{2}(I_1 + I_2')\omega_1^2 \tag{11.121}$$

Comparing both sides of Eq. (11.121), $I_2' = n^2 I_2$.

To derive the equivalent stiffness of the shaft in Fig. 11.23, its angle of twist is equated to that of the shaft system in Fig. 11.22. Let ϕ_{eq} be the angle of twist of the equivalent shaft and ϕ_1 and ϕ_2 the respective angles of twists of shafts of Fig. 11.22. Since the angle of twist of the equivalent shaft is as seen from the first shaft, the above angle of twists are related by

$$\phi_{eq} = \phi_1 + \frac{\phi_2}{n} \tag{11.122}$$

Since the angles of twists are related to the ratio of torque and torsional stiffness, the above angles can be expressed in terms of static torques T_1 and T_2 respectively in the two shafts, given by

$$\frac{T_1}{k_{t_{eq}}} = \frac{T_1}{k_{t_1}} + \frac{T_2}{nk_{t_2}} \tag{11.123}$$

Since $T_2 = T_1/n$, the above equation becomes,

$$\frac{T_1}{k_{t_{eq}}} = \frac{T_1}{k_{t_1}} + \frac{T_1}{n^2 k_{t_2}} \tag{11.124}$$

From Eq. (11.124), the torsional stiffness of the equivalent shaft to the original shaft system is related by

$$\frac{1}{k_{t_{eq}}} = \frac{1}{k_{t_1}} + \frac{1}{n^2 k_{t_2}} \tag{11.125}$$

Once any gear system is converted into an equivalent single shaft system, it can be easily solved as a multi-degree of freedom discrete system to solve for its torsional resonances.

11.7.3 Holzer's Method

This is one of the earliest methods used to determine the undamped natural frequencies and mode shapes of multiple degree of freedom semi-definite systems that is useful to study torsional oscillations. Since this involves simple computations, it can be easily programmed on a computer and is based on simple concepts. An example shown in Fig. 11.24 is used to demonstrate the basic concepts of this method. Although this example relates to MDOF linear vibrations, the model for torsional oscillations, however, is very similar. Masses will be replaced by moment of inertia, stiffness by torsional stiffness, and linear responses by angular response in case of torsional oscillations. Since it is easier to visualize a linear MDOF system, the above system was chosen to demonstrate Holzer's method.

The matrix differential equation of motion based on earlier discussions can be written as

$$\begin{bmatrix} m_1 & 0 & 0 \\ 0 & m_2 & 0 \\ 0 & 0 & m_3 \end{bmatrix} \begin{Bmatrix} \ddot{x}_1 \\ \ddot{x}_2 \\ \ddot{x}_3 \end{Bmatrix} + \begin{bmatrix} k_1 & -k_1 & 0 \\ -k_1 & k_1 + k_2 & -k_2 \\ 0 & -k_2 & k_2 \end{bmatrix} \begin{Bmatrix} x_1 \\ x_2 \\ x_3 \end{Bmatrix} = 0 \tag{11.126}$$

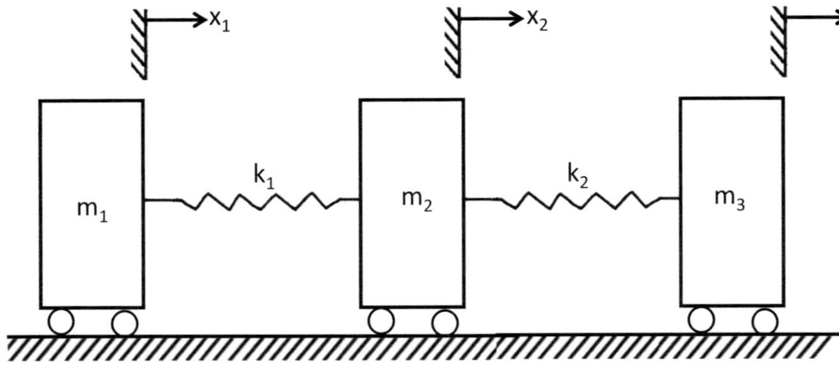

Fig. 11.24 Three-degree of freedom semi-definite undamped system

For a given set of initial conditions, the above system will vibrate at any or a linear combination of all of its natural frequencies. However, we can follow a single frequency ω and then prove that only three different values of this value, corresponding to the three-degree of freedom, can only satisfy the above equations and hence are the undamped natural frequencies of the system. Therefore, assuming that each degree of freedom is oscillating at a frequency ω, respectively with peak amplitudes X_1, X_2, and X_3, the above equation becomes

$$\begin{bmatrix} k_1 - m_1\omega^2 & -k_1 & 0 \\ -k_1 & k_1 + k_2 - m_2\omega^2 & -k_2 \\ 0 & -k_2 & k_2 - m_3\omega^2 \end{bmatrix} \begin{Bmatrix} X_1 \\ X_2 \\ X_3 \end{Bmatrix} = 0 \tag{11.127}$$

The above equations can be written as

$$\begin{aligned} &\left(k_1 - m_1\omega^2\right)X_1 - k_1X_2 = 0 \\ &-k_1X_1 + \left(k_1 + k_2 - m_2\omega^2\right)X_2 - k_2X_3 = 0 \\ &-k_2X_2 + \left(k_2 - m_3\omega^2\right)X_3 = 0 \end{aligned} \tag{11.128}$$

Rewriting the above equation in terms of difference between the peak amplitudes

$$\begin{aligned} &k_1(X_1 - X_2) = m_1\omega^2X_1 \\ &k_1(X_2 - X_1) + k_2(X_2 - X_3) = m_2\omega^2X_2 \\ &k_2(X_3 - X_2) = m_3\omega^2X_3 \end{aligned} \tag{11.129}$$

Equation (11.129) can be rewritten as

$$
\begin{aligned}
X_2 &= \frac{\left(k_1 - m_1\omega^2\right)X_1}{k_1} \\
X_3 &= \frac{\left\{k_1(X_2 - X_1) + \left(k_2 - m_2\omega^2\right)X_2\right\}}{k_2}
\end{aligned}
\tag{11.130}
$$

Since the above system is a semi-definite system, typical of many systems encountered in practice, the internal forces cancel with each other and therefore the sum of the inertia forces is equal to zero. Therefore, summing terms on the left and right sides of the Eq. (11.129)

$$
\sum_{i=1}^{3} m_i\omega^2 X_i = 0 \tag{11.131}
$$

As the undamped natural frequencies and the peak amplitudes are not known, we can write recursive equations starting with an assumed value of peak amplitude for a frequency and determine the rest of the peak amplitudes. These amplitudes are then substituted in Eq. (11.131). If the assumed value of the frequency is the natural frequency, then it results in zero. The above procedure is calculated for a number of frequencies and any sign change from Eq. (11.131) indicates the presence of an undamped natural frequency between those values.

Example 11.11

A three-degree of freedom semi-definite system as shown in Fig. 11.16 has to be analyzed for its undamped natural frequencies. Determine the undamped frequencies using the Holzer's method. $m_1 = 5$ kg, $m_2 = 10$ kg, $m_3 = 5$ kg and $k_1 = 10$ kN, $k_2 = 20$ kN.

The differential equation of motion for free vibration is given by

$$
\begin{bmatrix} m & 0 & 0 \\ 0 & 2m & 0 \\ 0 & 0 & m \end{bmatrix}
\begin{Bmatrix} \ddot{x}_1 \\ \ddot{x}_2 \\ \ddot{x}_3 \end{Bmatrix}
+
\begin{bmatrix} k & -k & 0 \\ -k & 3k & -2k \\ 0 & -2k & 2k \end{bmatrix}
\begin{Bmatrix} x_1 \\ x_2 \\ x_3 \end{Bmatrix} = 0 \tag{1}
$$

Equation (1) in terms of harmonic motions of peak amplitudes X_1, X_2, X_3, and frequency ω can be written as

$$
\begin{bmatrix} k - m\omega^2 & -k & 0 \\ -k & 3k - 2m\omega^2 & -2k \\ 0 & -2k & 2k - m\omega^2 \end{bmatrix}
\begin{Bmatrix} X_1 \\ X_2 \\ X_3 \end{Bmatrix} = 0 \tag{2}
$$

From Eq. (11.130)

$$X_1 = 1$$
$$X_2 = \frac{(k - m\omega^2)X_1}{k}$$
$$X_3 = \frac{\{k(X_2 - X_1) + 2(k - m\omega^2)X_2\}}{k} \tag{3}$$

Equation (3) is computed for various values of frequencies from 0 to 100 rad/s in steps of 10 rad/s and presented in Table 11.2.

From Fig. 11.25 and Table 11.2, it is clear that there is a change of signs between 40 to 50 rad/s and 80 to 90 rad/s. By linear interpolation, it is possible to determine the frequencies at which $\sum m_i X_i = 0$, from that gives the corresponding resonance frequency.

Table 11.2 Relative response at various frequencies using Holzer's method

Frequency (ω, rad/s)	X_1	X_2	X_3	$\sum m_i X_i$
0	1.00	1.00	−6.50	20.00
10	1.00	0.95	−6.30	18.89
20	1.00	0.80	−5.70	15.70
30	1.00	0.55	−4.70	10.89
40	1.00	0.20	−3.30	5.20
50	1.00	−0.25	−1.50	−0.31
60	1.00	−0.80	0.70	−4.30
70	1.00	−1.45	3.30	−5.11
80	1.00	−2.20	6.30	−0.80
90	1.00	−3.05	9.70	10.89
100	1.00	−4.00	13.50	32.50

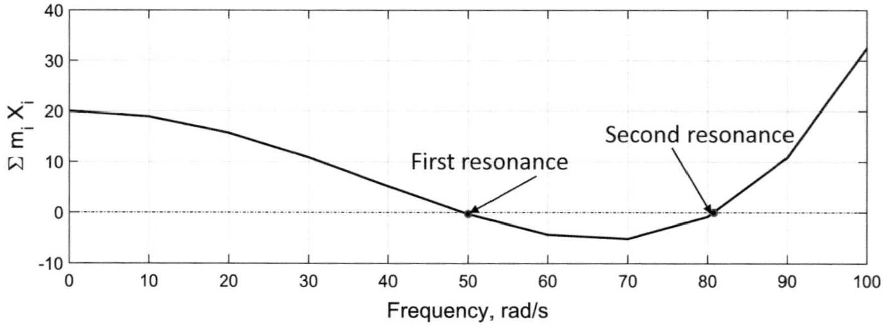

Fig. 11.25 Crossing points of $\sum m_i X_i$ for various frequencies shown in Table 11.2

Since the first natural frequency is zero due to the rigid body mode, natural frequencies are counted for increasing values of non-zero natural frequencies only. First, from Table 11.2, let us consider the crossover point between 40 to 50 rad/s, and let ω_x be the frequencies at which $\sum m_i X_i = 0$. A linear interpolation results in

$$\frac{\omega_x - 40}{50 - \omega_x} = \frac{5.20}{0.31} \tag{4}$$

Solving Eq. (4) gives $\omega_x = 49.44$ rad/s (first natural frequency) and it compares well with the exact value of 49.38 rad/s.

Second, from Table 11.2, let us consider the crossover point between 80 to 90 rad/s, and let ω_y be the frequencies at which $\sum m_i X_i = 0$. A linear interpolation results in

$$\frac{\omega_y - 80}{90 - \omega_y} = \frac{0.8}{10.89} \tag{5}$$

Solving Eq. (5) gives $\omega_y = 80.68$ rad/s (second natural frequency) and it compares well with the exact value of 81 rad/s.

If additional computations are carried at frequencies out closer to the crossover frequencies, more accurate values closer to the exact value can be obtained.

The mode shapes can be obtained by using the above natural frequencies in Eq. (11.130) and the mode shape for the first natural frequency at 49.44 rad/s is given by

$$\begin{aligned}
X_1 &= 1 \\
X_2 &= \frac{(k - m\omega^2)X_1}{k} = \frac{(10 \times 10^3 - 5 \times 49.44^2)X_1}{10 \times 10^3} = -0.222 \\
X_3 &= \frac{\{10 \times 10^3(-0.222 - 1) + 2(10 \times 10^3 - 5 \times 49.44^2)(-0.222)\}}{20 \times 10^3} = -0.5617
\end{aligned} \tag{6}$$

and for the second natural frequency at 80.68 rad/s, the mode shape is given by

$$\begin{aligned}
X_1 &= 1 \\
X_2 &= \frac{(k - m\omega^2)X_1}{k} = \frac{(10 \times 10^3 - 5 \times 80.68^2)X_1}{10 \times 10^3} = -2.2546 \\
X_3 &= \frac{\{10 \times 10^3(-2.2546 - 1) + 2(10 \times 10^3 - 5 \times 80.68^2)(-2.2546)\}}{20 \times 10^3} = 3.45
\end{aligned} \tag{7}$$

11.8 Conclusions

In this chapter, after providing a detailed discussion on the vibration of a single degree of freedom system, we have discussed many vibration problems encountered in typical machinery. Vibration is a very vast subject but only relevant portions were covered here that can help a practicing engineer to address vibration problems in machinery. It is, therefore, important to design machines by anticipating such vibration problems so costly breakdowns and modifications can be avoided at a later stage.

11.9 Exercise Problems

11.1 In order to restrain railway wagons at terminal stations, an arrangement shown in Fig. P11.1 is made in which a stiff spring is used to position the wagon in its final moments. If the wagon has a mass of 20,000 kg and a spring of stiffness 150 kN/m, and if the spring is deformed by 300 mm, (a) determine the velocity at which the wagon would have struck the spring. (b)For the same initial velocity of the wagon that strikes the spring, what would be its deformation if a viscous damper of damping coefficient 25 kN s/m is added in parallel?

11.2 A structure that can be modeled as a single degree of freedom spring-mass-damper system, when subjected to a static force of 500 N deflects 2 mm When it is given an initial deflection, the free oscillations decrease from 10 mm to 0.1 mm in 10 cycles and in one second it executes 7 oscillations. Determine (a) undamped natural frequency (b) damped natural frequency (c) damping factor (d) equivalent mass and stiffness of the frame.

11.3 As shown in Fig. P11.3, a ball of mass m = 50 g is dropped from a height of 1 m on a single degree of freedom spring-mass system of mass M = 1kg and stiffness, k = 2 kN/m. The mass M moves by 15 mm from its static equilibrium position due to the impact. Compute the height to which the ball will rebound.

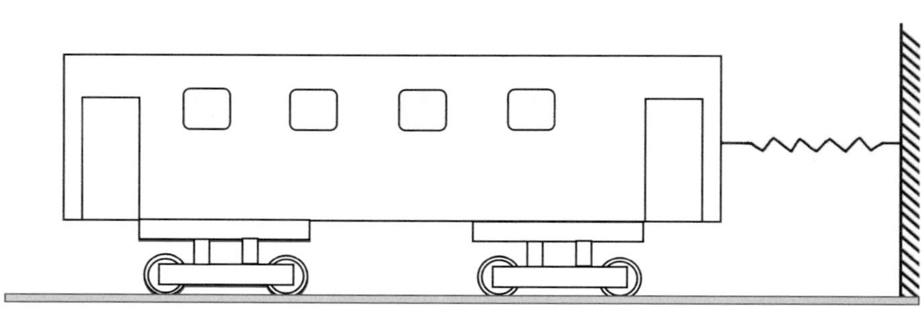

Fig. P11.1 SDOF model of a railway wagon stopping against a spring

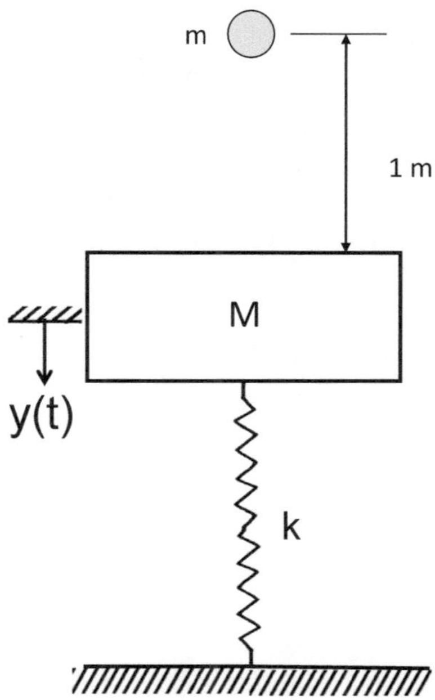

Fig. P11.3 Impact excitation of a SDOF by dropping a ball

11.4 A single degree of freedom spring-mass-damper system is hanging from a rigid structure. When the mass is given an initial displacement of 20 mm, the peak amplitude of oscillation reduces to 1 mm in 2 s after 5 cycles of oscillations in free vibration. After the oscillations in free vibrations come to a rest, the 5 kg mass is subjected to a force excitation, $F(t) = 10 \sin 2\pi ft, N$, where f = 5 Hz, determine (a) deflection of the mass (b) phase difference between deflection and force (c) amplitude of force transmitted to the structure supporting the system.

11.5 A press is modeled as an SDOF of the following parameters: mass 500 kg, spring stiffness 1.2 MN/m, viscous damping 5 kN s/m. If the above system is subjected to an impulse of 5000 N-s, what is the maximum displacement after the impulse is applied.

11.6 A single degree of freedom spring-mass-damper system is subjected to a force excitation, $25 \cos 30\pi t$. If m = 20 kg, k = 100 kN/m, and c = 600 N s/m, determine the peak amplitude of displacement, velocity, and acceleration and the phase angles between each of them.

11.7 An electric motor is mounted on a table and by indirect measurements, the effective mass of the combination is found to be 50 kg; the motor can be considered much heavier than the table. When the motor is mounted on the table, it deflects vertically by 5 mm. In free vibration, a displacement of 50 mm decays exponentially to 2 mm in 0.75 s; the motor has unbalanced rotating parts of 20 kg-mm. Determine (a) stiffness (b) damping coefficient (c) velocity response if the motor rotates at 1600 RPM.

11.8 Derive the equations of motion of the system shown in Fig. P11.8 and obtain its system transfer function.

11.9 Derive the equations of motion of the system shown in Fig. P11.9 and obtain its system transfer function.

11.10 Derive the equations of motion of the system shown in Fig. P11.10 and obtain its system transfer function.

11.11 An electro-magnetic exciter is used to apply a force excitation on a single degree of freedom spring-mass-damper system of spring stiffness 5 kN/m, mass 0.1 kg, and damping coefficient 5 N s/m that results in the displacement response measured using a transducer:$y(t) = 10\sin(40t - 5°) + 0.01\cos(700t - 170°)$ mm. Determine the corresponding time-domain equation for the force excitation applied by the shaker.

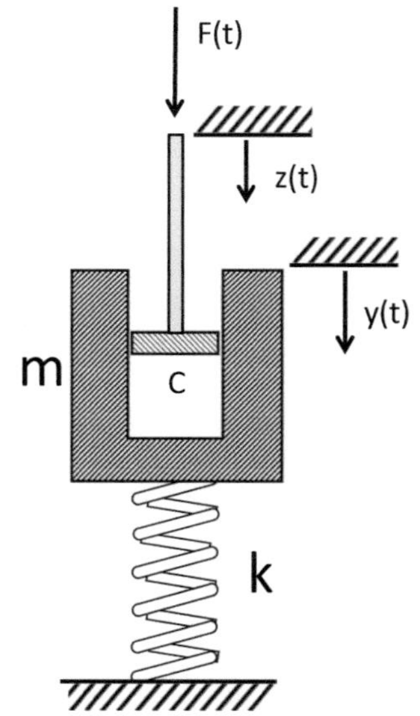

Fig. P11.8 SDOF with damper and spring connected in series

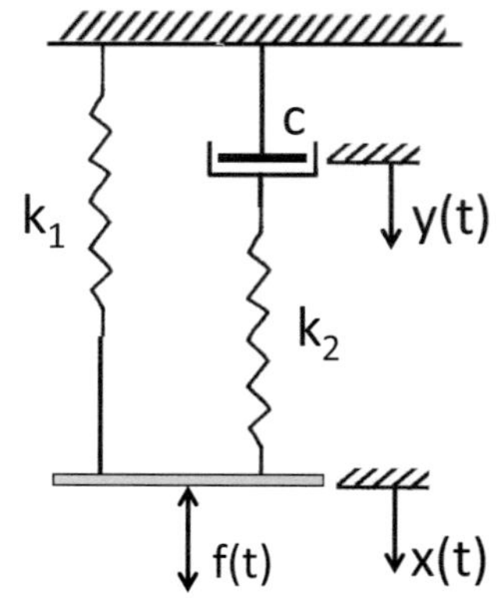

Fig. P11.9 Spring and damper in series connected with a spring in parallel

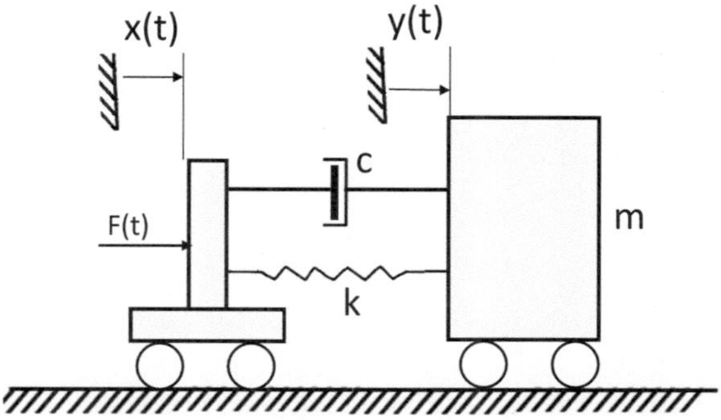

Fig. P11.10 Force excitation of a semi-definite system

11.12 A spring-mass system of stiffness 100 kN/m and mass 50 kg is subjected to a force of time-history shown in Fig. P11.12. Write down the equation for the response of the system for t > 10 s.

11.13 A spring-mass system of stiffness 100 kN/m and mass 50 kg is subjected to a force of time-history shown in Fig. P11.13. Write down the equation for the response of the system for t > 0.2 s.

11.14 A spring-mass system is subjected to a pulse as shown in Fig. P11.14. Determine the response of the system at $t = \frac{\pi}{\omega_n}$.

11.15 A spring-mass system of stiffness 100 kN/m and mass 50 kg is subjected to a force of time-history shown in Fig. P11.15. Determine the response at t = 0.02 s, 0.04 s, 0.06 s and 0.08 s.

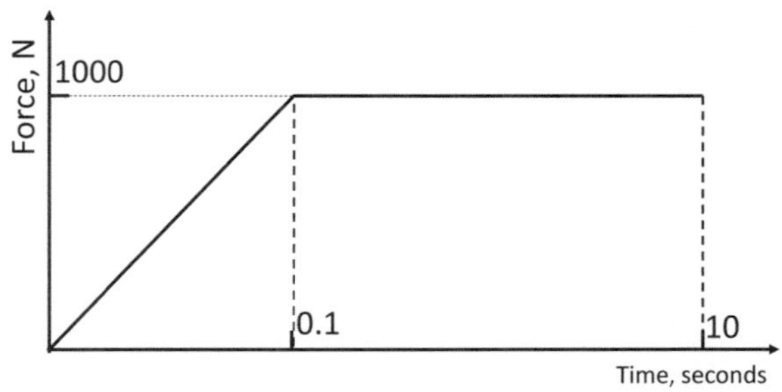

Fig. P11.12 Time history of a force applied to an SDOF

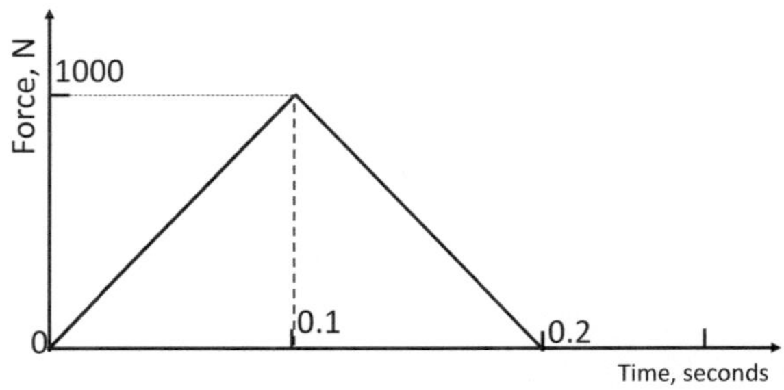

Fig. P11.13 Time history of a force applied to an SDOF

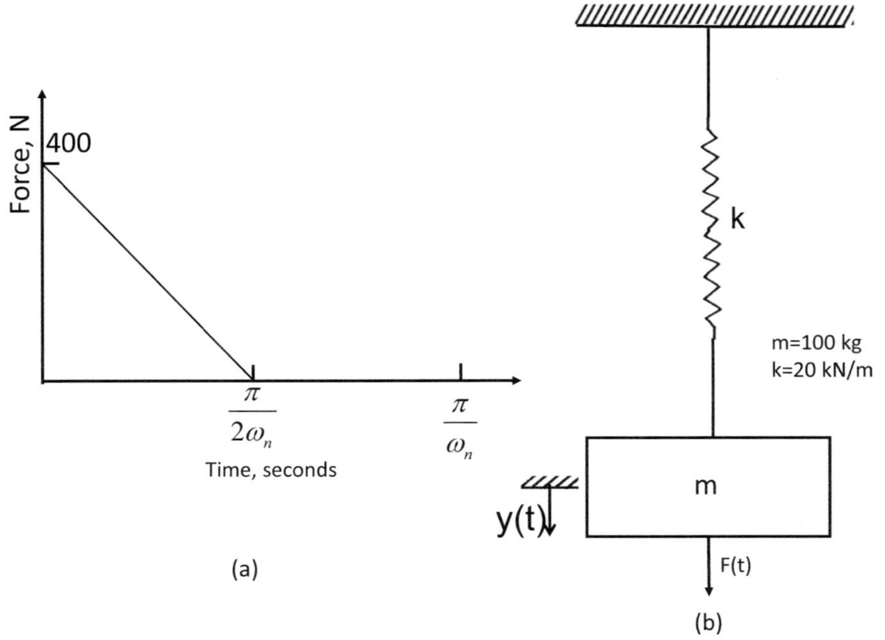

Fig. P11.14 **a** Time history of the force applied to a spring-mass of this figure. **b** Spring-mass system

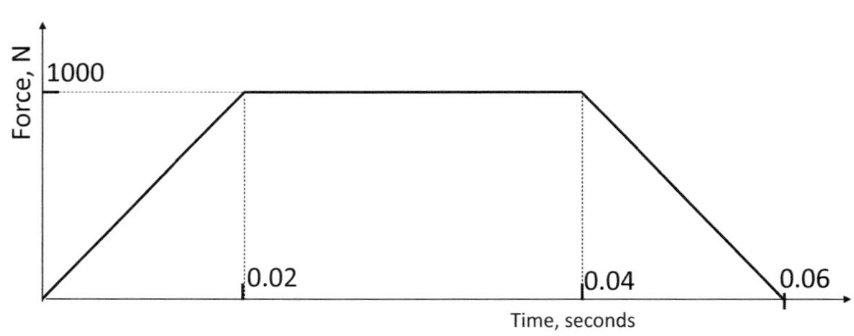

Fig. P11.15 Time history of a force applied to an SDOF

11.16 A single degree of freedom system of mass ***m***, stiffness ***k***, and damping coefficient ***c*** is subjected to a step input of magnitude F. Determine the response of the system using **Duhamel's integral**. What is the response of the system if there is no damping? Sketch the response in both cases.

11.17 A spring-mass-damper system is constrained to move along a smooth horizontal surface as shown in Fig. P11.17; it has mass M, stiffness k, and damping coefficient c and is subjected to a force excitation F(t). A slender rod of mass m and length ℓ is hinged to mass M as shown. Derive the equations of motion using Newton's law.

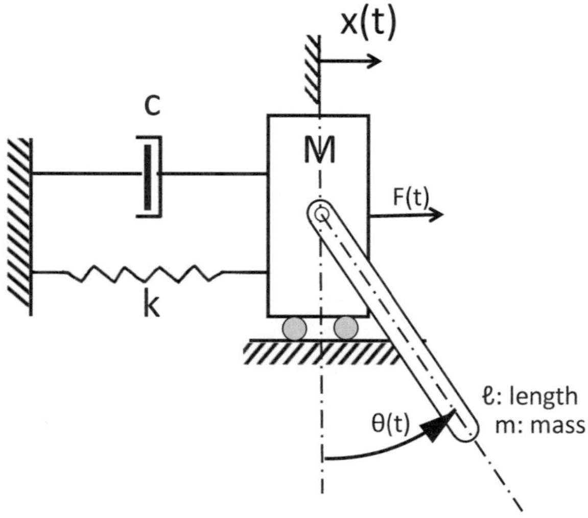

Fig. P11.17 SDOF coupled with a compound pendulum

11.18 A cylinder of mass *m*, radius *r*, and length ℓ is allowed to roll on a horizontal plane without slipping as shown in Fig. P11.18. Four springs, each of stiffness k/2, are attached to the center of the cylinder on its side faces such that the rolling of the cylinder does not interfere with the linear movement of the springs. (a) Obtain the differential equation of motion using Newton's laws and (b) the equation for the natural frequency of the system.

11.19 As shown in Fig. P11.19, a trolley of mass m_1 that has the coordinate x_1 is connected to springs of stiffness k/2 on either side, which in turn is attached to a fixed reference; the trolley moves on a smooth horizontal plane. A cylinder of mass m_2, radius r, and length ℓ that rolls horizontally on the trolly without slipping, is connected to the trolly by means of springs of stiffness k/2 on the center of its side faces; the mass center of the cylinder has the coordinate x_2. The springs, the trolley, and the cylinder are attached in such a way that they do not cause any obstruction to each other's movements.

(a) Derive the equations of motion using Newton's law (b)

Determine the natural frequencies, mode shapes, and the mass-normalized modal matrix, if $m_1 = m_2 = m$.

11.20 Determine the natural frequencies for the two-mass system shown in Fig. P11.20 and the corresponding mode shapes. Both masses move only in the vertical direction. If the lower mass is subjected to a unit impulse, determine the displacement response of both the masses as a function of time. You must use coordinate transformation using the mass-normalized modal matrix. $k = 5000$ N/m, $m = 1$ kg.

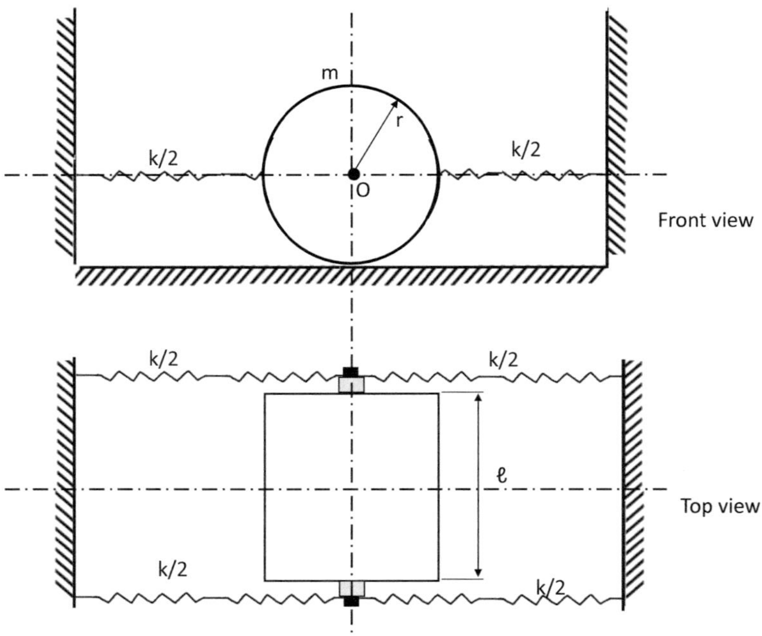

Fig. P11.18 A cylinder rolling without slipping attached to springs fixed at the other end

11.21 Determine the two natural frequencies and the corresponding mode shapes for the long slender bar of mass *m*, and length ℓ, supported by two springs as shown in Fig. P11.21, each having a stiffness, *k*. One spring is attached at the end of the bar, and the other is attached at a point which is at a variable distance for various values of n = 2, 3, 4, 5, 6.

11.22 A slender rod of mass m and length ℓ is pinned at the lower end and fixed to a ball of mass *M* at the upper end as shown in Fig. P11.22. It is horizontally supported by a spring of stiffness *k* at a height of *a* from the lower end. Determine the natural frequency of the system.

11.23 A horizontal slender rod of mass m and length ℓ is pinned at the left end and connected to a spring of stiffness k, a viscous damper of damping coefficient c, and a steel ball of mass M as shown in Fig. P11.23. Determine the natural frequency of the system.

11.24 A simplified model of an automobile is shown in Fig. P11.24. It consists of a larger mass M that is mounted on axles with equivalent spring stiffness at the front and rear and the entire automobile is resting on the tires that have an equivalent stiffness k_2 at the front and rear. The front and the back axle along with the tires each have a total mass m. Determine the natural frequencies and modeshapes assuming that all the masses move together in the vertical direction. M = 1500 kg, m = 200 kg, k_1 = 100 kN/m, k_2 = 1100 kN/m.

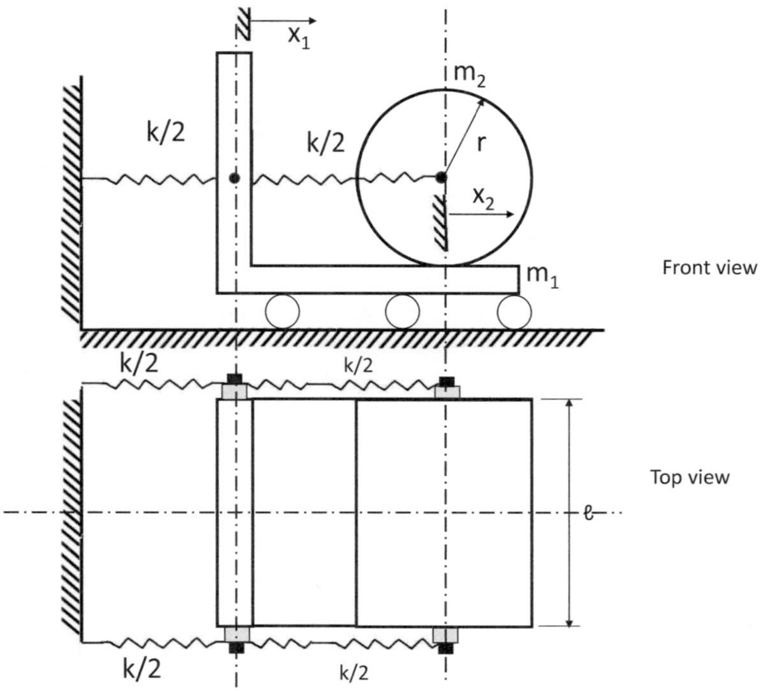

Fig. P11.19 A cylinder rolling without slipping attached to a moving trolly with springs, which is attached to springs fixed at the other end

11.25 A vertical shaft of moment inertia I about its rotational axis that is placed on a thrust bearing as shown in Fig. P11.25 is rotating at a constant angular velocity ω. A horizontal massless tube is attached to this vertical shaft as shown; two masses m_1 and m_2 connected by springs k_1 and k_2 as shown are placed inside the horizontal tube. The clearance between the masses and horizontal tube and the nature of the contact surfaces ensures there is no friction between them. Derive the equations of motion using Newton's law.

11.26 As shown in Fig. P11.26, An inverted pendulum consists of a slender rod of mass m and length ℓ is pinned at the lower end and has a concentrated mass at its tip. A spring of stiffness k and a viscous damper of damping coefficient c are attached as shown. Derive the equations of motion and its natural frequency.

11.27 Determine the natural frequencies and mode shapes of the system shown in Fig. P11.27.

Fig. P11.20 Two-degree of freedom spring mass system

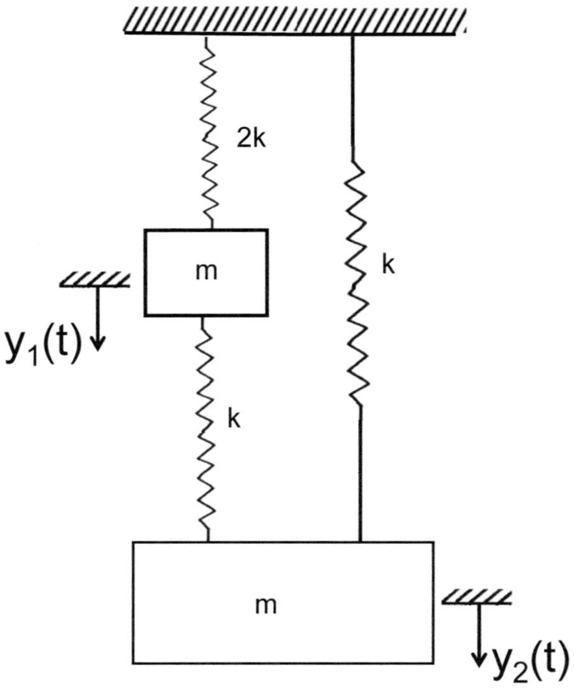

Fig. P11.21 Two parallel vertical springs attached to a horizontal bar

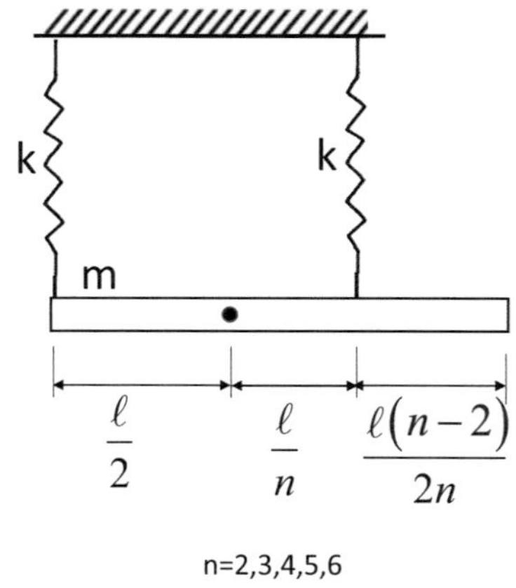

11.28 A one-fourth model of a car going over a speed breaker is shown in Fig. P11.28. Find the (i) critical speeds of the car in km/h and (ii) amplitude of the car body and axle vibration if it goes over the speed breaker at 45 km/h. $m_1 = 400$ kg, $m_2 = 80$ kg, $k_1 = 100$ kN/m, $k_2 = 1100$ kN/m. Assume the speed breaker shape to be a half-sinusoid.

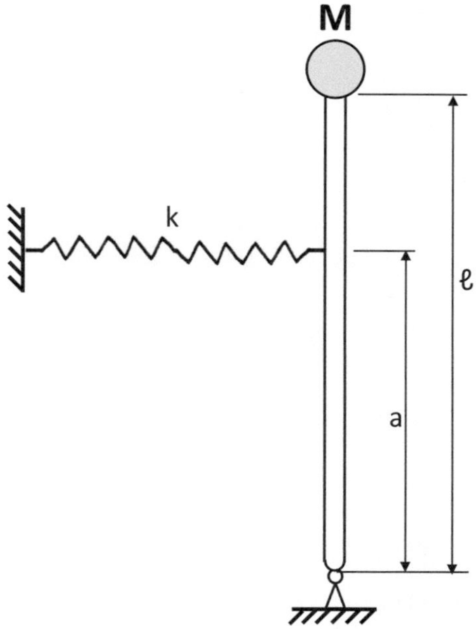

Fig. P11.22 A vertical inverted compound pendulum attached to a horizontal spring fixed at the other end

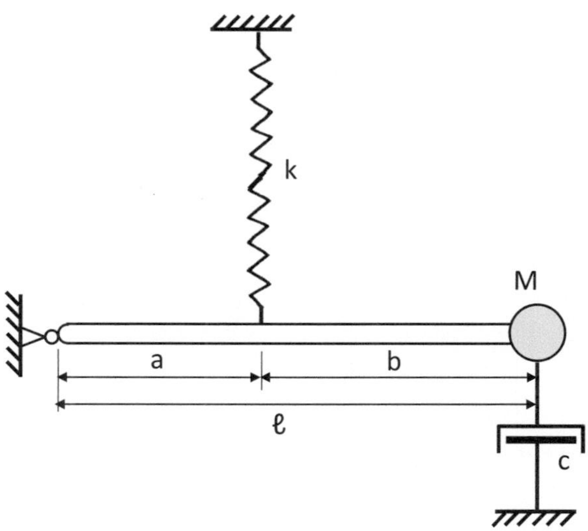

Fig. P11.23 Combined system of compound pendulum with spring and damper

11.29 A rigid cuboid of square cross-section of side q and height p, with mass m is resting on eight springs [of stiffness k_1(vertical) and k_2 (horizontal)], two at each of the bottom corners, as shown in Fig. P11.29, along the vertical and horizontal directions. Assume that the cuboid is capable of performing plane motion only in free vibration.

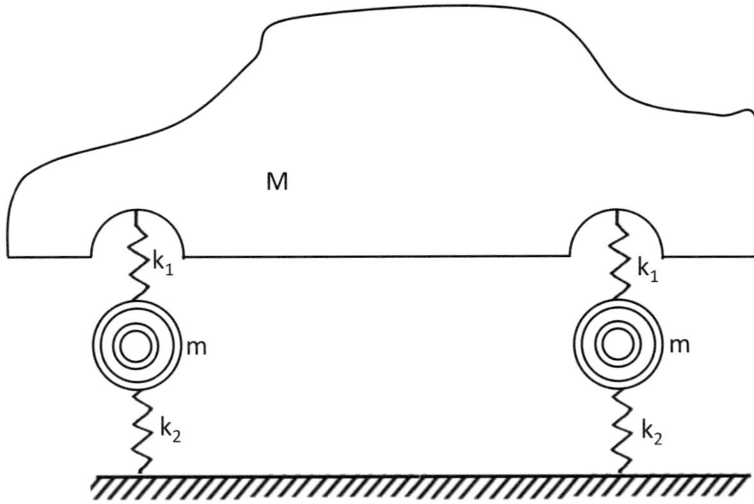

Fig. P11.24 Spring mass model of an automobile

Fig. P11.25 A vertical rotating shaft with a two-degree of freedom spring mass system moving along a horizontal tube

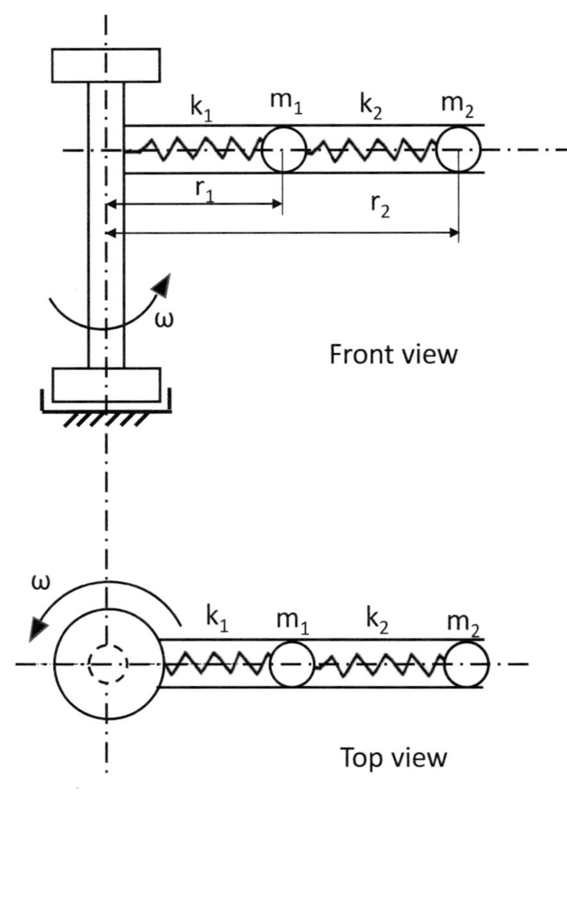

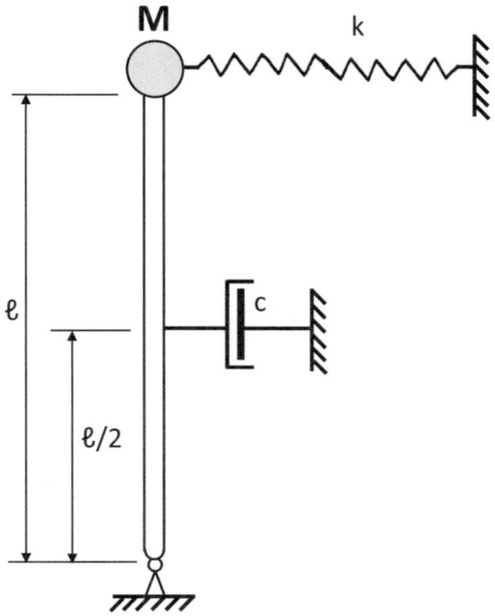

Fig. P11.26 A vertical inverted compound pendulum attached to a horizontal spring and damper fixed at the other end

(a) Derive the equations of motion,
(b) If m = 50 kg, p = 150 mm, q = 250 mm and k_1 = 30 kN/m, k_2 = 20 kN/m determine the natural frequencies of the system

11.30 If the lower mass (2m) of the two-degree of freedom system shown in Fig. P11.30 is subjected to a unit impulse, determine the response of both masses at t = 1 s. Assume k = 20 kN/m and m = 1 kg.

11.31 A machine that can be assumed as a concentrated mass of 250 kg operates between 500 and 1500 rpm during its normal operation. The machine produces an excitation harmonic force of constant amplitude 20,000 N at a frequency corresponding to the rpm. Design an isolation system so the maximum force transmitted to the foundation is 50,000 N during the starting phase and less than 2000 N during the normal operation.

11.32 A machine that can be modeled as a spring-mass system of stiffness k = 10 kN/m and mass M = 50 kg. The machine operates at 200 RPM with a rotating unbalance of 0.2 kg-m.

(a) Determine the rms value of response
(b) Determine the stiffness of an undamped absorber of mass 3 kg such that the vibrations are eliminated at 200 RPM.
(c) What are the natural frequencies of the machine with the absorber?
(d) What are the range of speeds for which the rms amplitude is less than 5 mm?

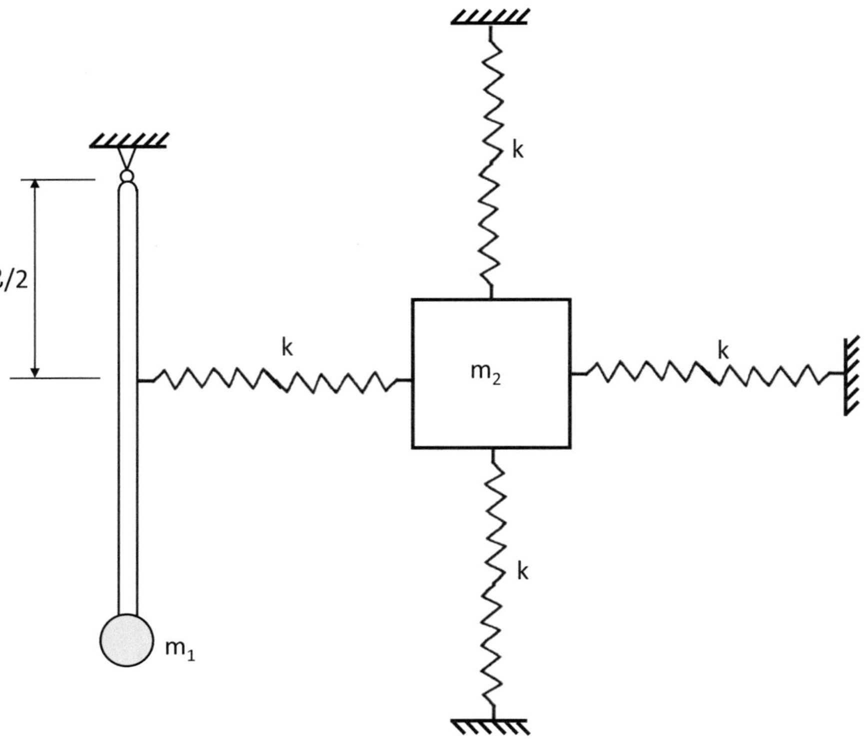

Fig. P11.27 A compound pendulum attached to a spring mass system with horizontal and vertical degrees of freedom

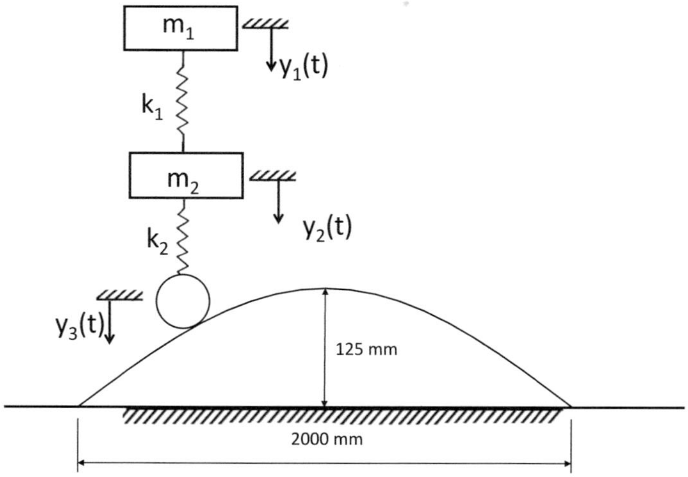

Fig. P11.28 One-fourth model of a car moving on a speed breaker

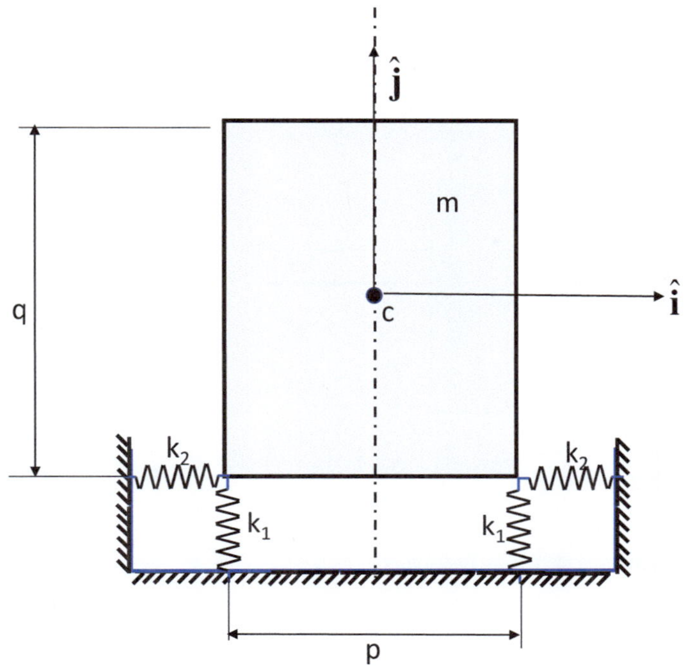

Fig. P11.29 A system of mass and springs with rotational and translational degrees of freedom

Fig. P11.30 Two degree of freedom spring mass system subjected to an impulse excitation

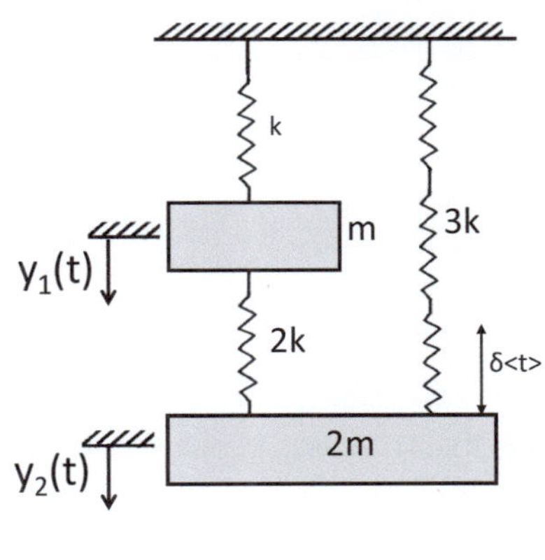

11.33 A shaft arrangement with two rotors is shown in Fig. P11.33. The two masses are : $m1 = 50$ kg and $m2 = 30$ kg. The influence coefficients are:$\alpha 11 = \alpha 22 = 3 \times 10^{-6}$ m/N, $\alpha 12 = \alpha 21 = 2 \times 10^{-6}$ m/N. Determine the first natural frequency using the Rayleigh's method and Dunkerley's method.

11.34 Figure P11.34 shows a rotor having a compressor wheel of mass $m1 = 35$ kg and a turbine wheel of mass $m2 = 22$ kg mounted on a steel shaft of uniform diameter. Determine the diameter of the shaft that will give a fundamental critical frequency that is 25% greater than the operating speed of 8000 rpm.

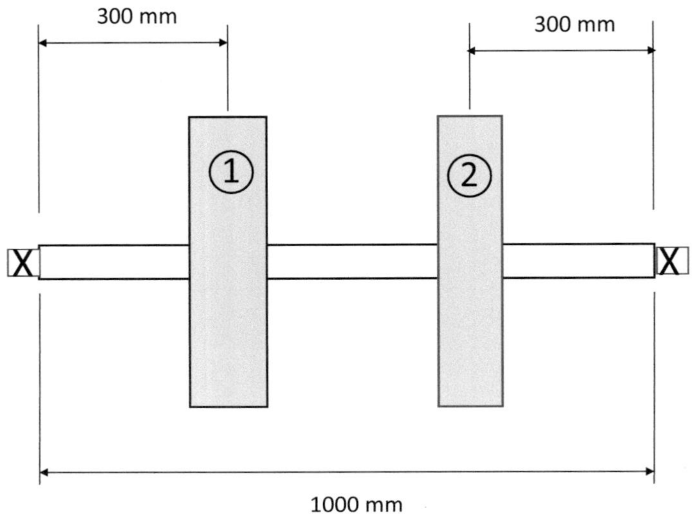

Fig. P11.33 A rotating shaft with two rotors

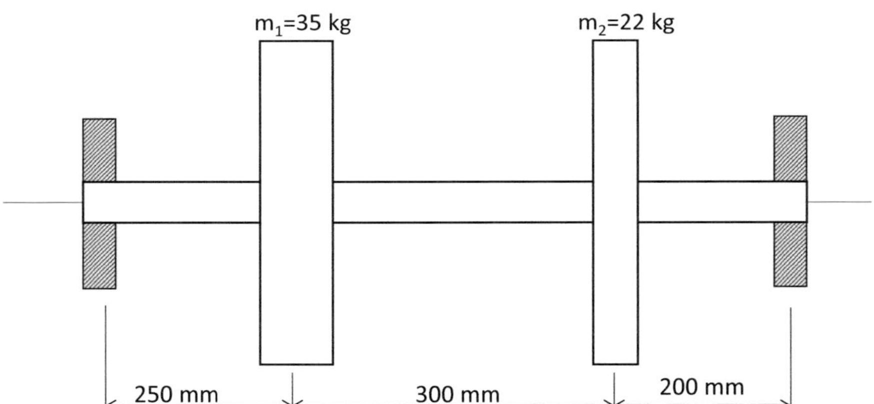

Fig. P11.34 A rotating shaft with two rotors

Use the following methods to determine the critical speed and choose the smallest among them to be the critical speed: (1) Rayleigh's method (2) Dunkerley's method.

11.35 A uniform circular steel shaft (E = 210 GPa) of 80 mm diameter and 3.5 m length is simply supported at its bearings as shown in Fig. P11.35. It has four rotors at the locations shown. Determine the lowest natural frequency of the shaft in lateral vibration (critical frequency) considering the weight of rotors as well as the shaft weight by using Dunkerley's method.

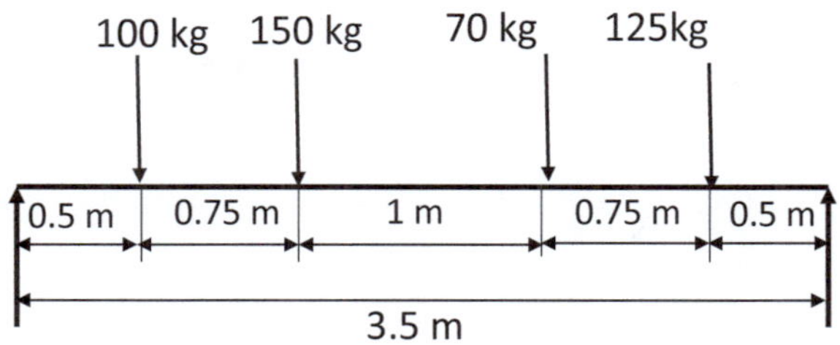

Fig. P11.35 A rotating shaft with two rotors

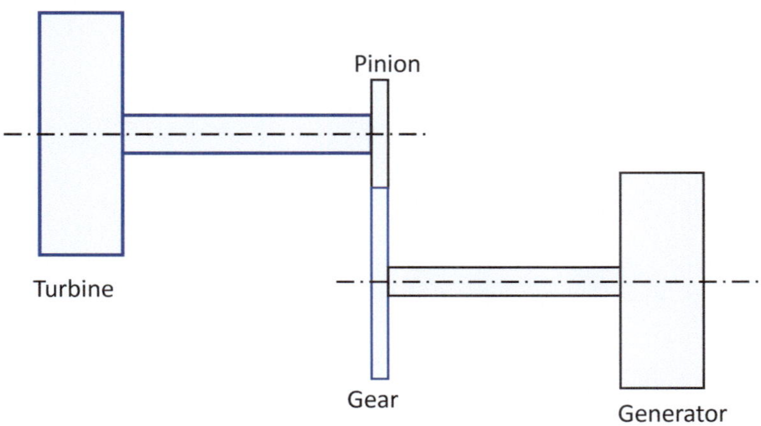

Fig. P11.36 A turbine generator system with a speed reduction geared system

11.36 A turbine ($J = 5000$ kg-m^2) rotating at 10,000 rpm is connected through a set of reduction gears to a generator ($J = 7000$ kg-m^2) so it runs at 2500 rpm as shown in Fig. P11.36; J_{pinion}=75 kg-m^2, J_{gear}=750 kg-m^2; stiffness of the shaft connecting the turbine and pinion is 2 MN/m and that of the shaft connecting the gear and generator is 3 MN/m Determine the natural frequencies of the system using Holzer's method.

Chapter 12
Gyroscopes

Gyroscopes are devices based on the simplest principle of mechanics-conservation of angular momentum. Hence, any spinning body in equilibrium would resist any disturbance to its current state. This is the driving principle behind a simple device like a spinning top or bicycle to the navigational systems that are formally named gyroscopes-the word gyro is derived from Greek that means turning. Of course, gyroscopes have an elaborate arrangement in which the spinning body is mounted on structures that are oriented in different directions so as to sense the resistance to the disturbance in various directions that can be accurately measured to obtain useful information.

In this chapter, however, we would not be interested in the operation of the gyroscopes. But we shall use their basic structure to analyze three-dimensional motion of mechanisms. In the earlier chapters, we had restricted our attention to the plane motion of bodies. However, from first principles, it is possible to derive equations for the angular momentum of a rotating body and equate its rate of change to the moments acting on it. We shall use these equations to obtain the reaction forces of a body moving in a three-dimensional space.

By using the above equations, we shall demonstrate the principle of precession that is encountered in many mechanisms and even in automobiles negotiating a curve.

12.1 Angular Velocity in Three-Dimensions

Consider a global coordinate system of unit vectors $\hat{\boldsymbol{I}}, \hat{\boldsymbol{J}}, \hat{\boldsymbol{K}}$ in a fixed frame $\mathcal{F}$ shown in Fig. 12.1 and a body $\mathcal{B}$ that is embedded in the coordinate system of unit vectors $\hat{\boldsymbol{i}}_0, \hat{\boldsymbol{j}}_0, \hat{\boldsymbol{k}}_0$; in the present position, both the coordinate systems are coincident. With three successive rotations, the body $\mathcal{B}$ can be moved to another position defined by $\hat{\boldsymbol{i}}, \hat{\boldsymbol{j}}, \hat{\boldsymbol{k}}$. These steps are described in the following paragraphs.

D. N. Manik, *Fundamentals of Mechanisms and Machines*,
https://doi.org/10.1007/978-981-97-1810-8_12

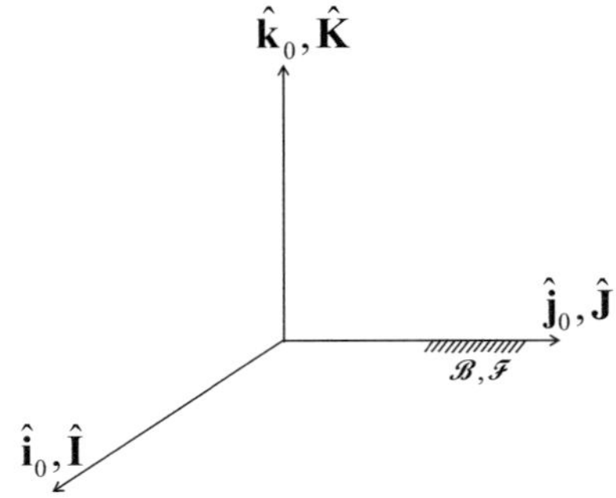

Fig. 12.1 Global coordinates and local coordinates in the initial position

In Fig. 12.2a, the first rotation through an angle ϕ is about the global $\hat{\boldsymbol{K}}$ unit vector. Hence, the unit vectors $\hat{\boldsymbol{i}}_0,\hat{\boldsymbol{j}}_0$ initially embedded on the body $\mathcal{B}$ move to their new position $\hat{\boldsymbol{i}}_1,\hat{\boldsymbol{j}}_1$, whereas the unit vector $\hat{\boldsymbol{k}}_0$ is still aligned with the global unit vector $\hat{\boldsymbol{K}}$ and is now named as $\hat{\boldsymbol{k}}_1$. The unit vectors of the body $\mathcal{B}$ are now in the intermediate position, $\hat{\boldsymbol{i}}_1,\hat{\boldsymbol{j}}_1,\hat{\boldsymbol{k}}_1$ and are in a frame $\mathcal{F}_1$ that has an angular velocity $\dot{\phi}\hat{\boldsymbol{K}}$ in $\mathcal{F}$.

Next, in Fig. 12.2b, the second rotation through an angle θ is about the unit vector $\hat{\boldsymbol{j}}_1$ embedded to the body. Hence, the unit vectors $\hat{\boldsymbol{i}}_1$, $\hat{\boldsymbol{k}}_1$ embedded on the body $\mathcal{B}$ move to their new position $\hat{\boldsymbol{i}}_2$, $\hat{\boldsymbol{k}}_2$, whereas the unit vector $\hat{\boldsymbol{j}}_1$ has not changed its position, but is now named as $\hat{\boldsymbol{j}}_2$. The unit vectors of the body $\mathcal{B}$ are now in the intermediate position, $\hat{\boldsymbol{i}}_2,\hat{\boldsymbol{j}}_2,\hat{\boldsymbol{k}}_2$ are in a frame $\mathcal{F}_2$ that has an angular velocity $\dot{\theta}\hat{\boldsymbol{j}}_1$ relative to the frame $\mathcal{F}_1$.

In the final position, as shown in Fig. 12.2c, the third rotation through an angle ψ is about the unit vector $\hat{\boldsymbol{k}}_2$ embedded to the body. Hence, the unit vectors $\hat{\boldsymbol{i}}_2,\hat{\boldsymbol{j}}_2$ embedded on the body $\mathcal{B}$ move to their final position $\hat{\boldsymbol{i}},\hat{\boldsymbol{j}}$, whereas the unit vector $\hat{\boldsymbol{k}}_2$ has not changed its position, but is now named as $\hat{\boldsymbol{k}}$. The unit vectors of the body in the final position, $\hat{\boldsymbol{i}},\hat{\boldsymbol{j}},\hat{\boldsymbol{k}}$ are in the body frame of $\mathcal{B}$ that has an angular velocity $\dot{\psi}\hat{\boldsymbol{k}}$ relative to the frame $\mathcal{F}_2$.

The angular velocity of $\mathcal{B}$ with respect to the fixed frame in $\mathcal{F}$ can be expressed as follows:

$$\begin{aligned}\omega_{\mathcal{B}/\mathcal{F}} &= \omega_{\mathcal{B}/\mathcal{F}_2} + \omega_{\mathcal{F}_2/\mathcal{F}_1} + \omega_{\mathcal{F}_1/\mathcal{F}} \\ &= \dot{\psi}\hat{\boldsymbol{k}} + \dot{\theta}\hat{\boldsymbol{j}}_2 + \dot{\phi}\hat{\boldsymbol{K}}\end{aligned} \tag{12.1}$$

The unit vector $\hat{\boldsymbol{j}}_2$ can be expressed in terms of $\hat{\boldsymbol{i}}$ and $\hat{\boldsymbol{j}}$ as

$$\hat{\boldsymbol{j}}_2 = \sin\psi\hat{\boldsymbol{i}} + \cos\psi\hat{\boldsymbol{j}} \tag{12.2}$$

The unit vector $\hat{\boldsymbol{i}}_2$ can be expressed in terms of $\hat{\boldsymbol{i}}$ and $\hat{\boldsymbol{j}}$ as

$$\hat{\boldsymbol{i}}_2 = \cos\psi\hat{\boldsymbol{i}} - \sin\psi\hat{\boldsymbol{j}} \tag{12.3}$$

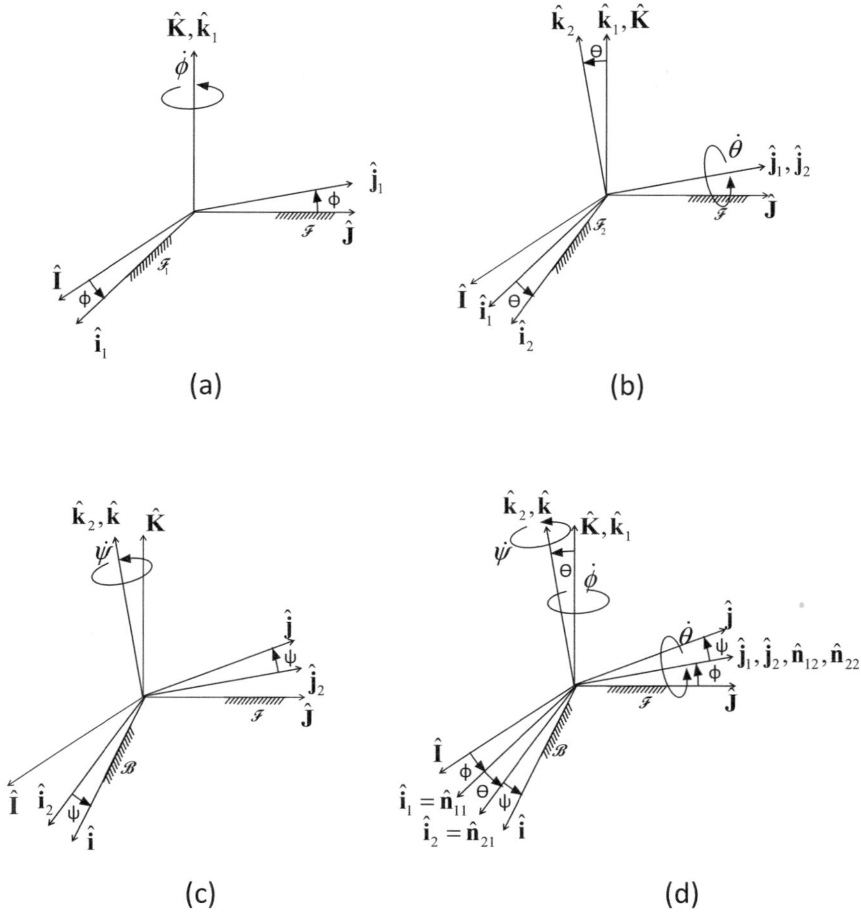

Fig. 12.2 **a** First Eulerian angle rotation, ϕ, about the $\hat{\boldsymbol{K}}$ global unit vector. **b** Second Eulerian angle rotation, Θ, about the $\hat{\boldsymbol{j}}_1$ body unit vector. **c** Third Eulerian angle rotation, ψ, about the $\hat{\boldsymbol{k}}_2$body unit vector. **d** Summary of all the three rotations

From Eq. (12.3), the global vector $\hat{\boldsymbol{K}}$ can be expressed in terms of unit vectors $\hat{\boldsymbol{i}}$,$\hat{\boldsymbol{j}}$ and $\hat{\boldsymbol{k}}$.

$$\begin{aligned}\hat{\boldsymbol{K}} &= -\sin\theta\hat{\boldsymbol{i}}_2 + \cos\theta\hat{\boldsymbol{k}} \\ &= -\sin\theta\left(\cos\psi\hat{\boldsymbol{i}} - \sin\psi\hat{\boldsymbol{j}}\right) + \cos\theta\hat{\boldsymbol{k}} \\ &= -\sin\theta\cos\psi\hat{\boldsymbol{i}} + \sin\theta\sin\psi\hat{\boldsymbol{j}} + \cos\theta\hat{\boldsymbol{k}}\end{aligned} \tag{12.4}$$

From Eqs. (12.1) through (12.4)

$$\begin{aligned}\omega_{\mathcal{B}/\mathcal{F}} &= \dot{\psi}\hat{\boldsymbol{k}} + \dot{\theta}\left(\sin\psi\hat{\boldsymbol{i}} + \cos\psi\hat{\boldsymbol{j}}\right) + \dot{\phi}\left(-\sin\theta\cos\psi\hat{\boldsymbol{i}} + \sin\theta\sin\psi\hat{\boldsymbol{j}} + \cos\theta\hat{\boldsymbol{k}}\right)\\ &= \left(\dot{\theta}\sin\psi - \dot{\phi}\sin\theta\cos\psi\right)\hat{\boldsymbol{i}} + \left(\dot{\theta}\cos\psi + \dot{\phi}\sin\theta\sin\psi\right)\hat{\boldsymbol{j}} + \left(\dot{\psi} + \dot{\phi}\cos\theta\right)\hat{\boldsymbol{k}}\end{aligned} \tag{12.5}$$

Alternatively, $\hat{\boldsymbol{j}}_2$ can be expressed in terms of the global vectors $\hat{\boldsymbol{I}}$ and $\hat{\boldsymbol{J}}$ as

$$\hat{\boldsymbol{j}}_2 = -\sin\phi\hat{\boldsymbol{I}} + \cos\phi\hat{\boldsymbol{J}} \tag{12.6}$$

and, $\hat{\boldsymbol{i}}_1$ can be expressed in terms of the global vectors $\hat{\boldsymbol{I}}$ and $\hat{\boldsymbol{J}}$ as

$$\hat{\boldsymbol{i}}_1 = \left(\cos\phi\hat{\boldsymbol{I}} + \sin\phi\hat{\boldsymbol{J}}\right) \tag{12.7}$$

Using Eq. (12.7), $\hat{\boldsymbol{k}}$ can be expressed in terms of $\hat{\boldsymbol{I}}$ and $\hat{\boldsymbol{J}}$ as

$$\begin{aligned}\hat{\boldsymbol{k}} &= \sin\theta\hat{\boldsymbol{i}}_1 + \cos\theta\hat{\boldsymbol{K}}\\ &= \sin\theta\left(\cos\phi\hat{\boldsymbol{I}} + \sin\phi\hat{\boldsymbol{J}}\right) + \cos\theta\hat{\boldsymbol{K}}\\ &= (\sin\theta\cos\phi)\hat{\boldsymbol{I}} + (\sin\theta\sin\phi)\hat{\boldsymbol{J}} + \cos\theta\hat{\boldsymbol{K}}\end{aligned} \tag{12.8}$$

From Eqs. (12.1), (12.6), and (12.8)

$$\begin{aligned}\omega_{\mathcal{B}/\mathcal{F}} &= \dot{\psi}\left[(\sin\theta\cos\phi)\hat{\boldsymbol{I}} + (\sin\theta\sin\phi)\hat{\boldsymbol{J}} + \cos\theta\hat{\boldsymbol{K}}\right] + \dot{\theta}\left(-\sin\phi\hat{\boldsymbol{I}} + \cos\phi\hat{\boldsymbol{J}}\right) + \dot{\phi}\hat{\boldsymbol{K}}\\ &= \left(\dot{\psi}\sin\theta\cos\phi - \dot{\theta}\sin\phi\right)\hat{\boldsymbol{I}} + \left(\dot{\psi}\sin\theta\sin\phi + \dot{\theta}\cos\phi\right)\hat{\boldsymbol{J}} + \left(\dot{\psi}\cos\theta + \dot{\phi}\right)\hat{\boldsymbol{K}}\end{aligned} \tag{12.9}$$

Alternatively, from Eqs. 12.1 and 12.8

$$\begin{aligned}\omega_{\mathcal{B}/\mathcal{F}} &= \dot{\psi}\left(\cos\theta\hat{\boldsymbol{K}} + \sin\theta\hat{\boldsymbol{i}}_1\right) + \dot{\theta}\hat{\boldsymbol{j}}_2 + \dot{\phi}\hat{\boldsymbol{K}}\\ &= \left(\dot{\psi}\sin\theta\right)\hat{\boldsymbol{i}}_1 + \dot{\theta}\hat{\boldsymbol{j}}_2 + \left(\dot{\psi}\cos\theta + \dot{\phi}\right)\hat{\boldsymbol{K}}\end{aligned} \tag{12.10}$$

12.2 Gyroscope

The equations derived in Sect. 12.1 can be directly used to relate the rate of change of angular velocity to the respective angular velocities in three dimensions. The construction of a typical gyroscope is shown in Fig. 12.3 in which a rotor (body $\mathcal{B}$) is embedded in axes $\hat{\boldsymbol{i}}, \hat{\boldsymbol{j}}, \hat{\boldsymbol{k}}$. The axis of rotation of the rotor is connected to the inner gimbal ($\mathcal{F}_2$) through bearings C and the inner gimbal is embedded in axes $\hat{\boldsymbol{i}}_2, \hat{\boldsymbol{j}}_2, \hat{\boldsymbol{k}}_2$; the inner gimbal ($\mathcal{F}_1$) is connected to the outer gimbal through bearings B

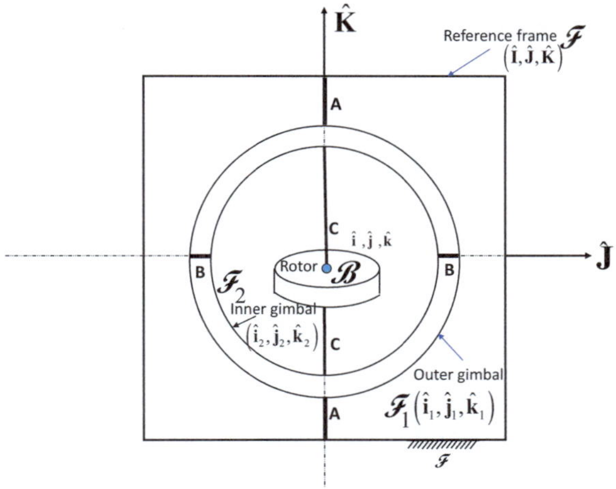

Fig. 12.3 Gyroscope

and is embedded in axes $\hat{\boldsymbol{i}}_1, \hat{\boldsymbol{j}}_1, \hat{\boldsymbol{k}}_1$; the outer gimbal is connected to the fixed frame ($\mathcal{F}_1$)through bearings A and is embedded in axes $\hat{\boldsymbol{I}}, \hat{\boldsymbol{J}}, \hat{\boldsymbol{K}}$.

The first rotation of the gyroscope, very similar to the action shown in Fig. 12.2a, using the outer gimbal is shown in Fig. 12.4. The outer gimbal that was earlier referred to as $\mathcal{F}_1$ about the $\hat{\boldsymbol{K}}$ axis is turned by an angle ϕ using bearings A at an angular velocity $\dot{\phi}$ in the counterclockwise direction. The inner gimbal and the rotor (body $\mathcal{B}$) move rigidly along with the outer gimbal and the rotor is not shown in this figure. The unit vector of the outer gimbal $\hat{\boldsymbol{i}}_1$ and$\hat{\boldsymbol{j}}_1$ moves toward the current position with respect to $\hat{\boldsymbol{I}}$ and $\hat{\boldsymbol{J}}$ vectors by an angle ϕ, as shown in Fig. 12.4, and $\hat{\boldsymbol{k}}_1$ remains aligned with the unit vector $\hat{\boldsymbol{K}}$ (Fig. 12.5).

In the second rotation, very similar to the rotation in Fig. 12.2b, the inner gimbal is rotated about the $\hat{\boldsymbol{j}}_1$ axis that is same as the $\hat{\boldsymbol{j}}_2$ by an angle Θ in the clockwise direction by tilting it with respect to the outer gimbal. The axes $\hat{\boldsymbol{i}}_1$ and $\hat{\boldsymbol{k}}_1$ are rotated clockwise by an angle Θ to respectively move into the new positions, $\hat{\boldsymbol{i}}_2$ and $\hat{\boldsymbol{k}}_2$. The inner gimbal rotates at an angular velocity $\dot{\theta}$ in the clockwise direction. The rotor (body $\mathcal{B}$) moves rigidly with the inner gimbal (Fig. 12.6).

The third rotation, very similar to the rotation in Fig. 12.2c, allows the rotor (body $\mathcal{B}$) to turn about $\hat{\boldsymbol{k}}_2$ axis through an angle ψ at an angular velocity $\dot{\psi}$ that is also its final position in the $\hat{\boldsymbol{k}}$ direction. Now the axis $\hat{\boldsymbol{i}}_2$ moves to $\hat{\boldsymbol{i}}$ and $\hat{\boldsymbol{j}}_2$ moves to $\hat{\boldsymbol{j}}$ directions respectively, each by an angle ψ. Now it should be clear that by moving through the Eulerian angles, a body in the three-dimensional space can be moved to any position with respect to the absolute frame of reference.

Equation 12.5 is reproduced as follows:

$$\omega_{\mathcal{B}/\mathcal{F}} = \left(\dot{\theta}\sin\psi - \dot{\phi}\sin\theta\cos\psi\right)\hat{\boldsymbol{i}} + \left(\dot{\theta}\cos\psi + \dot{\phi}\sin\theta\sin\psi\right)\hat{\boldsymbol{j}} + \left(\dot{\psi} + \dot{\phi}\cos\theta\right)\hat{\boldsymbol{k}} \tag{12.11}$$

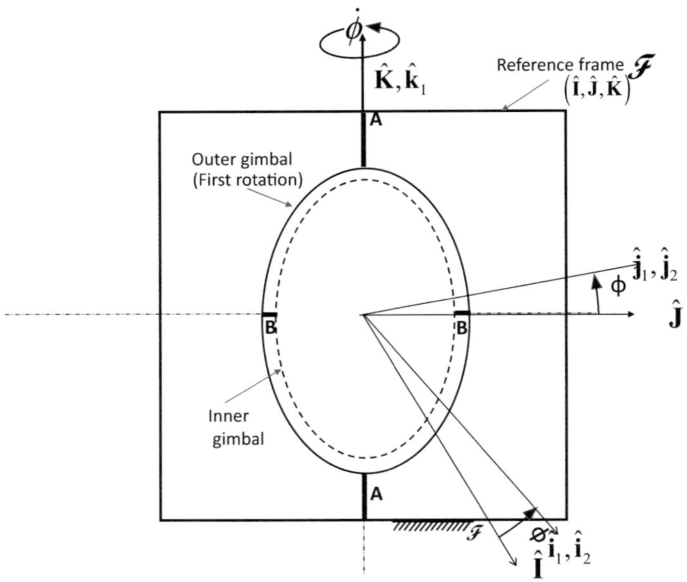

Fig. 12.4 First rotation of the gyroscope using the outer gimbal

The angular velocity of the body $\mathcal{B}$ can be written as:

$$\omega_{\mathcal{B}/\mathcal{F}} = \omega_1\hat{\boldsymbol{i}} + \omega_2\hat{\boldsymbol{j}} + \omega_3\hat{\boldsymbol{k}} \tag{12.12}$$

From Eqs. (12.11) and (12.12), the components of angular velocity are given by

$$\begin{aligned}\omega_1 &= \dot{\theta}\sin\psi - \dot{\phi}\sin\theta\cos\psi \\ \omega_2 &= \dot{\theta}\cos\psi + \dot{\phi}\sin\theta\sin\psi \\ \omega_3 &= \dot{\psi} + \dot{\phi}\cos\theta\end{aligned} \tag{12.13}$$

Equation (12.13) can be written in a matrix form as follows:

$$\begin{bmatrix}\sin\psi & -\sin\theta\cos\psi & 0 \\ \cos\psi & \sin\theta\sin\psi & 0 \\ 0 & \cos\theta & 1\end{bmatrix}\begin{Bmatrix}\dot{\theta} \\ \dot{\phi} \\ \dot{\psi}\end{Bmatrix} = \begin{Bmatrix}\omega_1 \\ \omega_2 \\ \omega_3\end{Bmatrix} \tag{12.14}$$

Solving Eq. (12.14)

$$\begin{aligned}\dot{\theta} &= \omega_1\sin\psi + \omega_2\cos\psi \\ \dot{\phi} &= \frac{\omega_2\sin\psi - \omega_1\cos\psi}{\sin\theta}\end{aligned}$$

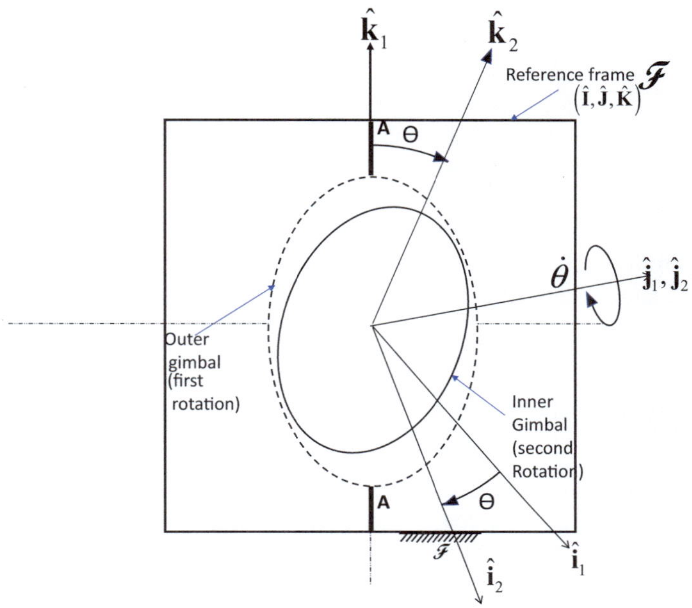

Fig. 12.5 Second rotation of the gyroscope using the inner gimbal

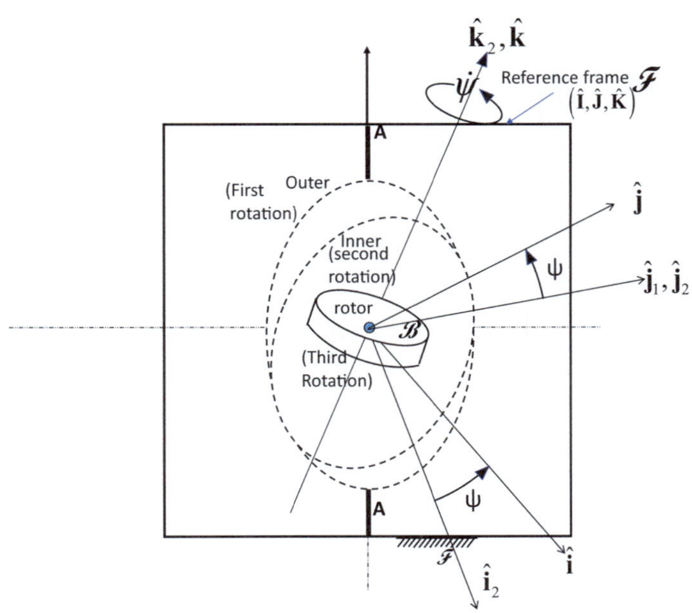

Fig. 12.6 Third rotation of the gyroscope using the rotor

$$\dot{\psi} = \frac{\omega_3 \sin\theta - \omega_2 \cos\theta \sin\psi + \omega_1 \cos\theta \cos\psi}{\sin\theta} \tag{12.15}$$

If the time-domain components of angular velocities are known, Eq. (12.15) can be integrated, at least numerically, but the problem of singularities due to the sinθ term in the denominator of the two Eq. (12.15) makes it difficult to obtain the precise position of an object moving in the three-dimensional space. In addition, the order in which the Eulerian angles is critical. Otherwise, the body will end up in a completely different position. By actually measuring the angular velocities using transducers, Eq. (12.15) can be used in the application of gyroscopes in navigation.

12.3 Rotation Matrices

The transformation between coordinates of a body in the three-dimensional space with respect to a fixed coordinate system can be easily obtained using rotational matrices. To begin with, consider a coordinate system shown in Fig. 12.7. It consists of cartesian coordinates in the plane of xy at with unit vectors $\hat{\boldsymbol{i}}_0$ and $\hat{\boldsymbol{j}}_0$. If this coordinate system is rotated counterclockwise by an angle θ_z in the xy plane, it moves to the new position with unit vectors $\hat{\boldsymbol{i}}_1$ and $\hat{\boldsymbol{j}}_1$. In the above action, the direction of the unit vector $\hat{\boldsymbol{k}}_o$, perpendicular to the xy plane, remains the same.

The above transformation can be summarized in the following equation:

$$\begin{Bmatrix} \hat{\boldsymbol{i}}_1 \\ \hat{\boldsymbol{j}}_1 \\ \hat{\boldsymbol{k}}_1 \end{Bmatrix} = \begin{bmatrix} \cos\theta_z & \sin\theta_z & 0 \\ -\sin\theta_z & \cos\theta_z & 0 \\ 0 & 0 & 1 \end{bmatrix} \begin{Bmatrix} \hat{\boldsymbol{i}}_0 \\ \hat{\boldsymbol{j}}_0 \\ \hat{\boldsymbol{k}}_0 \end{Bmatrix} \tag{12.16}$$

From Eq. 12.16, the transformation matrix for rotation in the xy plane is given by

$$\boldsymbol{T}_z = \begin{bmatrix} \cos\theta_z & \sin\theta_z & 0 \\ -\sin\theta_z & \cos\theta_z & 0 \\ 0 & 0 & 1 \end{bmatrix} \tag{12.17}$$

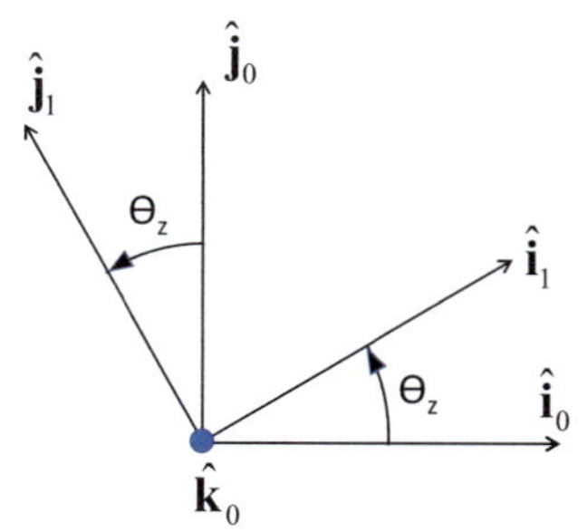

Fig. 12.7 Rotation about the z-axis

On similar lines, the rotation of the coordinates in the yz plane by θ_x results in the following transformation matrix

$$\boldsymbol{T}_x = \begin{bmatrix} 1 & 0 & 0 \\ 0 & \cos\theta_x & \sin\theta_x \\ 0 & -\sin\theta_x & \cos\theta_x \end{bmatrix} \tag{12.18}$$

and the rotation of the coordinates in the xz plane by θ_y results in the following transformation matrix

$$\boldsymbol{T}_y = \begin{bmatrix} \cos\theta_y & 0 & -\sin\theta_y \\ 0 & 1 & 0 \\ \sin\theta_y & 0 & \cos\theta_y \end{bmatrix} \tag{12.19}$$

By using the above transformation Eqs. (12.17)–(12.19), (12.11) can be derived. These equations are also useful whenever transformations from one coordinate system to another are required to be carried out.

12.4 Angular Momentum

The velocities of points P and Q are related by

$$\boldsymbol{v}_Q = \boldsymbol{v}_P + \boldsymbol{\omega} \times \boldsymbol{r}_{PQ} \tag{12.20}$$

where the angular velocity vector is given by

$$\boldsymbol{\omega} = \omega_x \hat{\boldsymbol{i}} + \omega_y \hat{\boldsymbol{j}} + \omega_z \hat{\boldsymbol{k}} \tag{12.21}$$

The acceleration of points P and Q are related by

$$\boldsymbol{a}_Q = \boldsymbol{a}_P + \boldsymbol{\alpha} \times \boldsymbol{r}_{PQ} + \boldsymbol{\omega} \times \boldsymbol{\omega} \times \boldsymbol{r}_{PQ} \tag{12.22}$$

where the angular acceleration vector is given by

$$\boldsymbol{\alpha} = \alpha_x \hat{\boldsymbol{i}} + \alpha_y \hat{\boldsymbol{j}} + \alpha_z \hat{\boldsymbol{k}} \tag{12.23}$$

The angular momentum of an element mass about a point is the cross product of its position vector with its linear momentum. Hence, the angular momentum of the rigid body shown in Fig. 12.8 about the point P is the volume integral of the angular moment of the element mass about P and is given by

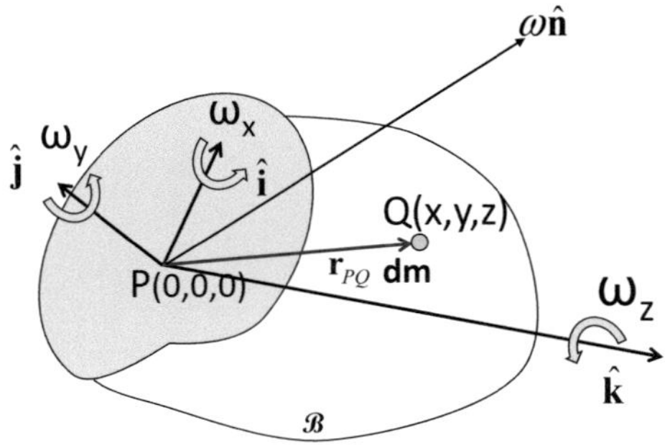

Fig. 12.8 A rigid body in three-dimensional motion

$$\boldsymbol{H}_P = \int_{\mathcal{B}} \boldsymbol{R} \times \boldsymbol{v}_Q dm \tag{12.24}$$

where $\boldsymbol{R} = \boldsymbol{r}_{PQ} = x\hat{\boldsymbol{i}} + y\hat{\boldsymbol{j}} + z\hat{\boldsymbol{k}}$ is the position vector of point Q with respect to the point P.

From Eq. (12.20), the angular momentum of the rigid body about the point P is given by

$$\boldsymbol{H}_P = \int_{\mathcal{B}} \boldsymbol{R} \times \left(\boldsymbol{v}_P + \boldsymbol{\omega} \times \boldsymbol{r}_{PQ}\right) dm \tag{12.25}$$

By substituting the position vector $\boldsymbol{r}_{PQ}$ of the elemental mass in Eq. (12.25), the angular momentum of the body about the point P is given by

$$\boldsymbol{H}_P = \int_{\mathcal{B}} \left(x\hat{\boldsymbol{i}} + y\hat{\boldsymbol{j}} + z\hat{\boldsymbol{k}}\right) \times \left(\boldsymbol{v}_P + \boldsymbol{\omega} \times \left[x\hat{\boldsymbol{i}} + y\hat{\boldsymbol{j}} + z\hat{\boldsymbol{k}}\right]\right) dm \tag{12.26}$$

Equation (12.26) can be written as

$$\boldsymbol{H}_P = \int_{\mathcal{B}} \left(x\hat{\boldsymbol{i}} + y\hat{\boldsymbol{j}} + z\hat{\boldsymbol{k}}\right) \times \boldsymbol{v}_P dm + \int_{\mathcal{B}} \left(x\hat{\boldsymbol{i}} + y\hat{\boldsymbol{j}} + z\hat{\boldsymbol{k}}\right) \times \left[\boldsymbol{\omega} \times \left(x\hat{\boldsymbol{i}} + y\hat{\boldsymbol{j}} + z\hat{\boldsymbol{k}}\right)\right] dm \tag{12.27}$$

In terms of the components of the angular velocity of Eq. (12.27), the angular momentum about the point P can be written as

$$\begin{aligned} \boldsymbol{H}_P = & \int_{\mathcal{B}} \left(x\hat{\boldsymbol{\imath}} + y\hat{\boldsymbol{\jmath}} + z\hat{\boldsymbol{k}}\right) \times \boldsymbol{v}_P dm + \\ & \int_{\mathcal{B}} \left[\omega_x\left(y^2 + z^2\right) - \omega_y xy - \omega_z xz\right]\hat{\boldsymbol{\imath}} dm + \int_{\mathcal{B}} \left[\omega_y\left(x^2 + z^2\right) - \omega_z yz - \omega_x xy\right]\hat{\boldsymbol{\jmath}} dm \\ & + \int_{\mathcal{B}} \left[\omega_z\left(x^2 + y^2\right) - \omega_x xz - \omega_y yz\right]\hat{\boldsymbol{k}} dm \end{aligned} \tag{12.28}$$

The various moments of inertia of the rigid body are expressed in the following equation:

$$\begin{aligned} I_{xx}^P &= \int_{\mathcal{B}} \left(y^2 + z^2\right) dm \\ I_{yy}^P &= \int_{\mathcal{B}} \left(x^2 + z^2\right) dm \\ I_{zz}^P &= \int_{\mathcal{B}} \left(x^2 + y^2\right) dm \\ I_{xy}^P &= -\int_{\mathcal{B}} xydm \\ I_{yz}^P &= -\int_{\mathcal{B}} yzdm \\ I_{xz}^P &= -\int_{\mathcal{B}} xzdm \end{aligned} \tag{12.29}$$

The first integral of Eq. (12.28) is given by

$$\int_{\mathcal{B}} \left(x\hat{\boldsymbol{\imath}} + y\hat{\boldsymbol{\jmath}} + z\hat{\boldsymbol{k}}\right) \times \boldsymbol{v}_P dm = m\boldsymbol{r}_{PC} \times \boldsymbol{v}_P \tag{12.30}$$

From Eqs. (12.28), (12.29), (12.30), the angular moment of the rigid body about the point P is given by

$$\begin{aligned} \boldsymbol{H}_P = m\boldsymbol{r}_{PC} \times \boldsymbol{v}_P + \left[\omega_x I_{xx}^P + \omega_y I_{xy}^P + \omega_z I_{xz}^P\right]\hat{\boldsymbol{\imath}} + \left[\omega_x I_{xy}^P + \omega_y I_{yy}^P + \omega_z I_{yz}^P\right]\hat{\boldsymbol{\jmath}} \\ + \left[\omega_x I_{xz}^P + +\omega_y I_{yz}^P + \omega_z I_{zz}^P\right]\hat{\boldsymbol{k}} \end{aligned} \tag{12.31}$$

The angular momentum about the mass center is given by (the first term of Eq. (12.31) will become zero)

$$\mathbf{H}_C = \left[\omega_x I_{xx}^C + \omega_y I_{xy}^C + \omega_z I_{xz}^C\right]\hat{\mathbf{i}} + \left[\omega_x I_{xy}^C + \omega_y I_{yy}^C + \omega_z I_{yz}^C\right]\hat{\mathbf{j}} + \left[\omega_x I_{xz}^C + +\omega_y I_{yz}^C + \omega_z I_{zz}^C\right]\hat{\mathbf{k}} \tag{12.32}$$

First, let us try to determine the sum of moments about the mass center; sum of moments about the mass center is equal to the time-derivative of the angular momentum about the mass center. It is given by taking the time-derivative of Eq. (12.32).

$$\sum \boldsymbol{M}_C = \frac{d\boldsymbol{H}_C}{dt} = \frac{d}{dt}\left\{\begin{array}{l}\left[\omega_x I_{xx}^C + \omega_y I_{xy}^C + \omega_z I_{xz}^C\right]\hat{i} + \left[\omega_x I_{xy}^C + \omega_y I_{yy}^C + \omega_z I_{yz}^C\right]\hat{j} \\ +\left[\omega_x I_{xz}^C + +\omega_y I_{yz}^C + \omega_z I_{zz}^C\right]\hat{k}\end{array}\right\} \tag{12.33}$$

The coordinates of Eq. (12.33) are on the moving body. Hence, we may recall our earlier discussion in Chap. 2 on how we could compute the time-derivative of vectors referenced to moving coordinates. If a vector $\boldsymbol{A}$ is referenced to the moving coordinate system on the rigid body rotating at an angular velocity $\boldsymbol{\omega}$, the time-derivative of this vector is given by

$$\dot{\boldsymbol{A}} = {}^{\mathcal{B}}\dot{\boldsymbol{A}} + \boldsymbol{\omega} \times (\boldsymbol{A}) \tag{12.34}$$

By using Eq. (12.34), the derivatives of Eq. (12.33) can be obtained by noting that moments of inertia do not change with respect to time, as the coordinate system is fixed on the moving body.

$$\begin{aligned}\sum \mathbf{M}_C &= \frac{d\mathbf{H}_C}{dt} = \frac{d}{dt}\left\{\begin{array}{l}\left[\omega_x I_{xx}^C + \omega_y I_{xy}^C + \omega_z I_{xz}^C\right]\hat{\mathbf{i}} + \left[\omega_x I_{xy}^C + \omega_y I_{yy}^C + \omega_z I_{yz}^C\right]\hat{j} \\ +\left[\omega_x I_{xz}^C + +\omega_y I_{yz}^C + \omega_z I_{zz}^C\right]\hat{\mathbf{k}}\end{array}\right\} \\ &= \left[\alpha_x I_{xx}^C + \alpha_y I_{xy}^C + \alpha_z I_{xz}^C\right]\hat{\mathbf{i}} + \left[\alpha_x I_{xy}^C + I_{yy}^C + I_{yz}^C\right]\hat{\mathbf{j}} \\ &\quad + \left[\alpha_x I_{xz}^C + \alpha_y I_{yz}^C + \alpha_z I_{zz}^C\right]\hat{\mathbf{k}} + \boldsymbol{\omega} \\ &\quad \times \left\{\begin{array}{l}\left[\omega_x I_{xx}^C + \omega_y I_{xy}^C + \omega_z I_{xz}^C\right]\hat{\mathbf{i}} + \left[\omega_x I_{xy}^C + \omega_y I_{yy}^C + \omega_z I_{yz}^C\right]\hat{j} \\ +\left[\omega_x I_{xz}^C + +\omega_y I_{yz}^C + \omega_z I_{zz}^C\right]\hat{\mathbf{k}}\end{array}\right\}\end{aligned} \tag{12.35}$$

Equation (12.35) expanded in terms of the components of angular velocity

$$\sum M_C = \left[\alpha_x I_{xx}^C + \alpha_y I_{xy}^C + \alpha_z I_{xz}^C\right]\hat{i} + \left[\alpha_x I_{xy}^C + I_{yy}^C + I_{yz}^C\right]\hat{j} + \left[\alpha_x I_{xz}^C + \alpha_y I_{yz}^C + \alpha_z I_{zz}^C\right]\hat{k}$$
$$+ \left[\omega_y\left(\omega_x I_{xz}^C + \omega_y I_{yz}^C + \omega_z I_{zz}^C\right) - \omega_z\left(\omega_x I_{xy}^C + \omega_y I_{yy}^C + \omega_z I_{yz}^C\right)\right]\hat{i} +$$
$$\left[\omega_z\left(\omega_x I_{xx}^C + \omega_y I_{xy}^C + \omega_z I_{xz}^C\right) - \omega_x\left(I_{xz}^C\omega_x + \omega_y I_{yz}^C + \omega_z I_{zz}^C\right)\right]\hat{j} +$$
$$\left[\omega_x\left(\omega_x I_{xy}^C + \omega_y I_{yy}^C + \omega_z I_{yz}^C\right) - \omega_y\left(\omega_x I_{xx}^C + \omega_y I_{xy}^C + \omega_z I_{xz}^C\right)\right]\hat{k} \tag{12.36}$$

Regrouping terms of Eq. (12.36)

$$\sum \mathbf{M}_C$$
$$= \left[\alpha_x I_{xx}^C + I_{xy}^C(\alpha_y - \omega_x\omega_z) + I_{xz}^C(\alpha_z + \omega_x\omega_y) - I_{yy}^C(\omega_y\omega_z) + I_{zz}^C(\omega_y\omega_z) + I_{yz}^C\left(\omega_y^2 - \omega_z^2\right)\right]\hat{\mathbf{i}}$$
$$+ \left[I_{xx}^C(\omega_x\omega_z) + I_{xy}^C(\alpha_x + \omega_y\omega_z) + I_{xz}^C\left(\omega_z^2 - \omega_x^2\right) + I_{yz}^C(\alpha_z - \omega_x\omega_y) + \alpha_y I_{yy}^C - I_{zz}^C\omega_x\omega_z\right]\hat{\mathbf{j}}$$
$$+ \left[I_{zz}^C\alpha_z + I_{xy}^C\left(\omega_x^2 - \omega_y^2\right) + I_{xz}^C(\alpha_x - \omega_y\omega_z) + I_{yz}^C(\alpha_y + \omega_x\omega_z) + I_{yy}^C\omega_x\omega_y - \omega_x\omega_y I_{xx}^C\right]\hat{\mathbf{k}} \tag{12.37}$$

Although Eq. (12.37) is a general equation relating moments about the mass center of a rigid body in three-dimensional motion to its angular velocity and acceleration, and moments of inertia, it should be used with caution by paying attention to the location of the coordinate system. Hence, it is safer to begin with the angular momentum Eq. (12.32) and then obtain the moment equation similar to the Eq. (12.36), by using the relevant angular velocity vector of the moving body on which the coordinate system is located. This is demonstrated in Example 12.1.

Example 12.1

As shown in Fig. 12.9, a massless rod of length 2.5 m rotates at a constant angular speed of 2 rad/s in the counterclockwise direction of a roller of diameter 0.35 m and mass 3 kg along a friction surface so the roller rotates without slipping. The coordinates are located at the mass center of the roller and hence assumed to located on the massless rod. If the mass center of roller moves at a constant velocity as it rotates about its axis, determine the friction forces experienced by the roller and the force exerted by the massless rod on the roller.

Assume that the roller rotates with an angular velocity ω_z with respect to the massless rod.

The angular velocity of the massless rod $\omega_y = 2$ rad/s.

Velocity of the center of mass of the roller, $\boldsymbol{v}_C = \omega_y\hat{\boldsymbol{j}} \times (-R\hat{\boldsymbol{k}}) = -2.5 \times 2\hat{\boldsymbol{i}} = -5\hat{\boldsymbol{i}}$ m/s.

The angular velocity of the roller is given by

$$\boldsymbol{\omega} = \omega_y\hat{\boldsymbol{j}} + \omega_z\hat{\boldsymbol{k}} = 2\hat{\boldsymbol{j}} + \omega_z\hat{\boldsymbol{k}} \tag{1}$$

If A represents a point of contact of the roller on the ground, the velocity of A will be zero at every instant of time. Relating the mass center velocity v_C to the velocity of point A

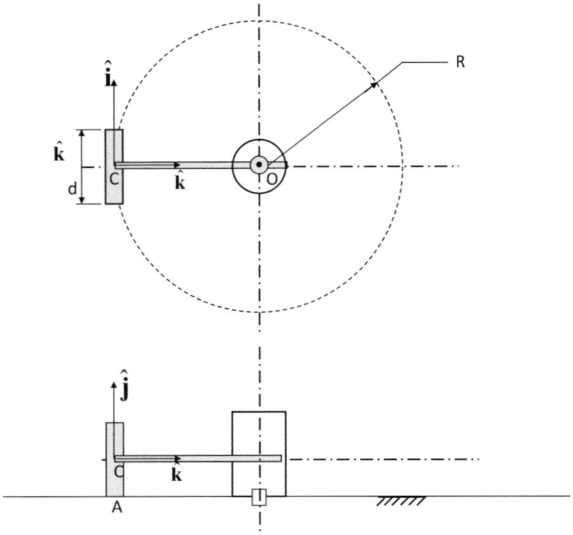

Fig. 12.9 Figure of Example 12.1

$$\begin{aligned} \boldsymbol{v}_C &= \boldsymbol{v}_A + \boldsymbol{\omega} \times \boldsymbol{r}_{AC} \\ -5\hat{\boldsymbol{i}} &= \frac{d}{2}\left(2\hat{\boldsymbol{j}} + \omega_z\hat{\boldsymbol{k}}\right) \times \hat{\boldsymbol{j}} \\ \Rightarrow \omega_z &= 28.57\,\text{rad/s} \end{aligned} \tag{2}$$

Acceleration of the roller mass center is given by

$$a_C = R\omega_y^2\hat{\boldsymbol{k}} = 2.5 \times 2^2\hat{\boldsymbol{k}} = 10\hat{\boldsymbol{k}}\,\text{m/s}^2 \tag{3}$$

The following forces are acting on the roller as shown in Fig. 12.10:

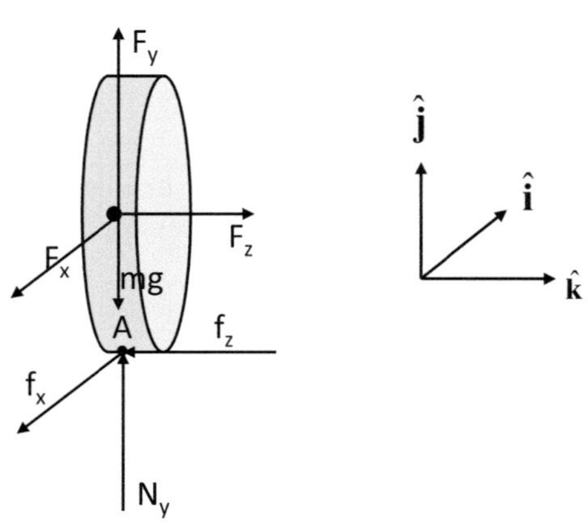

Fig. 12.10 Reaction forces of Example 12.1

$-f_z\hat{\boldsymbol{k}}$, the friction force resisting the centripetal force
$-f_x\hat{\boldsymbol{i}}$, the friction force resisting the rotation of the roller about its axis
$N_y\hat{\boldsymbol{j}}$, normal reaction of the surface to the roller
$F_z\hat{\boldsymbol{k}}, -F_x\hat{\boldsymbol{i}}, F_y\,\hat{\boldsymbol{j}}$ reaction forces at the roller massless rod interface
$-mg\hat{\boldsymbol{j}}$, the weight of the roller.

The net force acting on the roller is given by

$$\boldsymbol{F} = (-f_x - F_x)\hat{\boldsymbol{i}} + \left(N_y - mg + F_y\right)\hat{\boldsymbol{j}} + (F_z - fz)\hat{\boldsymbol{k}} \tag{4}$$

From Eqs. (3) and (4)

$$(-f_x - F_x)\hat{\boldsymbol{i}} + \left(N_y - mg + F_y\right)\hat{\boldsymbol{j}} + (F_z - fz)\hat{\boldsymbol{k}} = 30\hat{\boldsymbol{k}} \tag{5}$$

By comparing components of both sides of Eq. (5)

$$\begin{aligned} f_x + F_x &= 0 \\ N_y - mg + F_y &= 0 \\ F_z - fz &= 30 \end{aligned} \tag{6}$$

The forces that contribute toward the moment around the mass center are given by

$$\boldsymbol{F}_m = (-f_x)\hat{\boldsymbol{i}} + (-fz)\hat{\boldsymbol{k}} \tag{7}$$

Moment of the force $\boldsymbol{F}_m$ about the roller mass center is given by

$$\boldsymbol{M}_c = \boldsymbol{r}_{CA} \times \boldsymbol{F}_m = -\frac{d}{2}\hat{\boldsymbol{j}} \times \left((-f_x)\hat{\boldsymbol{i}} + (-f_z)\hat{\boldsymbol{k}}\right) = -\frac{d}{2}f_x\hat{\boldsymbol{k}} - \frac{d}{2}f_z\hat{\boldsymbol{i}} \tag{8}$$

The moments of inertia are given by

$$\begin{aligned} I_{yy}^C &= \frac{md^2}{16} = \frac{3(0.35)^2}{16} = 0.023 \text{ kg-m}^2 \\ I_{zz}^C &= \frac{md^2}{8} = \frac{3(0.35)^2}{8} = 0.0459 \text{ kg-m}^2 \end{aligned} \tag{9}$$

From Eq. 12.32, the angular momentum of the roller about its mass center is given by

$$\begin{aligned} \boldsymbol{H}_C &= \omega_y I_{yy}^C\hat{\boldsymbol{j}} + \omega_z I_{zz}^C\hat{\boldsymbol{k}} = 2 \times 0.0230\hat{\boldsymbol{j}} + 28.57 \times 0.0459\hat{\boldsymbol{k}} \\ &= 0.0459\hat{\boldsymbol{j}} + 1.3125\hat{\boldsymbol{k}} \text{ kg-m}^2\text{-rad/s} \end{aligned} \tag{10}$$

Since the coordinate system is located on the massless rod, the moment of forces about the mass center is given by

$$M_C = \omega_y \hat{\boldsymbol{j}} \times \boldsymbol{H}_c = 2\hat{\boldsymbol{j}} \times \left(0.0459\hat{\boldsymbol{j}} + 1.3125\hat{\boldsymbol{k}}\right) = 2.6250\hat{\boldsymbol{i}} \tag{11}$$

From Eqs. (8) and (11), $f_x = 0$ and $f_z = -15$. Since $f_x = 0$, $N_y = 0$, $F_y = mg = 3 \times 9.81 = 29.43$ N. Fz = 15 N.

12.5 Gyroscopic Forces in Mechanisms

All mechanisms that have rotating components that have a significant moment of inertia and angular speed, due to the principle of conservation of angular momentum, will resist any perturbation that will result in a couple that can result in additional reactions at the bearings. The perturbation will be mainly due to the change of direction of the axis of rotation, for example, when the mechanism is trying to negotiate a curve, that is very common. Therefore, it would be necessary to have an estimate of the resulting couple.

Consider a disk of moment of inertia I_{xx} spinning about its $\hat{\boldsymbol{i}}$ axis as shown in Fig. 12.11 with a constant angular velocity ω_x as shown. Before the disk reached this angular velocity, it was subjected to a torque T_x that resulted in an angular acceleration α_x. Once the disk reached the constant angular velocity ω_x, the torque was removed.

The torque T_x is given by

$$T_x = I_{xx}\alpha_x = I_{xx}\frac{d\omega_x}{dt} = \frac{d(I_{xx}\omega_x)}{dt} = \frac{d(H_x)}{dt} \tag{12.38}$$

The angular momentum is given by

$$H_x = I_{xx}\omega_x \tag{12.39}$$

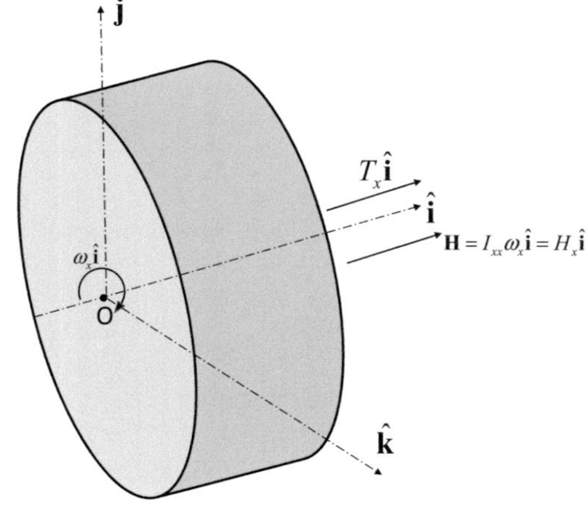

Fig. 12.11 A disk spinning about $\hat{\boldsymbol{i}}$ axis

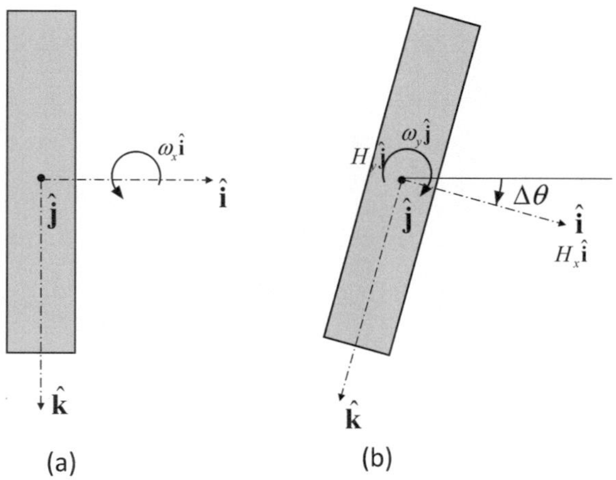

Fig. 12.12 Perturbing the rotating disk about $\hat{j}$ axis **a** Initial position **b** Perturbed position (tilted by $\Delta\theta$ in the clockwise direction)

Therefore, once the disk reaches its constant angular velocity, no additional torque is required and the angular momentum will be constant.

Figure 12.12 shows the perturbation of the rotating disk by $\Delta\theta$ in the clockwise direction about the $\hat{j}$ axis. Figure 12.12a shows the initial position that is same as the one represented by Fig. 12.12. Figure 12.12b is the perturbed position by rotating about the $\hat{j}$ axis. Although the disk continues to rotate about the $\hat{i}$ axis with the same angular speed and the corresponding angular momentum, there is an additional angular momentum introduced in the $\hat{j}$ direction due to a change in direction by $\Delta\theta$. From Eq. (12.37), the production of a torque due to this change of momentum, by neglecting all angular accelerations and cross moments of inertia, is given by

$$\sum \boldsymbol{M}_C = \omega_x \omega_y I_{xx}^C \hat{\boldsymbol{k}} = T_z \hat{\boldsymbol{k}} \tag{12.40}$$

Therefore, perturbing the rotating disk in the clockwise direction along the $\hat{j}$ axis results in a couple of magnitude T_z along the positive $\hat{k}$ axis. The angular velocity ω_y, the rate at which the spin axis is perturbed, is also called precession angular velocity, $\omega_p = \omega_y$. And the torque defined by Eq. (12.40) is known as the gyroscopic couple that is defined as

$$T_z = \omega_x \omega_p I_{xx} \tag{12.41}$$

Perturbing a rotating disk along a direction perpendicular to its axis of rotation at a precession angular velocity, ω_p, and the resulting gyroscopic couple, Tz, are shown in Fig. 12.13.

Since the angular velocity axes can be defined in any direction, a more general form of Eq. 12.40 can be written as

$$\boldsymbol{T} = \boldsymbol{\omega}_p \times I\boldsymbol{\omega} \tag{12.42}$$

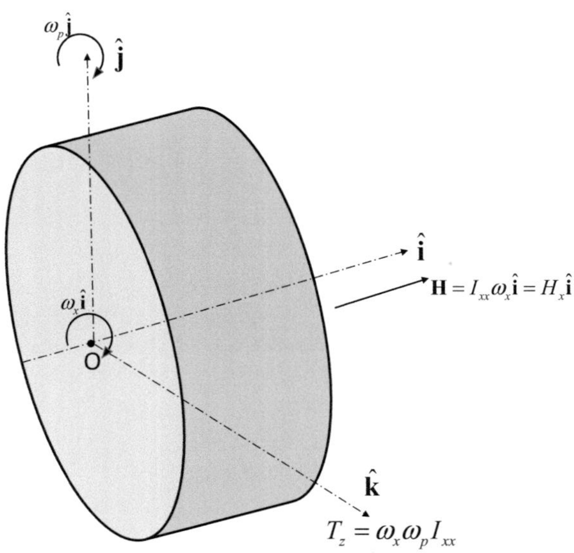

Fig. 12.13 Precession angular velocity and the gyroscopic couple

where ***T*** is the torque vector, *I* is moment of inertia of the rotating component about the spin axis, $\boldsymbol{\omega}$ is the angular velocity about the spin axis, and $\boldsymbol{\omega}_p$ is the precession angular velocity vector.

Example 12.2

Figure 12.14 shows the schematic of a single cylinder IC engine in which the crank is rotating counterclockwise at 3000 RPM direction and the crank is rotating about the x-axis that is also the moving direction of the automobile on which the above cylinder is mounted. If the automobile is traversing a curve of 500 m radius toward the left at a speed of 100 kmph. I_{xx}, the moment of inertia of the flywheel is 0.5 kg-m^2.

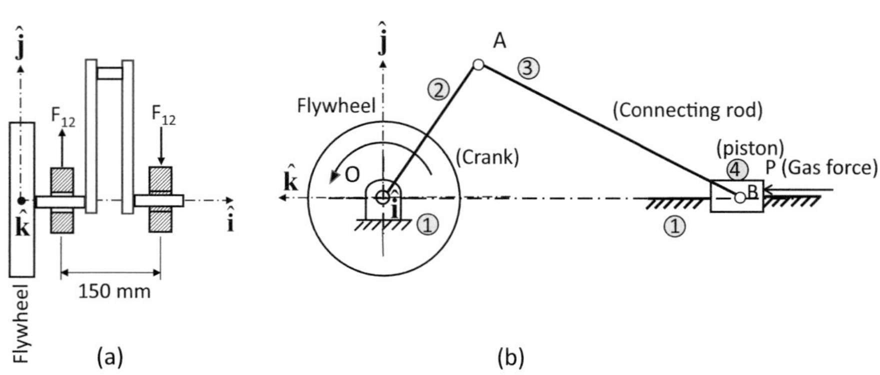

Fig. 12.14 Gyroscopic forces of a single cylinder IC engine **a** Side view **b** Front view

Angular velocity of the flywheel is given by

$$\omega = \frac{2\pi N}{60} = \frac{2\pi(3000)}{60} = 314.16\,\text{rad/s} \tag{1}$$

Angular velocity vector of the flywheel is given by

$$\boldsymbol{\omega} = 314.16\hat{\boldsymbol{i}}\ \text{rad/s} \tag{2}$$

Linear speed of the automobile is given by

$$V = \frac{100 \times 10^3}{3600} = 27.78\,\text{m/s} \tag{3}$$

If the vehicle is turning to the left while negotiating the curve of R = 500 m radius, the precision angular velocity will be in the counterclockwise direction, given by

$$\boldsymbol{\omega}_p = \frac{V}{R}\hat{\boldsymbol{j}} = \frac{27.78}{500}\hat{\boldsymbol{j}} = 0.0556\hat{\boldsymbol{j}}\,\text{rad/s} \tag{4}$$

From Eq. 12.42, the gyroscopic torque is given by

$$\boldsymbol{T} = \boldsymbol{\omega}_p \times I\boldsymbol{\omega} = 0.0556\hat{\boldsymbol{j}} \times (0.5)314.16\hat{\boldsymbol{i}} = -8.72\hat{\boldsymbol{k}}\,\text{Nm} \tag{5}$$

Since the gyroscopic torque is in the clockwise direction, the bearings in the directions as shown in Fig. 12.14 and their magnitude is given by

$$F_{12} = \frac{T_z}{L} = \frac{8.73}{0.175} = 49.89\,\text{N} \tag{6}$$

12.6 Conclusions

In this chapter, we have discussed three-dimensional motion of rigid bodies. Although gyroscopes were not the focus of study in this book, their study will provide insight into the three-dimensional motion of rigid bodies, which can be advantageously used in the kinetic and kinematic analysis of 3D mechanisms. Extending the same principles, we were able to discuss gyroscopic forces in mechanisms, which become important for designing bearings, if they are subjected to higher reaction forces due to these forces.

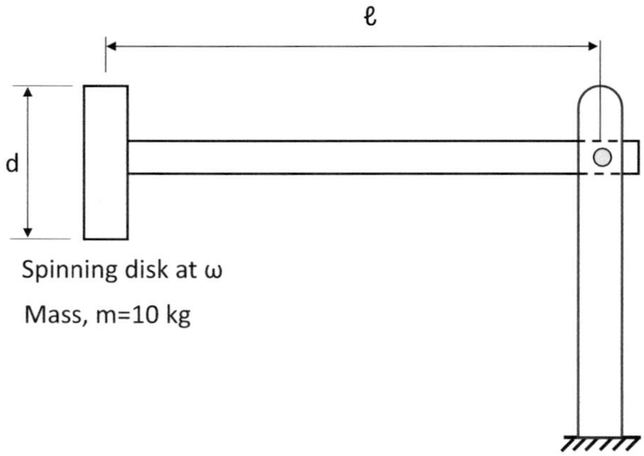

Fig. P12.1 Spinning disk in the vertical plane on a shaft rotating along a horizontal plane

12.7 Exercise Problems

12.1 Consider a disk of diameter of 250 mm spinning at an angular speed of 5000 RPM at the end of a shaft of length $\ell = 600$ mm, hinged as shown in Fig. P12.1. Mass of the disk is 10 kg. Determine the angular velocity of precession due to the torque introduced due to the gravitational force.

12.2 Consider the single engine of a jet engine shown in Fig. P12.2 rotating at a speed of 15,000 RPM. The entire engine has a mass of 700 kg and a radius of gyration of 250 mm. If the aircraft is negotiating a curve of 2 km radius at a speed of 900 kmph, determine the maximum bearing force.

12.3 Consider the gimbal mounting of a gyroscope shown in Fig. P12.3 Although the low friction bearings are used to prevent the precession of the gyro, some friction torque will be applied to cause precession. If the gyro spins at 15,000 RPM and a moment of inertia of about 0.00150 kg-m^2, how much friction torque must be applied to cause a precision of 0.75°/h?

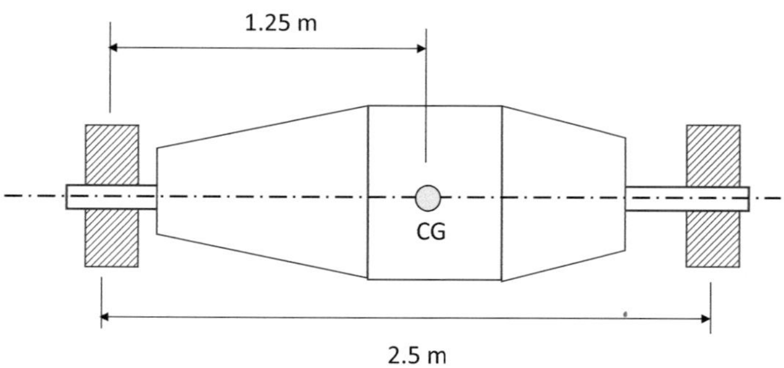

Fig. P12.2 A jet engine negotiating a curve

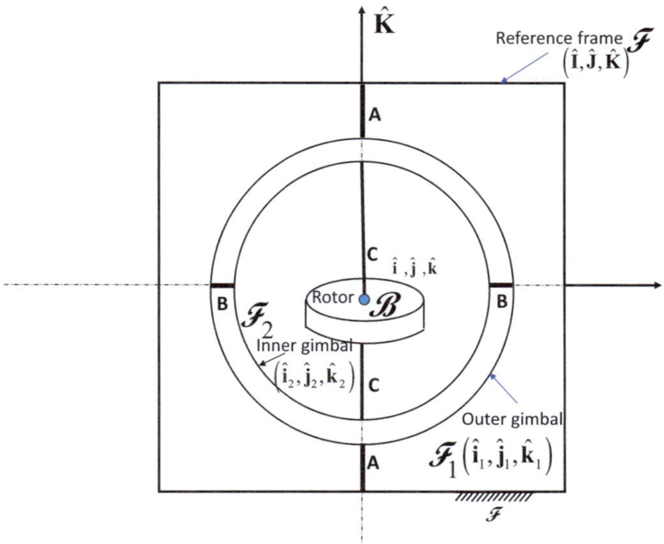

Fig. P12.3 Friction torque of a gyroscope

12.4 A 4 m diameter four-bladed propeller has a mass of 20 kg per blade. If the propeller is rotating at a speed of 1400 RPM, and one of the maneuvers causes an angular speed of 1.5 rad/s, resulting in a rotation of the aircraft about a horizontal axis along the wings, determine the gyroscopic couple. Assume a radius of gyration of 0.75 m for the propeller

12.5 Consider a single cylinder engine with a flywheel that is fitted on a motorcycle. As the motorcycle moves around a curve of 350 m radius at a speed of 100 kmph, determine the bearing reactions due to precession if the flywheel driven by the engine is rotating at a speed of 3500 RPM. Assume the moment of inertia of the flywheel as 0.5 kg-m^2.

12.6 A flywheel weighs 10 kg and its radius of gyration is 250 mm. It is rotating about a horizontal axis at a speed of 1000 RPM and the distance between the bearings and the plane of rotation is 200 mm. Determine the rate of precession.

12.7 The wheels of a motorcycle each have a moment of inertia of 2.5 kg-m^2 and the rotating engine parts have a moment of inertia 0.15 kg-m^2. The axis of rotation of the engine with the flywheel is parallel to the axis of rotation of the wheels. It is traveling in the third gear that has a reduction ratio of 1:5:1 at a speed of 40 kmph. If the diameter of the wheels is 600 mm, determine the magnitude and direction of the gyroscopic couple when it is negotiating a curve of 100 m radius.

12.8 The rotary engine of an airplane is rotating on axis that is parallel to the direction of its motion along a straight line; it weighs 500 kg with a radius of gyration of 0.4 m. The engine is rotating at a speed of 2000 RPM in the clockwise direction as seen from the front of the airplane. When the airplane is flying at a speed of 150 kmph and negotiating a curve of 200 m radius, determine the magnitude and direction of the gyroscopic couple acting on the aeroplane.

12.9 The turbine of a motor boat running at 1500 RPM weighs 650 kg and has a radius of gyration of 300 mm. The axis of rotation of the turbine is parallel to the longer dimension of the boat. If the boat pitches along an axis perpendicular to the engine axis at an angular speed of 0.5 rad/s, determine the magnitude and direction of the gyroscopic couple.

12.10 The turbine of a ship that weighs 6000 kg has a radius of gyration of 600 mm and the axis of rotating is parallel to the direction of motion of the ship. The ship is pitching along an axis perpendicular to the engine axis through an angle of 10° in a simple harmonic motion with a period of 25 s. Determine the magnitude and direction of the gyroscopic couple if the engine is rotating clockwise as seen from the front. What is the maximum angular acceleration of the ship during the pitching action?

Bibliography

1. Crossley FRE (1954) Dynamics in machines. The Ronald Press Company, New York
2. Hall AS (1961) Kinematics and linkage design. Prentice Hall, Englewood Cliffs, NJ
3. Hartenberg RS, Denavit J (1964) Kinematic synthesis of linkages. McGraw-Hill, New York
4. Hannah J, Stephens RC (1972) Mechanics and machines: advanced theory and examples. Edward Arnold, London
5. Mabie HH, Ocvirk FW (1978) Mechanisms and dynamics of machinery, 3rd edn. Wiley, New York
6. Steidel Jr RF (1979) An introduction to mechanical vibrations, 2nd edn. Wiley, New York
7. Patton WJ (1979) Kinematics. Reston Publishing, Reston, Virginia
8. Paul B (1979) Kinematics and dynamics of planar machinery. Prentice Hall, Englewood Cliffs, NJ
9. Martin GH (1982) Kinematics and dynamics of machines, 2nd edn. Waveland Press, Inc., Long Grove, IL
10. Mabie HH, Reinholtz CF (1986) Mechanisms and dynamics of machinery, 4th edn. Wiley, New York
11. McGill DJ, King WW (1989) An introduction to dynamics. PWS Kent, Boston
12. Shigley JE, Uicker Jr JJ (1995) Theory of machines and mechanisms, 2nd edn. McGraw-Hill International, New York
13. Rao SS (1995) Mechanical vibrations. Addison Wesley Publishing, New York
14. Meriam JL, Kraige LG (2001) Engineering mechanics, dynamics, 5th edn. New York
15. Wilson CW, Sadler JP (2003) Kinematics and dynamics of machinery, 3rd edn. Prentice Hall, New Jersey
16. Norton RL (2004) Design of machinery: an introduction to the synthesis and analysis of mechanisms and machines, 3rd edn. McGraw-Hill, New York
17. Bevan T (2010) Theory of machines, 3rd edn. Pearson, London
18. Russel K, Shen Q, Sodhi RS (2014) Mechanism design: visual and programmable approaches. CRC Press, Boca Raton

D. N. Manik, *Fundamentals of Mechanisms and Machines*,
https://doi.org/10.1007/978-981-97-1810-8

MIX
Papier aus verantwortungsvollen Quellen
Paper from responsible sources
FSC® C105338

If you have any concerns about our products,
you can contact us on
ProductSafety@springernature.com

In case Publisher is established outside the EU,
the EU authorized representative is:
Springer Nature Customer Service Center GmbH
Europaplatz 3, 69115 Heidelberg, Germany

Printed by Libri Plureos GmbH
in Hamburg, Germany